Volumetric Discrete Geometry

Discrete Mathematics and Its Applications

Handbook of Discrete and Computational Geometry, Third Edition
C. Toth, Jacob E. Goodman and Joseph O'Rourke

Handbook of Discrete and Combinatorial Mathematics, Second Edition
Kenneth H. Rosen

Crossing Numbers of Graphs
Marcus Schaefer

Graph Searching Games and Probabilistic Methods
Anthony Bonato and Paweł Prałat

Handbook of Geometric Constraint Systems Principles
Meera Sitharam, Audrey St. John, and Jessica Sidman,

Additive Combinatorics
Béla Bajnok

Algorithmics of Nonuniformity:
Tools and Paradigms
Micha Hofri and Hosam Mahmoud

Extremal Finite Set Theory
Daniel Gerbner and Balazs Patkos

Cryptology: Classical and Modern
Richard E. Klima and Neil P. Sigmon

Volumetric Discrete Geometry
Károly Bezdek and Zsolt Lángi

https://www.crcpress.com/Discrete-Mathematics-and-Its-Applications/book-series/CHDISMTHAPP?page=1&order=dtitle&size=12&view=list&status=published,forthcoming

Volumetric Discrete Geometry

Károly Bezdek
University of Calgary

Zsolt Lángi
Budapest University of Technology and Economics

CRC Press is an imprint of the
Taylor & Francis Group, an **informa** business
A CHAPMAN & HALL BOOK

CRC Press
Taylor & Francis Group
6000 Broken Sound Parkway NW, Suite 300
Boca Raton, FL 33487-2742

First issued in paperback 2022

ISBN 13: 978-1-03-247564-6 (pbk)
ISBN 13: 978-0-367-22375-5 (hbk)

DOI: 10.1201/9780429274572

Library of Congress Cataloging-in-Publication Data

Names: Bezdek, Károly, author. | Lángi, Zsolt, author.
Title: Volumetric discrete geometry / Károly Bezdek and Zsolt Lángi.
Description: Boca Raton : CRC Press, Taylor & Francis Group, 2019. | Includes bibliographical references.
Identifiers: LCCN 2018061556 | ISBN 9780367223755
Subjects: LCSH: Volume (Cubic content) | Geometry, Solid. | Discrete geometry.
Classification: LCC QC105 .B49 2019 | DDC 516/.11--dc23
LC record available at https://lccn.loc.gov/2018061556

Visit the Taylor & Francis Web site
at http://www.taylorandfrancis.com

and the CRC Press Web site
at http://www.crcpress.com

To Károly's wife, Éva, and Zsolt's wife, Kornélia,
for their exceptional and continuous support.

Contents

Preface xiii

Authors xvii

Symbols xix

I Selected Topics 1

1 Volumetric Properties of (m, d)-scribed Polytopes 3

1.1 The isoperimetric inequality . . . 3
1.2 Discrete isoperimetric inequalities: volume of polytopes circumscribed about a sphere . . . 5
1.3 Volume of polytopes inscribed in a sphere . . . 9
1.4 Polyhedra midscribed to a sphere . . . 14
1.5 Research Exercises . . . 20

2 Volume of the Convex Hull of a Pair of Convex Bodies 23

2.1 Volume of the convex hull of a pair of convex bodies in Euclidean space . . . 23
2.2 Volume of the convex hull of a pair of convex bodies in normed spaces . . . 29
2.3 Research Exercises . . . 35

3 The Kneser-Poulsen Conjecture Revisited 37

3.1 The Kneser-Poulsen conjecture . . . 37
3.2 The Kneser-Poulsen conjecture for continuous contractions of unions and intersections of balls . . . 38
3.3 The Kneser-Poulsen conjecture for contractions of unions and intersections of disks in $\mathbb{E}^2$ . . . 39
3.4 The Kneser-Poulsen conjecture for uniform contractions of r-ball polyhedra in $\mathbb{E}^d$, $\mathbb{S}^d$ and $\mathbb{H}^d$ . . . 41
3.5 The Kneser-Poulsen conjecture for contractions of unions and intersections of disks in $\mathbb{S}^2$ and $\mathbb{H}^2$ . . . 44
3.6 Research Exercises . . . 45

4 Volumetric Bounds for Contact Numbers **49**
4.1 Description of the basic geometric questions 49
4.2 Motivation from materials science 51
4.3 Largest contact numbers for congruent circle packings 52
4.3.1 The Euclidean plane 52
4.3.2 Spherical and hyperbolic planes 52
4.4 Largest contact numbers for unit ball packings in $\mathbb{E}^3$ 53
4.5 Upper bounding the contact numbers for packings by translates of a convex body in $\mathbb{E}^d$ 54
4.6 Contact numbers for digital and totally separable packings of unit balls in $\mathbb{E}^d$ 58
4.7 Bounds for contact numbers of totally separable packings by translates of a convex body in $\mathbb{E}^d$ with $d = 1, 2, 3, 4$ 60
4.7.1 Separable Hadwiger numbers 61
4.7.2 One-sided separable Hadwiger numbers 62
4.7.3 Maximum separable contact numbers 63
4.8 Appendix: Hadwiger numbers of topological disks 64
4.9 Research Exercises 65

5 More on Volumetric Properties of Separable Packings **67**
5.1 Solution of the contact number problem for smooth strictly convex domains in $\mathbb{E}^2$ 67
5.2 The separable Oler's inequality and its applications in $\mathbb{E}^2$ . . 70
5.2.1 Oler's inequality 70
5.2.2 An analogue of Oler's inequality for totally separable translative packings 72
5.2.3 On the densest totally separable translative packings . 73
5.2.4 On the smallest area convex hull of totally separable translative finite packings 74
5.3 Higher dimensional results: minimizing the mean projections of finite ρ-separable packings in $\mathbb{E}^d$ 75
5.4 Research Exercises 77

II Selected Proofs **81**

6 Proofs on Volumetric Properties of (m, d)-scribed Polytopes **83**
6.1 Proof of Theorem 3 83
6.2 Proofs of Theorems 10 and 11 86
6.3 Proof of Theorem 14 90
6.3.1 Preliminaries 90
6.3.2 Proof of Theorem 14 for $n \leq 6$ 92
6.3.3 Proof of Theorem 14 for $n = 7$ 92
6.3.4 Proof of Theorem 14 for $n = 8$ 93
6.4 Proofs of Theorems 16, 17 and 18 96

6.4.1 Proof of Theorem 16 and some lemmas for Theorems 17 and 18 96
6.4.2 Proofs of Theorems 17 and 18 99
6.5 Proof of Theorem 21 102
6.6 Proof of Theorem 22 105
6.7 Proof of Theorem 27 111
6.8 Proofs of Theorems 28 and 29 113
6.8.1 Preliminaries and the main idea of the proofs 113
6.8.2 The main lemma of the proofs 115
6.8.3 Proof of Theorem 28 118
6.8.4 Proof of Theorem 29 125

7 Proofs on the Volume of the Convex Hull of a Pair of Convex Bodies **131**
7.1 Proofs of Theorems 32 and 33 131
7.1.1 Proof of Theorem 32 131
7.1.2 Proof of Theorem 33 133
7.2 Proofs of Theorems 34, 36, 37 and 40 134
7.2.1 Preliminaries 134
7.2.2 Proofs of the Theorems 135
7.3 Proofs of Theorems 41 and 46 138
7.3.1 Proof of Theorem 41 138
7.3.2 Proof of Theorem 46 141
7.4 Proof of Theorem 53 142
7.5 Proof of Theorem 54. 145
7.6 Proofs of Theorems 57 and 58 148
7.6.1 Proof of Theorem 57 148
7.6.2 Proof of Theorem 58 151
7.7 Proofs of Theorems 59 and 60 151
7.7.1 The proof of the left-hand side inequality in (ii) 151
7.7.2 The proof of the right-hand side inequality in (ii) 155
7.7.3 The proofs of (i), (iii) and (iv) 156

8 Proofs on the Kneser-Poulsen Conjecture **159**
8.1 Proof of Theorem 67 159
8.2 Proof of Theorem 68 162
8.3 Proof of Theorem 69 165
8.4 Proof of Theorem 72 165
8.5 Proof of Theorem 73 168
8.5.1 Proof of (i) in Theorem 73 169
8.5.2 Proof of (ii) in Theorem 73 170
8.5.3 Proof of (iii) in Theorem 73 171
8.6 Proof of Theorem 74 173
8.6.1 The spherical leapfrog lemma 173
8.6.2 Smooth contractions via Schläfli's differential formula 174

8.6.3 From higher- to lower-dimensional spherical volume . 174
8.6.4 Putting pieces together 175
8.7 Proof of Theorem 75 176
8.8 Proof of Theorem 76 177
8.8.1 Proof of Part (i) of Theorem 76 177
8.8.2 Proof of Part (ii) of Theorem 76 179
8.9 Proof of Theorem 78 181
8.10 Proof of Theorem 79 182
8.10.1 Basic results on central sets of ball-polytopes 182
8.10.2 An extension theorem via piecewise isometries 184
8.10.3 Deriving Theorem 79 from the preliminary results . . 184

9 Proofs on Volumetric Bounds for Contact Numbers 187

9.1 Proof of Theorem 87 187
9.1.1 An upper bound for sphere packings: Proof of (i) . . . 187
9.1.2 An upper bound for the fcc lattice: Proof of (ii) 193
9.1.3 Octahedral unit sphere packings: Proof of (iii) 194
9.2 Proof of Theorem 88 196
9.3 Proof of Theorem 92 197
9.4 Proof of Theorem 93 199
9.5 Proof of Theorem 94 200
9.6 Proof of Theorem 95 203
9.7 Proof of Theorem 98 206
9.8 Proof of Theorem 99 211
9.9 Proof of Theorem 100 212
9.10 Proofs of Theorems 101, 102, and 103 214
9.10.1 Linearization, fundamental properties 214
9.10.2 Proofs of Theorems 101 and 102 217
9.10.3 Proof of Theorem 103 219
9.10.4 Remarks 223
9.11 Proof of Theorem 108 223
9.12 Proof of Theorem 110 226
9.13 Proof of Theorem 111 233

10 More Proofs on Volumetric Properties of Separable Packings 239

10.1 Proof of Theorem 113 239
10.2 Proof of Theorem 116 243
10.3 Proof of Corollary 117 246
10.4 Proof of Theorem 119 246
10.5 Proof of Theorem 123 249
10.6 Proof of Theorem 125 250
10.7 Proof of Theorem 126 250
10.8 Proof of Theorem 130 253

11 Open Problems: An Overview **259**
11.1 Chapter 1 259
11.2 Chapter 2 260
11.3 Chapter 3 261
11.4 Chapter 4 262
11.5 Chapter 5 263

Bibliography **265**

Index **283**

Preface

The volume of geometric objects has been studied by the ancient Greek mathematicians and it is quite remarkable that even today volume continues to play an important role in applied as well as pure mathematics. So, it did not come as a surprise to us that also in discrete geometry, which is a relatively new branch of geometry, volume played a significant role in generating interesting new topics for research. When writing this book, our goal was to demonstrate the more recent aspects of volume within discrete geometry. We found it convenient to split the book into two parts, with Part I consisting of survey chapters of selected topics on volume, and Part II consisting of chapters of selected proofs of theorems stated in Part I. This means also that the chapters can be read quite independently from each other. Moreover, our book provides a list of more than 30 open problems encouraging the interested reader to join in the recent efforts to progress research on the topics discussed. Also, there are more than 60 research exercises to help the reader with deeper understanding of the topics discussed. The prerequisite for Part I is rather modest, any advanced undergraduate student should be able to follow the developments in those chapters. However, Part II is of graduate level reading that intends to lead the interested reader to the frontiers of research in discrete geometry. In what follows, we give a brief description of the topics discussed in our book.

In Chapter 1 we investigate the volumetric properties of (m, d)-scribed polytopes, that is, d-dimensional polytopes whose every m-face touches a Euclidean ball. Starting with a short historical overview of the isoperimetric problem in Section 1.1, in Section 1.2 we examine a problem proposed by L'huilier in the 18th century, which asks about finding the minimum volume polytopes circumscribed about a ball. By the famous Lindelöf Condition in Theorem 4, these polytopes are the ones having maximal isoperimetric ratio and a fixed number of facets. In this section, besides describing the results regarding L'huilier's problem, we present a 'dynamic' version of Lindelöf's theorem. This version offers a possible explanation of the strange, elongated shape of the interstellar asteroid 'Oumuamua passing through the solar system. In Section 1.3 we investigate the 'dual' of L'huilier's problem, and deal with maximum volume polytopes inscribed in a ball. A consequence of the Circle Packing Theorem is that every combinatorial class of 3-dimensional convex polytope can be represented by a polytope, called Koebe polyhedron, whose every edge is tangent to a unit sphere. In Section 1.4 we collect the properties of Koebe polyhedra and, in particular, present results about the existence of Koebe polyhedra in every combinatorial class with additional requirements. Just as in other chapters, the goal of stating open questions and problems in Chapter 1 is to motivate further research on the topics discussed. The relevant selected proofs are discussed in Chapter 6.

In Chapter 2 we examine results about the volume of the convex hull of a pair of convex bodies. In Section 2.1 we introduce results of Rogers and Shephard, who in the 1950s in three subsequent papers defined various convex bodies associated to a convex body, and found the extremal values of their volumes on the family of convex bodies based on the convexity of some volume functional and Steiner symmetrization. After this we present a proof of a conjecture of Rogers and Shephard about the equality cases in their theorems, find a more general setting of their problems, and collect results about it. Section 2.2 describes a variant of the previous problems for normed spaces. In particular, using the notion of relative norm of a convex body, we collect results about the extremal values of four types of volume of the unit ball of a normed space. The problem of finding these values of one of these types leads to the famous Mahler Conjecture. We extend these notions to the translation bodies of convex bodies, defined in Section 2.1, and determine their extremal values in normed planes. Hopefully, the problems and conjectures of Chapter 2 will orient the reader towards new research directions. Chapter 7 contains the proofs of the theorems selected from Chapter 2.

The monotonicity of volume under contractions of arbitrary arrangements of spheres is a well-known fundamental problem in discrete geometry. The research on this topic started with the conjecture of Poulsen and Kneser in the late 1950s. In Chapter 3 we survey the status of the long-standing Kneser-Poulsen conjecture in Euclidean as well as in non-Euclidean spaces with emphases on the latest developments. With the update given on the Kneser-Poulsen conjecture in Chapter 3, our hope is to encourage the renewed interest in settling this old conjecture of discrete geometry in higher dimensions as well. The proofs of the most recent results of Chapter 3 are contained in Chapter 8.

The well-known kissing number problem asks for the largest number of non-overlapping unit balls touching a given unit ball in the Euclidean d-space $\mathbb{E}^d$. Generalizing the kissing number, the Hadwiger number or the translative kissing number $H(\mathbf{K})$ of a convex body $\mathbf{K}$ in $\mathbb{E}^d$ is the maximum number of non-overlapping translates of $\mathbf{K}$ that all touch $\mathbf{K}$. In Chapter 4 we study and survey the following natural extension of these problems. A finite translative packing of the convex body $\mathbf{K}$ in $\mathbb{E}^d$ is a finite family $\mathcal{P}$ of non-overlapping translates of $\mathbf{K}$ in $\mathbb{E}^d$. Furthermore, the contact graph $G(\mathcal{P})$ of $\mathcal{P}$ is the (simple) graph whose vertices correspond to the packing elements and whose two vertices are connected by an edge if and only if the corresponding two packing elements touch each other. The number of edges of $G(\mathcal{P})$ is called the contact number of $\mathcal{P}$. Finally, the contact number problem asks for the largest contact number, that is, for the maximum number $c(\mathbf{K}, n, d)$ of edges that a contact graph of n non-overlapping translates of $\mathbf{K}$ can have in $\mathbb{E}^d$. In the first half of Chapter 4 we survey the bounds proved for $c(\mathbf{K}, n, d)$ using volumetric methods. Then we turn our attention to an important subfamily of translative packings called totally separable packings. Here a packing of translates of a convex body $\mathbf{K}$ in $\mathbb{E}^d$ is called totally separable if any two packing elements can be separated by a hyperplane of $\mathbb{E}^d$ disjoint from the interior of every packing element. In the second half of Chapter 4 we study the analogues of the Hadwiger and contact numbers for totally separable translative packings of $\mathbf{K}$ labelled by $H_{\text{sep}}(\mathbf{K})$ and $c_{\text{sep}}(\mathbf{K}, n, d)$ and survey the bounds proved for $H_{\text{sep}}(\mathbf{K})$ as well as $c_{\text{sep}}(\mathbf{K}, n, d)$ using volumetric ideas. Along the way we call the reader's attention to a number of open problems. Chapter 9 contains our selection of proofs of some recent results of Chapter 4.

In Chapter 5 we continue our investigation of totally separable packings from a volumetric point of view. First, we outline the recent solution of the contact number problem for smooth strictly convex domains in $\mathbb{E}^2$. We discuss this approach in details based on angular measure, Birkhoff orthogonality, Birkhoff measure, (smooth) Birkhoff domains, and approximation by (smooth strictly convex) Auerbach domains, which are topics of independent interests as well. In the next part of Chapter 5, we connect the study of totally separable packings of discrete geometry to Oler's inequality of geometry of numbers. More concretely, we discuss an analogue of Oler's inequality for totally separable translative packings in $\mathbb{E}^2$ and then use it for finding the highest density of totally separable translative packings (resp., the smallest area convex hull of totally separable packings by n translates) of an arbitrary convex domain in $\mathbb{E}^2$. Finally, as a local version of totally separable packings, we introduce the family of ρ-separable translative packings of $\mathbf{o}$-symmetric convex bodies in $\mathbb{E}^d$. In particular, we investigate the fundamental problem of minimizing the mean i-dimensional projection of the convex hull of n non-overlapping translates of an $\mathbf{o}$-symmetric convex body $\mathbf{C}$ forming a ρ-separable packing in $\mathbb{E}^d$ for given $d > 1, n > 1$, and $\mathbf{C}$. We encourage the reader to take a closer look of the open problems of Chapter 5. Last but not least we encourage the reader to study the selected proofs included in Chapter 10.

Budapest and Calgary, January 12, 2019 Károly Bezdek and Zsolt Lángi

Authors

Károly Bezdek[1]

Canada Research Chair (Tier 1)
Department of Mathematics
University of Calgary, Canada

and

Professor
Department of Mathematics
University of Pannonia, Veszprém, Hungary

Zsolt Lángi[2]

Senior Research Fellow
Morphodynamics Research Group
Hungarian Academy of Sciences

and

Associate Professor
Department of Geometry
University of Technology, Budapest, Hungary

---

1 

Partially supported by a Natural Sciences and Engineering Research Council of Canada Discovery Grant

2 

Partially supported by the National Research, Development and Innovation Office, NKFI, K-119245 and K-119670, the János Bolyai Research Scholarship of the Hungarian Academy of Sciences, and grants BME FIKP-VÍZ and UNKP-18-4 New National Excellence Program by EMMI

Symbols

Symbol Description

Symbol	Description
$\mathbb{E}^d$	d-dimensional Euclidean space.
$\mathbf{o}$	The origin of $\mathbb{E}^d$.
$\mathbb{H}^d$	d-dimensional hyperbolic space.
$\mathrm{bd}(\cdot)$	Boundary of a set.
$\mathrm{int}(\cdot)$	Interior of a set.
$\mathrm{relbd}(\cdot)$	Relative boundary of a set.
$\mathrm{card}(\cdot)$, $\vert\cdot\vert$	Cardinality of a set.
$[\mathbf{p},\mathbf{q}]$	Closed segment with endpoints $\mathbf{p}$, $\mathbf{q}$.
$(\mathbf{p},\mathbf{q})$	Open segment with endpoints $\mathbf{p}$, $\mathbf{q}$.
$\Vert\mathbf{p}\Vert$	Euclidean norm of $\mathbf{p}$.
$\mathbf{B}^d$	Closed unit ball in $\mathbb{E}^d$ with $\mathbf{o}$ as its center.
$\mathbb{S}^{d-1}$	The boundary of $\mathbf{B}^d$; i.e., the $(d-1)$-dimensional spherical space.
$\mathrm{conv}(\cdot)$	Convex hull of a set in $\mathbb{E}^d$.
$\mathrm{aff}(\cdot)$	Affine hull of a set in $\mathbb{E}^d$.
$\mathrm{lin}(\cdot)$	Linear hull of a set in $\mathbb{E}^d$.
$\mathrm{vol}_d(\cdot)$	Volume of a set in $\mathbb{E}^d$.
$\mathrm{area}(\cdot)$	Area of a set in $\mathbb{E}^2$.
$\mathrm{svol}_{d-1}(\cdot)$	Surface volume of a set in $\mathbb{E}^d$.
$\mathrm{perim}(\cdot)$	Perimeter of a set in $\mathbb{E}^2$.
$\mathrm{Svol}_{d-1}(\cdot)$	Spherical volume of a set in $\mathbb{S}^{d-1}$.
κ_d	The volume of $\mathbf{B}^d$.
$\mathbf{K}^\circ$	The polar of the set $\mathbf{K}$.
$\mathrm{Sym}(\mathbf{S})$	The symmetry group of the set $\mathbf{S}$.
$M_\star^{d-1}(\cdot)$	The $(d-1)$-dimensional Minkowski content.
$L^d(\cdot)$	The d-dimensional Lebesgue measure.
$V(\cdot,\dots,\cdot)$	Mixed volume.
$W_i(\mathbf{K})$	The ith quermassintegral of $\mathbf{K}$.
$\mathbf{I_B}$	The rotated copy of $\mathbf{B}^\circ$ by $\frac{\pi}{2}$.
$I(\mathbf{K})$	Isoperimetric ratio of $\mathbf{K}$.
I_d^n	Maximum isoperimetric ratio of all convex polytopes in $\mathbb{E}^d$ with at most n facets.
$N(\mathbf{M})$	The set of outer unit normal vectors at the smooth points of $\mathbf{M}$.
$F(\mathbf{M})$	The form body of $\mathbf{M}$.
$V(\mathbf{P})$	The vertex set of the polytope $\mathbf{P}$.
$\mathcal{P}_d(n)$	The family of polytopes with n vertices and inscribed in $\mathbb{S}^{d-1}$.
$v_d(n)$	The maximum of $\mathrm{vol}_d(\mathbf{P})$ on the family $\mathcal{P}_d(n)$.
$\mathbf{C}_d(n)$	The cyclic polytope defined in Remark 20.
$\mathrm{skel}_k(\mathbf{P})$	The k-skeleton of the polytope $\mathbf{P}$ in $\mathbb{E}^d$.
$\mathrm{cm}_k(\mathbf{P})$	The center of mass of the k-skeleton of the polytope $\mathbf{P}$.
$\mathrm{cc}(\mathbf{P})$	The center of the smallest ball containing $\mathbf{P}$.
$\mathrm{IC}(\mathbf{P})$	The set of the centers of the largest balls contained in $\mathbf{P}$.
$\mathrm{ccm}(\mathbf{P})$	The circumcenter of mass of the simplicial polytope $\mathbf{P}$.
$\rho_T(\mathbf{C})$	The radius of the image of the spherical cap $\mathbf{C}$ under the Möbius transformation T.
$\mathbf{c}_T(\mathbf{C})$	The center of the image of the spherical cap $\mathbf{C}$ under the Möbius transformation T.
$S_H(\mathbf{K})$	The Steiner symmetrization of $\mathbf{K}$ with respect to the hyperplane H.
$V^*(\mathbf{K},\mathbf{L})$	Maximum of the volume of the convex hull of two intersecting translates of $\mathbf{K}$ and $\mathbf{L}$.
$R_\mathbf{p}(\mathbf{K})$	The reflection body of $\mathbf{K}$ with respect to $\mathbf{p}$.
$R(\mathbf{K})$	A reflection body of $\mathbf{K}$ minimizing $\mathrm{vol}_d(R_\mathbf{p}(\mathbf{K}))$ for all $\mathbf{p}\in\mathbf{K}$.
$R^*(\mathbf{K})$	A reflection body of $\mathbf{K}$ maximizing $\mathrm{vol}_d(R_\mathbf{p}(\mathbf{K}))$ for all $\mathbf{p}\in\mathbf{K}$.
$T_\mathbf{p}(\mathbf{K})$	The translation body of $\mathbf{K}$ with vector $\mathbf{p}$.
$T(\mathbf{K})$	A translation body of $\mathbf{K}$ minimizing $T_\mathbf{p}(\mathbf{K})$ under the condition $\mathbf{K}\cap(\mathbf{p}+\mathbf{K})\neq\emptyset$.
$C(\mathbf{K})$	The associated body of $\mathbf{K}$.
$c(\mathbf{K},\mathbf{L})$	Maximum volume of the convex hull of two intersecting congruent copies of $\mathbf{K}$ and $\mathbf{L}$.
$c(\mathbf{K}\vert\mathcal{S})$	Maximum volume of the convex hull of $\mathbf{K}$ and an intersecting congruent copy $\sigma(\mathbf{K})$ with $\sigma\in\mathcal{S}$.
$c_i(\mathbf{K}))$	The value of $c(\mathbf{K}\vert\mathcal{S})$ if $\mathcal{S}$ is the set of all reflections about i-flats.
$c^{tr}(\mathbf{K}))$	The value of $c(\mathbf{K}\vert\mathcal{S})$ if $\mathcal{S}$ is the set of all translations.
$c^{co}(\mathbf{K}))$	The value of $c(\mathbf{K}\vert\mathcal{S})$ if $\mathcal{S}$ is the set of all isometries.
$\mathrm{vol}_\mathbf{M}^{Bus}(\cdot)$	Busemann volume in the norm with unit ball $\mathbf{M}$.
$\mathrm{vol}_\mathbf{M}^{HT}(\cdot)$	Holmes-Thompson volume in the norm with unit ball $\mathbf{M}$.
$\mathrm{vol}_\mathbf{M}^{m}(\cdot)$	Gromov's mass in the norm with unit ball $\mathbf{M}$.
$\mathrm{vol}_\mathbf{M}^{m*}(\cdot)$	Gromov's mass* in the norm with unit ball $\mathbf{M}$.
$\mathbf{c}_{tr}^{Bus}(\mathbf{K})$	Maximal Busemann volume of the translation bodies of $\mathbf{K}$ in its relative norm.
$\mathbf{c}_{tr}^{HT}(\mathbf{K})$	Maximal Holmes-Thompson volume of the translation bodies of $\mathbf{K}$ in its relative norm.
$\mathbf{c}_{tr}^{m}(\mathbf{K})$	Maximal Gromov's mass of the translation bodies of $\mathbf{K}$ in its relative norm.
$\mathbf{c}_{tr}^{m*}(\mathbf{K})$	Maximal Gromov's mass* of the translation bodies of $\mathbf{K}$ in its relative norm.
$\mathbf{B}^d[\mathbf{p}_i,r_i]$	Closed Euclidean ball of radius r_i and centered at $\mathbf{p}_i$.
$\mathbf{B}_{\mathbb{M}^d}[\mathbf{x},r]$	Closed ball of radius r_i and centered at $\mathbf{p}_i$ in the space $\mathbb{M}^d$.
X^r	The r-dual body of X in the space $\mathbb{M}^d$.
$k(d)$	Kissing number of $\mathbf{B}^d$.
$c(n,d)$	Maximum contact number in all packings of n unit balls in $\mathbb{E}^d$.
$c(\mathbf{K},n,d)$	Maximum contact number in all packings of n translates of the convex body $\mathbf{K}$ in $\mathbb{E}^d$.
$c_{\mathrm{sep}}(n,d)$	Maximum contact number in all totally separable packings of n unit balls in $\mathbb{E}^d$.
$c_{\mathrm{fcc}}(n)$	Maximum contact number in all packings of n unit diameter balls in $\mathbb{E}^d$ whose centers are points in the face-centered cubic lattice.
$\mathrm{deg}_{\mathrm{avr}}(\cdot)$	Average degree in the contact graph of a packing.
$\mathrm{iq}(\mathbf{K})$	Isoperimetric quotient of $\mathbf{K}$.
$H(\mathbf{K})$	Hadwiger number of $\mathbf{K}$.
$h(\mathbf{K})$	One-sided Hadwiger number of $\mathbf{K}$.

$\delta(\mathbf{K})$	Maximum density of a packing of translates of **K**.
δ_d	Maximum density of a packing of unit balls in $\mathbb{E}^d$.
$\delta_d(n, \lambda)$	Maximum density of a packing of n translates of $\mathbf{B}^d$ with respect to their outer parallel domain of outer radius λ.
$\delta_d(\lambda)$	Upper limit of $\delta_d(n, \lambda)$ as $n \to \infty$.
$c_{\mathbb{Z}}(n, d)$	Maximum contact number in all digital packings of n unit balls in $\mathbb{E}^d$.
$H_{\rm sep}(\mathbf{K})$	Separable Hadwiger number of **K**.
$h_{\rm sep}(\mathbf{K})$	One-sided separable Hadwiger number of **K**.
$c_{\rm sep}(\mathbf{K}, n, d)$	Maximum contact number in all totally separable packings of n translates of **K** in $\mathbb{E}^d$.
$H_{\rm sep}(d)$	Maximum of $H_{\rm sep}$ over the family of **o**-symmetric, strictly convex, smooth bodies in $\mathbb{E}^d$.
$h_{\rm sep}(d)$	Maximum of $h_{\rm sep}$ over the family of **o**-symmetric, strictly convex, smooth bodies in $\mathbb{E}^d$.
$\mathbf{x} \dashv_{\mathbf{K}} \mathbf{y}$	**x** is Birkhoff orthogonal to **y** in the norm of **K**.
$\mathrm{cir}(\mathbf{A})$	Circular pieces of the A-convex domain **A**.
$e(\cdot)$	Euclidean angle measure.
$m(\cdot)$	The angle measure defined by (5.2).
$h(\mathbf{K}, \mathbf{L})$	Hausdorff distance of the convex bodies **K** and **L**.
$\|\|\cdot\|\|_{\mathbf{K}}$	The norm of a vector in the normed space with unit ball **K**.
$M_{\mathbf{K}}(G)$	The length of G measured in the relative norm of **K**.
$\diamond(\mathbf{K})$	A minimum area hexagon circumscribed about **K**.
$\square(\mathbf{K})$	A minimum area parallelogram circumscribed about **K**.
$\delta_{\rm sep}(\mathbf{K})$	Maximum density of all totally separable translative packings of **K**.
$\delta_{\rm sep}(\rho, \mathbf{K})$	Maximum density of all ρ-separable translative packings of **K**.
$M_i(\mathbf{C})$	Mean volume of the i-dimensional projections of **C**.
$R(\mathbf{C})$	Radius of a smallest Euclidean ball containing **C**.
$r(\mathbf{C})$	Radius of a largest Euclidean ball contained in **C**.

Part I

Selected Topics

1

Volumetric Properties of (m, d)-scribed Polytopes

Summary. In this chapter we investigate the volumetric properties of (m, d)-scribed polytopes, that is, d-dimensional polytopes whose every m-face touches a Euclidean ball. Starting with a short historical overview of the isoperimetric problem in Section 1.1, in Section 1.2 we examine a problem proposed by L'huilier in the 18th century, which asks about finding the minimum volume polytopes circumscribed about a ball. By the famous Lindelöf Condition in Theorem 4, these polytopes are the ones having maximal isoperimetric ratio and a fixed number of facets. In this section, besides describing the results regarding L'huilier's problem, we present a 'dynamic' version of Lindelöf's theorem. This version offers a possible explanation of the strange, elongated shape of the interstellar asteroid 'Oumuamua passing through the solar system. In Section 1.3 we investigate the 'dual' of L'huilier's problem, and deal with maximum volume polytopes inscribed in a ball. A consequence of the Circle Packing Theorem is that every combinatorial class of 3-dimensional convex polytope can be represented by a polytope, called Koebe polyhedron, whose every edge is tangent to a unit sphere. In Section 1.4 we collect the properties of polyhedra and, in particular, present results about the existence of Koebe polyhedra in every combinatorial class with additional requirements.

1.1 The isoperimetric inequality

One of the fundamental inequalities in the theory of convex bodies is the so-called *isoperimetric inequality.*

Theorem 1 (Isoperimetric Inequality) *Among all convex bodies in $\mathbb{E}^d$ of equal volumes, Euclidean balls have minimal surface volume.*

The planar case of this problem was essentially solved by Zenodorus, who showed that a circle has larger area than any polygon with the same perimeter. Clearly, in terms of rigorosity, his proof reflects the standards of mathematics at his time. Zenodorus's solution was lost; we know his work through Pappos and Theon of Alexandria [234]. In *Aeneid*, Virgil describes a variant of this question, which might be one of the first instances of a mathematical problem appearing in a literary work written for the general public. In his epic, Dido, the daughter of the king of Tyre, after fleeing from her city, landed on the north coast of Africa, and bargained for as much land as she could enclose with an oxhide. She cut the hide into thin strips, and arranged them roughly into a semicircle.

Later, Steiner gave five different proofs for the planar case [224, 225]. Whereas all his proofs implicitly assume the existence of an optimal solution, his work led to the discovery of Steiner symmetrization, which presently plays an important role in optimization procedures in convex geometry. The first exact proof of the planar case was found in the 19th century, due to Edler [94]. Since then, many different proofs of this statement appeared in the literature (see, e.g., the paper of Carathéodory and Study [74], or that of Lawlor [167]).

Nowadays the d-dimensional isoperimetric inequality is usually proved via the Brunn-Minkowski inequality [102, 194]. In [102], a more general version appears, which states that for any set $\mathbf{S} \subset \mathbb{E}^d$ whose closure has finite Lebesgue measure,

$$d\kappa_d^{\frac{1}{d}} L^d(\mathbf{S})^{\frac{d-1}{d}} \leq M_\star^{d-1}(\mathrm{bd}\mathbf{S}),$$

where $M_\star^{d-1}$ is the $(d-1)$-dimensional Minkowski content, L^d is the d-dimensional Lebesgue measure, and κ_d is the volume of the unit ball in $\mathbb{E}^d$. Other proofs of the isoperimetric inequality were given, for example, by Schmidt [215] and Gromov in the Appendix of [183].

A significant generalization of the isoperimetric inequality follows from a result of Alexandrov and Fenchel [69].

Theorem 2 (Alexandrov-Fenchel Inequality) *For any convex bodies* $\mathbf{K}_1, \mathbf{K}_2, \dots, \mathbf{K}_d \subset \mathbb{E}^d$, *we have*

$$V(\mathbf{K}_1, \mathbf{K}_2, \dots, \mathbf{K}_d)^2 \geq V(\mathbf{K}_1, \mathbf{K}_1, \mathbf{K}_3 \dots, \mathbf{K}_d)V(\mathbf{K}_2, \mathbf{K}_2, \mathbf{K}_3 \dots, \mathbf{K}_d), \quad (1.1)$$

where $V(\mathbf{K}_1, \mathbf{K}_2, \mathbf{K}_3, \dots, \mathbf{K}_d)$ *denotes the mixed volume of the convex bodies* $\mathbf{K}_1, \mathbf{K}_2, \dots, \mathbf{K}_d$.

The necessary and sufficient conditions for equality in (1.1) is known only in some special cases [217]. The inequality (1.1) implies that the sequence $\left\{\frac{W_{i+1}(\mathbf{K})}{W_i(\mathbf{K})}\right\}$ is decreasing for $i = 0, 1, \dots, n-1$, where the quantity $W_i(\mathbf{K})$, called *ith quermassintegral* of the convex body $\mathbf{K} \subset \mathbb{E}^d$, is defined as $W_i(\mathbf{K}) = V\left(\overbrace{\mathbf{K}, \dots, \mathbf{K}}^{d-i}, \overbrace{\mathbf{B}^d, \dots, \mathbf{B}^d}^{i}\right)$. A consequence of this fact is that for any $0 \leq i < j \leq q$, among convex bodies with a given ith quermassintegral, Euclidean balls have minimum jth quermassintegral. Here we note that the surface volume of a convex body $\mathbf{K}$ is $dW_{d-1}(\mathbf{K})$, $W_0(\mathbf{K}) = \mathrm{vol}_d(\mathbf{K})$, and $W_d(\mathbf{K}) = \mathrm{vol}_d(\mathbf{B}^d)$.

Schmidt [216] proved the isoperimetric inequality for convex bodies in spherical or hyperbolic d-space. Variants for arbitrary Riemannian geometries can be found, e.g., in [186]. In [71], Busemann proved the following isoperimetric inequality in normed planes. Let $M_{\mathbf{B}}(L)$ denote the perimeter of the simple, continuous curve L measured in the norm with unit ball $\mathbf{B}$ for some origin-symmetric plane convex body $\mathbf{B}$. Let $\mathbf{I_B}$ denote the rotated copy of the polar reciprocal $\mathbf{B}^\circ$ of $\mathbf{B}$ about the origin, with angle $\frac{\pi}{2}$. Then, for any convex body $\mathbf{K}$ with a given area in the normed plane with unit ball $\mathbf{B}$, the Minkowski perimeter $M_{\mathbf{B}}(\mathrm{bd}\mathbf{K})$ is minimal if and only if $\mathbf{K}$ is a positive homothetic copy of $\mathbf{I_B}$. A variant of this statement holds in d-dimensional normed spaces for $d > 2$, with a suitable definition of surface volume [72].

It is clear that there are convex bodies in $\mathbb{E}^d$ with fixed volume but arbitrarily large surface volume. Nevertheless, Ball [16] proved the following very interesting theorem, which was the starting point of the flourishing area of affine isoperimetric inequalities.

Theorem 3 *Let $\mathbf{K}$ be a convex body in $\mathbb{E}^d$. Then $\mathbf{K}$ has an affine image $\tilde{\mathbf{K}}$ such that the quantity $\frac{\mathrm{vol}_d(\tilde{\mathbf{K}})}{(\mathrm{svol}_{d-1}(\tilde{\mathbf{K}}))^{\frac{d}{d-1}}}$ is not smaller than the corresponding quantity for a regular simplex in $\mathbb{E}^d$.*

1.2 Discrete isoperimetric inequalities: volume of polytopes circumscribed about a sphere

Before proceeding further, following Pisanski et al. [201], we define the *isoperimetric ratio* of a convex body in the following way.

Definition 1 *Let $\mathbf{K} \subset \mathbb{E}^d$ be a convex body. The* isoperimetric ratio *of K is the quantity*

$$I(\mathbf{K}) = \frac{\mathrm{vol}_d(\mathbf{K})}{(\mathrm{svol}_{d-1}(\mathbf{K}))^{\frac{d}{d-1}}} \cdot \frac{(\mathrm{svol}_{d-1}(\mathbf{B}^d))^{\frac{d}{d-1}}}{\mathrm{vol}_d(\mathbf{B}^d)}.$$

We note that for every convex body $\mathbf{K}$, $0 < I(\mathbf{K}) \leq 1$, where $I(\mathbf{K}) = 1$ if and only if $\mathbf{K}$ is a Euclidean ball.

It is a natural question to ask about the maximum value of $I(\mathbf{K})$ on subfamilies of the family of d-dimensional convex bodies. The first result is due to Zenodorus, who proved that among convex n-gons in $\mathbb{E}^2$ of a given perimeter, the regular one has the largest area. It was probably L'huilier in the 18th century who suggested to find the convex polyhedra in $\mathbb{E}^3$ with maximal volume among those with a given surface area and a given number of faces. This question is particularly interesting for polyhedra whose number of faces is equal to that of one of the five platonic solids.

A significant step towards solving this question was made by Lindelöf [171], which was later generalized by Klee [153] for arbitrary dimensions.

Theorem 4 *Let* $\mathbf{P}$ *be a convex polytope in* $\mathbb{E}^d$ *with outer unit facet normal vectors* $\mathbf{x}_1, \mathbf{x}_2, \ldots, \mathbf{x}_n$. *Let* $\mathbf{Q}$ *be the convex polytope circumscribed about the unit ball* $\mathbf{B}^d$ *with the same outer unit face normal vectors. Then* $I(\mathbf{P}) \leq I(\mathbf{Q})$.

A consequence of Theorem 4 is that to maximize $I(\cdot)$ among convex polyhedra with n faces, it is sufficient to minimize the volume (or surface area) of polyhedra with n faces circumscribed about the unit ball $\mathbf{B}^3$.

Consider a convex polyhedron $\mathbf{Q}$ circumscribed about $\mathbf{B}^3$. If the center of mass of a face of $\mathbf{Q}$ does not coincide with the point of tangency on the face, then a slight deformation of the face yields a convex polyhedron $\mathbf{Q}'$, combinatorially equivalent to $\mathbf{Q}$ and circumscribed about $\mathbf{B}^3$, with $\mathrm{vol}_d(\mathbf{Q}') < \mathrm{vol}_d(\mathbf{Q})$. This simple observation, appeared in [153], combined with Theorem 4, implies that among tetrahedra in $\mathbb{E}^3$, the isoperimetric ratio is maximal for the regular ones. A similar argument can be applied to show the same statement for simplices in $\mathbb{E}^d$ with $d > 3$.

Theorem 5 *For any* $n \geq d+1$, *there is a convex polytope* $\mathbf{P}$ *in* $\mathbb{E}^d$ *with* n *facets such that* $I(\mathbf{P}) \geq I(\mathbf{Q})$ *for any convex polytope* $\mathbf{Q}$ *in* $\mathbb{E}^d$ *with* n *facets. Furthermore, any such polytope* $\mathbf{P}$ *is circumscribed about a Euclidean ball, and the point of tangency on each facet of* $\mathbf{P}$ *coincides with the center of mass of the facet.*

Theorem 6 *Among simplices in* $\mathbb{E}^d$, *the regular ones have maximal isoperimetric ratio.*

L. Fejes Tóth [109] proved the following theorem (see also [106]).

Theorem 7 *Let* F *and* V *denote the surface area and the volume of a convex polyhedron in* $\mathbb{E}^3$ *with* n *faces, and set* $\omega_n = \frac{n}{n-2}\frac{\pi}{6}$. *Then*

$$\frac{F^3}{V^2} \geq 54(n-2)\tan(\omega_n)\left(4\sin^2\omega_n - 1\right),$$

where equality holds only for the regular tetrahedron, hexahedron and dodecahedron.

By Theorem 4, this result is based on considering only polyhedra circumscribed about $\mathbf{B}^3$ and applying the so-called 'moment lemma' after centrally projecting each face of the polyhedron to $\mathbb{S}^2$.

Theorem 8 (Moment lemma) *Let* $\varphi(\rho)$ *be a strictly increasing function for* $0 \leq \rho < \frac{\pi}{2}$. *Let* $\mathbf{p}_1, \mathbf{p}_2, \ldots, \mathbf{p}_n$ *be* n *points of* $\mathbb{S}^2$ *not all of them lying on a hemisphere, and let* $\rho_{\mathbf{p}} = \min\{\mathbf{p}\mathbf{p}_1, \mathbf{p}\mathbf{p}_1, \ldots, \mathbf{p}\mathbf{p}_n\}$ *denote the spherical* *distance of a variable point* $\mathbf{p}$ *of* $\mathbb{S}^2$ *from a closest point of the system* $\{\mathbf{p}_i\}$. *Then, for the surface-integral of* $\varphi(\rho_{\mathbf{p}})$ *extended over* $\mathbb{S}^2$, *we have*

$$\int_{\mathbb{S}^2} \varphi(\rho_{\mathbf{p}})\, d\omega \geq (2n-4) \int_{\bar{\mathbf{\Delta}}} \varphi(\bar{\rho}_{\mathbf{p}})\, d\omega,$$

where $\bar{\mathbf{\Delta}}$ *is an equilateral spherical triangle with vertices* $\bar{\mathbf{p}}_1, \bar{\mathbf{p}}_2, \bar{\mathbf{p}}_3$ *having area* $\frac{2\pi}{n-2}$ *and* $\bar{\rho}_{\mathbf{p}} = \min\{\mathbf{p}\bar{\mathbf{p}}_1, \mathbf{p}\bar{\mathbf{p}}_2, \mathbf{p}\bar{\mathbf{p}}_3\}$ *is the spherical distance of* $\mathbf{p}$ *from a closest vertex of* $\bar{\mathbf{\Delta}}$.

Equality holds only if the system $\{\mathbf{p}_i\}$ *coincides with the system of vertices of a regular tetrahedron, octahedron or icosahedron.*

It is worth noting that by results of Lindelöf [171] and Goldberg [121], a regular triangle, respectively pentagon, based cylinder has maximal isoperimetric ratio among convex polyhedra in $\mathbb{E}^3$ with 5, respectively 7, faces, but the problem is still open for $n > 7$, $n \neq 12$ if $d = 3$, and for $n > d+1$ if $d > 3$. The interested reader can find a list on possibly optimal convex polyhedra in $\mathbb{E}^3$ on the webpage: http://schoengeometry.com/a_poly.html, in [191] and in [170].

An asymptotic estimate for the isoperimetric ratio for any $d \geq 2$ can be found in [129].

Theorem 9 *Let* I_d^n, *where* $n \geq d+1$, *denote the maximum of the isoperimetric ratio of the family of all convex polytopes in* $\mathbb{E}^d$ *with at most* n *facets. Then there is a constant* $\delta > 0$ *depending only on* d *such that*

$$I_d^n \sim \frac{\delta}{2d^{\frac{d}{d-1}}} \frac{\operatorname{svol}_{d-1}(\mathbf{B}^n)^{\frac{2d+1}{d-1}}}{\operatorname{vol}_d(\mathbf{B}^n)} \frac{1}{n^{\frac{2}{d-1}}}, \quad \text{as } n \to \infty.$$

The isoperimetric ratio plays an important role, among other things, in the theory of geometric partial differential equations (PDEs), in particular in curvature-driven flows, cf. e.g., [115, 137, 148]. We present a result which yields a 'dynamic' version of Theorem 4 based on the monotonicity of the isoperimetric ratio under the so-called Eikonal equation. The Eikonal equation is a non-linear PDE describing, among other things, the evolution of surfaces. Restricting it to the evolution $\mathbf{K}(t)$ of a convex body $\mathbf{K}(0) = \mathbf{K} \subset \mathbb{E}^d$, where $d > 1$, we may define it as

$$\mathbf{v}(\mathbf{p}) = 1, \tag{1.2}$$

where $\mathbf{p} \in \operatorname{bd}\mathbf{K}(t)$, and $\mathbf{v}(\mathbf{p})$ is the speed by which $\mathbf{p}$ moves in the direction of the inward surface normal of $\mathbf{K}(t)$ at $\mathbf{p}$. We note that by Alexandrov's theorem, stating that the boundary of a convex body is C^2-differentiable almost everywhere, it can be shown that the condition in (1.2) uniquely determines $\mathbf{K}(t)$. The next two theorems were proven by Domokos and Lángi [89], where we use the observation, explained more thoroughly in Section 6.2, that the set $\mathbf{K}(t)$ evolving under (1.2) is a convex body if and only if, $t < r(\mathbf{K})$, where

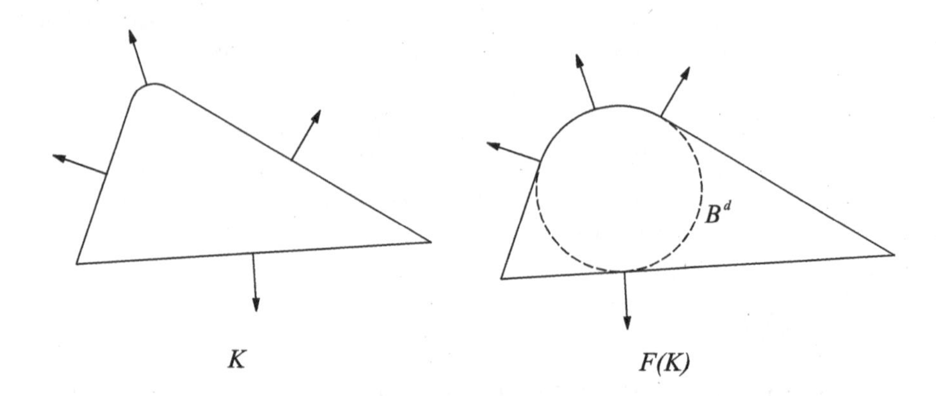

FIGURE 1.1
The form body of a convex body.

$r(\mathbf{K})$ is the radius of a largest sphere contained in $\mathbf{K}$. We note that Theorem 10 offers a possible explanation of the strange, elongated shape of the interstellar asteroid 'Oumuamua passing through the solar system [90].

Theorem 10 *For any convex body $\mathbf{K} \subset \mathbb{E}^d$, where $d > 1$, for the family $\mathbf{K}(t)$ evolving under (1.2), $I(\mathbf{K}(t))$ is either strictly decreasing on $[0, r(\mathbf{K}))$, or there is some value $t^\star \in [0, r(\mathbf{K}))$ such that $I(\mathbf{K}(t))$ is strictly decreasing on $[0, t^\star]$ and is a constant on $[t^\star, r(\mathbf{K}))$. Furthermore, in the latter case, for any $t > t^\star$, $\mathbf{K}(t)$ is homothetic to $\mathbf{K}(t^\star)$.*

Definition 2 *Let $\mathbf{M} \subset \mathbb{E}^d$ be a convex body, where $d > 1$. Let $N(\mathbf{M}) \subset \mathbb{S}^{d-1}$ denote the set of unit vectors which are external unit normal vectors of $\mathbf{M}$ at a smooth point of* $\mathrm{bd}\mathbf{M}$. *Then we call the set*

$$F(\mathbf{M}) = \{\mathbf{p} \in \mathbb{E}^d : \langle \mathbf{p}, \mathbf{x} \rangle \leq 1 \, \text{for every} \, \mathbf{x} \in N(\mathbf{M})\}$$

the form body *of $\mathbf{M}$ [217] (cf. Figure 1.1).*

Like in case of (1.2), by Alexandrov's theorem it follows that $F(\mathbf{M})$ is a convex body for every convex body $\mathbf{M} \subset \mathbb{E}^d$. Furthermore, if $\mathbf{M}$ is a convex polytope, then $F(\mathbf{M})$ is the convex polytope, circumscribed about $\mathbf{B}^d$, whose set of external unit facet normal vectors coincides with that of $\mathbf{M}$.

Theorem 11 *Let $\mathbf{K} \subset \mathbb{E}^d$ be a convex body, where $d > 1$. Then the quantity $I(\mathbf{K} + tF(\mathbf{K}))$ is an increasing function of $t \in [0, \infty)$.*

The proof of the fact that among spherical or hyperbolic n-gons of a given area, the regular ones have smallest perimeter can be found in [70]. Chakerian solved the same problem for arbitrary normed planes in [75]. Nevertheless, almost nothing is known about extremal polytopes in these spaces in dimensions $d > 2$. We ask the following questions.

Question 12 *Can Theorem 4 be modified for convex polytopes in $\mathbb{S}^d$ or $\mathbb{H}^d$?*

Open Problem 1 *Prove or disprove that among simplices in $\mathbb{S}^d$ or $\mathbb{H}^d$ of a given volume, where $d > 2$, the regular ones have smallest surface volume.*

We note that like in the Euclidean case, Theorem 8 implies that among tetrahedra in $\mathbb{S}^3$ or in $\mathbb{H}^3$ circumscribed about a ball, the regular ones have smallest volume.

Open Problem 2 *Prove or disprove that among simplices in $\mathbb{S}^d$ or $\mathbb{H}^d$ circumscribed about a ball, where $d > 3$, the regular ones have smallest volume.*

1.3 Volume of polytopes inscribed in a sphere

The dual of L'huilier's problem can be put as follows.

Open Problem 3 *Among convex polytopes in $\mathbb{E}^d$, inscribed in $\mathbb{S}^{d-1}$ and having n vertices, which polytope has the largest volume?*

This problem, with $d = 3$, was first mentioned by L. Fejes Tóth in [106] in 1964, who proved the following.

Theorem 13 *Let $\mathbf{P}$ be a convex polyhedron in $\mathbb{E}^3$ with n vertices, insribed in $\mathbf{B}^3$, and let $\omega_n = \frac{n}{n-2}\frac{\pi}{6}$. Then*

$$\mathrm{vol}_3(\mathbf{P}) \leq \frac{1}{6}(n-2)\cot\omega_n\left(3 - \cot^2\omega_n\right).$$

Here, equality holds if and only if $\mathbf{P}$ is a regular tetrahedron, a regular octahedron, or a regular icosahedron.

A systematic investigation of this question was started with the paper [20] of Berman and Hanes in 1970, who found a Lindelöf-type necessary condition for optimal polyhedra, and determined those with $n \leq 8$ vertices. The optimal configurations are listed below.

Theorem 14 *Let $\mathbf{P}$ be a maximal volume polyhedron with n vertices and inscribed in $\mathbb{S}^2$.*

(i) *If $n \leq 7$, then $\mathbf{P}$ is a double pyramid, with a regular $(n-2)$-gon centered at the origin as its base, and its two apexes lying on the line through the origin and perpendicular to its base.*

(ii) If $n = 8$, then $\mathbf{P}$ is congruent to the polyhedron with the vertices:

$$\mathbf{p}_1 = (\sin 3\varphi, 0, \cos 3\varphi), \quad \mathbf{p}_5 = (0, -\sin 3\varphi, -\cos 3\varphi),$$
$$\mathbf{p}_2 = (\sin \varphi, 0, \cos \varphi), \quad \mathbf{p}_6 = (0, -\sin \varphi, -\cos \varphi),$$
$$\mathbf{p}_3 = (-\sin \varphi, 0, \cos \varphi), \quad \mathbf{p}_7 = (0, \sin \varphi, -\cos \varphi),$$
$$\mathbf{p}_4 = (-\sin 3\varphi, 0, \cos 3\varphi), \quad \mathbf{p}_8 = (0, \sin 3\varphi, -\cos 3\varphi),$$

where $0 < \varphi < \frac{\pi}{2}$ and $\cos \varphi = \sqrt{\frac{15+\sqrt{145}}{40}}$.

Note that the average valence of a vertex of a convex polyhedron in $\mathbb{E}^3$ with n vertices and triangular faces is $6 - \frac{12}{n}$.

Definition 3 *Let $\mathbf{P}$ be a convex polyhedron in $\mathbb{E}^3$ with n vertices and triangular faces. We say that $\mathbf{P}$ is* medial *if the valence of every vertex of $\mathbf{P}$ is at least $\lfloor 6 - \frac{12}{n} \rfloor$ and at most $\lceil 6 - \frac{12}{n} \rceil$.*

The next conjecture is due to Goldberg [121].

Conjecture 15 *Let $\mathbf{P}$ be a convex polyhedron with n vertices and inscribed in $\mathbf{B}^3$. If there is a medial polyhedron satisfying these properties, then $\mathrm{vol}_3(\mathbf{P})$ is maximal under these conditions for a medial polyhedron.*

It is an elementary exercise to check that all polyhedra in Theorem 14 are medial. Thus, Conjecture 15 is satisfied for $n \leq 8$.

In 2016, G. Horváth and Lángi [145] extended the necessary condition for optimal polytopes in d-dimensional Euclidean space in the following form. Here, for a d-dimensional polytope $\mathbf{P}$, we denote the set of vertices of $\mathbf{P}$ by $V(\mathbf{P})$. We denote the family of d-dimensional polytopes, with n vertices and inscribed in the unit sphere $\mathbb{S}^{d-1}$, by $\mathcal{P}_d(n)$. We set $v_d(n) = \max\{\mathrm{vol}_d(\mathbf{P}) : \mathbf{P} \in \mathcal{P}_d(n)\}$, and note that by compactness, $v_d(n)$ exists for any values of d and n.

Definition 4 *Let $\mathbf{P} \in \mathcal{P}_d(n)$ be a polytope with vertex set $V(\mathbf{P}) = \{\mathbf{p}_1, \mathbf{p}_2, \dots, \mathbf{p}_n\}$. If for each i, there is an open set $\mathbf{U}_i \subset \mathbb{S}^{d-1}$ such that $\mathbf{p}_i \in \mathbf{U}_i$, and for any $\mathbf{q} \in \mathbf{U}_i$, we have*

$$\mathrm{vol}_d\left((\mathrm{conv}\left(V(\mathbf{P}) \setminus \{\mathbf{p}_i\}\right)) \cup \{\mathbf{q}\}\right) \leq \mathrm{vol}_d\left(\mathbf{P}\right),$$

then we say that $\mathbf{P}$ satisfies Property Z.

Let $\mathbf{P} \in \mathcal{P}_d(n)$ and let $V(\mathbf{P}) = \{\mathbf{p}_1, \mathbf{p}_2, \dots, \mathbf{p}_n\}$. Let $\mathcal{C}(\mathbf{P})$ be a simplicial complex with the property that its support is $|\mathcal{C}(\mathbf{P})| = \mathrm{bd}\mathbf{P}$, and that the vertices of $\mathcal{C}(\mathbf{P})$ are exactly the points of $V(\mathbf{P})$. We orient $\mathcal{C}(\mathbf{P})$ in such a way that for each $(d-1)$-simplex $(\mathbf{p}_{i_1}, \mathbf{p}_{i_2}, \dots, \mathbf{p}_{i_d})$ in $\mathcal{C}(\mathbf{P})$, the determinant $|\mathbf{p}_{i_1}, \dots, \mathbf{p}_{i_d}|$ is positive; and call the d-simplex $\mathrm{conv}\{\mathbf{o}, \mathbf{p}_{i_1}, \dots, \mathbf{p}_{i_d}\}$ a *facial simplex* of $\mathbf{P}$. We call the $(d-1)$-dimensional simplices of $\mathcal{C}(\mathbf{P})$ the *facets* of $\mathcal{C}(\mathbf{P})$.

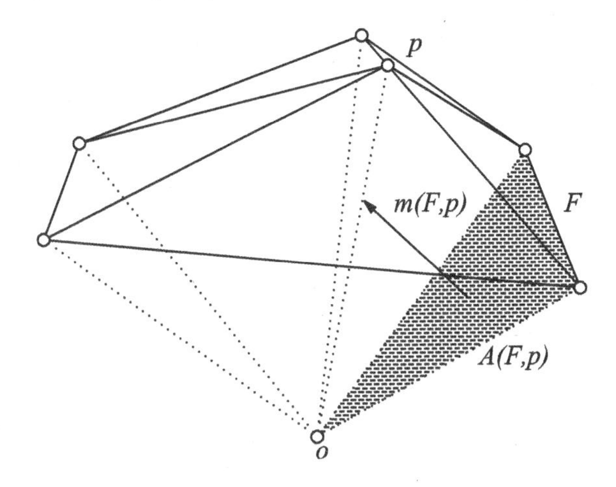

FIGURE 1.2
Notations for Theorem 16.

Theorem 16 *Consider a d-polytope $\mathbf{P}$ inscribed in $\mathbb{S}^{d-1}$ and satisfying Property Z. For any $\mathbf{p} \in V(\mathbf{P})$, let $\mathcal{F}_{\mathbf{p}}$ denote the family of the facets of $\mathcal{C}(\mathbf{P})$ containing $\mathbf{p}$. For any $\mathbf{F} \in \mathcal{F}_{\mathbf{p}}$, set $A(\mathbf{F}, \mathbf{p}) = \mathrm{vol}_{d-1}(\mathrm{conv}((V(\mathbf{F}) \cup \{\mathbf{o}\}) \setminus \{\mathbf{p}\}))$, and let $\mathbf{m}(\mathbf{F}, \mathbf{p})$ be the unit normal vector of the hyperplane, spanned by $(V(\mathbf{F}) \cup \{\mathbf{o}\}) \setminus \{\mathbf{p}\}$, pointing in the direction of the half space containing $\mathbf{p}$ (cf. Figure 1.2). Then*

(i) we have

$$\mathbf{p} = \frac{\mathbf{m}}{||\mathbf{m}||}, \text{ where } \mathbf{m} = \sum_{\mathbf{F} \in \mathcal{F}_{\mathbf{p}}} A(\mathbf{F}, \mathbf{p}) \mathbf{m}(\mathbf{F}, \mathbf{p}), \text{ and}$$

(ii) $\mathbf{P}$ is simplicial.

Here we note that it has been long conjectured, but never proved, that for every $n \geq 4$, every minimal volume convex polyhedron with n faces and circumscribed about $\mathbf{B}^3$ is simple.

By means of Theorem 16, we prove the following. Before stating the result, we remark that Theorem 16 yields possible candidates for an optimal polytope only within a fixed combinatorial class. Since the combinatorial classes of d-polytopes with at least $d + 4$ vertices are unknown [239], it seems that to treat this problem for polytopes with at least $d + 4$ vertices a new approach is needed.

Theorem 17 *Let $\mathbf{P} \in \mathcal{P}_d(d + 2)$ have maximal volume over $\mathcal{P}_d(d + 2)$. Then $\mathbf{P} = \mathrm{conv}(\mathbf{P}_1 \cup \mathbf{P}_2)$, where $\mathbf{P}_1$ and $\mathbf{P}_2$ are regular simplices of dimensions $\lfloor \frac{d}{2} \rfloor$*

and $\lceil \frac{d}{2} \rceil$*, respectively, inscribed in* $\mathbb{S}^{d-1}$*, and contained in orthogonal linear subspaces of* $\mathbb{E}^d$*. Furthermore,*

$$v_d(d+2) = \frac{1}{d!} \cdot \frac{(\lfloor d/2 \rfloor + 1)^{\frac{\lfloor d/2 \rfloor + 1}{2}} \cdot (\lceil d/2 \rceil + 1)^{\frac{\lceil d/2 \rceil + 1}{2}}}{\lfloor d/2 \rfloor^{\frac{\lfloor d/2 \rfloor}{2}} \cdot \lceil d/2 \rceil^{\frac{\lceil d/2 \rceil}{2}}}$$

Recall that a d-polytope with n vertices is *cyclic*, if it is combinatorially equivalent to the convex hull of n points on the moment curve $\gamma(t) = (t, t^2, \ldots, t^d)$, $t \in \mathbb{R}$.

Theorem 18 *Let* $\mathbf{P} \in \mathcal{P}_d(d+3)$ *satisfy Property Z. If* d *is even, assume that* $\mathbf{P}$ *is not cyclic. Then* $\mathbf{P} = \operatorname{conv}\{\mathbf{P}_1 \cup \mathbf{P}_2 \cup \mathbf{P}_3\}$*, where* $\mathbf{P}_1$*,* $\mathbf{P}_2$ *and* $\mathbf{P}_3$ *are regular simplices inscribed in* $\mathbb{S}^{d-1}$ *and contained in three mutually orthogonal linear subspaces of* $\mathbb{E}^d$*. Furthermore:*

(i) If d *is odd and* $\mathbf{P}$ *has maximal volume over* $\mathcal{P}_d(d+3)$*, then the dimensions of* $\mathbf{P}_1$*,* $\mathbf{P}_2$ *and* $\mathbf{P}_3$ *are* $\lfloor d/3 \rfloor$ *or* $\lceil d/3 \rceil$*. In particular, in this case we have*

$$(v_d(d+3) =) \operatorname{vol}_d(\mathbf{P}) = \frac{1}{d!} \cdot \prod_{i=1}^{3} (k_i+1)^{\frac{k_i+1}{2}} k_i^{\frac{k_i}{2}},$$

where $k_1 + k_2 + k_3 = d$ *and for every* i*, we have* $k_i \in \left\{ \lfloor \frac{d}{3} \rfloor, \lceil \frac{d}{3} \rceil \right\}$*.*

(ii) The same holds for the dimensions of $\mathbf{P}_1$*,* $\mathbf{P}_2$ *and* $\mathbf{P}_3$*, if* d *is even, and* $\mathbf{P}$ *has maximal volume over the family of* not cyclic *elements of* $\mathcal{P}_d(d+3)$ *satisfying Property Z.*

In light of Theorem 18, it seems interesting to find the maximum volume cyclic polytopes in $\mathcal{P}_d(d+3)$, with d even. Using the method of the proof of Theorem 18, the following can be shown.

Remark 19 *Let* $d = 2m$ *be even and let* $\mathbf{P} \in \mathcal{P}_d(d+3)$ *be a cyclic polytope satisfying Property Z. Then, for a suitable labeling* $\mathbf{p}_1, \mathbf{p}_2, \ldots, \mathbf{p}_{d+3}$ *of the vertices of* $\mathbf{P}$*, the value of* $||p_{i+1} - p_i||$ *is independent of* i*.*

Unfortunately, this method cannot be used to prove that for a suitable labeling $\mathbf{p}_1, \mathbf{p}_2, \ldots, \mathbf{p}_{d+3}$ of the vertices of $\mathbf{P}$, the value of $||\mathbf{p}_{i+k} - \mathbf{p}_i||$ is independent of i for any value of k different from 1. Nevertheless, in light of Remark 19, it seems reasonable to consider the possibility that if a cyclic polytope $\mathbf{P}$ with $d+3$ vertices in an even dimensional space $\mathbb{E}^d$ and inscribed in $\mathbb{S}^{d-1}$ has maximal volume within its combinatorial class, then $\mathbf{P}$ satisfies the above property for all values of k; or in other words, $\mathbf{P}$ has a dihedral symmetry of order $d+3$.

A possible candidate for this property is the polytope $\mathbf{C}_d(d+3)$ in Remark 20, obtained by choosing suitable points on the trigonometric moment curve $t \mapsto \left(\cos t, \sin t, \cos 2t, \sin 2t, \ldots, \cos\left(\frac{dt}{2}\right), \sin\left(\frac{dt}{2}\right)\right)$, where d is even.

Remark 20 *Let $d \geq 2$ be even, and $n \geq d+3$. Let*

$$\mathbf{C}_d(n) = \operatorname{conv}\{\mathbf{q}_i : i = 0, 1, \ldots, n-1\},$$

where

$$\mathbf{q}_i = \sqrt{\frac{2}{d}} \left(\cos\frac{i\pi}{n}, \sin\frac{i\pi}{n}, \cos\frac{2i\pi}{n}, \ldots, \cos\frac{di\pi}{2n}, \sin\frac{di\pi}{2n} \right)$$

Then $\mathbf{C}_d(n)$ is a cyclic d-polytope inscribed in $\mathbb{S}^{d-1}$, and its group of symmetries $\operatorname{Sym}(\mathbf{C}_d(n))$ is isomorphic to the dihedral group D_n.

It is a straightforward computation to check that $\mathbf{C}_4(7)$ satisfies Property Z. Furthermore, letting $\mathbf{P}_4 \in \mathcal{P}_4(7)$ be the convex hull of a regular triangle and two diameters of $\mathbb{S}^3$, in mutually orthogonal linear subspaces, and $\mathbf{P}_6 \in \mathcal{P}_6(9)$ be the convex hull of three regular triangles in $\mathbb{S}^5$ in mutually orthogonal linear subspaces, it is easy to check that

$$\operatorname{vol}_4(\mathbf{P}_4) = \frac{\sqrt{3}}{4} = 0.43301\ldots,$$

$$\operatorname{vol}_4(\mathbf{C}_4(7)) = \frac{49}{192}\left(\cos\frac{\pi}{7} + \cos\frac{2\pi}{7}\right) = 0.38905\ldots,$$

$$\operatorname{vol}_6(\mathbf{P}_6) = \frac{9\sqrt{3}}{640} = 0.02435\ldots$$

and

$$\operatorname{vol}_6(\mathbf{C}_6(9)) = \frac{7}{576}\sin\frac{\pi}{9} - \frac{7}{2880}\sin\frac{4\pi}{9} + \frac{7}{1152}\sin\frac{2\pi}{9} = 0.01697\ldots,$$

implying that

$$\operatorname{vol}_4(\mathbf{P}_4) > \operatorname{vol}_4(\mathbf{C}_4(7)) \text{ and } \operatorname{vol}_6(\mathbf{P}_6) > \operatorname{vol}_{d6}(\mathbf{C}_6(9)).$$

These inequalities suggest that the optimal polytopes in $\mathcal{P}_d(d+3)$ are not cyclic for any value of d.

A characterization of d-polytopes, with n vertices and symmetry group D_n can be found in [163]. For stating it, we introduce some notation. Let $d \geq 2$, $n > d$, $m = \lfloor d/2 \rfloor$, $0 < i_1 < i_2 < \ldots < i_s < \frac{k}{2}$, and for $k = 1, 2, \ldots, n$, let

$$\mathbf{q}_k = \frac{1}{\sqrt{m}} \left(\cos\frac{2i_1 k\pi}{n}, \sin\frac{2i_1 k\pi}{n}, \ldots, \cos\frac{k i_m \pi}{n}, \sin\frac{2k i_m \pi}{n} \right) \in \mathbb{E}^d$$

if d is even, and

$$\mathbf{q}_k = \frac{1}{\sqrt{m+1}} \left(\cos\frac{2i_1 k\pi}{n}, \sin\frac{2i_1 k\pi}{n}, \ldots, \sin\frac{2k i_s \pi}{n}, (-1)^k \right) \in \mathbb{E}^d$$

if d is odd. Furthermore, set $\mathbf{Q}(i_1, i_2, \ldots, i_n) = \operatorname{conv}\{\mathbf{q}_1, \mathbf{q}_2, \ldots, \mathbf{q}_n\}$. Note that as $||\mathbf{q}_k|| = 1$ for every value of k, the vertices of $\mathbf{Q}(i_1, i_2, \ldots, i_m)$ are $\mathbf{q}_1, \mathbf{q}_2, \ldots, \mathbf{q}_n$.

Theorem 21 *Let $d \geq 2$, and $\mathbf{P} \subset \mathbb{E}^d$ be a d-dimensional convex polytope with vertices $\mathbf{p}_1, \mathbf{p}_2, \ldots, \mathbf{p}_n$, where $n \geq 5$, and $n > d$. Then the following are equivalent.*

(i) For $k, i = 1, 2, \ldots, n$, the value of $||\mathbf{p}_{k+i} - \mathbf{p}_k||$ is independent of the value of i.

(ii) There is some $\phi \in \mathrm{Sym}(\mathbf{P})$ such that $\phi(\mathbf{p}_i) = \mathbf{p}_{i+1}$ for $i = 1, 2, \ldots, n$.

(iii) There are some $0 < i_1 < i_2 < \ldots < i_{\lfloor d/2 \rfloor} < \frac{n}{2}$ such that $\mathbf{P}$ is similar to $\mathbf{Q}(i_1, i_2, \ldots, i_{\lfloor d/2 \rfloor})$.

Unlike in the case of circumscribed simplices of minimal volume, it was proved by Böröczky [57] that among simplices in the spherical space $\mathbb{S}^d$ inscribed in a spherical ball, the regular ones have maximal volume. The same statement for simplices in the hyperbolic space $\mathbb{H}^d$ was proved by Peyerimhoff [200].

We finish this section with a question of G. Fejes Tóth, which appeared in [32].

Open Problem 4 *By Steiner symmetrization, it can be easily shown that the maximum volume of the intersection of a fixed ball in $\mathbb{E}^d$ and a variable simplex of given volume V is attained when the simplex is regular and concentric with the ball. Show that the above statement holds true in spherical and hyperbolic space as well.*

1.4 Polyhedra midscribed to a sphere

In the last two sections we examined, in particular, convex polyhedra in $\mathbb{E}^3$ whose every k-face, where $k = 0$ or $k = 2$, is tangent to the unit sphere $\mathbb{S}^2$. In this section we deal with the missing case $k = 1$, and introduce one of the 'gems' in the treasury of geometric literature.

The famous Circle Packing Theorem [197] states that every simple, connected plane graph can be realized as the intersection graph of a circle packing in the Euclidean plane, or equivalently, using a stereographic projection, on the sphere. Here, by intersection graph we mean a graph whose vertices are the centers of some mutually non-overlapping circles, and two vertices are connected if the corresponding circles are tangent.

This theorem was first proved by Koebe [156], and was later rediscovered by Thurston [235], who noted that this result also follows from the work of Andreev [11, 12]. The theorem has induced a significant interest in circle packings in many different settings, and has been generalized in many directions. One of the most known variants is due to Brightwell and Scheinerman [68]. By this result, any polyhedral graph (i.e., any simple, 3-connected planar graph

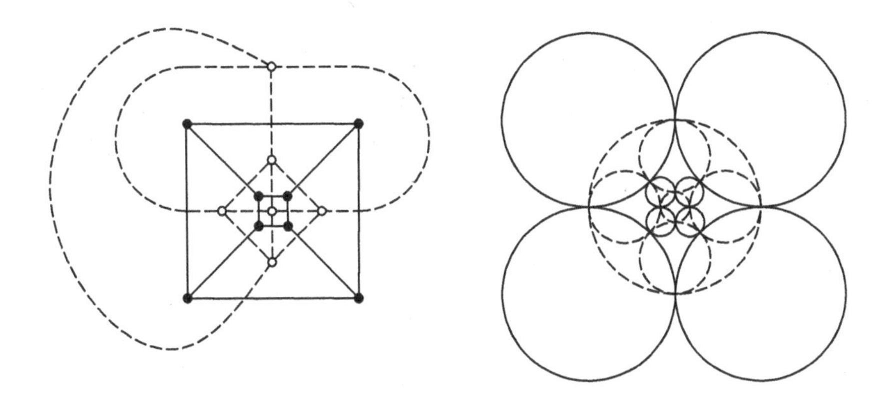

FIGURE 1.3
Representation of a polyhedral graph and its dual as intersection graphs of circle packings.

[228, 230]), together with its dual graph, can be realized simultaneously as intersection graphs of two circle packings with the property that each point of tangency belongs to two pairs of tangent circles which are orthogonal to each other.

Figure 1.3 shows an example of this representation. Here, on the left, the edge graph of a cube $\mathbf{C}$ is shown; filled circles are the vertices of the cube, edges connecting vertices are represented by continuous curves. The vertices and the edges of the edge graph of the dual of $\mathbf{C}$, an octahedron, are represented by empty circles and dashed curves, respectively. The corresponding circle packings are shown on the right. Here, circles drawn with continuous and dashed lines correspond to the vertices of $\mathbf{C}$ and its dual, respectively. Note that even though here a dashed circle contains all other dashed circles in its interior, using a suitable stereographic projection to $\mathbb{S}^2$, the corresponding spherical cap does not overlap the images of the other dashed circles.

Such a pair of families of circles on the unit sphere $\mathbb{S}^2$ centered at the origin $\mathbf{o}$ generate a convex polyhedron *midscribed* to the sphere; that is, having all edges tangent to it. In this polyhedron, members of one family, called *face circles*, are the incircles of the faces of the polyhedron, and members of the other family, called *vertex circles*, are circles passing through all edges starting at a given vertex. This yields the following theorem [68].

Theorem 22 *The combinatorial class of every convex polyhedron has a representative midscribed to the unit sphere* $\mathbb{S}^2$.

Such representatives of combinatorial classes are called Koebe polyhedra. By Mostow's rigidity theorem [187], these representations are unique up to

Möbius transformations of the sphere. Theorem 22 was significantly strengthened by Schramm [218], who proved the same statement for the boundary of any strictly convex, smooth body **K** playing the role of $\mathbb{S}^2$.

It is an interesting question to ask if Theorem 22 holds for representatives of polyhedral classes inscribed in or circumscribed about $\mathbb{S}^2$, or if not, which classes can be represented in this way. These two problems are equivalent, as the polar of any convex polyhedron inscribed in $\mathbb{S}^2$ is circumscribed about it and vice versa. These questions were posed by Steiner [226]. The first part of this question was solved only one hundred years later by Steinitz [229], who, using a variational principle and Lindelöf's result in Theorem 4 gave the first example, an infinite family of convex polyhedra, whose combinatorial classes do not contain representatives inscribed in $\mathbb{S}^2$. His result is a consequence of Theorem 23 from [229].

Theorem 23 *Let* **P** *be a convex polyhedron in* $\mathbb{E}^3$ *such that its vertices are colored black and white vertices such that it has more black vertices than white, and no two black vertices are adjacent. Then* **P** *cannot be inscribed in a sphere.*

Observe that the condition about **P** in Theorem 23 is purely combinatorial, and thus it holds for every polyhedron in the combinatorial class of **P**. We give an example of a convex polyhedron satisfying this condition, described also in [114].

Example 1 *Consider an octahedron* $\mathbf{P}_0$ *and color its vertices white. Attach a tetrahedron to each face of* $\mathbf{P}_0$*, and color the new vertices black. The so-obtained polyhedron* **P** *has six white and eight black vertices, and no two black vertices are consecutive.*

Whereas the condition in Theorem 23 is necessary for inscribability, it is not sufficient; it can be shown, for example, that the triakis tetrahedron, obtained by attaching a tetrahedron to each face of a tetrahedron, is not inscribable. Sufficient conditions for inscribability were given, e.g., by Dillencour and Smith [88], who proved in particular, that every 4-*connected* planar graph is the edge-graph of an inscribable convex polyhedron. The second part of Steiner's problem was completely solved by Rivin [206] in 1996, who characterized convex polyhedra inscribed in $\mathbb{S}^2$.

Higher dimensional analogues of Theorem 22 were examined by Schulte [219]. He called a convex polytope **P** in $\mathbb{E}^d$ (m, d)-scribable for some $0 \leq m \leq d-1$, if there is a convex d-polytope $\mathbf{P}'$ combinatorially equivalent to **P** whose every m-face is tangent to some Euclidean $(d-1)$-sphere. His result states that if $d \geq 4$, then for every value of m, there is a convex polytope **P** in $\mathbb{E}^d$ which is not (m, d)-scribable. Thus, the property in Theorem 22 holds only for 3-dimensional polytopes whose edges are tangent to a unit sphere. On the other hand, Padrol and Ziegler [198] proved the inscribability of certain families of convex polytopes in $\mathbb{E}^d$ with $d \geq 4$.

We note that using the projective ball model for $\mathbb{H}^3$ and central projection from $\mathbb{E}^3$ to $\mathbb{S}^3$ for $\mathbb{S}^3$, Theorem 22 can be stated in the same form for the combinatorial classes of convex polyhedra not only in Euclidean, but also for hyperbolic or spherical spaces.

Remark 24 *The question for midscribed polyhedra analoguous to those for circumscribed and inscribed polyhedra would be the following: For any value $e \geq 6$, find the polyhedra of maximal/minimal volume with e edges and midscribed to $\mathbb{S}^2$. Nevertheless, as we will see, the method of the proof of Theorem 28 yields that in every combinatorial class there is a Koebe polyhedron with arbitrarily small, and also one with arbitrarily large volume.*

In light of Remark 24, after proposing some related problems, we investigate Koebe polyhedra in a different direction. In their famous paper [18], among other things, Bárány and Füredi gave an upper bound on the radii of Euclidean balls in $\mathbb{E}^d$, centered at a point of a simplex $\mathbf{S}$ contained in $\mathbf{B}^d$ and satisfying the property that the k-skeleton of $\mathbf{S}$ is disjoint from the interior of the ball. They remarked that it is very likely that the best upper bound is attained if $\mathbf{S}$ is a regular simplex with $\mathbf{o}$ as its center, and the ball is centered at $\mathbf{o}$ and touches each k-face of $\mathbf{S}$.

Open Problem 5 *Prove or disprove that if $\mathbf{S}$ is a simplex contained in $\mathbf{B}^d$, and for some $\mathbf{x} \in \mathbf{S}$, $\mathbf{x} + \rho\mathbf{B}^d$ is a Euclidean ball whose interior is disjoint from the k-skeleton of $\mathbf{S}$, then $\rho \leq \sqrt{\frac{d-k}{d(k+1)}}$.*

Question 25 *Does the statement in Problem 5 hold for spherical or hyperbolic simplices?*

Question 26 *Consider a ball $\mathbf{S}$ in $\mathbb{S}^3$ or in $\mathbb{H}^3$. What is the infimum and the supremum of volumes of the tetrahedra midscribed to $\mathbf{S}$?*

Mani [175] strengthened Theorem 22 in a different direction. He proved that up to *Euclidean isometries*, every combinatorial class can be uniquely represented by a polyhedron midscribed to $\mathbb{S}^2$ such that the barycenter of the tangency points is the origin (cf. also [239, p.118] and [131, p.296a]). This result is an example of the problem of 'centering' via Möbius transformations, which was examined, from an algorithmic point of view in [22]. Springborn [222] gave an elegant different proof of Mani's statement, based on the application of the following theorem.

Theorem 27 *For any mutually distinct points $\mathbf{v}_1, \mathbf{v}_2, \ldots, \mathbf{v}_n$ on the d-dimensional unit sphere $\mathbb{S}^d$ centered at the origin $\mathbf{o}$, where $n \geq 3$ and $d \geq 2$, there is a Möbius transformation T of $\mathbb{S}^d$ such that $\sum_{i=1}^n T(\mathbf{v}_i) = o$. Furthermore, if $\tilde{T}$ is another such Möbius transformation, then $\tilde{T} = RT$, where R is an isometry of $\mathbb{S}^d$.*

A slightly more general approach of this problem can be found in [14], where the authors prove that the discrete measure in Theorem 27 can be replaced by any reasonable measure defined on $\mathbb{S}^2$. Their argument seems to work also for measures defined on $\mathbb{S}^d$. We note that in all the mentioned results some configurations *on* $\mathbb{S}^d$ are centered.

Lángi [162] examined a variant of this problem, namely the problem of centering Koebe polyhedra. It is worth mentioning that this problem cannot be regarded as centering a suitable measure on $\mathbb{S}^2$. Indeed, intuitively, whereas the measure on a certain subset of $\mathbb{S}^d$ is invariant under Möbius transformations, the same cannot be stated about Koebe polyhedra. This difference is represented in Remark 30.

To state the results in [162] we introduce some notations.

Let $\mathbf{P}$ be a convex polyhedron in $\mathbb{E}^3$. For $k = 0, 1, 2, 3$, let $\mathrm{skel}_k(\mathbf{P})$ denote the k-skeleton of $\mathbf{P}$. Then the *center of mass of* $\mathrm{skel}_k(\mathbf{P})$, which we denote by $\mathrm{cm}_k(\mathbf{P})$, is defined in the usual way as

$$\mathrm{cm}_k(\mathbf{P}) = \frac{\int_{\mathbf{p}\in \mathrm{skel}_k(\mathbf{P})} \mathbf{p}\, dv_k}{\int_{\mathbf{p}\in \mathrm{skel}_k(\mathbf{P})} dv_k},$$

where v_k denotes k-dimensional Lebesgue measure. By $\mathrm{cc}(\mathbf{P})$ and $\mathrm{IC}(\mathbf{P})$, we denote the center of the (unique) smallest ball containing $\mathbf{P}$ and the set of the centers of the largest balls contained in $\mathbf{P}$.

The next concept was defined for polygons in [1] and for simplicial polytopes in [231] (see also [3]). Before introducing it, we point out that, unlike in the other parts of the book, by the circumcenter of a non-degenerate simplex we mean the center of the unique sphere containing all vertices of the simplex, which may be different from the center of the smallest ball containing the simplex.

Definition 5 *Let* $\mathbf{P}$ *be an oriented simplicial polytope, and let* $\mathbf{o}$ *be a given reference point not contained in any of the facet hyperplanes of* $\mathbf{P}$*. Triangulate* $\mathbf{P}$ *by simplices whose bases are the facets of* $\mathbf{P}$ *and whose apex is* $\mathbf{o}$*. Let* $\mathbf{p}_i$ *and* m_i *denote, respectively, the circumcenter and the volume of the* i*th such simplex. Then the* circumcenter of mass *of* $\mathbf{P}$ *is defined as*

$$\mathrm{ccm}(\mathbf{P}) = \frac{\sum_i m_i \mathbf{p}_i}{\sum_i m_i}.$$

The authors of [231] show that the circumcenter of mass of a simplicial polytope $\mathbf{P}$ is

(i) independent of the choice of the reference point,

(ii) remains invariant under triangulations of $\mathbf{P}$ if no new vertex is chosen from the boundary of $\mathbf{P}$,

(iii) satisfies Archimedes' lemma: if we decompose $\mathbf{P}$ into two simplicial polytopes $\mathbf{Q}_1$ and $\mathbf{Q}_2$ in a suitable way, then $\operatorname{ccm}(\mathbf{P})$ is the weighted average of $\operatorname{ccm}(\mathbf{Q}_1)$ and $\operatorname{ccm}(\mathbf{Q}_2)$, where the weights are the volumes of $\mathbf{Q}_1$ and $\mathbf{Q}_2$,

(iv) if $\mathbf{P}$ is inscribed in a sphere, then its circumcenter of mass coincides with its circumcenter.

In addition, they use this point to define the *Euler line* of a simplicial polytope as the affine hull of the center of mass $\operatorname{cm}_3(\mathbf{P})$ of $\mathbf{P}$ and $\operatorname{ccm}(\mathbf{P})$. This definition is a generalization of the same concept defined for simplices. They show that for polygons, any notion of 'center' satisfying some elementary properties (i.e., it depends analytically on the vertices of the polygon, commutes with dilatations and satisfies Archimedes's lemma) is necessarily a point of the Euler line.

In the following theorem, with a little abuse of notation, if $\mathbf{P}$ is a Koebe polyhedron and T is a Möbius transformation, by $T(\mathbf{P})$ we mean the polyhedron defined by the images of the face circles and the vertex circles of $\mathbf{P}$ under T.

Theorem 28 *Let $S = \{\operatorname{cc}(\cdot), \operatorname{cm}_0(\cdot), \operatorname{cm}_1(\cdot), \operatorname{cm}_2(\cdot)\}$, and let $\mathbf{P}$ be a Koebe polyhedron. Then, for any $g(\cdot) \in S$, there is some Möbius transformation T_g such that $g(T_g(\mathbf{P})) = \mathbf{o}$. Furthermore, there is a Möbius transformation T_{ic} with $\mathbf{o} \in \operatorname{IC}(T_{\mathrm{ic}}(\mathbf{P}))$, and if $\mathbf{P}$ is simplicial, then for every $\lambda \in [0,1)$, there is a Möbius transformation T_λ satisfying $\lambda \operatorname{cm}_3(T_\lambda(\mathbf{P})) + (1-\lambda)\operatorname{ccm}(T_\lambda(\mathbf{P})) = \mathbf{o}$.*

In the next theorem, by a spherical cap on $\mathbb{S}^d$ we mean a d-dimensional closed spherical ball of spherical radius $0 < \rho < \frac{\pi}{2}$. Furthermore, if $T : \mathbb{S}^d \to \mathbb{S}^d$ is a Möbius transformation and $\mathbf{C}$ is a spherical cap, then by $\rho_T(\mathbf{C})$ and $\mathbf{c}_T(\mathbf{C})$ we denote the center and the spherical radius of the spherical cap $T(\mathbf{C})$, respectively.

Theorem 29 *Let $\mathbf{C}_1, \mathbf{C}_2, \ldots, \mathbf{C}_n \subset \mathbb{S}^d$ be spherical caps such that the union of their interiors is disconnected. For $i = 1, 2, \ldots, n$, let $w_i : \left(0, \frac{\pi}{2}\right) \to (0, \infty)$ be C^∞-class functions satisfying $\lim_{t \to \frac{\pi}{2} - 0} w_i(t) = \infty$ for all values of i. For any point $\mathbf{q} \in \mathbb{S}^d$, let $I(\mathbf{q})$ denote the set of the indices of the spherical caps whose boundary contains $\mathbf{q}$, and assume that for any $\mathbf{q} \in \mathbb{S}^d$, we have*

$$\lim_{t \to \frac{\pi}{2} - 0} \sum_{i \in I(\mathbf{q})} w_i(t) \cos t < \lim_{t \to 0+0} \sum_{i \notin I(\mathbf{q})} w_i(t). \tag{1.3}$$

Then there is a Möbius transformation $T : \mathbb{S}^d \to \mathbb{S}^d$ such that

$$\sum_{i=1}^{n} w_i(\rho_T(\mathbf{C}_i)) \mathbf{c}_T(\mathbf{C}_i) = \mathbf{o}. \tag{1.4}$$

Remark 30 *Let* $g(\cdot) \in \{\mathrm{cc}(\cdot), \mathrm{cm}_0(\cdot), \mathrm{cm}_1(\cdot), \mathrm{cm}_2(\cdot), \mathrm{cm}_3(\cdot)\}$ *and let* $\mathbf{P}$ *be a Koebe polyhedron. Then there is a Möbius transformation* $T : \mathbb{S}^2 \to \mathbb{S}^2$ *such that* $g(T(P)) \notin \mathbf{B}^3$. *Furthermore, if* $\mathbf{P}$ *is simplicial, the same statement holds for* $g(\cdot) = \mathrm{ccm}(\cdot)$.

The proof of Remark 30 can be found in [162]. We note that a similar argument works for configurations satisfying the conditions in Theorem 29.

Remark 31 *Using the idea of the proof of Theorem 28 for points of the Euler line, it is possible to prove the following, stronger statement: Let* $\mathbf{P}$ *be a Koebe polyhedron, and let* $g(\cdot) = \lambda_0 \,\mathrm{cm}_0(\cdot) + \lambda_1 \,\mathrm{cm}_1(\cdot) + \lambda_2 \,\mathrm{cm}_2(\cdot) + \lambda_3 \,\mathrm{cm}_3(\cdot)$, *where* $\sum_{i=0}^{3} \lambda_i = 1$, $\lambda_i \geq 0$ *for all values of* i *and* $\lambda_i > 0$ *for some* $i \neq 3$. *Then there is a Möbius transformation* T *such that* $g(T(\mathbf{P})) = \mathbf{o}$. *Furthermore, if* $\mathbf{P}$ *is simplicial, the same statement holds for the convex combination* $g(\cdot) = \lambda_0 \,\mathrm{cm}_0(\cdot) + \lambda_1 \,\mathrm{cm}_1(\cdot) + \lambda_2 \,\mathrm{cm}_2(\cdot) + \lambda_3 \,\mathrm{cm}_3(\cdot) + \lambda_4 \,\mathrm{ccm}(\cdot)$ *under the same conditions.*

Open Problem 6 *Prove or disprove that every combinatorial class of convex polyhedra contains a Koebe polyhedron whose center of mass is the origin.*

Open Problem 7 *Is it possible to prove variants of Theorem 29 if the weight functions* w_i *in (1.4) depend not only on* $\rho_T(\mathbf{C}_i)$, *but also on the radii of the other spherical caps as well?*

Open Problem 8 *As we already remarked in this section, Schramm [218] proved that if* $\mathbf{K}$ *is any smooth, strictly convex body in* $\mathbb{E}^3$, *then every combinatorial class of convex polyhedra contains a representative midscribed about* $\mathbf{K}$. *If* $\mathbf{K}$ *is symmetric to the origin, does this statement remain true with the additional assumption that the barycenter of the tangency points of this representative is the origin? Can the barycenter of the tangency points be replaced by other centers of the polyhedron?*

1.5 Research Exercises

Exercise 1.1 (Zenodorus) *Prove that if* $\mathbf{P}$ *is a convex* n*-gon with unit area and minimal perimeter for some* $n \geq 3$, *then* $\mathbf{P}$ *is equilateral (all its sides are of equal length) and equiangular (all its angles are equal).*

Exercise 1.2 (Dowker [91]) *For any* $n \geq 3$, *let* A_n *denote the minimum of the areas of convex* n*-gons circumscribed about the unit circle* $\mathbb{S}^1$.

(i) *Prove that the sequence* $\{A_n\}$ *is convex; i.e., for every* $n \geq 4$, *we have* $A_{n-1} + A_{n+1} \geq 2A_n$.

(ii) Prove that the value A_n is attained, e.g., for regular n-gons circumscribed about $\mathbb{S}^1$.

(iii) Use this statement to prove that among convex polygons with at most n vertices, regular n-gons have maximal isoperimetric ratio.

Exercise 1.3 (Dowker [91] and L. Fejes Tóth [106]) *Formulate and prove variants of the statements in the previous exercise in which area is replaced by perimeter, and/or circumscribed polygons are replaced by inscribed polygons.*

Exercise 1.4 *Derive a formula between the volume and the surface volume of a convex polytope in $\mathbb{E}^d$ circumscribed about $\mathbb{S}^{d-1}$. Use this formula to show that among convex polytopes with a given number of facets and circumscribed about $\mathbb{S}^{d-1}$, the ones with minimal volume coincide with the ones with minimal surface volume.*

Exercise 1.5 *Let $\mathbf{K} \subset \mathbb{E}^d$ be a convex body. Let $r(\mathbf{K})$ and $R(\mathbf{K})$ denote the radius of a largest ball contained in $\mathbf{K}$, and the radius of the smallest ball containing $\mathbf{K}$, respectively. Prove that*

$$r(\mathbf{K}) \leq \frac{d\mathrm{vol}_d(\mathbf{K})}{\mathrm{svol}_{d-1}(\mathbf{K})} \leq R(\mathbf{K}).$$

Exercise 1.6 *Prove that the average valence of a simplicial convex polyhedron in $\mathbb{E}^3$ with n vertices is equal to $6 - \frac{12}{n}$.*

Exercise 1.7 (Gale [116]) *Let $\gamma(t_i)$, where $i = 1, 2, \ldots, n$, $t_0 = t_n$, and $\gamma(t) = (t, t^2, \ldots, t^d)$ be the vertices of a d-dimensional cyclic polytope $\mathbf{P}$. Prove Gale's evenness condition, i.e., the vertices $\gamma(t_i)$, $t_i \in I$, $\mathrm{card}(I) = d$ are the vertices of a facet of $\mathbf{P}$ if and only if for any two indices* not *in I, there are even number of vertices in I that separate them.*

Exercise 1.8 *Using the properties of Steiner symmetrization, prove that the maximum volume of the intersection of a fixed ball in $\mathbb{E}^d$ and a variable simplex of given volume V is attained when the simplex is regular and concentric with the ball (cf. Problem 4).*

Exercise 1.9 *Prove that any six-neighbor circle packing, that is, a packing of circles each being tangent to at least six other circles, in the plane contains infinitely many circles.*

Exercise 1.10 (L. Fejes Tóth [111]) *Let $\mathbf{C}, \mathbf{C}_1, \mathbf{C}_2, \ldots, \mathbf{C}_n$ be mutually non-overlapping circular disks with radii $r, r_1, r_2, \ldots, r_n$. Assume that for $i = 1, 2, \ldots, n$, $\mathbf{C}_i$ touches exactly $\mathbf{C}, \mathbf{C}_{i-1}$ and $\mathbf{C}_{i+1}$. Prove that in this case*

$$\frac{1}{n}\sum_{i=1}^{n} r_i^d \geq \left(\frac{\sin\frac{\pi}{n} r}{1 - \sin\frac{\pi}{n}}\right)^d,$$

with equality if and only if $r_1 = r_2 = \ldots = r_n$.

Exercise 1.11 (Österreicher-Linhart [196]) *Prove that there is no finite four-neighbor circle packing in the plane in which all circles are congruent.*

2

Volume of the Convex Hull of a Pair of Convex Bodies

Summary. The aim of this chapter is to present results about the volume of the convex hull of a pair of convex bodies. In Section 2.1 we introduce results of Rogers and Shephard, who in the 1950s in three subsequent papers defined various convex bodies associated to a convex body, and found the extremal values of their volumes on the family of convex bodies based on the convexity of some volume functional and Steiner symmetrization. After this we present a proof of a conjecture of Rogers and Shephard about the equality cases in their theorems, find a more general setting of their problems, and collect results about it. Section 2.2 describes a variant of the previous problems for normed spaces. In particular, using the notion of relative norm of a convex body, we collect results about the extremal values of four types of volume of the unit ball of a normed space. The problem of finding these values of one of these types leads to the famous Mahler Conjecture. We extend these notions to the translation bodies of convex bodies, defined in Section 2.1, and determine their extremal values in normed planes.

2.1 Volume of the convex hull of a pair of convex bodies in Euclidean space

The volume of the convex hull of two convex bodies in the Euclidean d-space $\mathbb{E}^d$ has been the focus of research since the 1950s. One of the first results in this area is due to Fáry and Rédei [99], who proved that if one of the bodies is translated on a line at a constant velocity, then the volume of their convex hull is a convex function of time. This result was later reproved by Ahn et al. [2], and is used in the literature in various settings (see, e.g., the result of Alexander, Fradelizi and Zvavitch [4] on the maximal volume product of d-dimensional polytopes with d + 2 vertices). Rogers and Shephard [209] proved a similar result, stating that for any convex body $\mathbf{K}$ in $\mathbb{E}^d$,

$$\mathrm{vol}_d(\mathbf{K}-\mathbf{K}) \leq \binom{2d}{d} \mathrm{vol}_d(\mathbf{K}), \tag{2.1}$$

with equality if and only if $\mathbf{K}$ is a d-dimensional simplex. Their proof was based on the observation that for any convex body $\mathbf{K} \subset \mathbb{E}^d$, the property that for any $\mathbf{p} \in \mathbf{K} - \mathbf{K}$, $\mathbf{K} \cap (\mathbf{p} + \mathbf{K})$ is a homothetic copy of $\mathbf{K}$ is equivalent to the fact that $\mathbf{K}$ is a d-dimensional simplex. Here, it is worth noting that from the Brunn-Minkowski Inequality, stating that for any two convex bodies $\mathbf{K}, \mathbf{L}$ in $\mathbb{E}^d$, $(\mathrm{vol}_d(\mathbf{K} + \mathbf{L}))^{\frac{1}{d}} \geq (\mathrm{vol}_d(\mathbf{K}))^{\frac{1}{d}} + (\mathrm{vol}_d(\mathbf{L}))^{\frac{1}{d}}$, with equality if and only if $\mathbf{K}$ and $\mathbf{L}$ are homothetic, it immediately follows that $\mathrm{vol}_d(\mathbf{K}) \geq 2^d \mathrm{vol}_d(\mathbf{K})$, with equality if and only if $\mathbf{K}$ is centrally symmetric.

In two subsequent papers [210, 211], Rogers and Shephard generalized the result of Fáry and Rédei to prove volume inequalities on the convex hull of various convex bodies associated to a given convex body $\mathbf{K} \subset \mathbb{E}^d$. To state their two main tools to find lower bounds, we first need to introduce some concepts.

Definition 6 *Let $\mathcal{P} = \{\mathbf{p}_i : i \in I, \mathbf{p}_i \in \mathbb{E}^d\}$ be a set of points with some arbitrary set of indices I, and let $L = \{\lambda_i : i \in I, \lambda_i \in \mathbb{R}\}$. Assume that both $\mathcal{P}$ and L are bounded, and let $\mathbf{e} \in \mathbb{E}^d$. For every $t \in \mathbb{R}$, the convex set $\mathbf{K}(t) = \mathrm{conv}\{\mathbf{p}_i + t\lambda_i \mathbf{e} : i \in I\}$ is called a* linear parameter system.

Theorem 32 *The volume of a linear parameter system $\mathbf{K}(t)$ is a convex function of t.*

Definition 7 *Let $\mathbf{K}$ be a convex body and H be a hyperplane in $\mathbb{E}^d$. For any line L perpendicular to H, let d_L denote the length of the intersection $\mathbf{K} \cap L$. Let $[\mathbf{p}_L, \mathbf{q}_L]$ denote the segment in L whose length is d_L and which intersects H at $\frac{1}{2}(\mathbf{p}_L + \mathbf{q}_L)$. Then the set*

$$S_H(\mathbf{K}) = \bigcup \{[\mathbf{p}_L, \mathbf{q}_L] : L \textit{ is perpendicular to } H \textit{ and intersects } \mathbf{K}\}$$

is called the Steiner symmetrization *of $\mathbf{K}$ with respect to H (cf. Figure 2.1).*

This symmetrization procedure, introduced by Steiner in 1838, has the property that for any convex body $\mathbf{K}$ and hyperplane H in $\mathbb{E}^d$, $S_H(\mathbf{K})$ is a convex body in $\mathbb{E}^d$ of volume equal to $\mathrm{vol}_d(\mathbf{K})$ and symmetric about the hyperplane H. We note that one way to prove the isoperimetric inequality is to use the facts that surface volume does not increase under Steiner symmetrizations, and that applying subsequent, suitably chosen Steiner symmetrizations to any given convex body $\mathbf{K}$, the sequence of bodies obtained in this way approaches a Euclidean ball of the same volume (cf. e.g. Theorem 10.3.2 [217], or Lemma 164 in Section 7.4). The latter fact and Theorem 32 can be used to prove Theorem 33, where, for any convex body $\mathbf{K} \subset \mathbb{E}^d$, $\mathbf{B}_{\mathbf{K}}$ denotes a Euclidean ball of volume $\mathrm{vol}_d(\mathbf{K})$.

Definition 8 *Let $\mathbf{K}$ and $\mathbf{L}$ be convex bodies in $\mathbb{E}^d$. By $V^*(\mathbf{K}, \mathbf{L})$ we denote the supremum of the quantity $\mathrm{vol}_d(\mathrm{conv}(\mathbf{K} \cup (\mathbf{p} + \mathbf{L})))$ under the condition that $K \cap (\mathbf{p} + \mathbf{L})$ is not empty.*

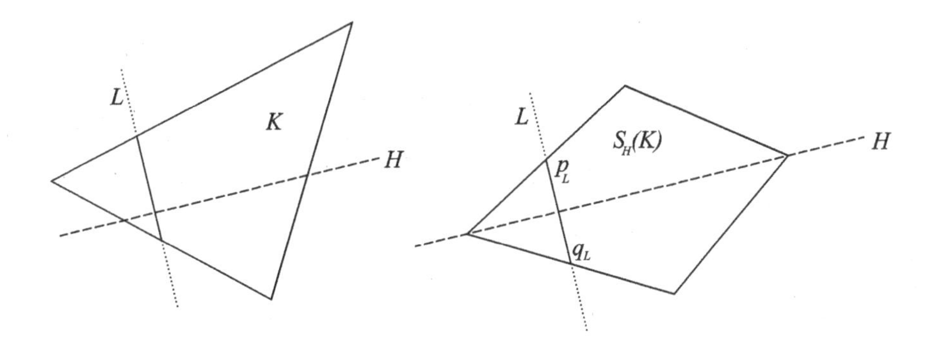

FIGURE 2.1
Steiner symmetrization of a triangle $\mathbf{K}$ with respect to the line H.

Theorem 33 *Let $\mathbf{K}$ and $\mathbf{L}$ be convex bodies and let H be a hyperplane in $\mathbb{E}^d$. Then*

(i) $V^*(\mathbf{K}, \mathbf{L}) \geq V^*(S_H(\mathbf{K}), S_H(\mathbf{L}))$, *and*

(ii) $V^*(\mathbf{K}, \mathbf{L}) \geq V^*(\mathbf{B_K}, \mathbf{B_L})$.

In [210, 211], Rogers and Shephard used Theorem 33 to examine the volumes of various convex bodies associated to a given convex body $\mathbf{K} \subset \mathbb{E}^d$, which we define in Definition 9, and present the corresponding results in Theorems 34-40.

Definition 9 *Let $\mathbf{K}$ be a convex body in $\mathbb{E}^d$, and let $\mathbf{p} \in \mathbb{E}^d$ be a point.*

(i) The convex body $\operatorname{conv}(\mathbf{K} \cup (2\mathbf{p} - \mathbf{K}))$, *denoted by $R_{\mathbf{p}}(\mathbf{K})$, is called the* reflection body *of $\mathbf{K}$ with respect to $\mathbf{p}$. If a reflection body $R_{\mathbf{p}}(\mathbf{K})$ with respect to $\mathbf{p}$ minimizes or maximizes* $\operatorname{vol}_d(R_{\mathbf{p}}(\mathbf{K}))$ *under the condition that $\mathbf{p} \in \mathbf{K}$, then we denote it by $R(\mathbf{K})$ and $R^*(\mathbf{K})$, respectively.*

(ii) The convex body $\operatorname{conv}(\mathbf{K} \cup (\mathbf{p} + \mathbf{K}))$, *denoted by $T_{\mathbf{p}}(\mathbf{K})$, is called the* translation body *of $\mathbf{K}$ with vector $\mathbf{p}$. If a translation body $T_{\mathbf{p}}(\mathbf{K})$ with vector $\mathbf{p}$ maximizes* $\operatorname{vol}_d(T_{\mathbf{p}}(\mathbf{K}))$ *under the condition that $\mathbf{K} \cap (\mathbf{p} + \mathbf{K}) \neq \emptyset$, then we denote it by $T^*(\mathbf{K})$.*

(iii) Imagine $\mathbb{E}^d$ as the hyperplane $\{x_{d+1} = 0\}$ in $\mathbb{E}^{d+1}$, and denote by $\mathbf{e}_{d+1}$ the last vector of the standard orthonormal basis of $\mathbb{E}^{d+1}$. The $(d+1)$-dimensional convex body $\operatorname{conv}(K \cup (\mathbf{e}_{d+1} - \mathbf{K}))$ *is called the* associated $(d+1)$-dimensional convex body *of $\mathbf{K}$. We denote it by $C(\mathbf{K})$.*

We note that some of the convex bodies in Definition 9, namely $R(\mathbf{K})$, $R^*(\mathbf{K})$ and $T^*(\mathbf{K})$, may not be well-defined. Nevertheless, we examine only their volumes, which are determined by $\mathbf{K}$.

Theorem 34 *For any convex body* $\mathbf{K} \subset \mathbb{E}^d$ *and* $\mathbf{p} \in \mathbf{K}$,

$$1 \leq \frac{\mathrm{vol}_d(R_{\mathbf{p}}(\mathbf{K}))}{\mathrm{vol}_d(\mathbf{K})} \leq 2^d$$

with equality on the left if and only if $\mathbf{K}$ *is symmetric about* $\mathbf{p}$, *and on the right if and only if* $\mathbf{K}$ *is a d-dimensional simplex with* $\mathbf{p}$ *as one of its vertices.*

By this result, the inequalities $1 \leq \frac{\mathrm{vol}_d(R(\mathbf{K}))}{\mathrm{vol}_d(\mathbf{K})} \leq 2^d$ clearly hold for any convex body $\mathbf{K}$. While here the left-hand side inequality is sharp if $\mathbf{K}$ is centrally symmetric, the exact quantity on the right-hand side is not known; according to a conjecture of Fáry and Rédei [99], the maximum of $R(\mathbf{K})$ is attained if $\mathbf{K}$ is a d-simplex. An elementary computation shows that in this special case

$$\frac{\mathrm{vol}_d(R(\mathbf{K}))}{\mathrm{vol}_d(\mathbf{K})} = \binom{d}{\lfloor \frac{d}{2} \rfloor} \sim \frac{2^d}{\pi\sqrt{d}}.$$

Conjecture 35 *Let* $\mathbf{K}$ *be a convex body in* $\mathbb{E}^d$. *Then there is a point* $\mathbf{x} \in \mathbf{K}$ *such that*

$$\frac{\mathrm{vol}_d(\mathrm{conv}(\mathbf{K} \cup (2\mathbf{x} - \mathbf{K})))}{\mathrm{vol}_d(\mathbf{K})} \leq \binom{d}{\lfloor \frac{d}{2} \rfloor}.$$

Recall that κ_d denotes the volume of the d-dimensional Euclidean unit ball.

Theorem 36 *For any convex body* $\mathbf{K} \subset \mathbb{E}^d$,

$$1 + \frac{2\kappa_{d-1}}{\kappa_d} \leq \frac{\mathrm{vol}_d(R^*(\mathbf{K}))}{\mathrm{vol}_d(\mathbf{K})} \leq 2^d \tag{2.2}$$

with equality on the left if $\mathbf{K}$ *is an ellipsoid, and on the right if and only if* $\mathbf{K}$ *is a simplex. Furthermore, if* $\mathbf{K}$ *is centrally symmetric, then*

$$1 + \frac{2\kappa_{d-1}}{\kappa_d} \leq \frac{\mathrm{vol}_d(R^*(\mathbf{K}))}{\mathrm{vol}_d(\mathbf{K})} \leq d+1, \tag{2.3}$$

with equality on the left if $\mathbf{K}$ *is an ellipsoid, and on the right if* $\mathbf{K}$ *is a centrally symmetric double pyramid on a convex base.*

Theorem 37 *For any convex body* $\mathbf{K} \subset \mathbb{E}^d$,

$$1 + \frac{2\kappa_{d-1}}{\kappa_d} \leq \frac{\mathrm{vol}_d(T^*(\mathbf{K}))}{\mathrm{vol}_d(\mathbf{K})} \leq d+1, \tag{2.4}$$

with equality on the left if $\mathbf{K}$ *is an ellipsoid, and on the right if and only if* $\mathbf{K}$ *is a simplex.*

We note that by [211], equality occurs for some convex body $\mathbf{K} \subset \mathbb{E}^d$ on the right in 2.3 or in 2.4 if and only if $\mathbf{K}$ is a centrally symmetric pseudo-double pyramid, or a pseudo-double pyramid, respectively, as defined in [211].

The next conjecture was made by Rogers and Shephard in [211].

Conjecture 38 *Equality holds for some convex body $\mathbf{K} \subset \mathbb{E}^d$ on the left in any of 2.2, 2.3 or 2.4 only if $\mathbf{K}$ is an ellipsoid.*

Conjecture 38 was proved almost 50 years later by Martini and Mustafaev [177], using measures associated to normed spaces. In Subsection 7.3.1, we give a simpler proof of this conjecture, published in [144] by G. Horváth and Lángi. The left-hand side inequality in Theorem 37 can be reformulated in a different way.

Theorem 39 *Let $\mathbf{K} \subset \mathbb{E}^d$ be a convex body. Then there is a cylinder $\mathbf{F}$ circumscribed about $\mathbf{K}$ such that $\frac{\mathrm{vol}_d(\mathbf{F})}{\mathrm{vol}_d(\mathbf{K})} \geq \frac{2\kappa_{d-1}}{\kappa_d}$.*

Theorem 40 *For any convex body $\mathbf{K} \subset \mathbb{E}^d$, we have*

$$1 \leq \frac{\mathrm{vol}_{d+1}(C(\mathbf{K}))}{\mathrm{vol}_d(\mathbf{K})} \leq \frac{2^d}{d+1},$$

with equality on the left if and only if $\mathbf{K}$ is centrally symmetric, and on the right if and only if $\mathbf{K}$ is a d-dimensional simplex.

Apart from the equality cases, we present the proofs of Theorems 34, 36, 37 and 40 in Chapter 7.

In [144], the authors examined a more general variant of this problem. To present it, we extend Definition 9.

Definition 10 *For two convex bodies $\mathbf{K}$ and $\mathbf{L}$ in $\mathbb{E}^d$, let*

$$c(\mathbf{K}, \mathbf{L}) = \max\{\mathrm{vol}_d(\mathrm{conv}(\mathbf{K}' \cup \mathbf{L}')) : \mathbf{K}' \cong \mathbf{K}, \mathbf{L}' \cong \mathbf{L} \text{ and } \mathbf{K}' \cap \mathbf{L}' \neq \emptyset\}$$

where $\cong$ denotes congruence. Furthermore, if $\mathcal{S}$ is a set of isometries of $\mathbb{E}^d$, we set

$$c(\mathbf{K}|\mathcal{S}) = \frac{\max\{\mathrm{vol}_d(\mathrm{conv}(\mathbf{K} \cup \sigma(\mathbf{K}))) : \mathbf{K} \cap \sigma(\mathbf{K}) \neq \emptyset, \sigma \in \mathcal{S}\}}{\mathrm{vol}_d(\mathbf{K})}.$$

Definition 11 *For any $i = 0, 1 \ldots, d-1$ and convex body $\mathbf{K} \subset \mathbb{E}^d$, we denote by $c_i(\mathbf{K})$ the value of $c(\mathbf{K}|\mathcal{S})$ if $\mathcal{S}$ is the set of all reflections about the i-dimensional affine subspaces of $\mathbb{E}^d$. Similarly, if $\mathcal{S}$ is the set of translations or the set of all isometries, we denote $c(\mathbf{K}|\mathcal{S})$ by $c^{tr}(\mathbf{K})$ or by $c^{co}(\mathbf{K})$, respectively.*

The first result, which we prove in Section 7.3, proves the conjecture of Rogers and Shephard for both translation and reflection bodies.

Theorem 41 *Let $\mathbf{K} \subset \mathbb{E}^d$ be a convex body, where $d \geq 2$. Then*

(i) $c^{tr}(\mathbf{K}) \geq 1 + \frac{2\kappa_{d-1}}{\kappa_d}$,

(ii) $c_0(\mathbf{K}) \geq 1 + \frac{2\kappa_{d-1}}{\kappa_d}$,

(iii) $c_{d-1}(\mathbf{K}) \geq 1 + \frac{2\kappa_{d-1}}{\kappa_d}$.

Furthermore, if we have equality in any of the inequalities in (i)-(ii), then **K** *is an ellipsoid, and if we have equality in (iii), then* **K** *is a Euclidean ball.*

This result readily implies the following.

Corollary 42 *For any convex body* $\mathbf{K} \subset \mathbb{E}^d$, $c^{co}(\mathbf{K}) \geq 1+\frac{2\kappa_{d-1}}{\kappa_d}$, *with equality if and only if* **K** *is a Euclidean ball.*

Corollary 43 *For any convex body* $K \subset \mathbb{E}^d$, *there is a direction* $\mathbf{m} \in \mathbb{S}^{d-1}$ *such that the right cylinder* $\mathbf{F_K}(\mathbf{m})$, *circumscribed about* **K** *and with generators parallel to* **m** *has volume*

$$\frac{\mathrm{vol}_d(\mathbf{F_K}(\mathbf{m}))}{\mathrm{vol}_d(\mathbf{K})} \geq \frac{2\kappa_{d-1}}{\kappa_d}. \tag{2.5}$$

Furthermore, if **K** *is not a Euclidean ball, then the inequality sign in (2.5) is a strict inequality for some direction* **m**.

Corollary 44 *For any convex body* $K \subset \mathbb{E}^d$, *there is a direction* $\mathbf{m} \in \mathbb{S}^{d-1}$ *such that any cylinder* $\mathbf{F_K}(\mathbf{m})$, *circumscribed about* **K** *and with generators parallel to* **m** *has volume*

$$\frac{\mathrm{vol}_d(\mathbf{F_K}(\mathbf{m}))}{\mathrm{vol}_d(\mathbf{K})} \geq \frac{2\kappa_{d-1}}{\kappa_d}. \tag{2.6}$$

Furthermore, if **K** *is not an ellipsoid, then the inequality sign in (2.6) is a strict inequality for some direction* **m**.

Conjecture 45 *Let* $d \geq 3$ *and* $1 < i < d-1$. *Prove that for any convex body* $\mathbf{K} \subset \mathbb{E}^d$, $c_i(\mathbf{K}) \geq 1 + \frac{2\kappa_{d-1}}{\kappa_d}$. *Is it true that equality holds only for Euclidean balls?*

Since the quantity $\frac{\mathrm{vol}_d(T_\mathbf{p}(\mathbf{K}))}{\mathrm{vol}_d(\mathbf{K})}$ does not change if we apply an affine transformation to both **K** and **p**, it follows that for any ellipsoid **E**, the quantity $\mathrm{vol}_d(T_\mathbf{p}(\mathbf{E}))$ is a constant under the condition that **E** and $\mathbf{p}+\mathbf{E}$ touch each other. It is an interesting question which other convex bodies satisfy this property.

Definition 12 *If, for a convex body* $\mathbf{K} \subset \mathbb{E}^d$, *we have that* $\mathrm{vol}_d(\mathrm{conv}((\mathbf{p}+\mathbf{K}) \cup (\mathbf{q}+\mathbf{K})))$ *has the same value for any touching pair of translates, we say that* **K** *satisfies the* translative constant volume property.

To state the corresponding result from [144], we note that an **o**-symmetric planar convex curve is a *Radon curve*, if, for the convex hull **K** of a suitable affine image of the curve, it holds that its polar K° is a rotated copy of **K** by $\frac{\pi}{2}$ (cf. [179]). Furthermore, a norm is a *Radon norm* if the boundary of its unit disk is a Radon curve.

Theorem 46 *For any plane convex body* $\mathbf{K} \subset \mathbb{E}^2$ *the following are equivalent.*

(i) $\mathbf{K}$ *satisfies the translative constant volume property.*

(ii) The boundary of $\frac{1}{2}(\mathbf{K} - \mathbf{K})$ *is a Radon curve.*

(iii) $\mathbf{K}$ *is a body of constant width in a Radon norm.*

It is known (cf. [7] or [179]) that for $d \geq 3$, if every planar section of a d-dimensional normed space is Radon, then the space is Euclidean; that is, its unit ball is an ellipsoid. This leads to Conjecture 47.

Conjecture 47 *Let* $d \geq 3$. *If some o-symmetric convex body* $\mathbf{K} \subset \mathbb{E}^d$ *satisfies the translative constant volume property, then* $\mathbf{K}$ *is an ellipsoid.*

We finish this section with a related problem.

Open Problem 9 *For any* $d \geq 3$ *and* $1 \leq i \leq d-1$, *find the least upper bound of* $c_i(\mathbf{K})$ *on the family of convex bodies in* $\mathbb{E}^d$. *Furthermore, find the least upper bound on* $c^{co}(\mathbf{K})$ *on the same family.*

2.2 Volume of the convex hull of a pair of convex bodies in normed spaces

Let us recall the well-known fact that any finite dimensional real normed space can be equipped with a Haar measure, and that it is unique up to multiplication of the standard Lebesgue measure by a scalar [132]. As a consequence, any 'meaningful' notion of volume is a scalar multiple of standard Euclidean volume. Nevertheless, depending on the choice of this scalar, one may define more than one version of normed volume. There are four variants that are regularly used in the literature [9, 10].

The *Busemann* and *Holmes-Thompson volume* of a set $\mathbf{S}$ in a d-dimensional normed space with unit ball $\mathbf{M}$, is defined as

$$\mathrm{vol}_{\mathbf{M}}^{Bus}(\mathbf{S}) = \frac{\kappa_d}{\mathrm{vol}_d(\mathbf{M})}\mathrm{vol}_d(\mathbf{S}) \text{ and } \mathrm{vol}_{\mathbf{M}}^{HT}(\mathbf{S}) = \frac{\mathrm{vol}_d(\mathbf{M}^\circ)}{\kappa_d}\mathrm{vol}_d(\mathbf{S}), \quad (2.7)$$

respectively. Note that the Busemann volume of the unit ball is equal to that of a Euclidean unit ball. For *Gromov's mass*, the scalar is chosen in such a way that the volume of a maximal volume cross-polytope, inscribed in the unit ball $\mathbf{M}$ is equal to $\frac{2^d}{d!}$, and for *Gromov's mass** (or Benson's definition of volume), the volume of a smallest volume parallelotope, circumscribed about $\mathbf{M}$, is equal to 2^d (cf. [9]). We denote the two latter quantities by $\mathrm{vol}_{\mathbf{M}}^{m}(\mathbf{S})$ and $\mathrm{vol}_{\mathbf{M}}^{m*}(\mathbf{S})$, respectively.

In light of the previous paragraphs, it is clear that for any fixed normed space and volume, the Euclidean result in Theorems 37 and 41 can be immediately applied.

Theorem 48 *Let $\mathcal{M}$ be a normed space with volume* $\mathrm{vol}_{\mathbf{M}}$. *Then, for any convex body* $\mathbf{K}$ *in* $\mathcal{M}$, *we have*

$$1 + \frac{2\kappa_{d-1}}{\kappa_d} \leq \frac{\max\{\mathrm{vol}_{\mathbf{M}}(\mathrm{conv}(\mathbf{K} \cup (\mathbf{p}+\mathbf{K}))) : (\mathbf{p}+\mathbf{K}) \cap \mathbf{K} \neq \emptyset\}}{\mathrm{vol}_{\mathbf{M}}(\mathbf{K})} \tag{2.8}$$

We observe that there is equality on the left if and only if $\mathbf{K}$ is an ellipsoid (cf. [144]), and on the right if and only if $\mathbf{K}$ is a pseudo-double pyramid (cf. [211]).

In the remaining part, we use a different approach. Before we present it, we collect estimates about the area and the perimeter of the unit disk of a normed plane.

The first result of this kind is due to Gołab [120] in 1932. We note that for any rectifiable curve γ in a normed plane with unit ball $\mathbf{M}$, the arclength of γ with respect to the norm can be defined in the usual way, i.e., as the supremum of the normed lengths of all polygonal curves with vertices on γ. We denote this quantity by $M_{\mathbf{M}}(\gamma)$.

Theorem 49 *Let* $\mathbf{M}$ *be any* $\mathbf{o}$*-symmetric plane convex body. Then* $6 \leq M_{\mathbf{M}}(\mathrm{bd}\mathbf{M}) \leq 8$. *There is equality on the left if and only if* $\mathbf{M}$ *is an affinely regular hexagon, and on the right if and only if* $\mathbf{M}$ *is a parallelogram.*

This result can be generalized for arbitrary convex bodies using the notion of relative norm [214] and a result of Fáry and Makai [101]. For any convex body $\mathbf{K}$ in $\mathbb{E}^d$, we say that the *relative norm* of $\mathbf{K}$ is the norm with the central symmetrization $\frac{1}{2}(\mathbf{K}-\mathbf{K})$ of $\mathbf{K}$ as its unit ball (cf. also [164] or [160]). Observe that, up to multiplication by a scalar, the relative norm of $\mathbf{K}$ is the unique norm in which $\mathbf{K}$ is a body of constant width. In the following, if $\mathbf{K}$ is not $\mathbf{o}$-symmetric, for any rectifiable curve γ, by $M_{\mathbf{K}}(\gamma)$ we denote the arclength of γ measured in the relative norm of $\mathbf{K}$.

Theorem 50 *Let* $\mathbf{M}$ *be an* $\mathbf{o}$*-symmetric plane convex body. Then* $M_{\mathbf{M}}(\mathrm{bd}\mathbf{K}) = M_{\mathbf{M}}\left(\mathrm{bd}\left(\frac{1}{2}(\mathbf{K}-\mathbf{K})\right)\right)$ *for any plane convex body* $\mathbf{K}$.

Corollary 51 *Let* $\mathbf{K}$ *be a plane convex body, and let* $\mathbf{M} = \frac{1}{2}(\mathbf{K}-\mathbf{K})$. *Then* $6 \leq M_{\mathbf{M}}(\mathrm{bd}\mathbf{K}) \leq 8$.

The next result is due to Schäffer [214].

Theorem 52 *For any* $\mathbf{o}$*-symmetric plane convex body* $\mathbf{M}$, *we have* $M_{\mathbf{M}}(\mathrm{bd}\mathbf{M}) = M_{\mathbf{M}^\circ}(\mathrm{bd}\mathbf{M}^\circ)$.

Note that by definition, if $\mathbf{M}$ is the unit ball of a d-dimensional normed space, $\mathrm{vol}_{\mathbf{M}}^{Bus}(\mathbf{M}) = \kappa_d$. The following theorem was proven by Blaschke [54] in the plane, and was generalized for higher dimensions by Santaló [213]. The case of equality was proven, e.g., by Saint-Raymond [212]. Apart from the case of equality, we prove Theorem 53 in Section 53.

Theorem 53 (Blaschke-Santaló Inequality) *If* $\mathbf{M}$ *is the unit ball of a* d*-dimensional normed space, then* $\mathrm{vol}_{\mathbf{M}}^{HT}(\mathbf{M}) \leq \kappa_d$*, with equality if and only if* $\mathbf{M}$ *is an ellipsoid.*

The best lower bound for the planar case was determined by Mahler [174]. Its generalization for arbitrary dimensions is one of the most important conjectures of convex geometry. In Section 7.5 we present a proof of the inequality in Theorem 54, but we examine equality only on the family of $\mathbf{o}$-symmetric convex polygons. This proof can be found in another paper of Mahler [173] (see also [142]), preceding [174].

Theorem 54 *If* $\mathbf{M}$ *is the unit disk of a normed plane, then* $\mathrm{vol}_{\mathbf{M}}^{HT}(\mathbf{M}) \geq \frac{8}{\pi}$*, with equality if and only if* $\mathbf{M}$ *is a parallelogram.*

Conjecture 55 (Mahler Conjecture) *Prove that among* d*-dimensional,* $\mathbf{o}$*-symmetric convex bodies, the minimum of* $\mathrm{vol}_{\mathbf{M}}^{HT}(\mathbf{M})$ *is attained, e.g., if* $\mathbf{M}$ *is a* d*-cube.*

The next result can be found in [9], and the estimates for the planar case in [10].

Theorem 56 *Let* $\mathbf{M}$ *be a* d*-dimensional,* $\mathbf{o}$*-symmetric convex body. Then*

$$\frac{2^d}{d!} \leq \mathrm{vol}_{\mathbf{M}}^{m}(\mathbf{M}) \leq \mathrm{vol}_{\mathbf{M}}^{m*}(\mathbf{M}) \leq 2^d,$$

where there is equality on the left if $\mathbf{M}$ *is a cross-polytope, and on the right if* $\mathbf{M}$ *is a parallelotope.*

We note that the best upper bound on $\mathrm{vol}_{\mathbf{M}}^{m}(\mathbf{M})$ and the best lower bound on $\mathrm{vol}_{\mathbf{M}}^{m*}(\mathbf{M})$ are not known for $d > 2$.

Theorem 57 *Let* $\mathbf{M}$ *be an* $\mathbf{o}$*-symmetric plane convex body. Then*

$$2 \leq \mathrm{vol}_{\mathbf{M}}^{m}(\mathbf{M}) \leq \pi,$$

with equality on the left if and only if $\mathbf{M}$ *is a parallelogram, and on the right if and only if* $\mathbf{M}$ *is an ellipse. Furthermore,*

$$3 \leq \mathrm{vol}_{\mathbf{M}}^{m*}(\mathbf{M}) \leq 4,$$

with equality on the left if and only if $\mathbf{M}$ *is an affinely regular hexagon, and on the right if and only if* $\mathbf{M}$ *is a parallelogram.*

If we restrict our examination to Radon norms, we obtain other interesting questions [10].

Theorem 58 *Let* $\mathbf{M}$ *be the unit disk of a Radon norm. Then*

(i) $6 \leq M_{\mathbf{M}}(\mathrm{bd}\mathbf{M}) \leq 2\pi$,

(ii) $\frac{9}{\pi} \leq \mathrm{vol}_{\mathbf{M}}^{HT}(\mathbf{M}) \leq \pi$,

(iii) $3 \leq \mathrm{vol}_{\mathbf{M}}^{m}(\mathbf{M}) \leq \pi$,

(iv) $3 \leq \mathrm{vol}_{\mathbf{M}}^{m*}(\mathbf{M}) \leq \pi$.

Here, in each case, we have equality on the left if and only if $\mathbf{M}$ *is an affinely regular hexagon, and on the right if and only if* $\mathbf{M}$ *is an ellipse.*

It seems to be an interesting problem to examine the extremal values of these quantities for the volume of any convex body $\mathbf{K}$, measured in the relative norm of $\mathbf{K}$. Nevertheless, we do not know of any such result in the literature.

To continue our investigation on the volume of the convex hull of two convex bodies, we introduce the following quantities.

Definition 13 *Let* $\mathbf{K}$ *be a d-dimensional convex body and* $\mathcal{M}$ *be the space with its relative norm. For* $\tau \in \{Bus, HT, m, m^*\}$, *let*

$$c_{tr}^{\tau}(\mathbf{K}) = \max\{\mathrm{vol}_{\mathbf{M}}^{\tau}(\mathrm{conv}(\mathbf{K} \cup (\mathbf{p} + \mathbf{K}))) : (\mathbf{p} + \mathbf{K}) \cap \mathbf{K} \neq \emptyset, \mathbf{p} \in \mathcal{M}\}. \quad (2.9)$$

Observe that the quantities in Definition 13 do not change under affine transformations. These quantities were examined in [161]. To present the related results, first we need to define the following plane convex body.

Consider the square $\mathbf{S}_0$ with vertices $\left(\pm\frac{1}{\sqrt{2}}, \pm\frac{1}{\sqrt{2}}\right)$ in a Cartesian coordinate system. Replace the two horizontal edges of $\mathbf{S}_0$ by the corresponding arcs of the ellipse with equation

$$\frac{x^2}{a^2} + \frac{y^2}{b^2} = 1,$$

where $a = 1.61803\ldots$, $b = \frac{a}{\sqrt{2a^2-1}}$. An elementary computation shows that the vertices of $\mathbf{S}_0$ are points of this ellipse. Replace the vertical edges of $\mathbf{S}_0$ by rotated copies of these elliptic arcs by $\frac{\pi}{2}$. We denote the plane convex body, obtained in this way and bounded by four congruent elliptic arcs, by $\mathbf{M}_0$ (cf. Figure 2.2). We remark that the value of a is obtained as a root of a transcendent equation, and has the property that the value of $\mathrm{area}(\mathbf{M}_0)\,(\mathrm{area}(\mathbf{M}_0) + 4)$ is maximal for all possible values of $a > 1$.

Theorem 59 *Let* $\mathbf{K}$ *be a plane convex body. Then*

(i) We have $2\pi \leq c_{tr}^{Bus}(\mathbf{K}) \leq 3\pi$, *with equality on the left if and only if* $\mathbf{K}$ *is a triangle, and on the right if and only if* $\mathbf{K}$ *is a parallelogram.*

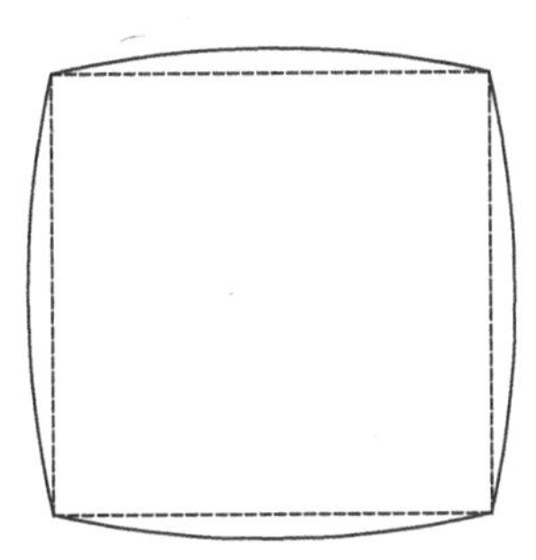

FIGURE 2.2
The plane convex body $\mathbf{M}_0$.

(ii) *We have* $\frac{18}{\pi} \leq c_{tr}^{HT}(\mathbf{K}) \leq 7.81111\ldots$, *with equality on the left if and only if* $\mathbf{K}$ *is a triangle, and on the right if* $\mathbf{K}$ *is an affine image of* $\mathbf{M}_0$.

(iii) *We have* $6 \leq c_m^{Bus}(\mathbf{K}) \leq \pi + 4$, *with equality on the left if and only if* $\mathbf{K}$ *is a (possibly degenerate) convex quadrilateral, and on the right if and only if* $\mathbf{K}$ *is an ellipse.*

(iv) *We have* $6 \leq c_m^{Bus}(\mathbf{K}) \leq 12$, *with equality on the left if and only if* $\mathbf{K}$ *is a triangle, and on the right if and only if* $\mathbf{K}$ *is a parallelogram.*

It is a natural question to ask for the extremal values of these four quantities over the family of centrally symmetric plane convex bodies. This question is answered in the next theorem.

Theorem 60 *Let* $\mathbf{M}$ *be an* $\mathbf{o}$*-symmetric plane convex body. Then*

(i) *We have* $\pi + 4 \leq c_{tr}^{Bus}(\mathbf{M}) \leq 3\pi$, *with equality on the left if and only if* $\mathbf{M}$ *is an ellipse, and on the right if and only if* $\mathbf{M}$ *is a parallelogram.*

(ii) *We have* $\frac{21}{\pi} \leq c_{tr}^{HT}(\mathbf{M}) \leq 7.81111\ldots$, *with equality on the left if and only if* $\mathbf{M}$ *is an affinely regular hexagon, and on the right if* $\mathbf{M}$ *is an affine image of* $\mathbf{M}_0$.

(iii) *We have* $6 \leq c_{tr}^{m}(\mathbf{M}) \leq \pi + 4$, *with equality on the left if and only if* $\mathbf{M}$ *is a parallelogram, and on the right if and only if* $\mathbf{M}$ *is an ellipse.*

(iv) *We have* $7 \leq c_{tr}^{m}(\mathbf{M}) \leq 12$, *with equality on the left if and only if* $\mathbf{M}$ *is an affinely regular hexagon, and on the right if and only if* $\mathbf{M}$ *is a parallelogram.*

Observe that in Theorems 59 and 60, the plane convex bodies for which $c_{tr}^{HT}(\cdot)$ is maximal are not determined. This naturally leads to the following problem.

Open Problem 10 *Prove or disprove that if $c_{tr}^{HT}(\mathbf{K})$ is maximal for some plane convex body $\mathbf{K}$, then $\mathbf{K}$ is an affine image of $\mathbf{M}_0$.*

Remark 61 *For any d-dimensional convex body $\mathbf{K}$ and direction $\mathbf{m} \in \mathbb{S}^{d-1}$, let $d_{\mathbf{m}}(\mathbf{K})$ denote the length of a maximal chord of $\mathbf{K}$ in the direction $\mathbf{m}$, and let $\mathbf{K}|\mathbf{m}^{\perp}$ be the orthogonal projection of $\mathbf{K}$ onto the hyperplane, through $\mathbf{o}$, that is perpendicular to $\mathbf{m}$. Then the maximal volume of the convex hull of two intersecting translates of $\mathbf{K}$ (that is, the numerator in the definition of $c^{tr}(\mathbf{K})$), is*

$$\mathrm{vol}_d(\mathbf{K}) + \max\{d_{\mathbf{m}}(\mathbf{K})\mathrm{vol}_{d-1}(K|\mathbf{m}^{\perp}) : \mathbf{m} \in \mathbb{S}^{d-1}\} \tag{2.10}$$

Note that for any $\mathbf{m} \in \mathbb{S}^{d-1}$, the central symmetrization of $\mathbf{K}|\mathbf{m}^{\perp}$ is $\left(\frac{1}{2}(\mathbf{K}-\mathbf{K})\right)|\mathbf{m}^{\perp}$. Thus, by the Brunn-Minkowski Inequality, the expression in (2.10) does not decrease under central symmetrization, with equality if and only if $\mathbf{K}$ is centrally symmetric. This yields that if $c_{tr}^{\tau}(\mathbf{K})$ is maximal for some convex body $\mathbf{K}$ for any $\tau \in \{Bus, HT, m, m^*\}$, then $\mathbf{K}$ is centrally symmetric.

Remark 62 *By Remark 61, to find the maximal value of $c_{tr}^{Bus}(\mathbf{K})$, it suffices to find the maximum of $c^{tr}(\mathbf{K})$ over the family of d-dimensional $\mathbf{o}$-symmetric convex bodies. Thus, from Theorem 37 (see Theorem 3 of [124]) it follows that*

$$c_{tr}^{Bus}(\mathbf{K}) \leq d+1,$$

with equality if and only if $\mathbf{K}$ is a centrally symmetric pseudo-double pyramid in the sense of [211]. Similarly, by [211] and [144], over the family of d-dimensional $\mathbf{o}$-symmetric convex bodies, we have

$$c_{tr}^{Bus}(\mathbf{K}) \geq 1 + \frac{2\kappa_{d-1}}{\kappa_d},$$

with equality if and only if $\mathbf{K}$ is an ellipsoid.

Open Problem 11 *For $d \geq 3$ and $\tau \in \{HT, m, m^*\}$, find the maximal values of $c_{tr}^{\tau}(\mathbf{K})$ over the family of d-dimensional convex bodies.*

Open Problem 12 *For $d \geq 3$ and $\tau \in \{Bus, HT, m, m^*\}$, find the minimal values of $c_{tr}^{\tau}(\mathbf{K})$ over the family of d-dimensional convex bodies.*

When finding the minimal value of $c_{tr}^{Bus}(\mathbf{K})$ over the family of plane convex bodies, one had to examine the smallest area convex disks of constant width two in a fixed normed plane. Nevertheless, for $d = 3$, even for the Euclidean norm, this question has been open for a long while (cf. [150]).

Other problems arise if, instead of two translates of a convex body, we consider other families related to the body as described in Section 2.1. Since for any convex body $\mathbf{K}$, $\mathbf{K} - \mathbf{K}$ is twice the unit ball of the relative norm of $\mathbf{K}$, the variant of our problem applied to difference bodies has already been examined in Theorems 53-57. These results can be regarded as the normed variants of the inequality (eq:differencebody) of Rogers and Shephard.

Another possibility is to examine the reflection bodies of $\mathbf{K}$, defined in Definition 9.

2.3 Research Exercises

Exercise 2.1 (Steiner [223]) *Prove that the Steiner symmetrization of any convex body is a convex body.*

Exercise 2.2 *Prove that if* $\mathbf{K}$ *is a convex body in* $\mathbb{E}^d$*, and* H *is a hyperplane, then* $\operatorname{diam}(S_H(\mathbf{K})) \leq \operatorname{diam}(\mathbf{K})$*, where* $\operatorname{diam}(\mathbf{S})$ *stands for the diameter of the set* $\mathbf{S}$.

Exercise 2.3 *Let* $\mathbf{K}$ *be a convex body in* $\mathbb{E}^d$*, and let* L *be a linear subspace of dimension* k*, with orthogonal complement* $L^{\perp}$*. Define* $S_L(\mathbf{K})$ *in the following way: For any* $\mathbf{x} \in L_1$*, let* $B_{d-k}(\mathbf{x})$ *be the closed* $(d-k)$*-dimensional ball in* $\mathbf{x} + L^{\perp}$*, centered at* $\mathbf{x}$ *such that* $\operatorname{vol}_{d-k}(B_{d-k}(\mathbf{x})) = \operatorname{vol}_{d-k}(\mathbf{K} \cap (\mathbf{x} + L^{\perp}))$*, and let* $S_L(\mathbf{K}) = \bigcup_{\mathbf{x} \in L} B_{d-k}(\mathbf{x})$*. Prove that* $S_L(\mathbf{K})$ *is a convex body satisfying* $\operatorname{vol}_d(S_L(\mathbf{K})) = \operatorname{vol}_d(\mathbf{K})$*,* $R(S_L(\mathbf{K})) \leq R(\mathbf{K})$ *and* $\operatorname{diam}(S_L(\mathbf{K})) \leq \operatorname{diam}(\mathbf{K})$*, where* $R(\mathbf{S})$ *is the radius of the smallest ball containing* $\mathbf{S}$.

Exercise 2.4 *Prove that for any* $\mathbf{o}$*-symmetric plane convex body* $\mathbf{K}$*, the midpoints of the sides of any minimum area parallelogram containing* $\mathbf{K}$ *belong to* $\mathbf{K}$.

Exercise 2.5 *Prove that for any* $\mathbf{o}$*-symmetric plane convex body* $\mathbf{K}$*, and any boundary point* $\mathbf{p} \in \operatorname{bd}\mathbf{K}$*, there is an affinely regular hexagon* $\mathbf{H}$ *inscribed in* $\mathbf{K}$ *such that* $\mathbf{p}$ *is a vertex of* $\mathbf{H}$.

Exercise 2.6 *Prove that the previous property holds for not necessarily* $\mathbf{o}$*-symmetric plane convex bodies as well, for a suitably chosen boundary point* $\mathbf{p} \in \operatorname{bd}\mathbf{K}$.

Exercise 2.7 (Gołab [120]) *Using the results in Exercises 2.4 and 2.5, prove Theorem 49.*

Exercise 2.8 *Let* $\mathbf{P}$ *be a convex polygon with vertices* $\mathbf{p}_1, \mathbf{p}_2, \ldots, \mathbf{p}_n = \mathbf{p}_1$ *in counterclockwise order, where the indices are understood* $\bmod n$*. For simplicity, we regard* $\mathbf{P}$ *as a polygon with* $\frac{n}{2}$ *pairs of parallel sides, where a side may have zero length, i.e., some of the vertices may not be distinct (cf. Figure 2.3).*

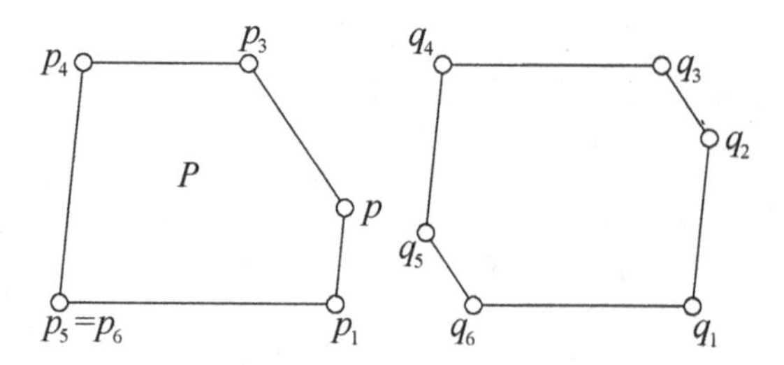

FIGURE 2.3
The polygon **P** in Exercise 2.8.

Let $\mathbf{q}_i = \frac{1}{2}(\mathbf{p}_i - \mathbf{p}_{i+n/2})$. *Prove that the vertices of the central symmetrization* $\frac{1}{2}(\mathbf{P} - \mathbf{P})$ *of* $\mathbf{P}$ *are* $\mathbf{q}_1, \mathbf{q}_2, \ldots, \mathbf{q}_n$ *in counterclockwise order.*

Exercise 2.9 (Fáry-Makai [101]) *Using the results of the previous exercise, prove Theorem 50.*

Exercise 2.10 (Estermann [97]) *Prove the following, slightly stronger form of Conjecture 35 in the plane: Let* $\mathbf{P} \subset \mathbb{E}^2$ *be a convex polygon of unit area, whose center of mass is* $\mathbf{o}$. *Let* $\mathbf{T}$ *be a triangle of unit area whose center of mass is* $\mathbf{o}$. *Then*

$$\operatorname{area}(\operatorname{conv}(\mathbf{P} \cup (-\mathbf{P}))) \leq \operatorname{area}(\operatorname{conv}(\mathbf{T} \cup (-\mathbf{T}))).$$

(Hint: use induction on the number of vertices, and the properties of linear parameter systems.)

Exercise 2.11 (Martini-Swanepoel [178]) *An* $\mathbf{o}$*-symmetric convex curve* Γ *is called* equiframed *if for any* $\mathbf{p} \in \Gamma$ *there is a minimal area parallelogram* $\mathbf{P}$ *circumscribed about* Γ *whose boundary contains* $\mathbf{p}$. *Prove that every Radon curve is equiframed, but not all equiframed curves are Radon.*

3

The Kneser-Poulsen Conjecture Revisited

Summary. The monotonicity of volume under contractions of arbitrary arrangements of spheres is a well-known fundamental problem in discrete geometry. The research on this topic started with the conjecture of Poulsen and Kneser in the late 1950s. We survey the status of the long-standing Kneser-Poulsen conjecture in Euclidean as well as in non-Euclidean spaces with emphases on the latest developments.

3.1 The Kneser-Poulsen conjecture

Recall that $\|\cdot\|$ denotes the standard Euclidean norm of the d-dimensional Euclidean space $\mathbb{E}^d$. So, if $\mathbf{p}_i, \mathbf{p}_j$ are two points in $\mathbb{E}^d$, then $\|\mathbf{p}_i - \mathbf{p}_j\|$ denotes the Euclidean distance between them. It is convenient to denote the (finite) point configuration consisting of the points $\mathbf{p}_1, \mathbf{p}_2, \ldots, \mathbf{p}_N$ in $\mathbb{E}^d$ by $\mathbf{p} = (\mathbf{p}_1, \mathbf{p}_2, \ldots, \mathbf{p}_N)$. Now, if $\mathbf{p} = (\mathbf{p}_1, \mathbf{p}_2, \ldots, \mathbf{p}_N)$ and $\mathbf{q} = (\mathbf{q}_1, \mathbf{q}_2, \ldots, \mathbf{q}_N)$ are two configurations of N points in $\mathbb{E}^d$ such that for all $1 \le i < j \le N$ the inequality $\|\mathbf{q}_i - \mathbf{q}_j\| \le \|\mathbf{p}_i - \mathbf{p}_j\|$ holds, then we say that $\mathbf{q}$ is a *contraction* of $\mathbf{p}$. If $\mathbf{q}$ is a contraction of $\mathbf{p}$, then there may or may not be a continuous motion $\mathbf{p}(t) = (\mathbf{p}_1(t), \mathbf{p}_2(t), \ldots, \mathbf{p}_N(t))$, with $\mathbf{p}_i(t) \in \mathbb{E}^d$ for all $0 \le t \le 1$ and $1 \le i \le N$ such that $\mathbf{p}(0) = \mathbf{p}$ and $\mathbf{p}(1) = \mathbf{q}$, and $\|\mathbf{p}_i(t) - \mathbf{p}_j(t)\|$ is monotone decreasing for all $1 \le i < j \le N$. When there is such a motion, we say that $\mathbf{q}$ is a *continuous contraction* of $\mathbf{p}$. Finally, let $\mathbf{B}^d[\mathbf{p}_i, r_i]$ denote the (closed) d-dimensional ball centered at $\mathbf{p}_i$ with radius r_i in $\mathbb{E}^d$ and let $\mathrm{vol}_d(\cdot)$ represent the d-dimensional volume (Lebesgue measure) in $\mathbb{E}^d$. In 1954 Poulsen [203] and in 1955 Kneser [155] independently conjectured the following for the case when $r_1 = \cdots = r_N$.

Conjecture 63 *If* $\mathbf{q} = (\mathbf{q}_1, \mathbf{q}_2, \ldots, \mathbf{q}_N)$ *is a contraction of* $\mathbf{p} = (\mathbf{p}_1, \mathbf{p}_2, \ldots, \mathbf{p}_N)$ *in* $\mathbb{E}^d$*, then*

$$\mathrm{vol}_d\left(\bigcup_{i=1}^{N} \mathbf{B}^d[\mathbf{p}_i, r_i]\right) \ge \mathrm{vol}_d\left(\bigcup_{i=1}^{N} \mathbf{B}^d[\mathbf{q}_i, r_i]\right).$$

A similar conjecture was proposed for intersections of balls by Klee and Wagon [154] in 1991.

Conjecture 64 *If* $\mathbf{q} = (\mathbf{q}_1, \mathbf{q}_2, \ldots, \mathbf{q}_N)$ *is a contraction of* $\mathbf{p} = (\mathbf{p}_1, \mathbf{p}_2, \ldots, \mathbf{p}_N)$ *in* $\mathbb{E}^d$, *then*

$$\mathrm{vol}_d\left(\bigcap_{i=1}^{N} \mathbf{B}^d[\mathbf{p}_i, r_i]\right) \le \mathrm{vol}_d\left(\bigcap_{i=1}^{N} \mathbf{B}^d[\mathbf{q}_i, r_i]\right).$$

For the sake of simplicity, we refer to Conjectures 63 and 64 as the Kneser-Poulsen conjecture.

3.2 The Kneser-Poulsen conjecture for continuous contractions of unions and intersections of balls

For a given point configuration $\mathbf{p} = (\mathbf{p}_1, \mathbf{p}_2, \ldots, \mathbf{p}_N)$ in $\mathbb{E}^d$ and radii $r_1, r_2, \ldots, r_N$ consider the following sets,

$$\mathbf{V}_i = \{\mathbf{x} \in \mathbb{E}^d \mid \text{for all } j,\ \|\mathbf{x} - \mathbf{p}_i\|^2 - r_i^2 \le \|\mathbf{x} - \mathbf{p}_j\|^2 - r_j^2\},$$

$$\mathbf{V}^i = \{\mathbf{x} \in \mathbb{E}^d \mid \text{for all } j,\ \|\mathbf{x} - \mathbf{p}_i\|^2 - r_i^2 \ge \|\mathbf{x} - \mathbf{p}_j\|^2 - r_j^2\}.$$

The set $\mathbf{V}_i$ (resp., $\mathbf{V}^i$) is called the *nearest (resp., farthest) point Voronoi cell* of the point $\mathbf{p}_i$. (For a detailed discussion on nearest as well as farthest point Voronoi cells, we refer the interested reader to [93] and [221].) We now restrict each of these sets as follows.

$$\mathbf{V}_i(r_i) = \mathbf{V}_i \cap \mathbf{B}^d[\mathbf{p}_i, r_i],$$

$$\mathbf{V}^i(r_i) = \mathbf{V}^i \cap \mathbf{B}^d[\mathbf{p}_i, r_i].$$

We call the set $\mathbf{V}_i(r_i)$ (resp., $\mathbf{V}^i(r_i)$) the *nearest (resp., farthest) point truncated Voronoi cell* of the point $\mathbf{p}_i$. For each $i \ne j$ let $W_{ij} = \mathbf{V}_i \cap \mathbf{V}_j$ and $W^{ij} = \mathbf{V}^i \cap \mathbf{V}^j$. The sets W_{ij} and W^{ij} are the *walls* between the nearest and farthest point Voronoi cells. Finally, it is natural to define the relevant *truncated walls* as follows.

$$\begin{aligned} W_{ij}(\mathbf{p}_i, r_i) &= W_{ij} \cap \mathbf{B}^d[\mathbf{p}_i, r_i] \\ &= W_{ij}(\mathbf{p}_j, r_j) = W_{ij} \cap \mathbf{B}^d[\mathbf{p}_j, r_j], \end{aligned}$$

$$\begin{aligned} W^{ij}(\mathbf{p}_i, r_i) &= W^{ij} \cap \mathbf{B}^d[\mathbf{p}_i, r_i] \\ &= W^{ij}(\mathbf{p}_j, r_j) = W^{ij} \cap \mathbf{B}^d[\mathbf{p}_j, r_j]. \end{aligned}$$

The following formula discovered by Csikós [80] proves Conjecture 63 as well as Conjecture 64 for continuous contractions in a straighforward way in any dimension. (Actually, the planar case of the Kneser-Poulsen conjecture under continuous contractions has been proved independently in [21, 55, 73, 79].)

Theorem 65 *Let $d \geq 2$ and let $\mathbf{p}(t), 0 \leq t \leq 1$ be a smooth motion of a point configuration in $\mathbb{E}^d$ such that for each t, the points of the configuration are pairwise distinct. Then*

$$\frac{d}{dt}\mathrm{vol}_d\left(\bigcup_{i=1}^{N}\mathbf{B}^d[\mathbf{p}_i(t), r_i]\right)$$

$$= \sum_{1\leq i<j\leq N}\left(\frac{d}{dt}d_{ij}(t)\right)\cdot \mathrm{vol}_{d-1}\left(W_{ij}(\mathbf{p}_i(t), r_i)\right),$$

$$\frac{d}{dt}\mathrm{vol}_d\left(\bigcap_{i=1}^{N}\mathbf{B}^d[\mathbf{p}_i(t), r_i]\right)$$

$$= \sum_{1\leq i<j\leq N} -\left(\frac{d}{dt}d_{ij}(t)\right)\cdot \mathrm{vol}_{d-1}\left(W^{ij}(\mathbf{p}_i(t), r_i)\right),$$

where $d_{ij}(t) = \|\mathbf{p}_i(t) - \mathbf{p}_i(t)\|$.

On the one hand, Csikós [82] managed to generalize his formula to configurations of balls called *flowers* which are sets obtained from balls with the help of operations $\cap$ and $\cup$. This work extends to hyperbolic as well as spherical space. On the other hand, Csikós [83] has succeeded in proving a Schläfli-type formula for polytopes with curved faces lying in pseudo-Riemannian Einstein manifolds, which can be used to provide another proof of Conjecture 63 as well as Conjecture 64 for continuous contractions (for more details see [83]).

3.3 The Kneser-Poulsen conjecture for contractions of unions and intersections of disks in $\mathbb{E}^2$

In [38] Bezdek and Connelly proved Conjecture 63 as well as Conjecture 64 in the Euclidean plane. Thus, we have the following theorem.

Theorem 66 *If $\mathbf{q} = (\mathbf{q}_1, \mathbf{q}_2, \ldots, \mathbf{q}_N)$ is a contraction of $\mathbf{p} = (\mathbf{p}_1, \mathbf{p}_2, \ldots, \mathbf{p}_N)$ in $\mathbb{E}^2$, then*

$$\mathrm{vol}_2\left(\bigcup_{i=1}^{N}\mathbf{B}^2[\mathbf{p}_i, r_i]\right) \geq \mathrm{vol}_2\left(\bigcup_{i=1}^{N}\mathbf{B}^2[\mathbf{q}_i, r_i]\right);$$

moreover,

$$\mathrm{vol}_2\left(\bigcap_{i=1}^{N}\mathbf{B}^2[\mathbf{p}_i,r_i]\right)\leq \mathrm{vol}_2\left(\bigcap_{i=1}^{N}\mathbf{B}^2[\mathbf{q}_i,r_i]\right).$$

In fact, the paper [38] contains a proof of an extension of the above theorem to flowers as well. In what follows we give an outline of the three-step proof published in [38] by phrasing it through a sequence of theorems each being higher-dimensional. Voronoi cells play an essential role in the proofs of Theorems 67 and 68.

Theorem 67 *Consider N moving closed d-dimensional balls $\mathbf{B}^d[\mathbf{p}_i(t),r_i]$ with $1\leq i\leq N, 0\leq t\leq 1$ in $\mathbb{E}^d, d\geq 2$. If $F_i(t)$ is the contribution of the ith ball to the boundary of the union $\bigcup_{i=1}^{N}\mathbf{B}^d[\mathbf{p}_i(t),r_i]$ (resp., of the intersection $\bigcap_{i=1}^{N}\mathbf{B}^d[\mathbf{p}_i(t),r_i]$), then*

$$\sum_{1\leq i\leq N}\frac{1}{r_i}\,\mathrm{svol}_{d-1}\left(F_i(t)\right)$$

decreases (resp., increases) in t under any analytic contraction $\mathbf{p}(t)$ of the center points, where $0\leq t\leq 1$ and $\mathrm{svol}_{d-1}(\dots)$ refers to the relevant $(d-1)$-dimensional surface volume.

Theorem 68 *Let the centers of the closed d-dimensional balls $\mathbf{B}^d[\mathbf{p}_i,r_i]$, $1\leq i\leq N$ lie in the $(d-2)$-dimensional affine subspace L of $\mathbb{E}^d, d\geq 3$. If F_i stands for the contribution of the ith ball to the boundary of the union $\bigcup_{i=1}^{N}\mathbf{B}^d[\mathbf{p}_i,r_i]$ (resp., of the intersection $\bigcap_{i=1}^{N}\mathbf{B}^d[\mathbf{p}_i,r_i]$), then*

$$\mathrm{vol}_{d-2}\left(\bigcup_{i=1}^{N}\mathbf{B}^{d-2}[\mathbf{p}_i,r_i]\right)=\frac{1}{2\pi}\sum_{1\leq i\leq N}\frac{1}{r_i}\,\mathrm{svol}_{d-1}(F_i)$$

$$\left(\textit{resp.,}\ \mathrm{vol}_{d-2}\left(\bigcap_{i=1}^{N}\mathbf{B}^{d-2}[\mathbf{p}_i,r_i]\right)=\frac{1}{2\pi}\sum_{1\leq i\leq N}\frac{1}{r_i}\,\mathrm{svol}_{d-1}(F_i)\right),$$

where $\mathbf{B}^{d-2}[\mathbf{p}_i,r_i]=\mathbf{B}^d[\mathbf{p}_i,r_i]\cap L, 1\leq i\leq N$.

Theorem 69 *If $\mathbf{q}=(\mathbf{q}_1,\mathbf{q}_2,\dots,\mathbf{q}_N)$ is a contraction of $\mathbf{p}=(\mathbf{p}_1,\mathbf{p}_2,\dots,\mathbf{p}_N)$ in $\mathbb{E}^d, d\geq 1$, then there is an analytic contraction $\mathbf{p}(t)=(\mathbf{p}_1(t),\dots,\mathbf{p}_N(t)), 0\leq t\leq 1$ in $\mathbb{E}^{2d}$ such that $\mathbf{p}(0)=\mathbf{p}$ and $\mathbf{p}(1)=\mathbf{q}$.*

Note that Theorems 67, 68, and 69 imply Theorem 66 in a straighforward way.

Also, we note that Theorem 69 (called the Leapfrog Lemma) cannot be improved; namely, it has been shown in [19] that there exist point configurations $\mathbf{q}$ and $\mathbf{p}$ in $\mathbb{E}^d$, actually constructed in the way suggested in [38], such that $\mathbf{q}$ is a contraction of $\mathbf{p}$ in $\mathbb{E}^d$ and there is no continuous contraction from $\mathbf{p}$ to $\mathbf{q}$ in $\mathbb{E}^{2d-1}$.

In order to describe a more complete picture of the status of the Kneser-Poulsen conjecture, we mention two additional corollaries obtained from the proof published in [38] and just outlined above. (For more details see [38].)

Theorem 70 *Let $\mathbf{p} = (\mathbf{p}_1, \mathbf{p}_2, \ldots, \mathbf{p}_N)$ and $\mathbf{q} = (\mathbf{q}_1, \mathbf{q}_2, \ldots, \mathbf{q}_N)$ be two point configurations in $\mathbb{E}^d$ such that $\mathbf{q}$ is a piecewise-analytic contraction of $\mathbf{p}$ in $\mathbb{E}^{d+2}$. Then the conclusions of Conjecture 63 as well as Conjecture 64 hold in $\mathbb{E}^d$.*

The following generalizes a result of Gromov [128], who proved it in the case $N \leq d+1$.

Theorem 71 *If $\mathbf{q} = (\mathbf{q}_1, \mathbf{q}_2, \ldots, \mathbf{q}_N)$ is an arbitrary contraction of $\mathbf{p} = (\mathbf{p}_1, \mathbf{p}_2, \ldots, \mathbf{p}_N)$ in $\mathbb{E}^d$ and $N \leq d+3$, then both Conjecture 63 and Conjecture 64 hold.*

Thus, it is natural to conclude this section with the following special and still open case of the Kneser-Poulsen conjecture.

Open Problem 13 *Prove the Kneser-Poulsen conjecture (i.e., Conjectures 63 and 64) for $d+4$ balls in $\mathbb{E}^d$, $d \geq 3$.*

3.4 The Kneser-Poulsen conjecture for uniform contractions of r-ball polyhedra in $\mathbb{E}^d$, $\mathbb{S}^d$ and $\mathbb{H}^d$

Let $\mathbb{M}^d$, $d > 1$ denote the d-dimensional Euclidean, hyperbolic, or spherical space, i.e., one of the simply connected complete Riemannian manifolds of constant sectional curvature. Since simply connected complete space forms, the sectional curvature of which have the same sign are similar, we may assume without loss of generality that the sectional curvature κ of $\mathbb{M}^d$ is $0, -1$, or 1. Let $\mathbf{R}_+$ denote the set of positive real numbers for $\kappa \leq 0$ and the half-open interval $(0, \frac{\pi}{2}]$ for $\kappa = 1$. Let $\mathrm{dist}_{\mathbb{M}^d}(\mathbf{x}, \mathbf{y})$ stand for the geodesic distance between the points $\mathbf{x} \in \mathbb{M}^d$ and $\mathbf{y} \in \mathbb{M}^d$. Furthermore, let $\mathbf{B}_{\mathbb{M}^d}[\mathbf{x}, r]$ denote the closed d-dimensional ball with center $\mathbf{x} \in \mathbb{M}^d$ and radius $r \in \mathbf{R}_+$ in $\mathbb{M}^d$, i.e., let $\mathbf{B}_{\mathbb{M}^d}[\mathbf{x}, r] := \{\mathbf{y} \in \mathbb{M}^d \,|\mathrm{dist}_{\mathbb{M}^d}(\mathbf{x}, \mathbf{y}) \leq r\}$. Now, we are ready to introduce the central notion of this section.

Definition 14 *For a set $X \subseteq \mathbb{M}^d$, $d > 1$ and $r \in \mathbf{R}_+$ let the* r-ball body *X^r generated by X be defined by $X^r := \bigcap_{\mathbf{x} \in X} \mathbf{B}_{\mathbb{M}^d}[\mathbf{x}, r]$. If $X \subseteq \mathbb{M}^d$ is a finite set, then we call X^r an* r-ball polyhedron.

We note that either $X^r = \emptyset$, or X^r is a point in $\mathbb{M}^d$, or $\mathrm{int}(X^r) \neq \emptyset$. Perhaps not surprisingly, r-ball bodies of $\mathbb{E}^d$ have already been investigated in a number of papers however, under various names such as "überkonvexe Menge" ([176]), "r-convex domain" ([107]), "spindle convex set" ([46], [158]), "ball convex set" ([164]), "hyperconvex set" ([113]), and "r-dual set" ([36]). r-ball bodies satisfy some basic identities such as

$$((X^r)^r))^r = X^r \text{ and } (X \cup Y)^r = X^r \cap Y^r,$$

which hold for any $X \subseteq \mathbb{M}^d$ and $Y \subseteq \mathbb{M}^d$. Clearly, also monotonicity holds namely, $X \subseteq Y \subseteq \mathbb{M}^d$ implies $Y^r \subseteq X^r$. Thus, there is a good deal of similarity between r-ball bodies and polar sets (resp., spherical polar sets) in $\mathbb{E}^d$ (resp., $\mathbb{S}^d$). In this section we explore further this similarity by investigating a volumetric relation between X^r and X in $\mathbb{M}^d$. For this reason let $V_{\mathbb{M}^d}(\cdot)$ denote the Lebesgue measure in $\mathbb{M}^d$, to which we are going to refer as volume in $\mathbb{M}^d$. Now, recall the recent theorem of Gao, Hug, and Schneider [117] stating that for any convex body of given volume in $\mathbb{S}^d$ the volume of the spherical polar body becomes maximal if the convex body is a ball. The following extension of their theorem has been proved by Bezdek [36].

Theorem 72 *Let $\mathbf{A} \subseteq \mathbb{M}^d$, $d > 1$ be a compact set of volume $V_{\mathbb{M}^d}(\mathbf{A}) > 0$ and $r \in \mathbf{R}_+$. If $\mathbf{B} \subseteq \mathbb{M}^d$ is a ball with $V_{\mathbb{M}^d}(\mathbf{A}) = V_{\mathbb{M}^d}(\mathbf{B})$, then $V_{\mathbb{M}^d}(\mathbf{A}^r) \leq V_{\mathbb{M}^d}(\mathbf{B}^r)$.*

Note that the Gao-Hug-Schneider theorem is a special case of Theorem 72 namely, when $\mathbb{M}^d = \mathbb{S}^d$ and $r = \frac{\pi}{2}$. As this theorem of [117] is often called a spherical counterpart of the Blaschke-Santaló inequality, one may refer to Theorem 72 as a Blaschke–Santaló-type inequality for r-ball bodies in $\mathbb{M}^d$.

From our point of view, the importance of Theorem 72 lies in the following application. For stating it in a proper way, we recall the following definition from [48]. Note that in what follows, labelled point sets play the role of point configurations whenever the contraction is a uniform one.

Definition 15 *We say that the (labeled) point set $\{\mathbf{q}_1, \ldots, \mathbf{q}_N\} \subset \mathbb{M}^d$ is a uniform contraction of the (labeled) point set $\{\mathbf{p}_1, \ldots, \mathbf{p}_N\} \subset \mathbb{M}^d$ with separating value $\lambda > 0$ in $\mathbb{M}^d$, $d > 1$ if*

$$\mathrm{dist}_{\mathbb{M}^d}(\mathbf{q}_i, \mathbf{q}_j) \leq \lambda \leq \mathrm{dist}_{\mathbb{M}^d}(\mathbf{p}_i, \mathbf{p}_j)$$

holds for all $1 \leq i < j \leq N$.

The following theorem published by Bezdek [36] can be summarized by saying that the volume of an r-ball polyhedron (generated by sufficiently many balls of radius r) increases under uniform contractions in $\mathbb{M}^d$. The more exact details are as follows.

Theorem 73

(i) Let $d \in \mathbf{Z}$ *and* $\delta, \lambda \in \mathbf{R}_+$ *be given such that* $d > 1$ *and* $0 < \lambda \leq \sqrt{2}\delta$. *If* $Q := \{\mathbf{q}_1, \dots, \mathbf{q}_N\} \subset \mathbb{E}^d$ *is a uniform contraction of* $P := \{\mathbf{p}_1, \dots, \mathbf{p}_N\} \subset \mathbb{E}^d$ *with separating value* λ *in* $\mathbb{E}^d$ *and* $N \geq (1+\sqrt{2})^d$, *then* $V_{\mathbb{E}^d}(P^\delta) < V_{\mathbb{E}^d}(Q^\delta)$.

(ii) Let $d \in \mathbf{Z}$ *and* $\delta, \lambda \in \mathbf{R}_+$ *be given such that* $d > 1, 0 < \delta < \frac{\pi}{2}$, *and* $0 < \lambda < \min\left\{\frac{2\sqrt{2}}{\pi}\delta, \pi - 2\delta\right\}$. *If* $Q := \{\mathbf{q}_1, \dots, \mathbf{q}_N\} \subset \mathbb{S}^d$ *is a uniform contraction of* $P := \{\mathbf{p}_1, \dots, \mathbf{p}_N\} \subset \mathbb{S}^d$ *with separating value* λ *in* $\mathbb{S}^d$ *and* $N \geq 2ed\pi^{d-1}\left(\frac{1}{2} + \frac{\pi}{2\sqrt{2}}\right)^d$, *then* $V_{\mathbb{S}^d}(P^\delta) < V_{\mathbb{S}^d}(Q^\delta)$.

(iii) Let $d, k \in \mathbf{Z}$ *and* $\delta, \lambda \in \mathbf{R}_+$ *be given such* $d > 1, k > 0$ *and* $0 < \frac{\sinh k}{\sqrt{2}k}\lambda \leq \delta < k$. *If* $Q := \{\mathbf{q}_1, \dots, \mathbf{q}_N\} \subset \mathbb{H}^d$ *is a uniform contraction of* $P := \{\mathbf{p}_1, \dots, \mathbf{p}_N\} \subset \mathbb{H}^d$ *with separating value* λ *in* $\mathbb{H}^d$ *and* $N \geq (\frac{\sinh 2k}{2k})^{d-1}(\frac{\sqrt{2}\sinh k}{k} + 1)^d$, *then* $V_{\mathbb{H}^d}(P^\delta) < V_{\mathbb{H}^d}(Q^\delta)$.

It is somewhat surprising that in spherical space for the specific radius of balls (i.e., spherical caps) one can find a proof of both Conjecture 63 and Conjecture 64 in all dimensions. The magic radius is $\frac{\pi}{2}$ and the following theorem of Bezdek and Connelly ([39]) describes the desired result in details, which one should compare to part (ii) of Theorem 73.

Theorem 74 *If a finite set of closed d-dimensional balls of radius* $\frac{\pi}{2}$ *(i.e., of closed hemispheres) in the d-dimensional spherical space* $\mathbb{S}^d, d \geq 2$ *is rearranged so that the (spherical) distance between each pair of centers does not increase, then the (spherical) d-dimensional volume of the intersection does not decrease and the (spherical) d-dimensional volume of the union does not increase.*

The method of the proof published by Bezdek and Connelly [39] can be described as follows. First, one can use a leapfrog lemma (similar to Theorem 69) to move one configuration to the other in an analytic and monotone way, but only in higher dimensions. Then the higher-dimensional balls have their combined volume (their intersections or unions) change monotonically, a fact that one can prove using Schläfli's differential formula. Then one can apply an integral formula to relate the volume of the higher-dimensional object to the volume of the lower-dimensional object, obtaining the volume inequality for the more general discrete motions.

Bezdek [30] has proved the following extension of Theorem 72 to intrinsic volumes in $\mathbb{E}^d$.

Theorem 75 *Let* $\mathbf{A} \subseteq \mathbb{E}^d$, $d > 1$ *be a compact set of volume* $V_d(\mathbf{A}) > 0$ *and let* $r > 0$. *If* $\mathbf{B} \subseteq \mathbb{E}^d$ *is a ball with volume* $V_d(\mathbf{A}) = V_d(\mathbf{B})$, *then*

$$V_k(\mathbf{A}^r) \leq V_k(\mathbf{B}^r) \tag{3.1}$$

holds for all $1 \leq k \leq d$, *where* $V_k(\cdot)$ *denotes the k-th intrinsic volume of the corresponding set in* $\mathbb{E}^d$.

Using Theorem 75, one can give a rather short proof ([30]) of the following theorem, part (i) of which has been published by Bezdek and Naszódi [48]. Furthermore, we note also that the following theorem is a strengthening of part (i) of Theorem 73.

Theorem 76 *Let $d > 1$, $\lambda > 0$, $r > 0$, and $1 \leq k \leq d$ be given and let $Q := \{\mathbf{q}_1, \ldots, \mathbf{q}_N\} \subset \mathbb{E}^d$ be a uniform contraction of $P := \{\mathbf{p}_1, \ldots, \mathbf{p}_N\} \subset \mathbb{E}^d$ with separating value λ in $\mathbb{E}^d$.*

(i) If $1 < d < 42$ and $N \geq (1+\sqrt{2})^d$, then

$$V_k(P^r) \leq V_k(Q^r) \tag{3.2}$$

(ii) If $d \geq 42$ and $N \geq \sqrt{\frac{\pi}{2d}}(1+\sqrt{2})^d + 1$, then (3.2) holds.

Theorem 76 leads to the following question (raised also in [48]).

Open Problem 14 *Let $1 \leq k \leq d$, $1 < N$, and $r > 0$. Prove or disprove that if $\mathbf{q} = (\mathbf{q}_1, \mathbf{q}_2, \ldots, \mathbf{q}_N)$ is any contraction of $\mathbf{p} = (\mathbf{p}_1, \mathbf{p}_2, \ldots, \mathbf{p}_N)$ in $\mathbb{E}^d$, then $V_k(P^r) \leq V_k(Q^r)$.*

In particular, the following conjecture has been put forward by Alexander [5].

Conjecture 77 *Under an arbitrary contraction of the center points of finitely many congruent disks in the plane, the perimeter of the intersection of the disks cannot decrease.*

For the sake of completeness, we conclude this section with the following statement, which is a natural dual of Theorem 76 for uniform contractions of unions of congruent balls in $\mathbb{E}^d$. Also, it improves Theorem 1.5 of [48].

Theorem 78 *Let $d > 1$, $\lambda > 0$, and $r > 0$ be given and let $Q := \{\mathbf{q}_1, \ldots, \mathbf{q}_N\} \subset \mathbb{E}^d$ be a uniform contraction of $P := \{\mathbf{p}_1, \ldots, \mathbf{p}_N\} \subset \mathbb{E}^d$ with separating value λ in $\mathbb{E}^d$. If $N \geq 2^d$, then $\mathrm{V}_d\left(\bigcup_{i=1}^N \mathbf{B}^d[\mathbf{p}_i, r]\right) \geq \mathrm{V}_d\left(\bigcup_{i=1}^N \mathbf{B}^d[\mathbf{q}_i, r]\right)$.*

3.5 The Kneser-Poulsen conjecture for contractions of unions and intersections of disks in $\mathbb{S}^2$ and $\mathbb{H}^2$

Let $\mathbb{M}^d$, $d > 1$ denote the d-dimensional Euclidean, hyperbolic, or spherical space. Applying results on central sets of unions of finitely many balls in $\mathbb{M}^d$, Gorbovickis ([124]) has proved the following new special cases of the Kneser-Poulsen conjecture in $\mathbb{M}^2$.

Theorem 79 *If the union of a finite set of closed disks in $\mathbb{S}^2$ (resp., $\mathbb{H}^2$) has a simply connected interior, then the area of the union of these disks cannot increase after any contractive rearrangement.*

The following statement published by Gorbovickis [124] is a straightforward corollary of Theorem 79.

Corollary 80

(i) If the intersection of a finite set of closed disks in $\mathbb{S}^2$ is connected, then the area of the intersection of these disks cannot decrease after any contractive rearrangement.

(ii) The area of the intersection of finitely many closed disks having radii at most $\frac{\pi}{2}$ in $\mathbb{S}^2$ cannot decrease after any contractive rearrangement.

One can regard part (ii) of Corollary 80 as an extension of Theorem 74 in $\mathbb{S}^2$. Finally, we mention the following recent result of Csikós and Horváth ([81]) that extends part (ii) of Corollary 80 to $\mathbb{H}^2$ by applying the method of Gorbovickis ([124]) to co-central sets, which one can regard as natural duals of central sets.

Theorem 81 *The area of the intersection of finitely many closed disks in $\mathbb{H}^2$ cannot decrease after any contractive rearrangement.*

3.6 Research Exercises

Exercise 3.1 *Prove that if $\mathbf{q} = (\mathbf{q}_1, \mathbf{q}_2, \ldots, \mathbf{q}_N)$ is a contraction of $\mathbf{p} = (\mathbf{p}_1, \mathbf{p}_2, \ldots, \mathbf{p}_N)$ in $\mathbb{E}^d$ such that $\mathbf{q}_i = \lambda \mathbf{p}_i$, $1 \leq i \leq N$ with $0 < \lambda < 1$, then the inequalities $\mathrm{vol}_d\left(\bigcup_{i=1}^N \mathbf{B}^d[\mathbf{p}_i, r_i]\right) \geq \mathrm{vol}_d\left(\bigcup_{i=1}^N \mathbf{B}^d[\mathbf{q}_i, r_i]\right)$ and $\mathrm{vol}_d\left(\bigcap_{i=1}^N \mathbf{B}^d[\mathbf{p}_i, r_i]\right) \leq \mathrm{vol}_d\left(\bigcap_{i=1}^N \mathbf{B}^d[\mathbf{q}_i, r_i]\right)$ hold.*

Exercise 3.2 (Kirszbraun-Alexander [6]) *If $\mathbf{q} = (\mathbf{q}_1, \mathbf{q}_2, \ldots, \mathbf{q}_N)$ is a contraction of $\mathbf{p} = (\mathbf{p}_1, \mathbf{p}_2, \ldots, \mathbf{p}_N)$ in $\mathbb{E}^d$ and $\bigcap_{i=1}^N \mathbf{B}^d[\mathbf{p}_i, r_i] \neq \emptyset$, then $\bigcap_{i=1}^N \mathbf{B}^d[\mathbf{q}_i, r_i] \neq \emptyset$.*

Exercise 3.3 (Bollobás [55]) *Prove that if $\mathbf{q} = (\mathbf{q}_1, \mathbf{q}_2, \ldots, \mathbf{q}_N)$ is a continuous contraction of $\mathbf{p} = (\mathbf{p}_1, \mathbf{p}_2, \ldots, \mathbf{p}_N)$ in $\mathbb{E}^2$, then $\mathrm{vol}_2\left(\bigcup_{i=1}^N \mathbf{B}^2[\mathbf{p}_i, r]\right) \geq \mathrm{vol}_2\left(\bigcup_{i=1}^N \mathbf{B}^2[\mathbf{q}_i, r]\right)$.* (Hint: Prove the analogue inequality for the perimeter.)

Exercise 3.4 (Capoyleas [73]) *Prove that if $\mathbf{q} = (\mathbf{q}_1, \mathbf{q}_2, \ldots, \mathbf{q}_N)$ is a continuous contraction of $\mathbf{p} = (\mathbf{p}_1, \mathbf{p}_2, \ldots, \mathbf{p}_N)$ in $\mathbb{E}^2$, then $\mathrm{vol}_2\left(\bigcap_{i=1}^N \mathbf{B}^2[\mathbf{p}_i, r]\right) \leq \mathrm{vol}_2\left(\bigcap_{i=1}^N \mathbf{B}^2[\mathbf{q}_i, r]\right)$.* (Hint: Prove the analogue inequality for the perimeter.)

Exercise 3.5 (Alexander [5]) *If* $\{\mathbf{q}_1, \mathbf{q}_2, \ldots, \mathbf{q}_N\}$ *and* $\{\mathbf{p}_1, \mathbf{p}_2, \ldots, \mathbf{p}_N\}$ *are given in* $\mathbb{E}^d$ *such that* $\|\mathbf{q}_i - \mathbf{q}_j\| \le c\|\mathbf{p}_i - \mathbf{p}_j\|$ *holds for all* $1 \le i < j \le N$ *with* $c > 0$, *then*

$$V_1 (\text{conv} (\{\mathbf{q}_i \mid 1 \le i \le N\})) \le c\, V_1 (\text{conv} (\{\mathbf{p}_i \mid 1 \le i \le N\})),$$

where $V_1(\cdot)$ *denotes the first intrinsic volume of the corresponding set in* $\mathbb{E}^d$.

Exercise 3.6 (Gorbovickis [122]) *Prove that if* $\mathbf{q} = (\mathbf{q}_1, \mathbf{q}_2, \ldots, \mathbf{q}_N)$ *is a contraction of* $\mathbf{p} = (\mathbf{p}_1, \mathbf{p}_2, \ldots, \mathbf{p}_N)$ *in* $\mathbb{E}^d$ *and* $\mathbf{q}$ *and* $\mathbf{p}$ *are not congruent, then*

$$V_1 (\text{conv} (\{\mathbf{q}_i \mid 1 \le i \le N\})) < V_1 (\text{conv} (\{\mathbf{p}_i \mid 1 \le i \le N\})).$$

Exercise 3.7 (Csikós-Horváth [81]) *Let the point set* $\{\mathbf{q}_1, \ldots, \mathbf{q}_N\}$ *be a contraction of the point set* $\{\mathbf{p}_1, \ldots, \mathbf{p}_N\}$ *in* $\mathbb{M}^2$ *(and in case* $\mathbb{M}^2 = \mathbb{S}^2$ *let* $\bigcup_{i=1}^N \mathbf{B}_{\mathbb{M}^2}[\mathbf{p}_i, r_i]$ *be contained in a closed hemisphere of* $\mathbb{S}^2$*). Then*

$$\text{per}\left(\text{conv}\left(\bigcup_{i=1}^N \mathbf{B}_{\mathbb{M}^2}[\mathbf{q}_i, r_i]\right)\right) \le \text{per}\left(\text{conv}\left(\bigcup_{i=1}^N \mathbf{B}_{\mathbb{M}^2}[\mathbf{p}_i, r_i]\right)\right),$$

where $\text{per}(\cdot)$ *denotes the perimeter of the corresponding set in* $\mathbb{M}^2$.

Exercise 3.8 (Bezdek-Connelly-Csikós [40]) *Prove that if* $\mathbf{q} = (\mathbf{q}_1, \mathbf{q}_2, \ldots, \mathbf{q}_N)$ *is a contraction of* $\mathbf{p} = (\mathbf{p}_1, \mathbf{p}_2, \ldots, \mathbf{p}_N)$ *for* $1 < N \le 4$ *in* $\mathbb{E}^2$, *then* $\text{per}\left(\bigcap_{i=1}^N \mathbf{B}^d[\mathbf{p}_i, r]\right) \le \text{per}\left(\bigcap_{i=1}^N \mathbf{B}^d[\mathbf{q}_i, r]\right)$ *holds.*

Exercise 3.9 (Gorbovickis [123]) *Prove that if* $\mathbf{q} = (\mathbf{q}_1, \mathbf{q}_2, \ldots, \mathbf{q}_N)$ *is a contraction of* $\mathbf{p} = (\mathbf{p}_1, \mathbf{p}_2, \ldots, \mathbf{p}_N)$ *in* $\mathbb{E}^d$, *then* $\text{vol}_d\left(\bigcup_{i=1}^N \mathbf{B}^d[\mathbf{p}_i, r_i]\right) \ge \text{vol}_d\left(\bigcup_{i=1}^N \mathbf{B}^d[\mathbf{q}_i, r_i]\right)$ *follows under the additional assumption that the intersection of every pair of balls from* $\{\mathbf{B}^d[\mathbf{q}_i, r_i] \mid 1 \le i \le N\}$ *has common interior points with no more than* $d+1$ *other balls from* $\{\mathbf{B}^d[\mathbf{q}_i, r_i] \mid 1 \le i \le N\}$.

Exercise 3.10 (Gorbovickis [122]) *Prove that if* $\mathbf{q} = (\mathbf{q}_1, \mathbf{q}_2, \ldots, \mathbf{q}_N)$ *is a contraction of* $\mathbf{p} = (\mathbf{p}_1, \mathbf{p}_2, \ldots, \mathbf{p}_N)$ *in* $\mathbb{E}^d$, *then there exists* r_0 *(depending on* $\mathbf{q}$ *and* $\mathbf{p}$*) such that for any* $r \ge r_0$ *the inequalities* $\text{vol}_d\left(\bigcup_{i=1}^N \mathbf{B}^d[\mathbf{p}_i, r_i]\right) \ge \text{vol}_d\left(\bigcup_{i=1}^N \mathbf{B}^d[\mathbf{q}_i, r_i]\right)$ *and* $\text{vol}_d\left(\bigcap_{i=1}^N \mathbf{B}^d[\mathbf{p}_i, r_i]\right) \le \text{vol}_d\left(\bigcap_{i=1}^N \mathbf{B}^d[\mathbf{q}_i, r_i]\right)$ *hold. Moreover, if* $\mathbf{q}$ *and* $\mathbf{p}$ *are not congruent, then the inequalities are strict.*

For the next problem, we fix an orthonormal basis (i.e., a Cartesian coordinate system) in $\mathbb{E}^d$ and refer to the coordinates of the point $\mathbf{x} \in \mathbb{E}^d$ by writing $\mathbf{x} = (x^{(1)}, \ldots, x^{(d)})$. Now, if $\mathbf{p} = (\mathbf{p}_1, \ldots, \mathbf{p}_N)$ and $\mathbf{q} = (\mathbf{q}_1, \ldots, \mathbf{q}_N)$ are two configurations of N points in $\mathbb{E}^d$ such that for all $1 \le k \le d$ and $1 \le i < j \le N$ the inequality $||q_i^{(k)} - q_j^{(k)}|| \le ||p_i^{(k)} - p_j^{(k)}||$ holds, then we say that $\mathbf{q}$ is a *strong contraction* of $\mathbf{p}$. Clearly, if $\mathbf{q}$ is a strong contraction of $\mathbf{p}$, then $\mathbf{q}$ is a contraction of $\mathbf{p}$ as well. Furthermore, recall that a convex body $\mathbf{K}$ is called an *unconditional* (or, 1-*unconditional*) convex body if for any $\mathbf{x} = (x^{(1)}, \dots, x^{(d)}) \in \mathbf{K}$ also $(\pm x^{(1)}, \dots, \pm x^{(d)}) \in \mathbf{K}$ holds. Clearly, if $\mathbf{K}$ is an unconditional convex body in $\mathbb{E}^d$, then $\mathbf{K}$ is symmetric about the origin $\mathbf{o}$ of $\mathbb{E}^d$.

Exercise 3.11 (Bezdek-Naszódi [48]) *Let $\mathbf{K}_1, \dots, \mathbf{K}_N$ be (not necessarily distinct) unconditional convex bodies in $\mathbb{E}^d$, $d \geq 2$. If $\mathbf{q} = (\mathbf{q}_1, \dots, \mathbf{q}_N)$ is a strong contraction of $\mathbf{p} = (\mathbf{p}_1, \dots, \mathbf{p}_N)$ in $\mathbb{E}^d$, then $\mathrm{vol}_d\left(\bigcup_{i=1}^N (\mathbf{p}_i + \mathbf{K}_i)\right) \geq \mathrm{vol}_d\left(\bigcup_{i=1}^N (\mathbf{q}_i + \mathbf{K}_i)\right)$ and $\mathrm{vol}_d\left(\bigcap_{i=1}^N (\mathbf{p}_i + \mathbf{K}_i)\right) \leq \mathrm{vol}_d\left(\bigcap_{i=1}^N (\mathbf{q}_i + \mathbf{K}_i)\right)$.*

Exercise 3.12 (Meyer-Reisner-Schmuckenschläger [181]) *Prove that if for an $\mathbf{o}$-symmetric convex body $\mathbf{K} \subset \mathbb{E}^d$ and some $\tau > 0$, $\mathrm{vol}_d\left(\mathbf{K} \cap (\mathbf{x} + \tau\mathbf{K})\right)$ depends on $\|\mathbf{x}\|_{\mathbf{K}}$ only (where $\|\cdot\|_{\mathbf{K}}$ denotes the norm of $\mathbb{E}^d$ generated by $\mathbf{K}$), then $\mathbf{K}$ is an ellipsoid.*

4

Volumetric Bounds for Contact Numbers

Summary. The well-known "kissing number problem" asks for the largest number of non-overlapping unit balls touching a given unit ball in the Euclidean d-space $\mathbb{E}^d$. Generalizing the kissing number, the Hadwiger number or the translative kissing number $H(\mathbf{K})$ of a convex body $\mathbf{K}$ in $\mathbb{E}^d$ is the maximum number of non-overlapping translates of $\mathbf{K}$ that all touch $\mathbf{K}$. In this chapter we study and survey the following natural extension of these problems. A finite translative packing of the convex body $\mathbf{K}$ in $\mathbb{E}^d$ is a finite family $\mathcal{P}$ of non-overlapping translates of $\mathbf{K}$ in $\mathbb{E}^d$. Furthermore, the contact graph $G(\mathcal{P})$ of $\mathcal{P}$ is the (simple) graph whose vertices correspond to the packing elements and whose two vertices are connected by an edge if and only if the corresponding two packing elements touch each other. The number of edges of $G(\mathcal{P})$ is called the contact number of $\mathcal{P}$. Finally, the "contact number problem" asks for the largest contact number, that is, for the maximum number $c(\mathbf{K}, n, d)$ of edges that a contact graph of n non-overlapping translates of $\mathbf{K}$ can have in $\mathbb{E}^d$. In the first half of this chapter, we survey the bounds proved for $c(\mathbf{K}, n, d)$ using volumetric methods. Then we turn our attention to an important subfamily of translative packings called totally separable packings. Here a packing of translates of a convex body $\mathbf{K}$ in $\mathbb{E}^d$ is called totally separable if any two packing elements can be separated by a hyperplane of $\mathbb{E}^d$ disjoint from the interior of every packing element. In the second half of this chapter, we study the analogues of the Hadwiger and contact numbers for totally separable translative packings of $\mathbf{K}$ labelled by $H_{\text{sep}}(\mathbf{K})$ and $c_{\text{sep}}(\mathbf{K}, n, d)$ and survey the bounds proved for $H_{\text{sep}}(\mathbf{K})$ as well as $c_{\text{sep}}(\mathbf{K}, n, d)$ using volumetric ideas. This chapter is a revised and strongly extended version of [41].

4.1 Description of the basic geometric questions

The well-known *"kissing number problem"* asks for the maximum number $k(d)$ of non-overlapping unit balls that can touch a unit ball in the d-dimensional Euclidean space $\mathbb{E}^d$. The problem originated in the 17th century from a disagreement between Newton and Gregory about how many 3-dimensional unit spheres without overlap could touch a given unit sphere. The former maintained that the answer was 12, while the latter thought it was 13. The question was finally settled many years later [220] when Newton was proved correct. The known values of $k(d)$ are $k(2) = 6$ (trivial), $k(3) = 12$ ([220]), $k(4) = 24$ ([189]), $k(8) = 240$ ([192]), and $k(24) = 196560$ ([192]). The problem of finding kissing numbers is closely connected to the more general problems of finding bounds for spherical codes and sphere packings. For old and new results on kissing numbers, we refer the interested reader to the recent survey article [64]. In this paper, we focus on a more general relative of kissing number called contact number.

Let $\mathbf{B}^d$ be the d-dimensional unit ball centered at the origin $\mathbf{o}$ in $\mathbb{E}^d$. As is well known, a *finite packing* of unit balls in $\mathbb{E}^d$ is a finite family of non-overlapping translates of $\mathbf{B}^d$ in $\mathbb{E}^d$. Furthermore, the *contact graph* of a finite unit ball packing in $\mathbb{E}^d$ is the (simple) graph whose vertices correspond to the packing elements and whose two vertices are connected by an edge if and only if the corresponding two packing elements touch each other. The number of edges of a contact graph is called the *contact number* of the underlying unit ball packing. The *"contact number problem"* asks for the largest contact number, that is, for the maximum number $c(n, d)$ of edges that a contact graph of n non-overlapping translates of $\mathbf{B}^d$ can have in $\mathbb{E}^d$.

The problem of determining $c(n, d)$ is equivalent to Erdős's repeated shortest distance problem, which asks for the largest number of repeated shortest distances among n points in $\mathbb{E}^d$. The planar case of this question was originally raised by Erdős in 1946 [95], with an answer conjectured by Reutter in 1972 and established by Harborth [139] in 1974, whereas the problem in its more general forms was popularized by Erdős and Ulam. Another way to look at the contact number problem is to think of it as the combinatorial analogue of the densest sphere packing problem, which dates back to the 17th century.

Let $\mathbf{K}$ be a convex body, i.e., a compact convex set with non-empty interior in $\mathbb{E}^d$. (If $d = 2$, then $\mathbf{K}$ is called a convex domain.) If $\mathbf{K}$ is symmetric about the origin $\mathbf{o}$ in $\mathbb{E}^d$, then one can regard $\mathbf{K}$ as the unit ball of a given norm in $\mathbb{R}^d$. In the same way as above, one can talk about the largest contact number of packings by n translates of $\mathbf{K}$ in $\mathbb{E}^d$ and label it by $c(\mathbf{K}, n, d)$. Here we survey the results on $c(n, d)$ as well as $c(\mathbf{K}, n, d)$.

The notion of total separability was introduced in [104] as follows: a packing of unit balls in $\mathbb{E}^d$ is called *totally separable* if any two unit balls can be separated by a hyperplane of $\mathbb{E}^d$ such that it is disjoint from the interior of each unit ball in the packing. Finding the densest totally separable unit ball packing is a difficult problem, which is solved only in dimensions two ([27, 104]) and three ([152]). As a close combinatorial relative, it is natural to investigate the maximum contact number $c_{\mathrm{sep}}(n, d)$ of totally separable packings of n unit balls in $\mathbb{E}^d$. In what follows, we survey the results on $c_{\mathrm{sep}}(n, d)$ and include in our overview the relevant aspects of totally separable translative packings as well.

4.2 Motivation from materials science

In addition to finding its origins in the works of pioneers like Newton, Erdős, Ulam and Fejes Tóth (see Section 4.3 for more on the role of the latter two), the contact number problem is also important from an applications point of view. Packings of hard sticky spheres - impenetrable spheres with short-range attractive forces - provide excellent models for the formation of several real-life materials such as colloids, powders, gels and glasses [140]. The particles in these materials can be thought of as hard spheres that self-assemble into small and large clusters due to their attractive forces. This process, called *self-assembly*, is of tremendous interest to materials scientists, chemists, statistical physicists and biologists alike.

Of particular interest are *colloids*, which consist of particles at micron scale, dispersed in a fluid and kept suspended by thermal interactions [176]. Colloidal matter occurs abundantly around us - for example in glue, milk and paint. Moreover, controlled colloid formation is a fundamental tool used in scientific research to understand the phenomena of self-assembly and phase transition.

From thermodynamical considerations, it is clear that colloidal particles assemble so as to minimize the potential energy of the cluster. Since the range of attraction between these particles is extremely small compared to their sizes, two colloidal particles do not exert any force on each other until they are infinitesimally close, at which point there is strong attraction between them. As a result, they stick together, are resistant to drift apart, but are strongly resistant to move any closer [13, 140]. Thus two colloidal particles experiencing an attractive force from one another in a cluster can literally be thought of as being in contact.

It can be shown that under the force law described above, the potential energy of a colloidal cluster at reasonably low temperatures is inversely proportional to the number of contacts between its particles [13, 143, 146]. Thus the particles are highly likely to assemble in packings that maximize the contact number. This has generated significant interest among materials scientists towards the contact number problem [13, 146] and has led to efforts in developing computer-assisted approaches to attack the problem.

4.3 Largest contact numbers for congruent circle packings

4.3.1 The Euclidean plane

Harborth [139] proved the following well-known result on the contact graphs of congruent circular disk packings in $\mathbb{E}^2$.

Theorem 82 $c(n,2) = \lfloor 3n - \sqrt{12n-3} \rfloor$, *for all* $n \geq 2$.

This result shows that an optimal way to pack n congruent disks to maximize their contacts is to pack them in a 'hexagonal arrangement'. The arrangement starts by packing 6 unit disks around a central disk in such a way that the centers of the surrounding disks form a regular hexagon. The pattern is then continued by packing hexagonal layers of disks around the first hexagon. Thus the hexagonal packing arrangement, which is known to be the densest congruent disk packing arrangement, also achieves the maximum contact number $c(n,2)$, for all n.

Interestingly, this also means that $c(n,2)$ equals the maximum number of sides that can be shared between n cells of a regular hexagon tiling of the plane. This connection was explored in [138], where isoperimetric hexagonal lattice animals of a given area n were explored.

In 1984, Ulam ([96]) proposed to investigate Erdős-type distance problems in normed spaces. Pursuing this idea, Brass [66] proved the following extension of Theorem 82 to normed planes.

Theorem 83 *Let* $\mathbf{K}$ *be a convex domain different from a parallelogram in* $\mathbb{E}^2$. *Then for all* $n \geq 2$, *one has* $c(\mathbf{K}, n, 2) = \lfloor 3n - \sqrt{12n-3} \rfloor$. *If* $\mathbf{K}$ *is a parallelogram, then* $c(\mathbf{K}, n, 2) = \lfloor 4n - \sqrt{28n-12} \rfloor$ *holds for all* $n \geq 2$.

4.3.2 Spherical and hyperbolic planes

An analogue of Harborth's theorem in the hyperbolic plane $\mathbb{H}^2$ was found by Bowen [62]. In fact, his method extends to the 2-dimensional spherical plane $\mathbb{S}^2$. We prefer to quote these results as follows.

Theorem 84 *Consider disk packings in* $\mathbb{H}^2$ *(resp.,* $\mathbb{S}^2$*) by finitely many congruent disks, which maximize the number of touching pairs for the given number of congruent disks and of given diameter* D. *Then such a packing must have all of its centers located on the vertices of a triangulation of* $\mathbb{H}^2$ *(resp.,* $\mathbb{S}^2$*) by congruent equilateral triangles of side length* D *provided that the equilateral triangle in* $\mathbb{H}^2$ *(resp.,* $\mathbb{S}^2$*) of side length* D *has each of its angles equal to* $\frac{2\pi}{N}$ *for some positive integer* $N \geq 3$.

In 1984, L. Fejes Tóth ([31]) raised the following attractive and related problem in $\mathbb{S}^2$: Consider an arbitrary packing $\mathcal{P}_r$ of disks of radius $r > 0$ in $\mathbb{S}^2$. Let $\deg_{\mathrm{avr}}(\mathcal{P}_r)$ denote the average degree of the vertices of the contact graph of $\mathcal{P}_r$. Then prove or disprove that $\limsup_{r\to 0}\left(\sup_{\mathcal{P}_r}\deg_{\mathrm{avr}}(\mathcal{P}_r)\right) < 5$. This problem was settled in [31].

Theorem 85 *Let $\mathcal{P}_r$ be an arbitrary packing of disks of radius $r > 0$ in $\mathbb{S}^2$. Then*

$$\limsup_{r\to 0}\left(\sup_{\mathcal{P}_r}\deg_{\mathrm{avr}}(\mathcal{P}_r)\right) < 5.$$

We conclude this section with the still open hyperbolic analogue of Theorem 85 which was raised in [31].

Conjecture 86 *Let $\mathcal{P}_r$ be an arbitrary packing of disks of radius $r > 0$ in $\mathbb{H}^2$. Then*

$$\limsup_{r\to 0}\left(\sup_{\mathcal{P}_r}\deg_{\mathrm{avr}}(\mathcal{P}_r)\right) < 5.$$

4.4 Largest contact numbers for unit ball packings in $\mathbb{E}^3$

Theorem 82 implies in a straightforward way that

$$\lim_{n\to+\infty}\frac{3n - c(n,2)}{\sqrt{n}} = \sqrt{12} = 3.464\ldots\ . \tag{4.1}$$

Although one cannot hope for an explicit formula for $c(n,3)$ in terms of n, there might be a way to prove a proper analogue of (4.1) in $\mathbb{E}^3$.

To this end, we know only what is stated in Theorem 87. In order to state these results we need an additional concept. Let us imagine that we generate packings of n unit balls in $\mathbb{E}^3$ in such a special way that each and every center of the n unit balls chosen is a lattice point of the face-centered cubic lattice with shortest non-zero lattice vector of length 2. Then let $c_{\mathrm{fcc}}(n)$ denote the largest possible contact number of all packings of n unit balls obtained in this way.

The motivation for considering $c_{\mathrm{fcc}}(n)$ is obvious. Since in the planar case, the densest disk packing arrangement also maximizes contacts between disks and the face-centered cubic lattice is the densest for sphere packings in $\mathbb{E}^3$ [135], it makes sense to consider $c_{\mathrm{fcc}}(n)$ as a candidate for $c(n,3)$. Moreover, it is easy to see that $c_{\mathrm{fcc}}(2) = c(2,3) = 1$, $c_{\mathrm{fcc}}(3) = c(3,3) = 3$ and $c_{\mathrm{fcc}}(4) = c(4,3) = 6$.

Theorem 87

(i) $c(n,3) < 6n - 0.926n^{\frac{2}{3}}$, *for all* $n \geq 2$.

(ii) $c_{\mathrm{fcc}}(n) < 6n - \frac{3\sqrt[3]{18\pi}}{\pi}n^{\frac{2}{3}} = 6n - 3.665\ldots n^{\frac{2}{3}}$, *for all* $n \geq 2$.

(iii) $6n - \sqrt[3]{486}n^{\frac{2}{3}} < 2k(2k^2-3k+1) \leq c_{\text{fcc}}(n) \leq c(n,3)$, *for all* $n = \frac{k(2k^2+1)}{3}$ *with* $k \geq 2$.

Recall that (*i*) was proved in [50] (using the method of [34]), while (*ii*) and (*iii*) were proved in [34]. Clearly, Theorem 87 implies that

$$0.926 < \frac{6n - c(n,3)}{n^{\frac{2}{3}}} < \sqrt[3]{486} = 7.862\ldots, \tag{4.2}$$

for all $n = \frac{k(2k^2+1)}{3}$ with $k \geq 2$.

4.5 Upper bounding the contact numbers for packings by translates of a convex body in $\mathbb{E}^d$

One of the main results of this section is an upper bound for the number of touching pairs in an arbitrary finite packing of translates of a convex body, proved in [33]. In order to state the theorem in question in a concise way, we need a bit of notation. Let $\mathbf{K}$ be an arbitrary convex body in $\mathbb{E}^d$, $d \geq 3$. Then let $\delta(\mathbf{K})$ denote the density of a densest packing of translates of the convex body $\mathbf{K}$ in $\mathbb{E}^d$, $d \geq 3$. Moreover, let

$$\mathrm{iq}(\mathbf{K}) := \frac{(\mathrm{svol}_{d-1}(\mathrm{bd}\mathbf{K}))^d}{(\mathrm{vol}_d(\mathbf{K}))^{d-1}}$$

be the isoperimetric quotient of the convex body $\mathbf{K}$, where $\mathrm{svol}_{d-1}(\mathrm{bd}\mathbf{K})$ denotes the $(d-1)$-dimensional surface volume of the boundary $\mathrm{bd}\mathbf{K}$ of $\mathbf{K}$ and $\mathrm{vol}_d(\mathbf{K})$ denotes the d-dimensional volume of $\mathbf{K}$. Furthermore, let $H(\mathbf{K})$ denote the Hadwiger number of $\mathbf{K}$, which is the largest number of non-overlapping translates of $\mathbf{K}$ that can all touch $\mathbf{K}$ (for more details, cf. Section 4.7). Finally, let the one-sided Hadwiger number $h(\mathbf{K})$ of $\mathbf{K}$ be the largest number of non-overlapping translates of $\mathbf{K}$ that touch $\mathbf{K}$ and that all lie in a closed supporting half space of $\mathbf{K}$. In [37], using the Brunn–Minkowski inequality, it is proved that $h(\mathbf{K}) \leq 2 \cdot 3^{d-1} - 1$, where equality is attained if and only if $\mathbf{K}$ is an affine d-cube. Let $\mathbf{K_o} := \frac{1}{2}(\mathbf{K} + (-\mathbf{K}))$ be the normalized (centrally symmetric) difference body assigned to $\mathbf{K}$.

Theorem 88 *Let* $\mathbf{K}$ *be an arbitrary convex body in* $\mathbb{E}^d$, $d \geq 3$. *Then*

$$c(\mathbf{K},n,d) \leq \frac{H(\mathbf{K_o})}{2}n - \frac{1}{2^d\delta(\mathbf{K_o})^{\frac{d-1}{d}}}\sqrt[d]{\frac{\mathrm{iq}(\mathbf{B}^d)}{\mathrm{iq}(\mathbf{K_o})}}\, n^{\frac{d-1}{d}} - (H(\mathbf{K_o}) - h(\mathbf{K_o}) - 1)$$

$$\leq \frac{3^d - 1}{2}n - \frac{\sqrt[d]{\kappa_d}}{2^{d+1}}n^{\frac{d-1}{d}},$$

where $\kappa_d = \frac{\pi^{\frac{d}{2}}}{\Gamma(\frac{d}{2}+1)} = \mathrm{vol}_d(\mathbf{B}^d)$.

Since we are interested in contact numbers of sphere packings as well, it is interesting to see the form Theorem 88 takes when $\mathbf{K} = \mathbf{B}^d$. Recall that $k(d)$ denotes the kissing number of a unit ball in $\mathbb{E}^d$. Let δ_d stand for the largest possible density for (infinite) packings of unit balls in $\mathbb{E}^d$. The following consequence of Theorem 88 was reported in [34].

Corollary 89 *Let $n > 1$ and $d \geq 3$ be positive integers. Then*

$$c(n,d) < \frac{1}{2}k(d)\, n - \frac{1}{2^d}\delta_d^{-\frac{d-1}{d}} n^{\frac{d-1}{d}}.$$

Now, recall the well-known theorem of Kabatjanskiĭ and Levensteĭn [149] that $k(d) \leq 2^{0.401d(1+o(1))}$ and $\delta_d \leq 2^{-0.599d(1+o(1))}$ as $d \to +\infty$. Together with Corollary 89 this gives

$$c(n,d) < \frac{1}{2}2^{0.401d(1+o(1))}n - \frac{1}{2^d}2^{0.599(1+o(1))(d-1)}n^{\frac{d-1}{d}},$$

for $n > 1$, as $d \to +\infty$.

In particular, for $d = 3$ we have $k(3) = 12$ [220] and $\delta_3 = \frac{\pi}{\sqrt{18}}$ [135]. Thus, by combining these with Corollary 89, we find that for $n > 1$,

$$c(n,3) < 6n - \frac{1}{8}\left(\frac{\pi}{\sqrt{18}}\right)^{-\frac{2}{3}} n^{\frac{2}{3}} = 6n - 0.152\ldots n^{\frac{2}{3}}.$$

The above upper bound for $c(n,3)$ was substantially improved, first in [34] and then further in [50]. The current best upper bound is stated in part (i) of Theorem 87.

In the proof of Theorem 88 published in [33], the following statement plays an important role that might be of independent interest and so we quote it as follows. For the sake of completeness, we wish to point out that Theorem 90 and Corollary 91 are actual strengthenings of Theorem 3.1 and Corollary 3.1 of [26] mainly because, in our case, the containers of the packings in question are highly non-convex.

Theorem 90 *Let $\mathbf{K_o}$ be a convex body in $\mathbb{E}^d$, $d \geq 2$ symmetric about the origin $\mathbf{o}$ of $\mathbb{E}^d$ and let $\{\mathbf{c}_1+\mathbf{K_o}, \mathbf{c}_2+\mathbf{K_o}, \ldots, \mathbf{c}_n+\mathbf{K_o}\}$ be an arbitrary packing of $n > 1$ translates of $\mathbf{K_o}$ in $\mathbb{E}^d$. Then*

$$\frac{n\mathrm{vol}_d(\mathbf{K_o})}{\mathrm{vol}_d(\bigcup_{i=1}^n(\mathbf{c}_i + 2\mathbf{K_o}))} \leq \delta(\mathbf{K_o}).$$

For a generalization, see Lemma 235. The following is an immediate corollary of Theorem 90.

Corollary 91 *Let $\mathcal{P}_n(\mathbf{K_o})$ be the family of all possible packings of $n > 1$ translates of the $\mathbf{o}$-symmetric convex body $\mathbf{K_o}$ in $\mathbb{E}^d$, $d \geq 2$. Moreover, let*

$$\delta(\mathbf{K_o}, n) := \max\left\{\frac{n\mathrm{vol}_d(\mathbf{K_o})}{\mathrm{vol}_d(\bigcup_{i=1}^n(\mathbf{c}_i + 2\mathbf{K_o}))} \,\middle|\, \{\mathbf{c}_1+\mathbf{K_o}, \ldots, \mathbf{c}_n+\mathbf{K_o}\} \in \mathcal{P}_n(\mathbf{K_o})\right\}.$$

Then

$$\limsup_{n\to\infty} \delta(\mathbf{K_o}, n) = \delta(\mathbf{K_o}).$$

Interestingly enough, one can interpret the contact number problem on the exact values of $c(n,d)$ as a volume minimization question. Here we give only an outline of that idea introduced and discussed in detail in [43].

Definition 16 *Let $\mathcal{P}^n := \{\mathbf{c}_i + \mathbf{B}^d \mid 1 \le i \le n$ with $\|\mathbf{c}_j - \mathbf{c}_k\| \ge 2$ for all $1 \le j < k \le n\}$ be an arbitrary packing of $n > 1$ unit balls in $\mathbb{E}^d$. The part of space covered by the unit balls of $\mathcal{P}^n$ is labelled by $\mathbf{P}^n := \bigcup_{i=1}^n (\mathbf{c}_i + \mathbf{B}^d)$. Moreover, let $C^n := \{\mathbf{c}_i \mid 1 \le i \le n\}$ stand for the set of centers of the unit balls in $\mathcal{P}^n$. Furthermore, for any $\lambda > 0$ let $\mathbf{P}^n_\lambda := \bigcup\{\mathbf{x} + \lambda\mathbf{B}^d \mid \mathbf{x} \in \mathbf{P}^n\} = \bigcup_{i=1}^n (\mathbf{c}_i + (1+\lambda)\mathbf{B}^d)$ denote the outer parallel domain of $\mathbf{P}^n$ having outer radius λ. Finally, let*

$$\delta_d(n,\lambda) := \max_{\mathcal{P}^n} \frac{n\kappa_d}{\mathrm{vol}_d(\mathbf{P}^n_\lambda)} = \frac{n\kappa_d}{\min_{\mathcal{P}^n} \mathrm{vol}_d\left(\bigcup_{i=1}^n (\mathbf{c}_i + (1+\lambda)\mathbf{B}^d)\right)}$$

and

$$\delta_d(\lambda) := \limsup_{n\to+\infty} \delta_d(n,\lambda).$$

Now, let $\mathcal{P} := \{\mathbf{c}_i + \mathbf{B}^d \mid i = 1, 2, \dots$ with $\|\mathbf{c}_j - \mathbf{c}_k\| \ge 2$ for all $1 \le j < k\}$ be an arbitrary infinite packing of unit balls in $\mathbb{E}^d$. Recall that the packing density δ_d of unit balls in $\mathbb{E}^d$ can be computed as follows:

$$\delta_d = \sup_{\mathcal{P}} \left(\limsup_{R\to+\infty} \frac{\sum_{\mathbf{c}_i + \mathbf{B}^d \subset R\mathbf{B}^d} \mathrm{vol}_d(\mathbf{c}_i + \mathbf{B}^d)}{\mathrm{vol}_d(R\mathbf{B}^d)} \right).$$

Hence, it is rather easy to see that $\delta_d \le \delta_d(\lambda)$ holds for all $\lambda > 0, d \ge 2$. On the other hand, it was proved in [33] (see also Corollary 91) that $\delta_d = \delta_d(\lambda)$ *for all* $\lambda \ge 1$ leading to the classical sphere packing problem. Now, we are ready to put forward the following question from [43].

Open Problem 15 *Determine $\delta_d(\lambda)$ for $d \ge 2$, $0 < \lambda < \sqrt{\frac{2d}{d+1}} - 1$.*

First, we note that $\frac{2}{\sqrt{3}} - 1 \le \sqrt{\frac{2d}{d+1}} - 1$ holds for all $d \ge 2$. Second, observe that as $\frac{2}{\sqrt{3}}$ is the circumradius of a regular triangle of side length 2, therefore if $0 < \lambda < \frac{2}{\sqrt{3}} - 1$, then for any unit ball packing $\mathcal{P}^n$ no three of the closed balls in the family $\{\mathbf{c}_i + (1+\lambda)\mathbf{B}^d \mid 1 \le i \le n\}$ have a point in common. In other words, for any λ with $0 < \lambda < \frac{2}{\sqrt{3}} - 1$ and for any unit ball packing $\mathcal{P}^n$, in the arrangement $\{\mathbf{c}_i + (1+\lambda)\mathbf{B}^d \mid 1 \le i \le n\}$ of closed balls of radii $1+\lambda$ only pairs of balls may overlap. Thus, computing $\delta_d(n,\lambda)$, i.e., minimizing $\mathrm{vol}_d(\mathbf{P}^n_\lambda)$ means maximizing the total volume of pairwise overlaps in the ball arrangement $\{\mathbf{c}_i + (1+\lambda)\mathbf{B}^d \mid 1 \le i \le n\}$ with the underlying packing $\mathcal{P}^n$. Intuition would suggest to achieve this by simply maximizing the number of touching pairs in the unit ball packing $\mathcal{P}^n$. Hence, Problem 15 becomes very close to the *contact number problem* of finite unit ball packings for $0 < \lambda < \frac{2}{\sqrt{3}} - 1$. Indeed, we have the following statement proved in [43].

Theorem 92 *Let $n > 1$ and $d > 1$ be given. Then there exists $\lambda_{d,n} > 0$ and a packing $\widehat{\mathcal{P}}^n$ of n unit balls in $\mathbb{E}^d$ possessing the largest contact number for the given n such that for all λ satisfying $0 < \lambda < \lambda_{d,n}$, $\delta_d(n, \lambda)$ is generated by $\widehat{\mathcal{P}}^n$, i.e., $\mathrm{vol}_d(\mathbf{P}^n_\lambda) \ge \mathrm{vol}_d(\widehat{\mathbf{P}}^n_\lambda)$ holds for every packing $\mathcal{P}^n$ of n unit balls in $\mathbb{E}^d$.*

Theorem 92 leads us to the problem of upper bounding $\delta_d(n, \lambda)$. The following statement proved in [43] gives a partial answer to that question.

Theorem 93 *Let d and λ be chosen satisfying $\sqrt[d]{d} - 1 \le \lambda \le \sqrt{2} - 1$. Then*

$$\delta_d(\lambda) \le \sup_n \delta_d(n, \lambda) \le \frac{2d+4}{(2-(1+\lambda)^2)\,d+4}(1+\lambda)^{-d} \le \frac{d+2}{2}(1+\lambda)^{-d} \le 1. \tag{4.3}$$

We note that Blichfeldt's upper bound $\frac{d+2}{2}2^{-\frac{d}{2}}$ for the packing density of unit balls in $\mathbb{E}^d$ can be obtained from the upper bound formula of Theorem 93 by making the substitution $\lambda = \sqrt{2} - 1$. We close this section by stating the following improvements on the estimates of Theorem 93 for $d = 2, 3$, which have been published in [43].

Theorem 94 *Let λ be chosen satisfying $0 < \lambda < \frac{2}{\sqrt{3}} - 1 = 0.1547\ldots$ and let $\mathbf{H}$ be a regular hexagon circumscribed about the unit disk $\mathbf{B}^2$ centered at the origin $\mathbf{o}$ in $\mathbb{E}^2$. Then*

$$\delta_2(\lambda) = \frac{\pi}{\mathrm{area}\,(\mathbf{H} \cap (1+\lambda)\mathbf{B}^2)}.$$

Definition 17 *Let $\mathbf{T}^d := \mathrm{conv}\{\mathbf{t}_1, \mathbf{t}_2, \ldots, \mathbf{t}_{d+1}\}$ be a regular d-simplex of edge length 2 in $\mathbb{E}^d, d \ge 2$ and let $0 < \lambda < \sqrt{\frac{2d}{d+1}} - 1$. Set*

$$\sigma_d(\lambda) := \frac{(d+1)\mathrm{vol}_d\left(\mathbf{T}^d \cap (\mathbf{t}_1 + \mathbf{B}^d)\right)}{\mathrm{vol}_d\left(\mathbf{T}^d \cap \left(\cup_{i=1}^{d+1} \mathbf{t}_i + (1+\lambda)\mathbf{B}^d\right)\right)} < 1.$$

An elementary computation yields that if $0 < \lambda < \frac{2}{\sqrt{3}} - 1$, then

$$\sigma_3(\lambda) = \frac{\pi - 6\phi_0}{\pi\lambda^3 + (3\pi - 9\phi_0)\,\lambda^2 + (3\pi - 18\phi_0 +)\,\lambda + \pi - 6\phi_0},$$

where $\phi_0 := \arctan \frac{1}{\sqrt{2}} = 0.615479\ldots$.

Theorem 95 *Let* $0 < \lambda < \frac{2}{\sqrt{3}} - 1 = 0.1547\ldots$ *Set* $\psi_0 := -\arctan\left(\sqrt{\frac{2}{3}}\tan(5\phi_0)\right) = 0.052438\ldots$ *Then*

$$\delta_3(\lambda) \leq \sup_n \delta_3(n, \lambda)$$

$$\leq \frac{\pi - 6\psi_0}{\pi - 6\psi_0 + (3\pi - 18\psi_0)\lambda - 18\psi_0\lambda^2 - (\pi + 6\psi_0)\lambda^3} < \sigma_3(\lambda). \tag{4.4}$$

Finally, we raise the following problem.

Conjecture 96 *Prove that* $\delta_d(n, \lambda) \leq \sigma_d(\lambda)$ *holds for all* $d \geq 3$, $0 < \lambda < \sqrt{\frac{2d}{d+1}} - 1$, *and* $n > 1$.

4.6 Contact numbers for digital and totally separable packings of unit balls in $\mathbb{E}^d$

In this section, we use the terms 'cube', 'sphere' and 'ball' to refer to two- and three-dimensional objects of these types. Consider the three-dimensional (resp. two-dimensional) integer lattice $\mathbb{Z}^3$ (resp. $\mathbb{Z}^2$), which can be thought of as an infinite space tiling array of unit cubes called *lattice cells.* For convenience, we imagine these cubes to be centered at the integer points, rather than having their vertices at these points. Two lattice cells are *connected* if they share a facet.

We refer to a packing of congruent unit diameter spheres centered at the points of $\mathbb{Z}^3$ (resp. $\mathbb{Z}^2$) as a *digital sphere packing.* These packings provide a natural means for generating totally separable sphere packings. We denote the maximal contact number of such a digital packing of n spheres by $c_{\mathbb{Z}}(n, 3)$ (resp. $c_{\mathbb{Z}}(n, 2)$). Clearly, $c_{\mathbb{Z}}(n, 2) \leq c_{\text{sep}}(n, 2)$ and $c_{\mathbb{Z}}(n, 3) \leq c_{\text{sep}}(n, 3)$. The question is how large the maximum digital contact number can be and whether it equals the corresponding maximum contact number of totally separable sphere packings.

A three-dimensional (resp. two-dimensional) *polyomino* is a finite collection of connected lattice cells of $\mathbb{Z}^3$ (resp. $\mathbb{Z}^2$). Considering the maximum volume ball contained in a cube, each polyomino corresponds to a digital sphere (circle) packing and vice versa. Moreover, since the ball (circle) intersects the cube (square) at 6 points (4 points), one on each facet, it follows that the number of facets shared between the cells of the polyomino equals the contact number of the corresponding digital packing.

It is easy to see that minimizing the surface area (resp., perimeter) of a three-dimensional (resp., two-dimensional) polyomino of volume n corresponds to finding the maximum contact number of a digital packing of n

spheres. Harary and Harborth [138] studied the problem of finding isoperimetric polyominoes of area n in 2-space. Their key insight was that n squares can be arranged in a square-like arrangement so as to minimize the perimeter of the resulting polyomino. The same construction appears in [8], but without referencing [138]. The three-dimensional case has a similar solution which first appeared in [8]. The proposed arrangement consists of forming a quasicube (an orthogonal box with one or two edges deficient by at most one unit) followed by attaching as many of the remaining cells as possible in the form of a quasisquare layer. The rest of the cells are then attached to the quasicube in the form of a row. The main results of [138] and [8] on isoperimetric polyominoes in $\mathbb{E}^2$ and $\mathbb{E}^3$ can be used to derive the following about the maximum digital contact numbers (see [51]).

Theorem 97 *Given $n \geq 2$, we have*

(i) $c_{\mathbb{Z}}(n,2) = \lfloor 2n - 2\sqrt{n} \rfloor$.

(ii) $c_{\mathbb{Z}}(n,3) = 3n - 3n^{\frac{2}{3}} - o(n^{\frac{2}{3}})$.

We now turn to the more general totally separable sphere packings in $\mathbb{E}^2$ and $\mathbb{E}^3$. The contact number problem for such packings was discussed in the very recent paper [51].

Theorem 98 *For all $n \geq 2$, we have*

(i) $c_{\text{sep}}(n,2) = \lfloor 2n - 2\sqrt{n} \rfloor$.

(ii) $3n - 3n^{\frac{2}{3}} - o(n^{\frac{2}{3}}) \leq c_{\text{sep}}(n,3) < 3n - 1.346n^{\frac{2}{3}}$.

Part (i) follows from a natural modification of Harborth's proof [139] of Theorem 82 (for details see [51]). The lower bound in (ii) comes from the fact that every digital sphere packing is totally separable. However, proving the upper bound in (ii) is more involved.

Theorem 98 can be used to generate the following analogues of relations (4.1) and (4.2).

$$\lim_{n \to +\infty} \frac{2n - c_{\text{sep}}(n,2)}{\sqrt{n}} = 2 \ . \tag{4.5}$$

$$1.346 < \frac{3n - c_{\text{sep}}(n,3)}{n^{\frac{2}{3}}} \leq 3 + o(1) \ . \tag{4.6}$$

Since the bounds in (4.6) are tighter than (4.2), it is reasonable to conjecture that the limit of $\frac{3n - c_{\text{sep}}(n,3)}{n^{2/3}}$ exists as $n \to +\infty$. In fact, it can be asked if this limit equals 3. Furthermore, a comparison of Theorem 97 and Theorem 98 shows that $c_{\text{sep}}(n,2) = c_{\mathbb{Z}}(n,2)$ holds for all positive integers n. Therefore, it is natural to raise the following open problem.

Open Problem 16 *Show that*

$$\lim_{n \to +\infty} \frac{3n - c_{\text{sep}}(n,3)}{n^{\frac{2}{3}}} = 3 .$$

In particular, is it the case that $c_{\text{sep}}(n,3) = c_{\mathbb{Z}}(n,3)$, *for all positive integers* n?

Let us imagine that we generate totally separable packings of unit diameter balls in $\mathbb{E}^d$ such that every center of the balls chosen is a lattice point of the integer lattice $\mathbb{Z}^d$ in $\mathbb{E}^d$. Then, as in Section 4.6, let $c_{\mathbb{Z}}(n,d)$ denote the largest possible contact number of all packings of n unit diameter balls obtained in this way.

Theorem 99 $c_{\mathbb{Z}}(n,d) \leq \lfloor dn - dn^{\frac{d-1}{d}} \rfloor$, *for all* $n > 1$ *and* $d \geq 2$.

Here we recall Theorem 97 and refer to [51] to note that the upper bound of Theorem 99 is sharp for $d = 2$ and all $n > 1$ and for $d \geq 3$ and all $n = k^d$ with $k > 1$. On the other hand, it is not a sharp estimate for example, for $d = 3$ and $n = 5$.

We close this section by stating the recent upper bounds of [51] for the contact numbers of totally separable unit ball packings in $\mathbb{E}^d$.

Theorem 100 $c_{\text{sep}}(n,d) \leq dn - \frac{1}{2d^{\frac{d-1}{2}}} n^{\frac{d-1}{d}}$, *for all* $n > 1$ *and* $d \geq 4$.

4.7 Bounds for contact numbers of totally separable packings by translates of a convex body in $\mathbb{E}^d$ with $d = 1, 2, 3, 4$

Generalizing the kissing number, the *Hadwiger number* or *the translative kissing number* $H(\mathbf{K})$ of a convex body $\mathbf{K}$ is the maximum number of non-overlapping translates of $\mathbf{K}$ that all touch $\mathbf{K}$. Given the difficulty of the kissing number problem, determining Hadwiger numbers is highly non-trivial with few exact values known for $d \geq 3$. The best general upper and lower bounds on $H(\mathbf{K})$ are due to Hadwiger [133] and Talata [232], respectively, and can be expressed as

$$2^{cd} \leq H(\mathbf{K}) \leq 3^d - 1, \tag{4.7}$$

where c is an absolute constant and equality holds in the right inequality if and only if $\mathbf{K}$ is an affine d-dimensional cube [127].

A packing of translates of a *convex domain*, that is, a convex body $\mathbf{K}$ in $\mathbb{E}^2$ is said to be *totally separable* if any two packing elements can be separated by a line of $\mathbb{E}^2$ disjoint from the interior of every packing element. This notion was introduced by G. Fejes Tóth and L. Fejes Tóth [104].

One can define a totally separable packing of translates of a d-dimensional convex body $\mathbf{K}$ in a similar way by requiring any two packing elements to be separated by a hyperplane in $\mathbb{E}^d$ disjoint from the interior of every packing element [51, 152].

Recall that the *contact graph* of a packing of translates of $\mathbf{K}$ is the simple graph whose vertices are the members of the packing, and whose two vertices are connected by an edge if and only if the two members touch each other. In the recent paper [49] the authors investigate the maximum vertex degree (called *separable Hadwiger number*), as well as the maximum number of edges (called the *maximum separable contact number*) of the contact graphs of totally separable packings by a given number of translates of a smooth or strictly convex body $\mathbf{K}$ in $\mathbb{E}^d$. This extends and generalizes the results of [42] and [51]. The details are the following.

4.7.1 Separable Hadwiger numbers

It is natural to introduce the totally separable analogue of the Hadwiger number as follows [42].

Definition 18 *Let* $\mathbf{K}$ *be a convex body in* $\mathbb{E}^d$. *We call a family of translates of* $\mathbf{K}$ *that all touch* $\mathbf{K}$ *and, together with* $\mathbf{K}$, *form a totally separable packing in* $\mathbb{E}^d$ *a* separable Hadwiger configuration *of* $\mathbf{K}$. *The* separable Hadwiger number $H_{\text{sep}}(\mathbf{K})$ *of* $\mathbf{K}$ *is the maximum size of a separable Hadwiger configuration of* $\mathbf{K}$.

Recall that the *Minkowski symmetrization* of the convex body $\mathbf{K}$ in $\mathbb{E}^d$ denoted by $\mathbf{K_o}$ is defined by $\mathbf{K_o} := \frac{1}{2}(\mathbf{K} + (-\mathbf{K})) = \frac{1}{2}(\mathbf{K} - \mathbf{K}) = \frac{1}{2}\{\mathbf{x} - \mathbf{y} : \mathbf{x}, \mathbf{y} \in \mathbf{K}\}$. Clearly, $\mathbf{K_o}$ is an $\mathbf{o}$-symmetric d-dimensional convex body. Minkowski [184] showed that if $\mathcal{P} = \{\mathbf{x}_1 + \mathbf{K}, \mathbf{x}_2 + \mathbf{K}, \dots, \mathbf{x}_n + \mathbf{K}\}$ is a packing of translates of $\mathbf{K}$, then $\mathcal{P}_\mathbf{o} = \{\mathbf{x}_1 + \mathbf{K_o}, \mathbf{x}_2 + \mathbf{K_o}, \dots, \mathbf{x}_n + \mathbf{K_o}\}$ is a packing as well. Moreover, the contact graphs of $\mathcal{P}$ and $\mathcal{P}_\mathbf{o}$ are the same. Using the same method, it is easy to see that Minkowski's above statement applies to totally separable packings as well. (See also [42].) Thus, from this point on, we only consider $\mathbf{o}$-symmetric convex bodies.

It is mentioned in [51] that based on [85] (see also [204] and [157]) it follows in a straightforward way that $H_{\text{sep}}(\mathbf{B}^d) = 2d$ for all $d \geq 2$. On the other hand, if $\mathbf{K}$ is an $\mathbf{o}$-symmetric convex body in $\mathbb{E}^d$, then each facet of the minimum volume circumscribed parallelotope of $\mathbf{K}$ touches $\mathbf{K}$ at the center of the facet and so, clearly $H_{\text{sep}}(\mathbf{K}) \geq 2d$. Thus,

$$2d \leq H_{\text{sep}}(\mathbf{K}) \leq H(\mathbf{K}) \leq 3^d - 1 \tag{4.8}$$

holds for any $\mathbf{o}$-symmetric convex body $\mathbf{K}$ in $\mathbb{E}^d$. Furthermore, the affine d-cube is the only $\mathbf{o}$-symmetric convex body in $\mathbb{E}^d$ with separable Hadwiger number $3^d - 1$ [127].

We investigate equality in the first inequality of (4.8). First, we note as an easy exercise that H_{sep} as a map from the set of convex bodies equipped with any reasonable topology to the reals is upper semi-continuous. Thus, for any d, if an **o**-symmetric convex body $\mathbf{K}$ in $\mathbb{E}^d$ is sufficiently close to the Euclidean ball $\mathbf{B}^d$ (say, $\mathbf{B}^d \subseteq \mathbf{K} \subseteq (1+\varepsilon_d)\mathbf{B}^d$, where $\varepsilon_d > 0$ depends on d only), then $H_{\text{sep}}(\mathbf{K}) = 2d$.

Hence, it is natural to ask whether the set of those **o**-symmetric convex bodies in $\mathbb{R}^d$ with $H_{\text{sep}}(\mathbf{K}) = 2d$ is dense. In [49], the authors investigate whether $H_{\text{sep}}(\mathbf{K}) = 2d$ holds for any **o**-symmetric smooth or strictly convex $\mathbf{K}$ in $\mathbb{E}^d$. The first main result (Theorem 1) of [49] is a partial answer to this question.

Definition 19 *An* Auerbach basis *of an* **o**-*symmetric convex body* $\mathbf{K}$ *in* $\mathbb{E}^d$ *is a set of* d *points on the boundary of* $\mathbf{K}$ *that form a basis of* $\mathbb{E}^d$ *with the property that the hyperplane through any one of them, parallel to the other* $d-1$ *supports* $\mathbf{K}$.

Theorem 101 *Let* $\mathbf{K}$ *be an* **o**-*symmetric convex body in* $\mathbb{E}^d$, *which is smooth* or *strictly convex. Then*

(i) For $d \in \{1,2,3,4\}$, *we have* $H_{\text{sep}}(\mathbf{K}) = 2d$ *and, in any separable Hadwiger configuration of* $\mathbf{K}$ *with* $2d$ *translates of* $\mathbf{K}$, *the half translation vectors are* d *pairs of opposite vectors, where picking one from each pair yields an Auerbach basis of* $\mathbf{K}$. *(ii)* $H_{\text{sep}}(\mathbf{K}) \leq 2^{d+1} - 3$ *for all* $d \geq 5$.

We note that part (i) of Theorem 101 was proved for $d = 2$ and smooth **o**-symmetric convex domains in [42]. It is natural to finish this subsection with the following question.

Open Problem 17 *Determine the largest value of* $H_{\text{sep}}(\mathbf{K})$ *for* **o**-*symmetric smooth strictly convex bodies in* $\mathbb{E}^d$, $d \geq 5$.

4.7.2 One-sided separable Hadwiger numbers

Recall that the one-sided Hadwiger number $h(\mathbf{K})$ of an **o**-symmetric convex body $\mathbf{K}$ in $\mathbb{E}^d$ has been defined in [37] as the maximum number of non-overlapping translates of $\mathbf{K}$ that can touch $\mathbf{K}$ and lie in a *closed* supporting half space of $\mathbf{K}$. It is proved in [37] that $h(\mathbf{K}) \leq 2 \cdot 3^{d-1} - 1$ holds for any **o**-symmetric convex body $\mathbf{K}$ in $\mathbb{E}^d$ with equality for affine d-cubes only.

One could consider the obvious extension of the one-side Hadwiger number to separable Hadwiger configurations. However, a more restrictive and slightly more technical definition serves our purposes better, the reason of which will become clear in Theorem 102 and Example 2.

Definition 20 *Let* $\mathbf{K}$ *be a smooth* **o**-*symmetric convex body in* $\mathbb{E}^d$. *The* one-sided separable Hadwiger number $h_{\text{sep}}(\mathbf{K})$ *of* $\mathbf{K}$ *is the maximum number* n *of translates* $2\mathbf{x}_1 + \mathbf{K}, \ldots, 2\mathbf{x}_n + \mathbf{K}$ *of* $\mathbf{K}$ *that form a separable Hadwiger configuration of* $\mathbf{K}$, *and the following holds. If* $f_1, \ldots, f_n$ *denote supporting linear* *functionals of* $\mathbf{K}$ *at the points* $\mathbf{x}_1, \dots, \mathbf{x}_n$*, respectively, then* $\mathbf{o} \notin \operatorname{conv}\{\mathbf{x}_1, \dots, \mathbf{x}_n\}$ *and* $\mathbf{o} \notin \operatorname{conv}\{f_1, \dots, f_n\}$.

Definition 21 *For a positive integer* d*, let*

$$h_{\text{sep}}(d) := \max\{h_{\text{sep}}(\mathbf{K}) \ : \ \mathbf{K} \text{ is an } \mathbf{o}\text{-symmetric, smooth and strictly convex body in } \mathbb{E}^d\},$$
$$H_{\text{sep}}(d) := \max\{H_{\text{sep}}(\mathbf{K}) \ : \ \mathbf{K} \text{ is an } \mathbf{o}\text{-symmetric, smooth and strictly convex body in } \mathbb{E}^d\},$$

and set $H_{\text{sep}}(0) = h_{\text{sep}}(0) = 0$.

The proof of part (i) of Theorem 101 relies on the following fact: for the smallest dimensional example $\mathbf{K}$ of an $\mathbf{o}$-symmetric, smooth and strictly convex body with $H_{\text{sep}}(\mathbf{K}) > 2d$, we have $h_{\text{sep}}(\mathbf{K}) > 2d$. More precisely, the following statement is proved in [49] (Theorem 2).

Theorem 102 *(i)* $h_{\text{sep}}(d) \le H_{\text{sep}}(d) \le \max\{2\ell + h_{\text{sep}}(d-\ell) \ : \ \ell = 0, \dots, d\}$. *(ii)* $h_{\text{sep}}(d) = d$ *for* $d \in \{1, 2, 3, 4\}$. *(iii)* $h_{\text{sep}}(\mathbf{B}^d) = d$ *for the* d*-dimensional Euclidean ball* $\mathbf{B}^d$ *with* $d \in \mathbb{Z}^+$.

According to Note 1, when bounding $H_{\text{sep}}(\mathbf{K})$ for a smooth *or* strictly convex body $\mathbf{K}$, it is sufficient to consider smooth *and* strictly convex bodies.

As a warning sign, it is shown in Example 2 (see Example 3.1 in [49]) that there is an $\mathbf{o}$-symmetric, smooth and strictly convex body $\mathbf{K}$ in $\mathbb{E}^5$, which has a set of 6 translates that form a separable Hadwiger configuration, and the origin is not in the convex hull of the translation vectors.

4.7.3 Maximum separable contact numbers

Let $\mathbf{K}$ be an $\mathbf{o}$-symmetric convex body in $\mathbb{E}^d$, and let $\mathcal{P} := \{\mathbf{x}_1 + \mathbf{K}, \dots, \mathbf{x}_n + \mathbf{K}\}$ be a packing of translates of $\mathbf{K}$. Recall that the number of edges in the contact graph of $\mathcal{P}$ is called the contact number of $\mathcal{P}$. Moreover, $c(\mathbf{K}, n, d)$ denotes the largest contact number of a packing of n translates of $\mathbf{K}$ in $\mathbb{E}^d$. It is proved in [33] (see also Theorem 88) that $c(\mathbf{K}, n, d) \le \frac{H(\mathbf{K})}{2} n - n^{\frac{d-1}{d}} g(\mathbf{K})$ holds for all $n > 1$, where $g(\mathbf{K}) > 0$ depends on $\mathbf{K}$ only.

Definition 22 *If* $d, n \in \mathbb{Z}^+$ *and* $\mathbf{K}$ *is an* $\mathbf{o}$*-symmetric convex body in* $\mathbb{E}^d$*, then let* $c_{\text{sep}}(\mathbf{K}, n, d)$ *denote the largest contact number of a totally separable packing of* n *translates of* $\mathbf{K}$.

According to Theorem 101, the maximum degree in the contact graph of a totally separable packing of a smooth convex body $\mathbf{K}$ is $2d$, and hence, $c_{\text{sep}}(\mathbf{K}, n, d) \le dn$, for $d \in \{1, 2, 3, 4\}$. The third main result (Theorem 3 of [49]) is the following stronger bound.

Theorem 103 *Let* $\mathbf{K}$ *be a smooth* $\mathbf{o}$*-symmetric convex body in* $\mathbb{E}^d$ *with* $d \in \{1, 2, 3, 4\}$. *Then*

$$c_{\text{sep}}(\mathbf{K}, n, d) \leq dn - n^{(d-1)/d} f(\mathbf{K})$$

for all $n > 1$, *where* $f(\mathbf{K}) > 0$ *depends on* $\mathbf{K}$ *only.*

In particular, if $\mathbf{K}$ *is a smooth* $\mathbf{o}$*-symmetric convex domain in* $\mathbb{E}^2$, *then*

$$c_{\text{sep}}(\mathbf{K}, n, 2) \leq 2n - \frac{\sqrt{\pi}}{8}\sqrt{n}$$

holds for all $n > 1$.

4.8 Appendix: Hadwiger numbers of topological disks

A well-known theorem of Hadwiger [133] states that the Hadwiger number of a plane convex body $\mathbf{K}$ satisfies $6 \leq H(\mathbf{K}) \leq 8$. Giving an affirmative answer to a question of Hadwiger, Grünbaum [130] strenghtened this result by proving the following theorem.

Theorem 104 *Let* $\mathbf{K}$ *be a plane convex body. If* $\mathbf{K}$ *is a parallelogram, then* $H(\mathbf{K}) = 8$. *Otherwise,* $H(\mathbf{K}) = 6$.

To generalize this problem, we define a *topological ball* as a subset of $\mathbb{E}^d$ homeomorphic to the Euclidean ball $\mathbf{B}^d$. Furthermore, we call a set $\mathbf{S}$ a *starlike ball* if it is a topological ball which is starlike relative to a point $\mathbf{p} \in \mathbf{S}$; that is, if for any point $\mathbf{q} \in \mathbf{S}$, we have $[\mathbf{p}, \mathbf{q}] \subset \mathbf{S}$. If $d = 2$, we may call a topological or starlike ball a topological or starlike *disk*, respectively. It is worth noting that the notions of overlapping and touching, and thus, that of Hadwiger number, can be generalized for topological balls in a natural way, and that Halberg et al. [134] proved that the Hadwiger number of every topological disk is at least six.

In 1995, Bezdek, Kuperberg and Kuperberg [29] proved the following related result.

Theorem 105 *For any topological disk* $\mathbf{S}$ *in* $\mathbb{E}^2$, *the maximum number of mutually touching translates of* $\mathbf{S}$ *is at most four.*

Furthermore, in this paper they proposed a conjecture and a question related to Hadwiger numbers. These problems can be found also in [67].

Conjecture 106 *The Hadwiger number of any starlike disk is at most* 8.

Question 107 *Is it true that the Hadwiger number of any topological disk is at most eight? If not, is there a universal constant* K *such that the Hadwiger number of any topological disk is at most* K*?*

Question 107 was answered in the negative, in the strong sense, by a clever construction of Cheong and Lee [77].

Theorem 108 *For any positive integer k, there is a topological disk* $\mathbf{S}$ *with* $H(\mathbf{S}) \geq k$.

The first result about the Hadwiger numbers of starlike disks, based on estimating the area of certain multiple packings, is due to Bezdek in [28].

Theorem 109 *The Hadwiger number of any starlike disk is at most* 75.

Lángi improved this upper bound both in the general [160], and in a special case [159].

Theorem 110 *The Hadwiger number of any centrally symmetric starlike disk is at most* 12.

Theorem 111 *The Hadwiger number of any starlike disk is at most* 35.

Nevertheless, Conjecture 106 is still open.

A variant of Question 107 can be found in [160], where the author puts a restriction on the number of the connected components of $(\mathrm{conv}\mathbf{S}) \setminus \mathbf{S}$, where $\mathbf{S}$ is a topological disk. In particular, he proved Theorem 112.

Theorem 112 *Let* $\mathbf{S}$ *be a topological disk such that* $(\mathrm{conv}\mathbf{S}) \setminus \mathbf{S}$ *is connected. Then* $H(\mathbf{S}) \leq 8$.

In this context, the following natural variant of Question 107 can be proposed [160].

Open Problem 18 *Is it true that for every positive integer k there is an integer $N(k)$ such that for any topological disk* $\mathbf{S}$, *if* $(\mathrm{conv}\mathbf{S}) \setminus \mathbf{S}$ *has at most k connected components, then* $H(\mathbf{S}) \leq N(k)$?

4.9 Research Exercises

Exercise 4.1 *Prove that for any convex body* $\mathbf{K} \subset \mathbb{E}^d$ *and point set* $\mathcal{C} \subset \mathbb{E}^d$, *the family* $\{\mathbf{x}+\mathbf{K} : \mathbf{x} \in \mathcal{C}\}$ *is a packing if and only if* $\{\mathbf{x} + \frac{1}{2}(\mathbf{K} - \mathbf{K}) : \mathbf{x} \in \mathcal{C}\}$ *is a packing.*

Exercise 4.2 (Grünbaum [130]) *Using Exercises 4.1 and 2.5, prove that the Hadwiger number of every convex domain is at least* 6 *in* $\mathbb{E}^2$. *Moreover, show that the Hadwiger number of every convex domain different from a parallelogram is* 6 *in* $\mathbb{E}^2$.

Exercise 4.3 *Prove that the separable Hadwiger number of every smooth convex domain is* 4 *in* $\mathbb{E}^2$.

Exercise 4.4 (Bezdek-Kuperberg-Kuperberg [29]) *Prove that for any positive integer* N *there is a starlike ball* $\mathbf{S}_N$ *in* $\mathbb{E}^3$ *such that* $N \leq H(\mathbf{S}_N)$.

Exercise 4.5 (Maehara [172]) *Prove that the kissing number (i.e., Hadwiger number) of a ball in* $\mathbb{E}^3$ *is* 12, *i.e.,* $k(3) = 12$.

Exercise 4.6 (Fejes Tóth [103]) *Prove that the one-sided kissing number (i.e., one-sided Hadwiger number) of a ball in* $\mathbb{E}^3$ *is* 9.

Exercise 4.7 (Talata [233]) *Prove that the Hadwiger number (resp., separable Hadwiger number) of a tetrahedron in* $\mathbb{E}^3$ *is* 18.

Exercise 4.8 (Larman-Zong [165]) *Prove that the Hadwiger number (resp., separable Hadwiger number) of an octahedron in* $\mathbb{E}^3$ *is* 18.

Exercise 4.9 (Musin [189]) *Prove that the kissing number of a ball in* $\mathbb{E}^4$ *is* 24, *i.e.,* $k(4) = 24$.

Exercise 4.10 (Musin [188]) *Prove that the one-sided kissing number of a ball in* $\mathbb{E}^4$ *is* 18.

Exercise 4.11 (Odlyzko-Sloane [192]) *Prove that* $k(8) = 240$ *and* $k(24) = 196560$.

Exercise 4.12 (Talata [232]) *Show that there exists an absolute constant* $c > 0$ *such that* $H(\mathbf{K}) \geq 2^{cd}$ *for every positive integer* d *and every convex body* $\mathbf{K}$ *in* $\mathbb{E}^d$.

Exercise 4.13 (Bezdek-Brass [37]) *Prove that* $h(\mathbf{K}) \leq 2 \cdot 3^{d-1} - 1$ *holds for any* $\mathbf{o}$*-symmetric convex body* $\mathbf{K}$ *in* $\mathbb{E}^d$ *with equality for affine* d*-cubes only.*

5

More on Volumetric Properties of Separable Packings

Summary. In this chapter we continue our investigation of totally separable packings from a volumetric point of view. First, we outline the recent solution of the contact number problem for smooth strictly convex domains in $\mathbb{E}^2$. We discuss this approach in details based on angular measure, Birkhoff orthogonality, Birkhoff measure, (smooth) Birkhoff domains, and approximation by (smooth strictly convex) Auerbach domains, which are topics of independent interests as well. In the next part of this chapter, we connect the study of totally separable packings of discrete geometry to Oler's inequality of geometry of numbers. More concretely, we discuss an analogue of Oler's inequality for totally separable translative packings in $\mathbb{E}^2$ and then use it for finding the highest density of totally separable translative packings (resp., for finding the smallest area convex hull of totally separable packings by n translates) of an arbitrary convex domain in $\mathbb{E}^2$. Finally, as a local version of totally separable packings, we introduce the family of ρ-separable translative packings of $\mathbf{o}$-symmetric convex bodies in $\mathbb{E}^d$. In particular, we investigate the fundamental problem of minimizing the mean i-dimensional projection of the convex hull of n non-overlapping translates of an $\mathbf{o}$-symmetric convex body $\mathbf{C}$ forming a ρ-separable packing in $\mathbb{E}^d$ for given $d > 1, n > 1$, and $\mathbf{C}$.

5.1 Solution of the contact number problem for smooth strictly convex domains in $\mathbb{E}^2$

This section is based on the results published in [42].

Definition 23 *Let* $\mathbf{K_o}$ *be an* $\mathbf{o}$*-symmetric convex body in* $\mathbb{E}^d$, $d \geq 2$. *A non-zero vector* $\mathbf{x}$ *in* $(\mathbb{R}^d, \|\cdot\|_{\mathbf{K_o}})$ *is said to be* Birkhoff orthogonal *to a non-zero vector* $\mathbf{y}$ *if* $\|\mathbf{x}\|_{\mathbf{K_o}} \leq \|\mathbf{x} + t\mathbf{y}\|_{\mathbf{K_o}}$, *for all* $t \in \mathbb{R}$ *[52], where* $\|\mathbf{x}\|_{\mathbf{K_o}} = \inf\{\lambda > 0 : \mathbf{x} \in \lambda\mathbf{K_o}\}$ *for every* $\mathbf{x} \in \mathbb{R}^d$. *We denote this by* $\mathbf{x} \dashv_{\mathbf{K_o}} \mathbf{y}$.

Note that in general, Birkhoff orthogonality is a non-symmetric relation, that is $\mathbf{x} \dashv_{\mathbf{K_o}} \mathbf{y}$ does not imply $\mathbf{y} \dashv_{\mathbf{K_o}} \mathbf{x}$.

Definition 24 *Let $\mathbf{K_o} \subseteq \mathbb{R}^2$ be an $\mathbf{o}$-symmetric convex domain in $\mathbb{E}^2$. An angular measure, also called an angle measure, in $(\mathbb{R}^2, \|\cdot\|_{\mathbf{K_o}})$ is a measure μ defined on $\mathrm{bd}\mathbf{K_o}$ that can be extended in a translation-invariant way to measure angles anywhere and satisfies the following properties [66]:*

(i) $\mu(\mathrm{bd}\mathbf{K_o}) = 2\pi$.

(ii) For any Borel set $X \subseteq \mathrm{bd}\mathbf{K_o}$, $\mu(X) = \mu(-X)$.

(iii) For each $\mathbf{x} \in \mathrm{bd}\mathbf{K_o}$, $\mu(\{\mathbf{x}\}) = 0$.

For any $\mathbf{x}, \mathbf{y} \in \mathrm{bd}\mathbf{K_o}$, we write $\mu([\mathbf{x}, \mathbf{y}]_{\mathbf{K_o}})$ for the measure of the angle subtended by the arc $[\mathbf{x}, \mathbf{y}]_{\mathbf{K_o}}$ at $\mathbf{o}$, where $[\mathbf{x}, \mathbf{y}]_{\mathbf{K}}$ denotes the smaller (in the norm $\|\cdot\|_{\mathbf{K_o}}$) of the two closed arcs on $\mathrm{bd}\mathbf{K_o}$ with endpoints $\mathbf{x}$ and $\mathbf{y}$. In [15, 92], angle measures are required to satisfy a fourth non-degeneracy condition, namely, for any $\mathbf{x} \neq \mathbf{y} \in \mathrm{bd}\mathbf{K_o}$, $\mu([\mathbf{x}, \mathbf{y}]_{\mathbf{K_o}}) > 0$. Here it suffices to adopt Brass's definition. We refer the interested reader to [15] for a very recent expository treatment of angle measures.

Note that the usual Euclidean angle measure in the plane satisfies these conditions. Moreover, for any angle measure in $(\mathbb{R}^2, \|\cdot\|_{\mathbf{K_o}})$, the sum of interior angles of any simple n-gon in $\mathbb{R}^2$ equals $(n-2)\pi$ [66].

Definition 25 *An angle measure μ in the plane $(\mathbb{R}^2, \|\cdot\|_{\mathbf{K_o}})$ is called a* Birkhoff measure *in short, B-*measure *[98] if for any $\mathbf{x}, \mathbf{y} \in \mathrm{bd}\mathbf{K_o}$, $\mathbf{x} \dashv_{\mathbf{K_o}} \mathbf{y}$ implies that $\mu([\mathbf{x}, \mathbf{y}]_{\mathbf{K_o}}) = \pi/2$.*

Definition 26 *Let $\mathbf{D} \subseteq \mathbb{E}^2$ be an $\mathbf{o}$-symmetric convex domain, then $\mathbf{D}$ is called a* Birkhoff domain *in short, B-*domain *if there is a B-measure defined in $(\mathbb{R}^2, \|\cdot\|_{\mathbf{D}})$.*

The following statement has been proved in [42].

Theorem 113 *If $\mathbf{D}$ is a smooth B-domain in $\mathbb{E}^2$ and $n \geq 2$, then we have*

$$c_{\mathrm{sep}}(\mathbf{D}, n, 2) = \left\lfloor 2n - 2\sqrt{n} \right\rfloor. \tag{5.1}$$

Definition 27 *Let $\mathbf{A} \subseteq \mathbb{E}^2$ be an $\mathbf{o}$-symmetric convex domain, $\mathbf{B}$ a circular disk centered at $\mathbf{o}$ and $c, -c, c', -c'$ non-overlapping arcs on $\mathrm{bd}\mathbf{B} \cap \mathrm{bd}\mathbf{A}$ such that for any $\mathbf{x} \in c$ there exists $\mathbf{x}' \in c'$ with $\mathbf{x} \dashv_{\mathbf{B}} \mathbf{x}'$ and vice versa. Then we call $\mathbf{A}$ an Auerbach domain, or simply an A-domain.*

Figure 5.1 illustrates Definition 27. Clearly, for any $\mathbf{x} \in c$ there exist antipodes $\mathbf{x}', -\mathbf{x}' \in \mathrm{bd}\mathbf{B}$ with $\mathbf{x} \dashv_{\mathbf{B}} \mathbf{x}'$ and $\mathbf{x} \dashv_{\mathbf{B}} -\mathbf{x}'$. From Definition 27, we must have $\mathbf{x}' \in c'$ and $-\mathbf{x}' \in -c'$. Moreover, an analogous statement holds for any $\mathbf{x}' \in c'$. Therefore, an A-domain $\mathbf{A}$ can be thought of as an $\mathbf{o}$-symmetric convex domain in $\mathbb{E}^2$ such that $\mathrm{bd}\mathbf{A}$ contains two pairs of antipodal circular arcs all lying on the same circle and with each pair being Birkhoff orthogonal to the other in $\mathbb{E}^2$. Note that this definition does not exclude the case when more than one set of such arcs occurs on $\mathrm{bd}\mathbf{A}$, in which case we choose the

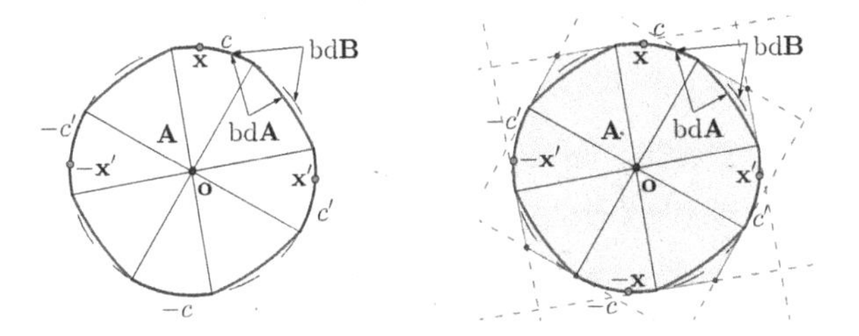

FIGURE 5.1
Definition 27 illustrated. An A-domain $\mathbf{A}$ with circular pieces c, c', $-c$ and $-c'$ lying on the boundary of the circular disk $\mathbf{B}$. Note that $\mathbf{A}$ is not necessarily contained in $\mathbf{B}$. However, due to convexity, $\mathbf{A}$ must lie in the shaded region determined by the tangent lines at the endpoints of the circular pieces.

set of four arcs arbitrarily, or even when $\mathbf{A}$ is a circular disk. Given an A-domain $\mathbf{A}$, we call the four circular arcs $c, -c, c', -c'$ chosen on its boundary, the *circular pieces* of $\mathbf{A}$ and write

$$\mathrm{cir}(\mathbf{A}) = c \cup (-c) \cup c' \cup (-c').$$

Clearly, if $\mathbf{x}, \mathbf{y} \in \mathrm{cir}(\mathbf{A})$, then $\mathbf{x} \dashv_{\mathbf{A}} \mathbf{y}$ holds if and only if $\mathbf{x} \dashv_{\mathbf{B}} \mathbf{y}$ holds. We observe that any $\mathbf{x} \in \mathrm{cir}(\mathbf{A})$ belongs to an Auerbach basis of $\mathbf{A}$. Furthermore, any Auerbach basis of $\mathbf{A}$ is either contained in $\mathrm{cir}(\mathbf{A})$ or $\mathrm{bd}\mathbf{A} \setminus \mathrm{cir}(\mathbf{A})$.

Let $\mathbf{A}$ be an A-domain with circular pieces c, $-c$, c' and $-c'$ lying on the boundary of a circular disk $\mathbf{B}$ and e denote the Euclidean angle measure. Then $e(c) = e(c') = e(-c) = e(-c')$ holds and we define an angle measure m on $\mathrm{bd}\mathbf{A}$ as follows. For any arc $a \subseteq \mathrm{bd}\mathbf{A}$, define

$$m(a) = 2\pi \frac{e(a \cap \mathrm{cir}(\mathbf{A}))}{e(\mathrm{cir}(\mathbf{A}))}. \tag{5.2}$$

Note that m assigns a measure of $\pi/2$ to each of the designated circular pieces on $\mathrm{bd}\mathbf{A}$ and a measure of 0 to the rest of $\mathrm{bd}\mathbf{A}$ (including any circular arcs not included among the circular pieces), that is, $m(c) = m(c') = m(-c) = m(-c') = \pi/2$ and $m(\mathrm{bd}\mathbf{A} \setminus \mathrm{cir}(\mathbf{A})) = 0$. It is easy to check (for more details see [42]) that m satisfies properties (i-iii) of Definition 24, as well as the following property.

Lemma 114 *Let $\mathbf{A}$ be an A-domain and m the angle measure defined on $\mathrm{bd}\mathbf{A}$ by (5.2). Also, let $\mathbf{x}, \mathbf{y} \in \mathrm{bd}\mathbf{A}$ be such that $\mathbf{x} \dashv_{\mathbf{A}} \mathbf{y}$. Then*

$$m([\mathbf{x}, \mathbf{y}]_{\mathbf{A}}) = \frac{\pi}{2}. \tag{5.3}$$

In other words, m is a B-measure in $(\mathbb{R}^2, \|\cdot\|_{\mathbf{A}})$ and every A-domain is a B-domain.

As $c_{\text{sep}}(\cdot, n, 2)$ is invariant under affine transformations, Theorem 113 and Lemma 114 give:

Corollary 115 *If* $\mathbf{A}$ *is (an affine image of) a smooth A-domain in* $\mathbb{E}^2$ *and* $n \geq 2$, *we have* $c_{\text{sep}}(\mathbf{A}, n, 2) = \lfloor 2n - 2\sqrt{n} \rfloor$.

Recall the definition of Hausdorff distance between two convex bodies.

Definition 28 *Given two (not necessarily* $\mathbf{o}$*-symmetric) convex bodies* $\mathbf{K}$ *and* $\mathbf{L}$ *in* $\mathbb{E}^d$, *the Hausdorff distance between them is defined as*

$$h(\mathbf{K}, \mathbf{L}) = \min\left\{\varepsilon : \mathbf{K} \subseteq \mathbf{L} + \varepsilon\mathbf{B}^d, \mathbf{L} \subseteq \mathbf{K} + \varepsilon\mathbf{B}^d\right\}.$$

It is well known that $h(\cdot, \cdot)$ is a metric on the set of all d-dimensional convex bodies [217, page 61]. The following results have been proved in [42].

Theorem 116 *Affine images of smooth strictly convex A-domains are dense (in the Hausdorff sense) in the space of smooth* $\mathbf{o}$*-symmetric strictly convex domains. Moreover, given any smooth* $\mathbf{o}$*-symmetric strictly convex domain* $\mathbf{K_o}$, *we can construct an affine image* $\mathbf{A}'$ *of a smooth strictly convex A-domain* $\mathbf{A}$ *such that the length of* $\mathrm{bd}\mathbf{A}' \cap \mathrm{bd}\mathbf{K_o}$ *can be made arbitrarily close to the length of* $\mathrm{bd}\mathbf{K_o}$.

Corollary 117 *Let* $\mathbf{K}$ *be an* $\mathbf{o}$*-symmetric smooth strictly convex domain in* $\mathbb{E}^2$ *and* $n \geq 2$. *Then* $c_{\text{sep}}(\mathbf{K}, n, 2) = \lfloor 2n - 2\sqrt{n} \rfloor$.

Open Problem 19 *One may wonder whether* $c_{\text{sep}}(\mathbf{K}, n, 2) = \lfloor 2n - 2\sqrt{n} \rfloor$ *holds for any* $\mathbf{o}$*-symmetric smooth convex domain* $\mathbf{K}$ *in* $\mathbb{E}^2$ *and* $n \geq 2$.

5.2 The separable Oler's inequality and its applications in $\mathbb{E}^2$

5.2.1 Oler's inequality

Our goal is to bridge totally separable packings of discrete geometry and Oler's inequality of geometry of numbers. Recall that the concept of totally separable packings was introduced by G. Fejes Tóth and L. Fejes Tóth [104] as follows. We say that a set of domains is totally separable if any two of them can be separated by a straight line avoiding the interiors of all domains. The main question investigated in [104] is to find the densest totally separable arrangement of congruent replicas of a given domain. The paper [104] generated a good deal of interest in the density problem of totally separable arrangements and led to further important publications such as [27] and [152]. Coming from this direction, the goal of Bezdek and Lángi [44] was to find the densest totally separable packing of translates of a given convex domain and then to extend that approach to the analogue question for finite totally separable packings. It turned out that an efficient method to achieve all that is based on a new version of Oler's classical inequality ([193]). So, next we introduce some basic terminology and then state Oler's inequality in the form which is most suitable for the presentation of the results of [44].

Let $\mathbf{K}$ be a *convex domain*, i.e., a compact convex set with non-empty interior in the Euclidean plane $\mathbb{E}^2$. A family $\mathcal{F}$ of n translates of $\mathbf{K}$ in $\mathbb{E}^2$ is called a *packing* if no two members of $\mathcal{F}$ have an interior point in common.

If $\mathbf{K}$ is an o-symmetric convex domain in $\mathbb{E}^2$, where $\mathbf{o}$ stands for the origin of $\mathbb{E}^2$, then let $\|\cdot\|_{\mathbf{K}}$ denote the *norm generated by* $\mathbf{K}$, i.e., let $\|\mathbf{x}\|_{\mathbf{K}} = \min\{\lambda : \mathbf{x} \in \lambda\mathbf{K}\}$ for any $\mathbf{x} \in \mathbb{E}^2$. The distance between the points $\mathbf{p}$ and $\mathbf{q}$ measured in the norm $\|\cdot\|_{\mathbf{K}}$ is denoted by $\|\mathbf{p}-\mathbf{q}\|_{\mathbf{K}}$. For the sake of simplicity, the Euclidean distance between the points $\mathbf{p}$ and $\mathbf{q}$ of $\mathbb{E}^2$ is denoted by $\|\mathbf{p}-\mathbf{q}\|$.

If $P = \bigcup_{i=1}^{n}[\mathbf{x}_{i-1}, \mathbf{x}_i]$ is a polygonal curve in $\mathbb{E}^2$ with $[\mathbf{x}_{i-1}, \mathbf{x}_i]$ standing for the closed line segment connecting $\mathbf{x}_{i-1}$ and $\mathbf{x}_i$, and $\mathbf{K}$ is an $\mathbf{o}$-symmetric plane convex domain, then the *Minkowski length* of P is defined as $M_{\mathbf{K}}(P) = \sum_{i=1}^{n}\|\mathbf{x}_i - \mathbf{x}_{i-1}\|_{\mathbf{K}}$. Based on this and using approximation by closed polygons one can define the Minkowski length $M_{\mathbf{K}}(G)$ of any rectifiable curve $G \subseteq \mathbb{E}^2$ in the norm $\|\cdot\|_{\mathbf{K}}$. If $\mathbf{K}$ is a not o-symmetric, by $M_{\mathbf{K}}(G)$ we mean the length of G in the *relative norm* of $\mathbf{K}$, i.e., in the norm defined by $\frac{1}{2}(\mathbf{K}-\mathbf{K})$ [193] (cf. also Section 2.2).

Finally, if $\mathbf{K}$ is an o-symmetric convex domain in $\mathbb{E}^2$, then let $\diamond(\mathbf{K})$ denote a minimal area circumscribed hexagon of $\mathbf{K}$.

Now, we are ready to state Oler's inequality ([193]) in the following form. Let $\mathbf{K}$ be an o-symmetric convex domain in $\mathbb{E}^2$. Let

$$\mathcal{F} = \{\mathbf{x}_i + \mathbf{K} : i = 1, 2, \ldots, n\}$$

be a packing of n translates of $\mathbf{K}$ in $\mathbb{E}^2$, and set $X = \{\mathbf{x}_1, \mathbf{x}_2, \ldots, \mathbf{x}_n\}$. Furthermore, let Π be a simple closed polygonal curve with the following properties:

(i) the vertices of Π are points of X

and

(ii) $X \subseteq \Pi^*$ with $\Pi^* = \Pi \cup \mathrm{int}\Pi$, where $\mathrm{int}\Pi$ refers to the interior of Π.

Then

$$\frac{\mathrm{area}(\Pi^*)}{\mathrm{area}\,(\diamond(\mathbf{K}))} + \frac{M_{\mathbf{K}}(\Pi)}{4} + 1 \geq n, \tag{5.4}$$

where area$(\cdot)$ denotes the area of the corresponding set. The formula (5.4) was conjectured by H. J. Zassenhaus and has a number of interesting aspects discussed in [238] (see also [26] and [60]).

5.2.2 An analogue of Oler's inequality for totally separable translative packings

We recall the following definitions from [44].

Definition 29 *A closed polygonal curve* $P = \bigcup_{i=1}^{m} [\mathbf{x}_{i-1}, \mathbf{x}_i]$, *where* $\mathbf{x}_0 = \mathbf{x}_m$, *is called* permissible *if there is a sequence of simple closed polygonal curves* $P^n = \bigcup_{i=1}^{m} [\mathbf{x}_{i-1}^n, \mathbf{x}_i^n]$, *where* $\mathbf{x}_0^n = \mathbf{x}_m^n$, *satisfying* $\mathbf{x}_i^n \to \mathbf{x}_i$ *for every value of* i. *The interior* $\mathrm{int}P$ *is defined as* $\lim_{n\to\infty} \mathrm{int}P^n$.

Remark 118 *By the properties of limits, if* $P = \bigcup_{i=1}^{m} [\mathbf{x}_{i-1}, \mathbf{x}_i]$ *is permissible and* P^n *and* Q^n *are sequences of simple closed polygonal curves with* $\lim_{n\to\infty} P^n = \lim_{n\to\infty} Q^n = P$, *then* $\lim_{n\to\infty} \mathrm{int}P^n = \lim_{n\to\infty} \mathrm{int}Q^n$, *i.e., the interior of a permissible curve is well defined.*

Definition 30 *Let* $\mathbf{K}$ *be a convex domain in* $\mathbb{E}^2$. *Then let* $\square(\mathbf{K})$ *denote a minimal area circumscribed parallelogram of* $\mathbf{K}$.

The following totally separable analogue of Oler's inequality has been published in [44].

Theorem 119 *Let* $\mathbf{K}$ *be an* $\mathbf{o}$*-symmetric convex domain in* $\mathbb{E}^2$. *Let*

$$\mathcal{F} = \{\mathbf{x}_i + \mathbf{K} : i = 1, 2, \ldots, n\}$$

be a totally separable packing of n *translates of* $\mathbf{K}$ *in* $\mathbb{E}^2$, *and set* $X = \{\mathbf{x}_1, \mathbf{x}_2, \ldots, \mathbf{x}_n\}$. *Furthermore, let* Π *be a permissible closed polygonal curve with the following properties:*

(i) the vertices of Π *are points of* X

and

(ii) $X \subseteq \Pi^*$ *with* $\Pi^* = \Pi \cup \mathrm{int}\Pi$.

Then

$$\frac{\mathrm{area}(\Pi^*)}{\mathrm{area}\,(\square(\mathbf{K}))} + \frac{M_{\mathbf{K}}(\Pi)}{4} + 1 \geq n. \tag{5.5}$$

Remark 120 *We note that unlike in Oler's original inequality, equality in (5.5) of Theorem 119 is attained in a variety of ways. This is illustrated in Fig. 5.2, where the polygon* Π *consists of blocks of zig-zags and simple closed polygons having sides parallel to the two sides of a chosen* $\square(\mathbf{K})$. *Furthermore, we note that characterizing the case of equality in (5.5) of Theorem 119 remains an open problem.*

Remark 121 *It is well known that the width of any convex body* $\mathbf{K}$ *in any direction is equal to the width of its central symmetrization* $\frac{1}{2}(\mathbf{K} - \mathbf{K})$ *in this direction. This readily implies that* $\square(\mathbf{K})$ *does not change under central symmetrization.*

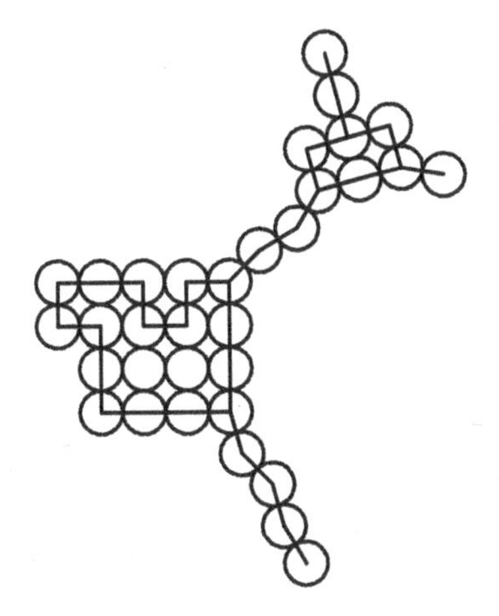

FIGURE 5.2
A totally separable packing of translates of $\mathbf{K}$ (with $\mathbf{K}$ being a circular disk for the sake of simplicity), which satisfies the conditions in Theorem 119 and for which there is equality in (5.5) of Theorem 119.

Remark 122 *Let $\mathcal{F} = \{\mathbf{x}_i + \mathbf{K} : i = 1, 2, \dots, n\}$ be a family of n translates of $\mathbf{K}$ in $\mathbb{E}^2$, where $\mathbf{K}$ is an o-symmetric convex domain of $\mathbb{E}^2$, and let $\mathbf{K}^*$ be a convex domain satisfying $\mathbf{K} = \frac{1}{2}(\mathbf{K}^* - \mathbf{K}^*)$ with $\mathbf{o} \in \mathrm{int}\mathbf{K}^*$, and let $\mathcal{F}^* = \{\mathbf{x}_i + \mathbf{K}^* : i = 1, 2, \dots, n\}$. Then $\mathcal{F}$ is a packing if and only if $\mathcal{F}^*$ is a packing, and $\mathcal{F}$ is a totally separable packing if and only if $\mathcal{F}^*$ is a totally separable packing. (For details see for example, [42].) Thus, Theorem 119 holds for any (not necessarily* **o***-symmetric) plane convex domain $\mathbf{K}^*$ (with $o \in \mathrm{int}\mathbf{K}^*$) as well.*

5.2.3 On the densest totally separable translative packings

Theorem 119 and Remark 122 imply the following statement published in [44], which was proved for **o**-symmetric convex domains in [104] with a weaker estimate than (5.6) for convex domains in general, namely, with $\square(\mathbf{K})$ standing for a minimal area circumscribed quadrangle of $\mathbf{K}$.

Theorem 123 *If $\delta_{sep}(\mathbf{K})$ denotes the largest (upper) density of totally separable translative packings of the convex domain $\mathbf{K}$ in $\mathbb{E}^2$, then*

$$\delta_{sep}(\mathbf{K}) = \frac{\mathrm{area}(\mathbf{K})}{\mathrm{area}(\square(\mathbf{K}))}. \tag{5.6}$$

Remark 124 *It is worth mentioning here that by (5.6) of Theorem 123, the densest totally separable translative packing of a convex domain is attained by a lattice arrangement.*

Open Problem 20 *Let* $\mathbf{K}$ *be a convex body in* $\mathbb{E}^d$, $d \geq 3$. *Prove or disprove that the highest density of totally separable translative packings of* $\mathbf{K}$ *in* $\mathbb{E}^d$ *is attained by the totally separable lattice packing of* $\mathbf{K}$ *generated by (any of) the smallest volume parallelotope (i.e., affine d-cube) circumscribed* $\mathbf{K}$.

Theorem 125 published in [44] is a totally separable analogue of the well-known theorem (which is a combination of the results published in [100], [110], [108], and [207]), stating that the maximal density of translative packings of a convex domain in $\mathbb{E}^2$ is minimal if and only if the domain is a triangle.

Theorem 125 *For any convex domain* $\mathbf{K}$ *in* $\mathbb{E}^2$, *we have*

$$\frac{1}{2} \leq \delta_{sep}(\mathbf{K}) \leq 1, \tag{5.7}$$

with equality on the left if and only if $\mathbf{K}$ *is a triangle, and on the right if and only if* $\mathbf{K}$ *is a parallelogram.*

5.2.4 On the smallest area convex hull of totally separable translative finite packings

The following area inequalities have been published in [44].

Theorem 126 *Let* $\mathcal{F} = \{\mathbf{c}_i + \mathbf{K} : i = 1, 2, \ldots, n\}$ *be a totally separable packing of* n *translates of the convex domain* $\mathbf{K}$ *in* $\mathbb{E}^2$. *Let* $\mathbf{C} = \mathrm{conv}\{\mathbf{c}_1, \mathbf{c}_2, \ldots, \mathbf{c}_n\}$.

(i) *Then we have*

$$\mathrm{area}\left(\mathrm{conv}\left(\bigcup_{i=1}^{n}(\mathbf{c}_i + \mathbf{K})\right)\right) = \mathrm{area}(\mathbf{C}+\mathbf{K}) \geq \frac{2}{3}(n-1)\mathrm{area}\left(\Box(\mathbf{K})\right)$$

$$+\mathrm{area}(\mathbf{K}) + \frac{1}{3}\mathrm{area}(\mathbf{C}).$$

(ii) *If* $\mathbf{K}$ *or* $\mathbf{C}$ *is centrally symmetric, then*

$$\mathrm{area}(\mathbf{C} + \mathbf{K}) \geq (n-1)\mathrm{area}\left(\Box(\mathbf{K})\right) + \mathrm{area}(\mathbf{K}).$$

Remark 127 *We note that equality is attained in (i) of Theorem 126 for the following totally separable translative packings of a triangle (cf. Figure 5.3). Let* $\mathbf{K}$ *be a triangle, with the origin* $\mathbf{o}$ *at a vertex, and* $\mathbf{u}$ *and* $\mathbf{v}$ *being the position vectors of the other two vertices, and let* $\mathbf{T} = m\mathbf{K}$, *where* $m > 1$ *is an integer. Let* $\mathcal{F}$ *be the family consisting of the elements of the lattice packing* $\{i\mathbf{u}+j\mathbf{v}+\mathbf{K} : i, j, \in \mathbb{Z}\}$ *contained in* $\mathbf{T}$. *Then* $\mathcal{F}$ *is a totally separable packing of* $n = \frac{m(m+1)}{2}$ *translates of* $\mathbf{K}$ *with* $\mathrm{conv}\left(\bigcup \mathcal{F}\right) = \mathbf{T} = \mathbf{C}+\mathbf{K}$, *where* $\mathbf{C} = (m-1)\mathbf{K}$. *Thus,* $\mathrm{area}(\mathbf{T}) = m^2\mathrm{area}(\mathbf{K}) = [\frac{2}{3}m(m+1) - \frac{1}{3} + \frac{1}{3}(m-1)^2]\mathrm{area}(\mathbf{K}) = \frac{4}{3}(n-1)\mathrm{area}(\mathbf{K})+\mathrm{area}(\mathbf{K})+\frac{1}{3}\mathrm{area}(\mathbf{C}) = \frac{2}{3}(n-1)\mathrm{area}(\Box(\mathbf{K}))+\mathrm{area}(\mathbf{K}) + \frac{1}{3}\mathrm{area}(\mathbf{C})$.

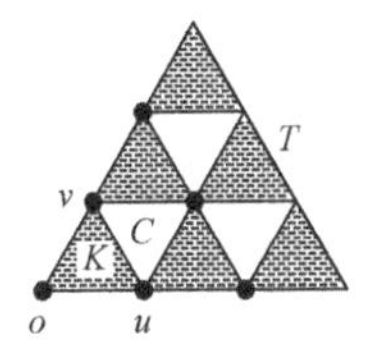

FIGURE 5.3
An example for equality in (i).

Remark 128 *In (ii) of Theorem 126 equality can be attained in a variety of ways shown in Figure 5.4 for both cases, namely, when* **C** *is centrally symmetric (and* **K** *is not centrally symmetric such as a triangle) and when* **K** *is centrally symmetric (such as a circular disk) without any assumption on the symmetry of* **C**.

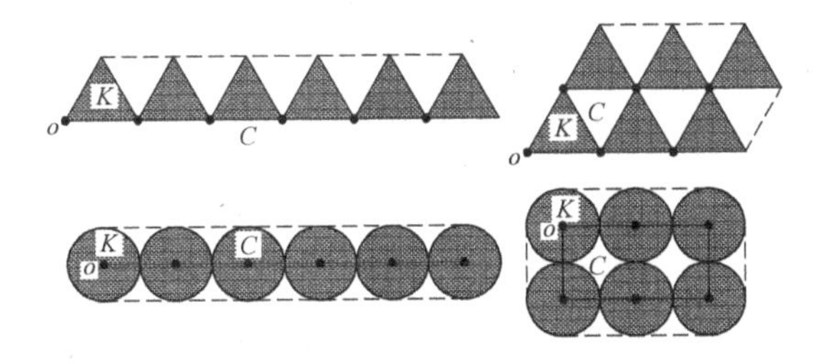

FIGURE 5.4
Totally separable translative packings of a triangle and a unit disk for which equality is attained in (ii) in Theorem 126.

We conclude this subsection with the following.

Open Problem 21 *Let* **C** *be an* **o***-symmetric convex body in* $\mathbb{E}^d$, $d \geq 3$ *and* $n > 1$. *Prove or disprove that the smallest volume of the convex hull of* n *translates of* **C** *forming a totally separable packing in* $\mathbb{E}^d$ *is obtained when the* n *translates of* **C** *form a sausage, that is, a linear packing.*

5.3 Higher dimensional results: minimizing the mean projections of finite ρ-separable packings in $\mathbb{E}^d$

The starting point is the following elegant theorem of Böröczky Jr. [58]: Consider the convex hull **Q** of n non-overlapping translates of an arbitrary convex body $\mathbf{C}$ in $\mathbb{E}^d$ with n being sufficiently large. If $\mathbf{Q}$ has minimal mean i-dimensional projection for given i with $1 \leq i < d$, then $\mathbf{Q}$ is approximately a d-dimensional ball. The main goal of [45] is to prove an extension of this theorem to ρ-separable translative packings of convex bodies in $\mathbb{E}^d$. Next, following [45] we define the concept of ρ-separable translative packings and then state the main result of [45].

Definition 31 *Let* $\mathbf{C}$ *be an* $\mathbf{o}$*-symmetric convex body of* $\mathbb{E}^d$*. Furthermore, let* $\|\cdot\|_{\mathbf{C}}$ *denote the norm generated by* $\mathbf{C}$*, i.e., let* $\|\mathbf{x}\|_{\mathbf{C}} := \inf\{\lambda \mid \mathbf{x} \in \lambda\mathbf{C}\}$ *for any* $\mathbf{x} \in \mathbb{E}^d$*. Now, let* $\rho \geq 1$*. We say that the packing*

$$\mathcal{P}_{\text{sep}} := \{\mathbf{c}_i + \mathbf{C} \mid i \in I \text{ with } \|\mathbf{c}_j - \mathbf{c}_k\|_{\mathbf{C}} \geq 2 \text{ for all } j \neq k \in I\}$$

of (finitely or infinitely many) non-overlapping translates of $\mathbf{C}$ *with centers* $\{\mathbf{c}_i \mid i \in I\}$ *is a* ρ-separable packing *in* $\mathbb{E}^d$ *if for each* $i \in I$ *the finite packing* $\{\mathbf{c}_j + \mathbf{C} \mid \mathbf{c}_j + \mathbf{C} \subseteq \mathbf{c}_i + \rho\mathbf{C}\}$ *is a totally separable packing (in* $\mathbf{c}_i + \rho\mathbf{C}$*). Finally, let* $\delta_{\text{sep}}(\rho, \mathbf{C})$ *denote the largest density of all* ρ*-separable translative packings of* $\mathbf{C}$ *in* $\mathbb{E}^d$*, i.e., let*

$$\delta_{\text{sep}}(\rho, \mathbf{C}) := \sup_{\mathcal{P}_{\text{sep}}} \left(\limsup_{\lambda \to +\infty} \frac{\sum_{\mathbf{c}_i + \mathbf{C} \subset \mathbf{W}_\lambda^d} \mathrm{vol}_d(\mathbf{c}_i + \mathbf{C})}{\mathrm{vol}_d(\mathbf{W}_\lambda^d)} \right),$$

where $\mathbf{W}_\lambda^d$ *denotes the d-dimensional cube of edge length* 2λ *centered at* $\mathbf{o}$ *in* $\mathbb{E}^d$ *having edges parallel to the coordinate axes of* $\mathbb{E}^d$ *and* $\mathrm{vol}_d(\cdot)$ *refers to the d-dimensional volume of the corresponding set in* $\mathbb{E}^d$*.*

Remark 129 *Let* $\delta(\mathbf{C})$ *(resp.,* $\delta_{\text{sep}}(\mathbf{C})$*) denote the supremum of the upper densities of all translative packings (resp., totally separable translative packings) of the* $\mathbf{o}$*-symmetric convex body* $\mathbf{C}$ *in* $\mathbb{E}^d$*. Clearly,* $\delta_{\text{sep}}(\mathbf{C}) \leq \delta_{\text{sep}}(\rho, \mathbf{C}) \leq \delta(\mathbf{C})$ *for all* $\rho \geq 1$*. Furthermore, if* $1 \leq \rho < 3$*, then any* ρ*-separable translative packing of* $\mathbf{C}$ *in* $\mathbb{E}^d$ *is simply a translative packing of* $\mathbf{C}$ *and therefore,* $\delta_{\text{sep}}(\rho, \mathbf{C}) = \delta(\mathbf{C})$*.*

Open Problem 22 *Let* $\mathbf{C}$ *be an* $\mathbf{o}$*-symmetric convex body* $\mathbf{C}$ *in* $\mathbb{E}^d$*,* $d \geq 2$*. Then prove or disprove that there exists* $\rho(\mathbf{C}) > 0$ *such that for any* $\rho \geq \rho(\mathbf{C})$ *one has* $\delta_{\text{sep}}(\mathbf{C}) = \delta_{\text{sep}}(\rho, \mathbf{C})$*.*

Recall that the mean i-dimensional projection $M_i(\mathbf{C})$ $(i = 1, 2, \ldots, d-1)$ of the convex body $\mathbf{C}$ in $\mathbb{E}^d$, can be expressed ([217]) with the help of a mixed volume via the formula

$$M_i(\mathbf{C}) = \frac{\kappa_i}{\kappa_d} V(\overbrace{\mathbf{C}, \ldots, \mathbf{C}}^{i}, \overbrace{\mathbf{B}^d, \ldots, \mathbf{B}^d}^{d-i}),$$

where κ_d is the volume of $\mathbf{B}^d$ in $\mathbb{E}^d$. Note that $M_i(\mathbf{B}^d) = \kappa_i$, and the surface volume of $\mathbf{C}$ is $\mathrm{svol}_{d-1}(\mathbf{C}) = \frac{d\kappa_d}{\kappa_{d-1}} M_{d-1}(\mathbf{C})$ and in particular, $\mathrm{svol}_{d-1}(\mathbf{B}^d) =$

$d\kappa_d$. Set $M_d(\mathbf{C}) := \mathrm{vol}_d(\mathbf{C})$. Finally, let $R(\mathbf{C})$ (resp., $r(\mathbf{C})$) denote the circumradius (resp., inradius) of the convex body $\mathbf{C}$ in $\mathbb{E}^d$, which is the radius of the smallest (resp., a largest) ball that contains (resp., is contained in) $\mathbf{C}$. The following is the main result of [45].

Theorem 130 *Let $d \geq 2$, $1 \leq i \leq d-1$, $\rho \geq 1$, and let $\mathbf{Q}$ be the convex hull of a ρ-separable packing of n translates of the $\mathbf{o}$-symmetric convex body $\mathbf{C}$ in $\mathbb{E}^d$ such that $M_i(\mathbf{Q})$ is minimal and $n \geq \frac{4^d d^{4d}}{\delta_{\mathrm{sep}}(\rho,\mathbf{C})^{d-1}} \cdot \left(\rho\frac{R(\mathbf{C})}{r(\mathbf{C})}\right)^d$. Then*

$$\frac{r(\mathbf{Q})}{R(\mathbf{Q})} \geq 1 - \frac{\omega}{n^{\frac{2}{d(d+3)}}}, \tag{5.8}$$

for $\omega = \lambda(d)\left(\frac{\rho R(\mathbf{C})}{r(\mathbf{C})}\right)^{\frac{2}{d+3}}$, where $\lambda(d)$ depends only on the dimension d. In addition,

$$M_i(\mathbf{Q}) = \left(1 + \frac{\sigma}{n^{\frac{1}{d}}}\right) M_i(\mathbf{B}^d)\left(\frac{\mathrm{vol}_d(\mathbf{C})}{\delta_{\mathrm{sep}}(\rho,\mathbf{C})\kappa_d}\right)^{\frac{i}{d}} \cdot n^{\frac{i}{d}},$$

where $-\frac{2.25R(\mathbf{C})\rho d i}{r(\mathbf{C})\delta_{\mathrm{sep}}(\rho,\mathbf{C})} \leq \sigma \leq \frac{2.1R(\mathbf{C})\rho i}{r(\mathbf{C})\delta_{\mathrm{sep}}(\rho,\mathbf{C})}$.

Remark 131 *It is worth restating Theorem 130 as follows: Consider the convex hull $\mathbf{Q}$ of n non-overlapping translates of an arbitrary $\mathbf{o}$-symmetric convex body $\mathbf{C}$ forming a ρ-separable packing in $\mathbb{E}^d$ with n being sufficiently large. If $\mathbf{Q}$ has minimal mean i-dimensional projection for given i with $1 \leq i < d$, then $\mathbf{Q}$ is approximately a d-dimensional ball.*

Open Problem 23 *The nature of the question analogue to Theorem 130 on minimizing $M_d(\mathbf{Q}) = \mathrm{vol}_d(\mathbf{Q})$ is very different. Namely, recall that Betke and Henk [24] proved L. Fejes Tóth's sausage conjecture for $d \geq 42$ according to which the smallest volume of the convex hull of n non-overlapping unit balls in $\mathbb{E}^d$ is obtained when the n unit balls form a sausage, that is, a linear packing (see also [25, 26]). As linear packings of unit balls are ρ-separable, therefore the above theorem of Betke and Henk applies to ρ-separable packings of unit balls in $\mathbb{E}^d$ for all $\rho \geq 1$ and $d \geq 42$. On the other hand, the problem of minimizing the volume of the convex hull of n unit balls forming a ρ-separable packing in $\mathbb{E}^d$ remains an interesting open problem for $\rho \geq 1$ and $2 \leq d < 42$. Last but not least, the problem of minimizing $M_d(\mathbf{Q})$ for $\mathbf{o}$-symmetric convex bodies $\mathbf{C}$ different from a ball in $\mathbb{E}^d$ seems to be wide open for $\rho \geq 1$ and $d \geq 2$.*

5.4 Research Exercises

Given an $\mathbf{o}$-symmetric convex domain $\mathbf{K_o}$ in $\mathbb{E}^2$, the normed plane $(\mathbb{R}^2, \|\cdot\|_{\mathbf{K_o}})$ is called a *Radon plane* if for any $\mathbf{x}, \mathbf{y} \in \mathrm{bd}\mathbf{K_o}$, $\mathbf{x} \dashv_{\mathbf{K_o}} \mathbf{y}$ implies $\mathbf{y} \dashv_{\mathbf{K_o}}$

$\mathbf{x}$. In other words, a Radon plane is one in which the relation of Birkhoff orthogonality is symmetric. We define a *Radon domain* as the closed unit disk of a Radon plane. If $\mathbf{K_o}$ is an $\mathbf{o}$-symmetric convex domain in $\mathbb{E}^2$ that is not necessarily a Radon domain, then an (non-trivial) arc $a \subseteq \mathrm{bd}\mathbf{K_o}$ is said to be a *Radon arc* if $\mathbf{x} \dashv_{\mathbf{K_o}} \mathbf{y}$ for any $\mathbf{x} \in a$ and $\mathbf{y} \in \mathrm{bd}\mathbf{K_o}$ implies $\mathbf{y} \dashv_{\mathbf{K_o}} \mathbf{x}$.

Exercise 5.1 (Fankhänel [98]) *Show that if the boundary of an* $\mathbf{o}$*-symmetric convex domain* $\mathbf{K_o}$ *contains a Radon arc, then* $(\mathbb{R}^2, \|\cdot\|_{\mathbf{K_o}})$ *possesses a B-measure. Moreover, if* $\mathbf{K_o}$ *is a smooth Radon domain, then* $(\mathbb{R}^2, \|\cdot\|_{\mathbf{K_o}})$ *possesses a strictly increasing B-measure.*

Exercise 5.2 *Show that the unit disk of the plane* $(\mathbb{R}^2, \|\cdot\|_{p,q})$*, where* $1 \leq p, q, \leq \infty$, $\frac{1}{p} + \frac{1}{q} = 1$, *and*

$$\|(x,y)\|_{p,q} = \begin{cases} (|x|^p + |y|^p)^{1/p}, & x, y \geq 0 \,\mathrm{or}\, x, y \leq 0, \\ (|x|^q + |y|^q)^{1/q}, & \mathrm{otherwise.} \end{cases}$$

is a Radon domain. Moreover, prove that a Radon domain is smooth if and only if it is strictly convex.

Let $\mathbf{K_o}$ be a smooth $\mathbf{o}$-symmetric convex domain in $\mathbb{E}^2$ and $\mathbf{P}$ any parallelogram (not necessarily of minimum area) circumscribing $\mathbf{K_o}$ such that $\mathbf{K_o}$ touches each side of $\mathbf{P}$ at its midpoint (and not at the corners of $\mathbf{P}$ as $\mathbf{K_o}$ is smooth). Let $\mathbf{x}$ and $\mathbf{y}$ be the midpoints of any two adjacent sides of $\mathbf{P}$. Then $-\mathbf{x}$ and $-\mathbf{y}$ are also points of intersection of $\mathbf{K_o}$ and $\mathbf{P}$. It is easy to see that $\{\mathbf{x}, \mathbf{y}\}$ is an Auerbach basis of the normed plane $(\mathbb{R}^2, \|\cdot\|_{\mathbf{K_o}})$. We call the lattice $\mathcal{L}_{\mathbf{P}}$ in $(\mathbb{R}^2, \|\cdot\|_{\mathbf{K_o}})$ with fundamental cell $\mathbf{P}$, an *Auerbach lattice* of $\mathbf{K_o}$ as we can think of $\mathcal{L}_{\mathbf{P}}$ as being generated by the Auerbach basis $\{\mathbf{x}, \mathbf{y}\}$ of $(\mathbb{R}^2, \|\cdot\|_{\mathbf{K_o}})$.

Exercise 5.3 (Bezdek-Khan-Oliwa [42]) *Let* $\mathbf{R}$ *be a smooth Radon domain and let* $n = \ell(\ell + \epsilon) + k \geq 4$ *be the decomposition of a positive integer* n *such that* $k \neq 1$. *Prove that if* $\mathcal{P}$ *is a totally separable packing of* n *translates of* $\mathbf{R}$ *with* $c_{\mathrm{sep}}(\mathbf{K}, n)(\mathbf{R}, n, 2) = \lfloor 2n - 2\sqrt{n} \rfloor$ *contacts, then* $\mathcal{P}$ *is a finite lattice packing lying on an Auerbach lattice of* $\mathbf{R}$.

A triplet of non-overlapping convex domains in $\mathbb{E}^2$ is said to be *separable* if there is a straight line not intersecting the interiors of them, but containing both sides at least one of them. Finally, we say that a packing of convex domains in $\mathbb{E}^2$ is *locally separable* if any triplet of the convex domains is separable.

Exercise 5.4 (Bezdek [27]) *Prove that the (upper) density of any locally separable packing of congruent circular disks in* $\mathbb{E}^2$ *is at most* $\frac{\pi}{4}$.

Exercise 5.5 (Groemer [126]) *Prove that if the compact convex set* $\mathbf{C}$ *contains the centers of* n *non-overlapping unit disks, then*

$$\frac{1}{2\sqrt{3}}\mathrm{area}(\mathbf{C}) + \frac{1}{4}\mathrm{per}(\mathbf{C}) + 1 \geq n$$

Exercise 5.6 (Wegner [237]) *Prove that if $\mathbf{D}_n$ is the convex hull of n non-overlapping unit disks in $\mathbb{E}^2$, then*

$$\text{area}(\mathbf{D}_n) \geq 2\sqrt{3}(n-1) + (2-\sqrt{3})\left\lceil \sqrt{12n-3} - 2 \right\rceil + \pi$$

Exercise 5.7 (Fejes Tóth-Fejes Tóth [104]) *Prove that if a totally separable packing of n congruent convex domains is contained in a convex quadrangle of area A and a is the area of a convex quadrangle of least area containing a domain, then $n \leq \frac{A}{a}$.*

Exercise 5.8 (Kertész [152]) *Show that if a cube of volume V contains a totally separable packing of N balls of radius r in $\mathbb{E}^3$, then $V \geq 8Nr^3$.*

Exercise 5.9 (Betke-Henk-Wills [25]) *Let $\mathbf{B}^d$ denote the d-dimensional unit ball centered at the origin $\mathbf{o}$ in $\mathbb{E}^d$, and for a positive integer n, let $C_n := \{\mathbf{x}_1, \ldots, \mathbf{x}_n\}$ be a packing set of $\mathbf{B}^d$, i.e., $\|\mathbf{x}_i - \mathbf{x}_j\| \geq 2$, $1 \leq i < j \leq n$. Then show that for every $\rho < \sqrt{2}$ a dimension $d(\rho)$ exists such that, for $d \geq d(\rho)$,*

$$\text{vol}_d\left(\text{conv}(C_n) + \rho\mathbf{B}^d\right) \geq \text{vol}_d\left(\text{conv}(S_n) + \rho\mathbf{B}^d\right),$$

where S_n is a minimal linear packing set of n unit balls, i.e., a line segment of length $2(n-1)$, holds.

Exercise 5.10 (Betke-Henk-Wills [25]) *Show that for every convex body $\mathbf{K}$ in $\mathbb{E}^d$ and $\rho < \frac{1}{32d^2}$,*

$$\text{vol}_d\left(\text{conv}(C_n) + \rho\mathbf{K}\right) \geq \text{vol}_d\left(\text{conv}(S_n) + \rho\mathbf{K}\right),$$

where C_n is a packing set with respect to n translates of $\mathbf{K}$ and S_n is a minimal linear packing set of n translates of $\mathbf{K}$, holds.

Part II

Selected Proofs

6

Proofs on Volumetric Properties of (m, d)-scribed Polytopes

Summary. In this chapter we present selected proofs of some theorems from Chapter 1 about the isoperimetric problem and the volume of polytopes. In Section 6.1 we prove Ball's famous reverse isoperimetric inequality. In Section 6.2 we prove the monotonicity of the isoperimetric ratio under the Eikonal equation, and a dynamic variant of Lindelöf's Condition. In Sections 6.3 and 6.4 we determine the largest volume of polytopes inscribed in the unit sphere of $\mathbb{E}^3$ and $\mathbb{E}^d$, respectively, with a small number of vertices. In Section 6.5 we characterize polytopes with n vertices whose symmetry group is isomorphic to the dihedral group D_n. Section 6.6 presents the proof of the realization of every combinatorial class of 3-dimensional convex polyhedra by a midscribed polyhedron, that is, by a Koebe polyhedron. In Section 6.7 we prove that this realization is unique, up to Euclidean isometries, under the additional assumption that the barycenter of the tangency points of the polyhedron is the center of the midsphere. Finally, in Section 6.8, we prove results about the general problem of centering Koebe polyhedra.

6.1 Proof of Theorem 3

The *volume ratio* of a convex body $\mathbf{K} \subset \mathbb{E}^d$ is defined as

$$\mathrm{vr}(\mathbf{K}) = \left(\frac{\mathrm{vol}_d(\mathbf{K})}{\mathrm{vol}_d(\mathbf{B})} \right)^{\frac{1}{d}},$$

where $\mathbf{B}$ is the largest volume ellipsoid contained in $\mathbf{K}$. We note that this ellipsoid, the so-called John *ellipsoid*, uniquely exists, and that $\mathrm{vr}(\cdot)$ does not change under affine transformations of $\mathbb{E}^d$.

First, we show how Theorem 3 follows from the next theorem. We remark that an elementary computation yields that the volume and the surface volume of a regular simplex $\mathbf{S}$ in $\mathbb{E}^d$, circumscribed about $\mathbf{B}^d$, are

$$\mathrm{vol}_d(\mathbf{S}) = \frac{d^{\frac{d}{2}} \cdot (d+1)^{\frac{d+1}{2}}}{d!}, \quad \mathrm{svol}_{d-1}(\mathbf{S}) = \frac{d^{\frac{d}{2}} \cdot (d+1)^{\frac{d+1}{2}}}{(d-1)!}. \tag{6.1}$$

Theorem 132 *The volume ratio of a convex body* $\mathbf{K}$ *is less than or equal to the volume ratio of a regular simplex circumscribed about* $\mathbf{B}^d$.

Proof of Theorem 3 Without loss of generality, we may assume that the largest volume ellipsoid contained in $\mathbf{K}$ is $\mathbf{B}^d$. Then,

$$\mathrm{svol}_{d-1}(\mathbf{K}) = \lim_{\varepsilon\to 0^+} \frac{\mathrm{vol}_d(\mathbf{K}+\varepsilon\mathbf{B}^d) - \mathrm{vol}_d(\mathbf{K})}{\varepsilon}$$

$$\leq \lim_{\varepsilon\to 0^+} \frac{\mathrm{vol}_d(\mathbf{K}+\varepsilon\mathbf{K}) - \mathrm{vol}_d(\mathbf{K})}{\varepsilon} = \mathrm{vol}_d(\mathbf{K}) \lim_{\varepsilon\to 0^+} \frac{(1+\varepsilon)^d - 1}{\varepsilon} = d\mathrm{vol}_d(\mathbf{K}).$$

Note that if $\mathbf{S}$ is a regular simplex circumscribed about $\mathbf{B}^d$, then the largest volume ellipsoid contained in $\mathbf{S}$ is $\mathbf{B}^d$. Thus, from Theorem 132 it follows that $\mathrm{vol}_d(\mathbf{K}) \leq \mathrm{vol}_d(\mathbf{S})$. Combining it with the previous inequality, we have

$$\frac{\mathrm{vol}_d(\mathbf{K})}{(\mathrm{svol}_{d-1}(\mathbf{K}))^{\frac{d}{d-1}}} \geq \frac{1}{d^{\frac{d}{d-1}}(\mathrm{vol}_d(\mathbf{K}))^{\frac{1}{d-1}}} \geq \frac{1}{d^{\frac{d}{d-1}}(\mathrm{vol}_d(\mathbf{S}))^{\frac{1}{d-1}}}$$

$$= \frac{\mathrm{vol}_d(\mathbf{S})}{(\mathrm{svol}_{d-1}(\mathbf{S}))^{\frac{d}{d-1}}}. \qquad \square$$

The proof of Theorem 132 is based on two lemmas. The first one determines the properties of John ellipsoids (cf., e.g., [141]).

Lemma 133 *Let* $\mathbf{K}$ *be a convex body in* $\mathbb{E}^d$, *and assume that the maximal volume ellipsoid contained in* $\mathbf{K}$ *is* $\mathbf{B}^d$. *Then there are contact points* $\mathbf{x}_1, \mathbf{x}_2, \ldots, \mathbf{x}_m$ *in* $(\mathrm{bd}\mathbf{K}) \cap \mathbf{B}^d$ *and positive values* $\lambda_1, \lambda_2, \ldots, \lambda_m$ *such that*

$$\sum_{i=1}^{m} \lambda_i \mathbf{x}_i = \mathbf{o}, \tag{6.2}$$

and

$$\sum_{i=1}^{m} \lambda_i \mathbf{x}_i \otimes \mathbf{x}_i = \mathrm{Id}_d\,. \tag{6.3}$$

Here Id_d denotes the $d \times d$ identity operator, and $\mathbf{x} \otimes \mathbf{x}$ is the rank-1 orthogonal projection of $\mathbb{E}^d$ onto the line spanned by $\mathbf{x}$. The equality of the traces on the two sides of (6.3) yields that $\sum_{i=1}^{m} \lambda_i = d$. The other lemma is a convolution inequality of Brascamp and Lieb [65].

Lemma 134 *Let* $\mathbf{x}_1, \mathbf{x}_2, \ldots, \mathbf{x}_m$ *be unit vectors in* $\mathbb{E}^d$, *and* $\lambda_1, \lambda_2, \ldots, \lambda_m$ *positive numbers such that*

$$\sum_{i=1}^{m} \lambda_i \mathbf{x}_i \otimes \mathbf{x}_i = \mathrm{Id}_d\,.$$

For each value of i, *let* $f_i : \mathbb{R} \to [0, \infty)$ *be an integrable function. Then*

$$\int_{\mathbb{E}^d} \prod_{i=1}^{m} \left(f_i\left(\langle \mathbf{x}_i, \mathbf{y}\rangle\right)\right)^{\lambda_i} d\mathbf{y} \leq \prod_{i=1}^{m} \left(\int_{\mathbb{R}} f_i(t)\,dt\right)^{\lambda_i}. \tag{6.4}$$

Proof of Theorem 132 Without loss of generality, we may assume that the largest volume ellipsoid contained in $\mathbf{K}$ is $\mathbf{B}^d$. Then we need to show that

$$\mathrm{vol}_d(\mathbf{K}) \leq \mathrm{vol}_d(\mathbf{S}) = \frac{d^{\frac{d}{2}} \cdot (d+1)^{\frac{d+1}{2}}}{d!}.$$

By Lemma 133, there are unit vectors $\mathbf{x}_1, \mathbf{x}_2, \ldots, \mathbf{x}_m \in \mathrm{bd}\mathbf{K}$ and positive numbers $\lambda_1, \lambda_2, \ldots, \lambda_m$ satisfying the conditions in (6.2) and (6.3). Let us define the convex body $\mathbf{C}$ as

$$\mathbf{C} = \left\{ \mathbf{y} \in \mathbb{E}^d : \langle \mathbf{x}_i, \mathbf{y} \rangle \leq 1, i = 1, 2, \ldots, m \right\}. \tag{6.5}$$

Clearly, $\mathbf{C} \subseteq \mathbf{K}$. We show that $\mathrm{vol}_d(\mathbf{C}) \leq \mathrm{vol}_d(\mathbf{S})$. We prove this inequality by applying Lemma 134 to a family of functions defined on $\mathbb{E}^{d+1}$.

Let us regard $\mathbb{E}^{d+1}$ as $\mathbb{E}^d \times \mathbb{R}$. For $i = 1, 2, \ldots, m$, set

$$\mathbf{y}_i = \sqrt{\frac{d}{d+1}} \left(-\mathbf{x}_i, \frac{1}{\sqrt{d}} \right),$$

and

$$\mu_i = \frac{d+1}{d} \lambda_i.$$

It is easy to see that for each value of i, $\mathbf{y}_i$ is a unit vector, $\sum_{i=1}^m \mu_i = d+1$, and that the conditions in (6.2) and (6.3) imply that

$$\sum_{i=1}^{m} \mu_i \mathbf{y}_i \otimes \mathbf{y}_i = \mathrm{Id}_{d+1}.$$

For $i = 1, 2, \ldots, m$, define the function $f_i : \mathbb{R} \to [0, \infty)$ as

$$f_i(t) = \begin{cases} e^{-t} & \text{if} \quad t \geq 0, \\ 0 & \text{if} \quad t < 0. \end{cases}$$

For any $\mathbf{z} \in \mathbb{E}^{d+1}$, set

$$F(\mathbf{z}) = \prod_{i=1}^{m} \left(f_i \left(\langle \mathbf{y}_i, \mathbf{z} \rangle \right) \right)^{\mu_i}.$$

By Lemma 134, we have

$$\int_{\mathbb{E}^{d+1}} F(\mathbf{z})\, d\mathbf{z} \leq \prod_{i=1}^{m} \left(\int_{\mathbb{R}} f_i(t)\, dt \right)^{\mu_i} = 1. \tag{6.6}$$

Consider some point $\mathbf{z} = (\mathbf{w}, r) \in \mathbb{E}^{d+1}$. Then, for each value of i, we have

$$\langle \mathbf{y}_i, \mathbf{z} \rangle = \frac{r}{\sqrt{d+1}} - \sqrt{\frac{d}{d+1}} \langle \mathbf{x}_i, \mathbf{w} \rangle.$$

Since $\sum_{i=1}^{m} \lambda_i \mathbf{x}_i = \mathbf{o}$, there is some index j such that $\langle \mathbf{x}_j, \mathbf{w} \rangle \geq 0$. Thus, if $r < 0$, then $\langle \mathbf{y}_j, \mathbf{z} \rangle < 0$ and the definition of F yields that $F(\mathbf{z}) = 0$. Furthermore, if $r \geq 0$, then $F(\mathbf{z})$ is non-zero if and only if for all values of i,

$$\langle \mathbf{y}_i, \mathbf{w} \rangle \leq \frac{r}{\sqrt{d}}.$$

In this case

$$F(\mathbf{z}) = \exp\left(-\sum_{i=1}^{m} \mu_i \left(\frac{r}{\sqrt{d+1}} - \sqrt{\frac{d}{d+1}} \langle \mathbf{y}_i, \mathbf{w} \rangle\right)\right)$$

$$= \exp\left(-\sqrt{d+1}r + \sqrt{\frac{d+1}{d}} \langle \sum_{i=1}^{m} \lambda_i \mathbf{x}_i, \mathbf{w} \rangle\right) = \exp\left(-\sqrt{d+1}r\right).$$

Hence, for any $r \geq 0$, the integral of F over the hyperplane $\{x_{d+1} = r\}$ is

$$e^{-\sqrt{d+1}r} \mathrm{vol}_d\left(\frac{r}{\sqrt{d}} \mathbf{C}\right) = e^{-\sqrt{d+1}r} \left(\frac{r}{\sqrt{d}}\right)^d \mathrm{vol}_d(\mathbf{C}).$$

Then (6.6) yields that

$$\mathrm{vol}_d(\mathbf{C}) \int_0^{\infty} e^{-\sqrt{d+1}r} \left(\frac{r}{\sqrt{d}}\right)^d dr = \frac{\mathrm{vol}_d(\mathbf{C}) d!}{d^{\frac{d}{2}} \cdot (d+1)^{\frac{d+1}{2}}} \leq 1,$$

from which the assertion readily follows. □

6.2 Proofs of Theorems 10 and 11

Before proving the theorems, we need to find the geometric interpretation of (1.2). In the literature, there are two different interpretations of this equation.

The *Eikonal wavefront model*, used for example in optics, starts with a smooth hypersurface at time $t = 0$. In this model, one obtains the evolving hypersurface at time t by translating every *point* of its boundary in the direction of the inward surface normal by a vector of length t (cf. Figure 6.1 (A)). If $\mathbf{K}(0)$ is a smooth, convex body with minimal curvature radius $r_{\min}(0)$ (i.e., the reciprocal of the maximal principal curvature), then in the Eikonal wavefront model the evolving hypersurface will exhibit its first singularity at $t = r_{\min}(0)$ and for $t > r_{\min}(0)$ it will develop self-intersecting, non-convex parts.

In the *Eikonal abrasion model*, used for example in the investigation of the abrasion of particles, starting with the boundary of a (not necessarily smooth) convex body $\mathbf{K}(0)$, one obtains the evolving hypersurface at time

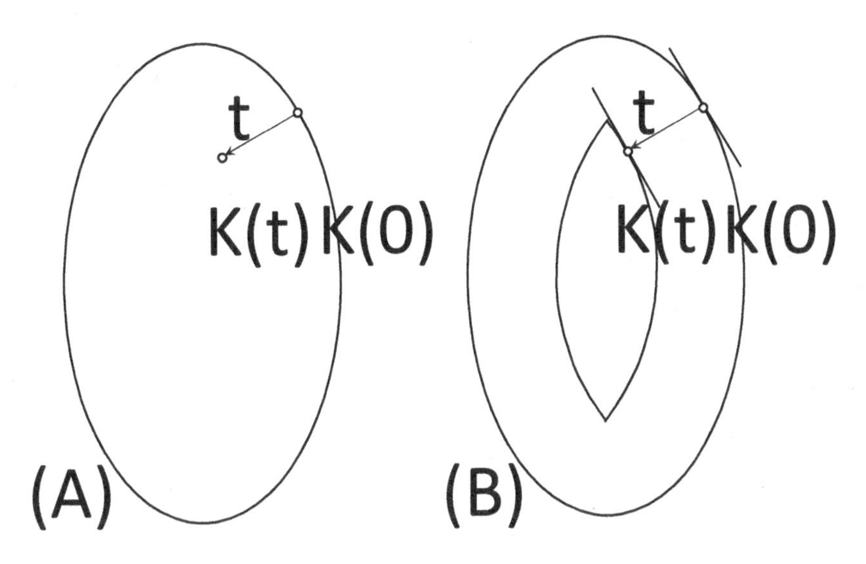

FIGURE 6.1
Alternative interpretations of the Eikonal equation (1.2): (A) The Eikonal Wavefront Model and (B) the Eikonal Abrasion Model.

t by translating *the supporting half space* at every point of its boundary in the direction of the inward surface normal by a vector of length t (cf. Figure 6.1 (B)). Obviously, in this model the evolving hypersurface $\mathbf{K}(t)$ will remain convex; however, initially smooth shapes will also develop singularities. The first such singularity appears, similarly to the wavefront model, at $t = r_{\min}(0)$. The singularities of the evolving hypersurface in the Eikonal abrasion model correspond to the self-intersections in the Eikonal wavefront model.

Since our examination is restricted to not necessarily smooth convex bodies, we use the Eikonal abrasion model. We start with introducing some notation.

Let $\mathbf{K}$ be a convex body in $\mathbb{E}^d$; that is, a compact, convex set with non-empty interior. Recall that the *support function* $h_{\mathbf{K}} : \mathbb{S}^{d-1} \to \mathbb{R}$ of $\mathbf{K}$ is defined by $h_{\mathbf{K}}(\mathbf{m}) = \sup\{\langle \mathbf{m}, \mathbf{p} \rangle : \mathbf{p} \in \mathbf{K}\}$ for all $\mathbf{m} \in \mathbb{S}^{d-1}$ [217]. For any $t \geq 0$, we let $\mathbf{H}_{\mathbf{K}}(\mathbf{m}, t)$ be the closed half space defined by $\{\mathbf{p} \in \mathbb{E}^d : \langle \mathbf{p}, \mathbf{m} \rangle \leq h_{\mathbf{K}}(\mathbf{m}) - t\}$. In the Eikonal abrasion model, the evolving convex body at time t is given by the formula

$$\mathbf{K}(t) = \bigcap_{\mathbf{m} \in S^{d-1}} \mathbf{H}_{\mathbf{K}}(\mathbf{m}, t).$$

We show that $\mathbf{K}(t)$ is the set of points in $\mathbf{K}$ at the distance at least t from bd$\mathbf{K}$; that is, the so-called *inner parallel body* of $\mathbf{K}$ at the distance t. Consider some point $\mathbf{p} \in \text{int}\mathbf{K}$, and let $\mathbf{q} \in \text{bd}\mathbf{K}$ be a point closest to $\mathbf{p}$ in bd$K$. Then the Euclidean ball $\mathbf{p} + ||\mathbf{q} - \mathbf{p}||\mathbf{B}^d$ is contained in $\mathbf{K}$. Let H be the hyperplane through $\mathbf{q}$ and perpendicular to $\mathbf{q}-\mathbf{p}$. Since H is the only hyperplane that supports the ball $\mathbf{p}+||\mathbf{q}-\mathbf{p}||\mathbf{B}^d$ at $\mathbf{q}$, and $\mathbf{p}+||\mathbf{q}-\mathbf{p}||\mathbf{B}^d \subseteq \mathbf{K}$, it follows that H is the unique supporting hyperplane of $\mathbf{K}$ at $\mathbf{q}$. Clearly, the distance of $\mathbf{p}$ and H is $||\mathbf{q}-\mathbf{p}||$. Since every closed supporting half space of $\mathbf{K}$ contains $\mathbf{K}$, it follows that the distance of $\mathbf{p}$ from every supporting hyperplane of $\mathbf{K}$ is at least $||\mathbf{q}-\mathbf{p}||$. Thus, the distance of $\mathbf{p}$ from $\mathrm{bd}\mathbf{K}$ is equal to the minimal distance of $\mathbf{p}$ from any of the supporting hyperplanes of $\mathbf{K}$, which implies that $\mathbf{K}(t)$ is indeed the inner parallel body of $\mathbf{K}$ at the distance t. This property, proved also in [166] together with some other elements of this proof, yields the following observation: If $r(\mathbf{K})$ is the radius of a largest ball contained in $\mathbf{K}$, then for any $0 \leq t < r(\mathbf{K})$, $\mathbf{K}(t)$ is a convex body, for $t = r(\mathbf{K})$, $\mathbf{K}(t)$ is a compact, convex set with no interior point, and for $t > r(\mathbf{K})$, $\mathbf{K}(t) = \emptyset$.

Now we turn to the proof of Theorem 10. As volume and surface volume are continuous with respect to Hausdorff distance and are strictly increasing with respect to containment in the family of convex bodies, both $\mathrm{vol}_d(\mathbf{K}(t))$ and $\mathrm{svol}_{d-1}(\mathbf{K}(t))$ are positive continuous, strictly decreasing functions of t on the interval $[0, r(\mathbf{K}))$. This implies that $I(\mathbf{K}(t))$ is also continuous on this interval.

Consider some $0 < t_0 < r(\mathbf{K})$. For brevity, we set $\mathbf{K}(t_0) = \mathbf{K}_0$, $N_0 = N(\mathbf{K}_0)$, $F(\mathbf{K}(t_0)) = \mathbf{F}_0$ and $I(\mathbf{K}(t)) = I(t)$. Let $\mathbf{p} \in \mathrm{bd}\mathbf{K}_0$. By our previous observation, for any $0 \leq t \leq t_0$ the minimum of the distances of $\mathbf{p}$ from the supporting hyperplanes of $\mathrm{bd}\mathbf{K}(t)$ is equal to $t_0 - t$. Since this minimum is attained at some point $\mathbf{q} \in \mathrm{bd}\mathbf{K}(t)$, the ball $\mathbf{p} + (t - t_0)\mathbf{B}^d$ touches $\mathrm{bd}\mathbf{K}(t)$ from inside. Thus, $\mathbf{K}_0 + (t_0 - t)\mathbf{B}^d \subseteq \mathbf{K}(t)$. On the other hand, we also have $h_{\mathbf{K}(t)}(\mathbf{m}) = h_{\mathbf{K}_0}(\mathbf{m}) + (t_0 - t)$ for any $\mathbf{m} \in N_0$, which yields that $\mathbf{K}(t) \subseteq \mathbf{K}_0 + (t_0 - t)\mathbf{F}_0$. Note that if $\mathbf{K}_0$ is smooth, then $\mathbf{K}(t) = \mathbf{K}_0 + (t_0 - t)\mathbf{B}^d$ for every $0 \leq t \leq t_0$, implying that $\mathbf{K}(t)$ is also smooth on this interval. More generally, $N(\mathbf{K}(t))$ decreases and $F(\mathbf{K}(t))$ increases in time with respect to containment; that is, for any $0 \leq t_1 < t_2 < r(\mathbf{K})$ we have $N(\mathbf{K}(t_2)) \subseteq N(\mathbf{K}(t_1))$ and $F(\mathbf{K}(t_1)) \subseteq F(\mathbf{K}(t_2))$.

Set $\mathbf{L}(t) = \mathbf{K}_0 + (t_0 - t)\mathbf{B}^d$, and $\mathbf{M}(t) = \mathbf{K}_0 + (t_0 - t)\mathbf{F}_0$. By Minkowski's theorem on mixed volumes [217], we have

$$\mathrm{vol}_d(\mathbf{L}(t)) = \sum_{j=0}^{d} \binom{d}{j} (t - t_0)^j W_j(\mathbf{K}_0), \text{ and} \tag{6.7}$$

$$\mathrm{vol}_d(\mathbf{M}(t)) = \sum_{j=0}^{d} \binom{d}{j} (t - t_0)^j V_j(\mathbf{K}_0), \tag{6.8}$$

where $W_j(\mathbf{K}_0)$ is the jth quermassintegral of $\mathbf{K}_0$, and we denote the mixed volume $V(\overbrace{\mathbf{K}_0, \ldots, \mathbf{K}_0}^{d-j}, \overbrace{\mathbf{F}_0, \ldots, \mathbf{F}_0}^{j})$ by $V_j(\mathbf{K}_0)$.

Observe that $W_0(\mathbf{K}_0) = V_0(\mathbf{K}_0) = \mathrm{vol}_d(\mathbf{K}_0)$, and that $dW_1(\mathbf{K}_0) = \mathrm{svol}_{d-1}(\mathbf{K}_0)$. We show that $dV_1(\mathbf{K}_0) = \mathrm{svol}_{d-1}(\mathbf{K}_0)$ as well. Since both mixed volumes and surface volume are continuous with respect to Hausdorff distance, it suffices to prove this equality for polytopes, and thus, assume for the moment that $\mathbf{K}_0$ is a polytope. In this case $\mathbf{F}_0$ is the polytope, circumscribed about $\mathbf{B}^d$, whose outer unit facet normal vectors coincide with those of $\mathbf{K}_0$. Thus, $\mathbf{M}(t)$ can be decomposed into $\mathbf{K}_0$, cylinders of height $t_0 - t$ with the facets of $\mathbf{K}_0$ as bases, and sets in the $(\rho(t_0 - t))$-neighborhood of the $(d-2)$-faces of $\mathbf{K}(t_0)$, where ρ is the diameter of $\mathbf{F}_0$. The volume of this set is

$$\mathrm{vol}_d(\mathbf{M}(t)) = \mathrm{vol}_d(\mathbf{K}_0) + (t_0 - t)\mathrm{svol}_{d-1}(\mathbf{K}_0) + O((t_0 - t)^2),$$

implying $-dV_1(\mathbf{K}_0) = \frac{d}{dt}\mathrm{vol}_d(\mathbf{M}(t))\big|_{t=t_0} = -\mathrm{svol}_{d-1}(\mathbf{K}(t_0))$.

Let us define the quantity

$$\underline{I}(t) = \frac{\mathrm{vol}_d(\mathbf{L}(t))}{\mathrm{svol}_{d-1}(\mathbf{M}(t))^{\frac{d}{d-1}}},$$

and set $\tau_d = \frac{\mathrm{vol}_d(\mathbf{B}^d)}{(\mathrm{svol}_{d-1}(\mathbf{B}^d))^{\frac{d}{d-1}}}$. We note that $\underline{I}(t)$ depends on t_0 and it is defined only for $0 \leq t \leq t_0$. Furthermore, as both volume and surface volume are strictly increasing with respect to inclusion, we have $\underline{I}(t) \leq I(t)\tau_d$. Differentiating this quantity, the formulas in (6.7) and (6.8), and their connection with $\mathrm{svol}_{d-1}(\mathbf{K}_0)$ yields that

$$\underline{I}'_-(t_0) = -\frac{d^2}{(\mathrm{svol}_{d-1}(\mathbf{K}_0))^{\frac{2d-1}{d-1}}}\left(V_1(\mathbf{K}_0)^2 - V_0(\mathbf{K}_0)V_2(\mathbf{K}_0)\right),$$

which is not positive by the Alexandrov-Fenchel inequality (cf. Theorem 2).

Even though there are only partial results to characterize the equality case in Theorem 2, these results permit us to show that if $\underline{I}'_-(t_0) = 0$, then $\mathbf{K}_0$ is homothetic to $\mathbf{F}_0$. To do it, we introduce some concepts from [217].

Let $\mathbf{p}$ be a boundary point of a convex body $\mathbf{Q}$ in $\mathbb{E}^d$. We denote the set of external unit normal vectors of the supporting hyperplanes of $\mathbf{Q}$ at $\mathbf{p}$ by $N_{\mathbf{Q}}(\mathbf{p})$. This set is a spherically convex set in $\mathbb{S}^{d-1}$ for any boundary point $\mathbf{p}$ of any convex body $\mathbf{Q}$. Furthermore, the relative interiors of these sets cover $\mathbb{S}^{d-1}$, and if two of them intersect, then they coincide. In other words, for any $\mathbf{m} \in \mathbb{S}^{d-1}$ there is a boundary point $\mathbf{p}$ of $\mathbf{Q}$ such that $\mathbf{m}$ is in the relative interior of $N_{\mathbf{Q}}(\mathbf{p})$, and if $\mathbf{m}$ is in the relative interiors of both $N_{\mathbf{Q}}(\mathbf{p}_1)$ and $N_{\mathbf{Q}}(\mathbf{p}_2)$, then $N_{\mathbf{Q}}(\mathbf{p}_1) = N_{\mathbf{Q}}(\mathbf{p}_2)$. If $\mathbf{m} \in \mathbb{S}^{d-1}$ is in the relative interior of $N_{\mathbf{Q}}(\mathbf{p})$, and the dimension of $N_{\mathbf{Q}}(\mathbf{p})$ is at most r, we say that $\mathbf{m}$ is an *r-extreme normal vector* of $\mathbf{Q}$. Note that the set of 0-extreme normal vectors of $\mathbf{Q}$ is exactly $N(\mathbf{Q})$, defined in Definition 2. A supporting hyperplane H of $\mathbf{Q}$ is called *r-extreme* if its outer unit normal vector is r-extreme.

A convex body $\mathbf{L}$, containing the convex body $\mathbf{K}$, is called an *r-tangential body* of $\mathbf{K}$ if every $(d-r-1)$-extreme support hyperplane of $\mathbf{L}$ is a support hyperplane of $\mathbf{K}$. By Theorem 2.2.10 in [217], if $\mathbf{K} \subseteq \mathbf{L}$ are convex bodies in $\mathbb{E}^d$, then $\mathbf{L}$ is an r-tangential body of $\mathbf{K}$ if and only if every supporting hyperplane of $\mathbf{L}$ that is not a supporting hyperplane of $\mathbf{K}$ contains only $(r-1)$-*singular points* of bd$\mathbf{L}$, that is, points $\mathbf{p}$ in bd$\mathbf{L}$ for which the spherical dimension of $N_{\mathbf{L}}(\mathbf{p})$ is at least $d-r$.

By Theorem 7.6.19 in [217], since both $\mathbf{K}_0$ and $\mathbf{F}_0$ are d-dimensional, $(V_1(\mathbf{K}_0))^2 = V_0(\mathbf{K}_0)V_2(\mathbf{K}_0)$ implies that $\mathbf{K}_0$ is homothetic to a $(d-2)$-tangential body of $\mathbf{F}_0$. More specifically, $\mathbf{K}_0$ has a homothetic copy $\mathbf{K}'$ such that $\mathbf{F}_0 \subseteq \mathbf{K}'$, and every supporting hyperplane of $\mathbf{K}'$ that does not support $\mathbf{F}_0$ contains only $(d-3)$-singular, or in particular, singular points of $\mathbf{K}'$. Hence, every supporting hyperplane of $\mathbf{K}'$ that contains a smooth point of bd$\mathbf{K}'$ supports $\mathbf{F}_0$ as well. Thus, the definition of $\mathbf{F}_0$ and the relation $\mathbf{F}_0 \subseteq \mathbf{K}'$ yields $\mathbf{F}_0 = \mathbf{K}'$. This means that if $\underline{I}'_-(t_0) = 0$, then $\mathbf{F}_0$ is homothetic to $\mathbf{K}_0$. Note that the reversed statement also holds: if $\mathbf{F}_0$ is homothetic to $\mathbf{K}_0$, then $\underline{I}'_-(t_0) = 0$, and even more, in this case $\mathbf{K}(t)$ is homothetic to $\mathbf{K}_0$ for any $t > t_0$.

Let $t^\star$ denote the smallest value of t such that $\underline{I}'_-(t) = 0$. Then $\mathbf{K}(t)$ is homothetic to $\mathbf{K}(t^\star)$ for any $t \in [t^\star, r(\mathbf{K}))$, and $I(t)$ is a constant on this interval. To finish the proof, we need to show that $I(t)$ strictly decreases on $[0, t^\star]$.

Since $\underline{I}'_-(t_0) < 0$ for any $0 < t_0 < t^\star$, for any such t_0 there is some $\varepsilon = \varepsilon(\mathbf{K}, t_0) > 0$ such that $I(t)\tau_d \geq \underline{I}(t) > I(t_0)\tau_d$ for all $t \in (t_0 - \varepsilon, t_0)$; that is, the function $I(t)$ is locally strictly decreasing from the left at every point. This and the continuity of $I(t)$ implies that $I(t)$ strictly decreases on this interval. Indeed, suppose for contradiction that for some $t_1 < t_2$ we have $I(t_1) \leq I(t_2)$. By continuity, $I(t)$ attains its global maximum on $[t_1, t_2]$ at some $t' \in [t_1, t_2]$. Clearly, since $I(t)$ is locally strictly decreasing from the left at t', it follows that $t' = t_1$ and $I(t_1) > I(t_2)$, a contradiction.

Now we prove Theorem 11. Let $\mathbf{K}$ be a convex body in $\mathbb{E}^d$, and let $t_0 > 0$ be arbitrary. Note that, in the Eikonal abrasion model, with $\mathbf{K} + t_0F(\mathbf{K})$ playing the role of the initial body, at time t the evolving body is $\mathbf{K} + (t_0 - t)F(\mathbf{K})$. Since $I(\mathbf{K} + (t_0 - t)F(\mathbf{K})$ is a decreasing function of $t_0 - t$ on the interval $[0, t_0]$, it follows that $I(\mathbf{K} + tF(\mathbf{K}))$ is an increasing function of t on the same interval. Since $t_0 > 0$ was arbitrary, we have that $I(\mathbf{K} + tF(\mathbf{K}))$ increases on $t \in [0, \infty)$.

6.3 Proof of Theorem 14

6.3.1 Preliminaries

Let $\mathbf{P} \in \mathcal{P}_3(n)$, and consider a triangulation $\mathcal{C}(\mathbf{P})$ of bd$\mathbf{P}$ such that all vertices of $\mathcal{C}(\mathbf{P})$ are vertices of $\mathbf{P}$. We call the convex hull of a triangle in $\mathcal{C}(\mathbf{P})$ and $\mathbf{o}$ a *facial tetrahedron* of $\mathbf{P}$. Note that the volume of the facial tetrahedron

$\mathrm{conv}\{\mathbf{o},\mathbf{p},\mathbf{q},\mathbf{r}\}$ of $\mathbf{P}$ is $\frac{1}{6}|\mathbf{p},\mathbf{q},\mathbf{r}|$, where $|\mathbf{p},\mathbf{q},\mathbf{r}|$ is the determinant whose columns are the vectors $\mathbf{p}$, $\mathbf{q}$ and $\mathbf{r}$.

For simplicity, for any points $\mathbf{p}_1,\mathbf{p}_2,\ldots,\mathbf{p}_n$, we introduce the notation $\mathbf{C}(\mathbf{p}_1,\mathbf{p}_2,\ldots,\mathbf{p}_n)=\mathrm{conv}\{\mathbf{p}_1,\mathbf{p}_2,\ldots,\mathbf{p}_n\}$. In this case for any $\mathbf{p}_i,\mathbf{p}_j$ we set $\mathbf{m}_{ij}=\frac{1}{6}(\mathbf{p}_i\times\mathbf{p}_j)$, $S_{ij}=[\mathbf{p}_i,\mathbf{p}_j]$, and $s_{ij}=||\mathbf{p}_i-\mathbf{p}_j||$. Finally, we note that for any fixed value of n, a polyhedron with n vertices and maximal volume inscribed in $\mathbb{S}^2$ satisfies Property Z in Definition 4.

We start the proof with two lemmas, which are the special cases of Theorem 16 and Lemma 142 for $d=3$, proved in Subsection 6.4.1. These lemmas were first proved for $d=3$ in [20].

Lemma 135 *Let* $\mathbf{P}=\mathbf{C}(\mathbf{p}_1,\mathbf{p}_2,\ldots,\mathbf{p}_n)\in\mathcal{P}_3(n)$ *satisfy Property Z. Let* $\mathcal{C}(\mathbf{P})$ *be a triangulation of* $\mathrm{bd}\mathbf{P}$ *such that all vertices of* $\mathcal{C}(\mathbf{P})$ *are vertices of* $\mathbf{P}$*. Assume that the vertices of* $\mathcal{C}(\mathbf{P})$ *adjacent to* $\mathbf{p}_1$ *are* $\mathbf{p}_2,\mathbf{p}_3,\ldots,\mathbf{p}_k$ *in cyclic order in* $\mathrm{bd}\mathbf{P}$.

(i) Then, $\mathbf{p}_1=\frac{\mathbf{z}}{||\mathbf{z}||}$, *where* $\mathbf{z}=\mathbf{m}_{23}+\mathbf{m}_{34}+\ldots+\mathbf{m}_{k2}$.

(ii) Furthermore, every face of $\mathcal{P}$ *is a triangle.*

Lemma 136 *Let* $\mathbf{P}=\mathbf{C}(\mathbf{p}_1,\mathbf{p}_2,\ldots,\mathbf{p}_n)\in\mathcal{P}_3(n)$ *satisfy Property Z. Let* $\mathcal{C}(\mathbf{P})$ *be a triangulation of* $\mathrm{bd}\mathbf{P}$ *such that all vertices of* $\mathcal{C}(\mathbf{P})$ *are vertices of* $\mathbf{P}$*. Assume that the vertices of* $\mathcal{C}(\mathbf{P})$ *adjacent to* $\mathbf{p}_1$ *are* $\mathbf{p}_2,\mathbf{p}_3,\ldots,\mathbf{p}_5$ *in cyclic order in* $\mathrm{bd}\mathbf{P}$*. Then* $s_{12}=s_{14}$ *and* $s_{13}=s_{15}$.

We also need the next remark and another lemma. Here, by a *double n-pyramid* we mean a double pyramid with n vertices.

Remark 137 *If, using the notation in Lemma 135, the points* $\mathbf{p}_2,\mathbf{p}_3,\ldots,\mathbf{p}_r$ *lie in a plane* H*, then by (i) of Lemma 135,* $\mathbf{p}_i$ *is one of the two tangent points of the supporting planes of* $\mathbb{S}^2$ *parallel to* H*. Thus, in this case* $s_{12}=s_{13}=\ldots=s_{1r}$.

Lemma 138 *Let* $\mathbf{P}=\mathbf{C}(\mathbf{p}_1,\mathbf{p}_2,\ldots,\mathbf{p}_n)\in\mathcal{P}_3(n)$ *satisfy Property Z. If* $\mathbf{P}$ *is combinatorially equivalent to a double n-pyramid, then* $\mathbf{P}$ *is a double n-pyramid, with a regular* $(n-2)$*-gon centered at* $\mathbf{o}$ *as its base, and the two tangent points of the supporting planes of* $\mathbb{S}^2$ *parallel to the plane of its base as apexes. Furthermore, in this case* $\mathrm{vol}_3(\mathbf{P})=\frac{n-2}{3}\sin\frac{2\pi}{n-2}$.

Proof of Lemma 138. Since $\mathbf{P}$ is combinatorially equivalent to a double n-pyramid, it contains $n-2$ vertices of valence 4, and two vertices of valence $n-2$. Without loss of generality, we may assume that the edges of $\mathbf{P}$ are $S_{(i(n-1))}$ and S_{in} for $i=1,2,\ldots,n-2$, and $S_{12},S_{23},\ldots,S_{(n-2)1}$. Then, by Lemma 136, we have $s_{i(n-1)}=s_{in}$ for all $n=1,2,\ldots,n-2$. Thus, for $i=1,2,\ldots,n-2$, $\mathbf{p}_i$ is contained in the plane equidistant from $\mathbf{p}_{n-1}$ and $\mathbf{p}_n$, implying that the points $\mathbf{p}_1,\mathbf{p}_2,\ldots,\mathbf{p}_{n-2}$ are the vertices of a planar convex $(n-2)$-gon P_0, where the plane of P_0 contains $\mathbf{o}$. Furthermore, by Lemma 136, it follows that $s_{12} = s_{23} = \ldots = s_{(n-2)1}$, or in other words, P_0 is a regular $(n-2)$-gon. Finally, Remark 137 implies that $\mathbf{p}_{n-1}$ and $\mathbf{p}_n$ are the two tangent points of the supporting planes of $\mathbb{S}^2$ parallel to the plane of P_0. This proves the first part of Lemma 138. The second part can be proved using elementary computations. □

6.3.2 Proof of Theorem 14 for $n \leq 6$

Consider the case that $n = 4$. Then $\mathbf{P}$ is a tetrahedron. Applying Remark 137 for $\mathbf{P}$, we obtain that all edges of $\mathbf{P}$ starting at any given vertex of $\mathbf{P}$ are of equal length. Thus, $\mathbf{P}$ is a regular tetrahedron inscribed in $\mathbb{S}^2$.

Now, assume that $n = 5$. Then $\mathbf{P}$ is a double 5-pyramid, and Theorem 14 immediately follows from Lemma 138. If $n = 6$, then it is combinatorially equivalent to a double 6-pyramid. Thus, Lemma 138 implies that $\mathbf{P}$ is a regular octahedron inscribed in $\mathbb{S}^2$.

6.3.3 Proof of Theorem 14 for $n = 7$

It was shown by Bowen and Fisk [63] that up to combinatorial isomorphism, the only polyhedron with $n = 7$ vertices and triangular faces, and having no 3-valent vertices, is a double 7-pyramid. By Lemma 138, there is a unique polyhedron satisfying Property Z and combinatorially equivalent to a double 7-pyramid, and the volume of this polyhedron is $\mathrm{vol}_3(\mathbf{P}) = \frac{5}{3}\sin\frac{2\pi}{5} = 1.58510\ldots$. Thus, it is sufficient to prove that if $\mathbf{P}$ has a 3-valent vertex, then its volume is less than $\frac{5}{3}\sin\frac{2\pi}{5}$.

Let $\mathbf{P} = \mathbf{C}(\mathbf{p}_1, \mathbf{p}_2, \ldots, \mathbf{p}_7) \in \mathcal{P}_3(7)$, and assume that $\mathrm{vol}_3(\cdot)$ is maximal at $\mathbf{P}$ on the family $\mathcal{P}_3(7)$. For contradiction, suppose that $\mathbf{p}_1$ is a 3-valent vertex of $\mathbf{P}$. Let the vertices of $\mathbf{P}$ adjacent to $\mathbf{p}_1$ be $\mathbf{p}_2, \mathbf{p}_3, \mathbf{p}_4$. Since $\mathbf{P}$ satisfies Property Z, the line through $\mathbf{o}$ and $\mathbf{p}_1$ contains the circumcenter of the triangle $T = \mathrm{conv}\{\mathbf{p}_2, \mathbf{p}_3, \mathbf{p}_4\}$. Clearly, since $\mathbf{P}$ has maximal volume, it contains $\mathbf{o}$, and thus, T contains its circumcenter. Thus, by [106, p. 267], the area of T is at most $\frac{3\sqrt{3}}{4}\left(1 - \frac{1}{3}\tan^2\frac{2\pi-\tau}{6}\right)$, where τ is the area of the central projection of T onto $\mathbb{S}^2$. Let $\theta = 4\pi - \tau$. Then the central projection of the seven facial tetrahedra of $\mathbf{P}$ not incident to $\mathbf{p}_1$ has total area θ. By [106, p. 275], the total volume of these seven facial tetrahedra is at most

$$\frac{7}{4}\tan\frac{2\pi - \frac{\theta}{7}}{6}\left(1 - \frac{1}{3}\tan^2\frac{2\pi - \frac{\theta}{7}}{6}\right) = \frac{7}{4}\tan\frac{10\pi + \tau}{42}\left(1 - \frac{1}{3}\tan^2\frac{10\pi + \tau}{42}\right).$$

Thus, $\mathrm{vol}_3(\mathbf{P}) \leq f(\tau)$, where

$$f(\tau) = \frac{\sqrt{3}}{4}\left(1 - \frac{1}{3}\tan^2\frac{2\pi - \tau}{6}\right) + \frac{7}{4}\tan\frac{10\pi + \tau}{42}\left(1 - \frac{1}{3}\tan^2\frac{10\pi + \tau}{42}\right).$$

A direct computation shows that f is concave, and its maximum is less than $\frac{5}{3}\sin\frac{2\pi}{5}$. Thus, if $\mathbf{P}$ has a 3-valent vertex, then its volume is not maximal on the family $\mathcal{P}_3(7)$.

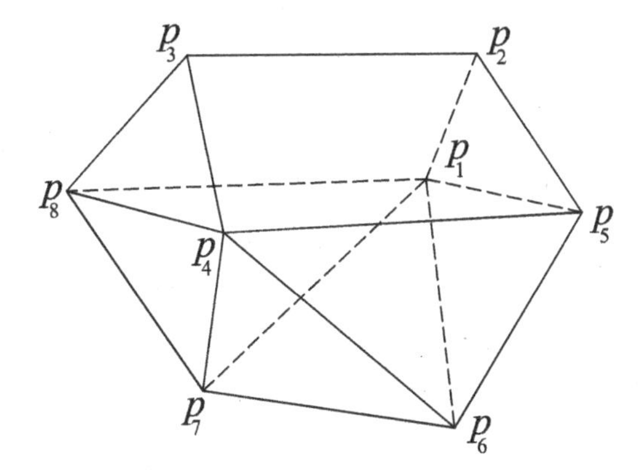

FIGURE 6.2
A medial polyhedron with 8 vertices.

6.3.4 Proof of Theorem 14 for $n = 8$

If $n = 8$ then the average valence of any polyhedron $\mathbf{P}$ in $\mathcal{P}_3(8)$ is 4.5. Bowen and Fisk [63] showed that up to combinatorial isomorphism, there are two simplical polyhedra with $n = 8$ vertices that have no 3-valent vertices. One of these is a double 8-pyramid, and the other one has four 5-valent, and four 4-valent vertices, and thus, it is medial. A schematic view of the second polyhedron can be seen on Figure 6.2. By Lemma 138, if $\mathbf{P}$ is combinatorially equivalent to a double 8-pyramid and satisfies Property Z, then it is uniquely determined up to isometries, and its volume is $\sqrt{3}$. In the next lemma, we consider the other case.

Lemma 139 *Let $\mathbf{P} \in \mathcal{P}_3(8)$ be a medial polyhedron satisfying Property Z. Then $\mathbf{P}$ is congruent to the polyhedron in (ii) of Theorem 14, and its volume is $\sqrt{\frac{475+29\sqrt{145}}{250}} = 1.815716\ldots$.*

Proof of Lemma 139. We denote the vertices of $\mathbf{P}$ with $\mathbf{p}_1, \mathbf{p}_2, \ldots, \mathbf{p}_8$ in such a way that it is consistent with the notation in Figure 6.2. Without loss of generality, we may assume that the vertices are distinct. Lemma 135 implies that

$$\mathbf{p}_1 = c_1 (\mathbf{m}_{25} + \mathbf{m}_{56}, +\mathbf{m}_{67} + \mathbf{m}_{78} + \mathbf{m}_{82}),$$

$$\mathbf{p}_4 = c_4 (\mathbf{m}_{38} + \mathbf{m}_{87}, +\mathbf{m}_{76} + \mathbf{m}_{65} + \mathbf{m}_{53}),$$

$$\mathbf{p}_5 = c_5 (\mathbf{m}_{61} + \mathbf{m}_{12}, +\mathbf{m}_{23} + \mathbf{m}_{34} + \mathbf{m}_{46}),$$

where $c_1, c_4, c_5 > 0$. Taking the cross-product of the first expression with $\mathbf{p}_1$ yields that

$$\langle \mathbf{p}_1, \mathbf{p}_5 - \mathbf{p}_8 \rangle \mathbf{p}_2 + \langle \mathbf{p}_1, \mathbf{p}_6 - \mathbf{p}_2 \rangle \mathbf{p}_5 + \langle \mathbf{p}_1, \mathbf{p}_7 - \mathbf{p}_5 \rangle \mathbf{p}_6 \qquad (6.9)$$

$$+\langle \mathbf{p}_1, \mathbf{p}_8 - \mathbf{p}_6\rangle \mathbf{p}_7 + \langle \mathbf{p}_1, \mathbf{p}_2 - \mathbf{p}_7\rangle \mathbf{p}_8 = \mathbf{o}.$$

Similarly, by taking the cross-product of the second expression with $\mathbf{p}_4$, we obtain that

$$\langle \mathbf{p}_4, \mathbf{p}_8 - \mathbf{p}_5\rangle \mathbf{p}_3 + \langle \mathbf{p}_4, \mathbf{p}_3 - \mathbf{p}_6\rangle \mathbf{p}_5 + \langle \mathbf{p}_4, \mathbf{p}_5 - \mathbf{p}_7\rangle \mathbf{p}_6 \tag{6.10}$$

$$+\langle \mathbf{p}_4, \mathbf{p}_6 - \mathbf{p}_8\rangle \mathbf{p}_7 + \langle \mathbf{p}_4, \mathbf{p}_7 - \mathbf{p}_3\rangle \mathbf{p}_8 = \mathbf{o}.$$

From Lemma 136, it follows that $\langle \mathbf{p}_1, \mathbf{p}_2\rangle = \langle \mathbf{p}_2, \mathbf{p}_3\rangle = \langle \mathbf{p}_3, \mathbf{p}_4\rangle$, and that $\langle \mathbf{p}_5, \mathbf{p}_6\rangle = \langle \mathbf{p}_6, \mathbf{p}_7\rangle = \langle \mathbf{p}_7, \mathbf{p}_8\rangle$. Furthermore, we obtain in the same way that $\langle \mathbf{p}_2, \mathbf{p}_5\rangle = \langle \mathbf{p}_2, \mathbf{p}_8\rangle$, $\langle \mathbf{p}_3, \mathbf{p}_5\rangle = \langle \mathbf{p}_3, \mathbf{p}_8\rangle$, $\langle \mathbf{p}_1, \mathbf{p}_6\rangle = \langle \mathbf{p}_4, \mathbf{p}_6\rangle$ and $\langle \mathbf{p}_1, \mathbf{p}_7\rangle = \langle \mathbf{p}_4, \mathbf{p}_7\rangle$. Using these equalities for the sum of (6.9) and (6.10), we have

$$\langle \mathbf{p}_1, \mathbf{p}_5 - \mathbf{p}_8\rangle \mathbf{p}_2 + \langle \mathbf{p}_4, \mathbf{p}_8 - \mathbf{p}_5\rangle \mathbf{p}_3 + \tag{6.11}$$

$$+\langle \mathbf{p}_5, \mathbf{p}_4 - \mathbf{p}_1\rangle \mathbf{p}_6 + \langle \mathbf{p}_8, \mathbf{p}_1 - \mathbf{p}_4\rangle \mathbf{p}_7 = \mathbf{o}.$$

Note that the sum of the coefficients in (6.11) is zero, which implies that if at least one of them is non-zero, then the vectors $\mathbf{p}_2, \mathbf{p}_3, \mathbf{p}_6$ and $\mathbf{p}_7$ are affinely dependent, and thus, they lie on a plane H.

Assume that there are no zero coefficients in (6.11). Since $\mathbf{p}_1, \mathbf{p}_4, \mathbf{p}_5, \mathbf{p}_8$ are 5-valent vertices, at most one of them is contained in H. Thus, if all of them are contained in one of the two closed half spaces bounded by H, then S_{26} or S_{27} is an edge, a contradiction. If one of them, say $\mathbf{p}_8$, is contained in one open half space bounded by H, and the other three vertices are not, then S_{68} is an edge, a contradiction. Thus, exactly two of them are contained in one open half space, and the other two are in the other open half space. Note that each such pair is connected by an edge, implying that $\mathbf{p}_5$ and $\mathbf{p}_8$, and similarly $\mathbf{p}_1$ and $\mathbf{p}_4$, are contained in different half spaces.

Without loss of generality, assume that $\mathbf{p}_1$ and $\mathbf{p}_8$ are on the same side of H. If S_{36} are adjacent vertices of the quadrilateral $\operatorname{conv}\{\mathbf{p}_2, \mathbf{p}_3, \mathbf{p}_6, \mathbf{p}_7\}$, then S_{13} or S_{68} is an edge. Hence, S_{36} is a diagonal of this quadrilateral. Consider the case that $\mathbf{p}_1$, $\mathbf{p}_8$ are contained in a component of $\mathbb{S}^2 \setminus H$ not greater than a hemisphere. Then $\langle \mathbf{p}_1, \mathbf{p}_2\rangle = \langle \mathbf{p}_2, \mathbf{p}_3\rangle$ implies that $s_{18} \leq s_{15}$; that is, that $\langle \mathbf{p}_1, \mathbf{p}_8\rangle \leq \langle \mathbf{p}_1, \mathbf{p}_5\rangle$. Thus, the coefficient of $\mathbf{p}_2$ in 6.11 is negative. Since S_{27} is a diagonal of the quadrilateral of $\operatorname{conv}\{\mathbf{p}_2, \mathbf{p}_3, \mathbf{p}_6, \mathbf{p}_7\}$, the coefficient of $\mathbf{p}_7$ is also negative; that is, $\langle \mathbf{p}_1, \mathbf{p}_8\rangle < \langle \mathbf{p}_4, \mathbf{p}_8\rangle$. From the equality $\langle \mathbf{p}_6, \mathbf{p}_7\rangle = \langle \mathbf{p}_7, \mathbf{p}_8\rangle$ it similarly follows that $\langle \mathbf{p}_1, \mathbf{p}_8\rangle > \langle \mathbf{p}_4, \mathbf{p}_8\rangle$; a contradiction. If $\mathbf{p}_1$, $\mathbf{p}_8$ are contained in a component of $\mathbb{S}^2 \setminus H$ greater than a hemisphere, an analogous argument gives a contradiction. Hence, at least one of the coefficients in (6.11) is zero.

Consider the case that exactly one coefficient in (6.11) is zero. Without loss of generality, let $\langle \mathbf{p}_1, \mathbf{p}_8\rangle = \langle \mathbf{p}_4, \mathbf{p}_8\rangle$. Since the sum of the coefficients is zero, we have that $\mathbf{p}_2, \mathbf{p}_4, \mathbf{p}_6$ are collinear, which contradicts the assumption that they are distinct unit vectors. Assume that exactly two coefficients are zero, then the remaining two vectors are antipodal. If, say, the coefficients of $\mathbf{p}_6$ and $\mathbf{p}_7$ are zero, then $\mathbf{p}_2$ and $\mathbf{p}_3$ are antipodal, which, together with the equality

$s_{12} = s_{13}$ implies that $\mathbf{p}_1 = \mathbf{p}_3$; a contradiction. If, say, the coefficients of $\mathbf{p}_2$ and $\mathbf{p}_7$ are zero, then $\langle \mathbf{p}_1, \mathbf{p}_5 \rangle = \langle \mathbf{p}_1, \mathbf{p}_8 \rangle = \langle \mathbf{p}_4, \mathbf{p}_8 \rangle$. Since in this case $\mathbf{p}_3$ and $\mathbf{p}_6$ are antipodal, from (6.11) it follows that the coefficient of $\mathbf{p}_6$ is also zero. If three coefficients are zero, then so is the fourth. Hence, we have

$$\langle \mathbf{p}_1, \mathbf{p}_5 \rangle = \langle \mathbf{p}_1, \mathbf{p}_8 \rangle = \langle \mathbf{p}_4, \mathbf{p}_8 \rangle = \langle \mathbf{p}_4, \mathbf{p}_5 \rangle.$$

This implies that $\mathbf{p}_1, \mathbf{p}_2, \mathbf{p}_3, \mathbf{p}_4$ lie in the plane bisecting S_{58}, and that $\mathbf{p}_5, \mathbf{p}_6, \mathbf{p}_7, \mathbf{p}_8$ lie in the plane bisecting S_{14}. Thus, $\mathbf{p}_1, \mathbf{p}_2, \mathbf{p}_3, \mathbf{p}_4$ and $\mathbf{p}_5, \mathbf{p}_6, \mathbf{p}_7, \mathbf{p}_8$ lie in orthogonal planes through $\mathbf{o}$. Without loss of generality, we may assume that for some $0 < \varphi, \tau < \frac{\pi}{3}$, we have

$$\mathbf{p}_1 = (\sin 3\varphi, 0, \cos 3\varphi), \qquad \mathbf{p}_5 = (0, -\sin 3\tau, -\cos 3\tau),$$

$$\mathbf{p}_2 = (\sin \varphi, 0, \cos \varphi), \qquad \mathbf{p}_6 = (0, -\sin \tau, -\cos \tau),$$

$$\mathbf{p}_3 = (-\sin \varphi, 0, \cos \varphi), \qquad \mathbf{p}_7 = (0, \sin \tau, -\cos \tau),$$

$$\mathbf{p}_4 = (-\sin 3\varphi, 0, \cos 3\varphi), \qquad \mathbf{p}_8 = (0, \sin 3\tau, -\cos 3\tau).$$

Substituting these into (6.9), we obtain

$$(\cos 2\varphi + \cos 3\varphi \cos \tau) \sin 3\tau - \cos 3\varphi \sin \tau (\cos 3\tau - \cos \tau) = 0,$$

or

$$3 \cos 3\varphi \sin 2\tau + 2 \cos 2\varphi \sin 3\tau = 0.$$

Applying Lemma 135 for $\mathbf{p}_5$ yields an expression similar to (6.9), from which we have

$$3 \cos 3\tau \sin 2\varphi + 2 \cos 2\tau \sin 3\varphi = 0.$$

Rewriting these expressions we obtain

$$3 \cos \varphi (4 \cos^2 \varphi - 3) \cos \tau + (2 \cos^2 \varphi - 1)(4 \cos^2 \tau - 1) = 0, \tag{6.12}$$

$$3 \cos \tau (4 \cos^2 \tau - 3) \cos \varphi + (2 \cos^2 \tau - 1)(4 \cos^2 \varphi - 1) = 0.$$

Subtracting and factoring yields

$$(2 \cos \varphi \cos \tau + 1)(\cos^2 \varphi - \cos^2 \tau) = 0.$$

From this, $\varphi = \tau$, and the solution of this equation is $\cos \varphi = \sqrt{\frac{15+\sqrt{145}}{40}}$.
□

Now we prove Theorem 14 for $n = 8$.

Let $\mathbf{P} = \mathbf{C}(\mathbf{p}_1, \mathbf{p}_2, \ldots, \mathbf{p}_8)$ be a polyhedron in $\mathcal{P}_3(8)$ that satisfies Property Z. Then $\mathbf{P}$ contains $\mathbf{o}$ and is simplicial. By the argument in the first paragraph of this subsection and by Lemma 139, if $\mathbf{P}$ is a polyhedron of maximal volume in $\mathcal{P}_3(8)$ not having 3-valent vertices, then $\mathbf{P}$ is congruent to the polyhedron in (ii) of Theorem 14, and its volume is $\sqrt{\frac{475+29\sqrt{145}}{250}} = 1.815716\ldots$. Thus, assume that at least one vertex of $\mathbf{P}$ is 3-valent. Using the idea of the proof of Theorem 14 for $n = 7$, we have that

$$\mathrm{vol}_3(\mathbf{P}) \leq \frac{\sqrt{3}}{4}\left(1 - \frac{1}{3}\tan^2\frac{2\pi - \tau}{6}\right) + \frac{9}{4}\tan\frac{14\pi + \tau}{54}\left(1 - \frac{1}{3}\tan^2\frac{14\pi + \tau}{54}\right).$$

This function is concave, and its maximum is strictly less than $\sqrt{\frac{475+29\sqrt{145}}{250}}$.

6.4 Proofs of Theorems 16, 17 and 18

First, in Subsection 6.4.1, we prove Theorem 16 and derive some lemmas from it that will help us prove Theorems 17 and 18. Then, in Subsection 6.4.2 we prove these two theorems.

6.4.1 Proof of Theorem 16 and some lemmas for Theorems 17 and 18

Proof of Theorem 16. We first recall some notation. Consider some d-polytope $\mathbf{P}$ inscribed in $\mathbb{S}^{d-1}$ and satisfying Property Z, that is, assume that no perturbation of any of its vertices increases its volume. Let $\mathbf{p} \in V(\mathbf{P})$ be a vertex of $\mathbf{P}$, and let $\mathcal{F}_{\mathbf{p}}$ denote the family of the facets of the triangulation $\mathcal{C}(\mathbf{P})$ of bd$\mathbf{P}$, containing $\mathbf{p}$. For any facet $\mathbf{F} \in \mathcal{F}_{\mathbf{p}}$, we set $A(\mathbf{F}, \mathbf{p}) = \mathrm{vol}_{d-1}(\mathrm{conv}(V(\mathbf{F}) \cup \{\mathbf{o}\} \setminus \{\mathbf{p}\}))$, and let $\mathbf{m}(\mathbf{F}, \mathbf{p})$ be the unit normal vector of the hyperplane, spanned by $V(\mathbf{F}) \cup \{\mathbf{o}\} \setminus \{\mathbf{p}\}$, pointing in the direction of the half space containing $\mathbf{p}$. Recall that a facial simplex of $\mathcal{C}(\mathbf{P})$ is the convex hull of $\mathbf{o}$ and a facet of $\mathcal{C}(\mathbf{P})$.

To prove the first part of the theorem, we need to prove that $\mathbf{p} = \frac{\mathbf{m}}{||\mathbf{m}||}$, where $\mathbf{m} = \sum_{\mathbf{F} \in \mathcal{F}_{\mathbf{p}}} A(\mathbf{F}, \mathbf{p})\mathbf{m}(\mathbf{F}, \mathbf{p})$. Using the triangulation $\mathcal{C}(\mathbf{P})$, we have that for any $\mathbf{p} \in V(\mathbf{P})$, the volume of $\mathbf{P}$ can be written as

$$\mathrm{vol}_d(\mathbf{P}) = V + \sum_{\mathbf{F} \in \mathcal{F}_{\mathbf{p}}} \mathrm{vol}_d\left(\mathrm{conv}(\mathbf{F} \cup \{\mathbf{o}\})\right) = V + \frac{1}{d}\sum_{\mathbf{F} \in \mathcal{F}_{\mathbf{p}}} A(\mathbf{F}, \mathbf{p})\langle \mathbf{m}(\mathbf{F}, \mathbf{p}), \mathbf{p}\rangle,$$

where V is the sum of the volumes of the facial simplices of $\mathcal{C}(\mathbf{P})$, *not* containing $\mathbf{p}$. From this it follows that

$$\mathrm{vol}_d(\mathbf{P}) = V + \frac{1}{d}\langle \mathbf{p}, \mathbf{m}\rangle.$$

Now, since $\mathbf{p} \in \mathbb{S}^{d-1}$, if $\langle \mathbf{p}, \mathbf{m}\rangle \; ||\mathbf{m}||$, then for any open set $\mathbf{U} \subset \mathbb{S}^{d-1}$ containing $\mathbf{p}$, there is a point $\mathbf{q} \in \mathbf{U}$ such that $\langle \mathbf{p}, \mathbf{m}\rangle < \langle \mathbf{q}, \mathbf{m}\rangle$, implying that

$$\mathrm{vol}_d(\mathbf{P}) < V + \frac{1}{d}\langle \mathbf{q}, \mathbf{m}\rangle \leq \mathrm{vol}_d\left(\mathrm{conv}\left(((V(\mathbf{P}) \cup \{\mathbf{q}\}) \setminus \{\mathbf{p}\})\right)\right),$$

which contradicts our assumption that $\mathbf{P}$ satisfies Property Z. Thus, we have $\langle \mathbf{p}, \mathbf{m} \rangle = ||\mathbf{m}||$, or in other words, $\mathbf{p} = \frac{\mathbf{m}}{||\mathbf{m}||}$.

Now we prove the second part, and assume that $\mathbf{P}$ is not simplicial, that is, that some facet $\mathbf{F}$ of $\mathbf{P}$ is not a simplex. First, we consider the case that $\mathbf{F}$ has $d+1$ vertices, and, as a $(d-1)$-polytope, it is simplicial. Then $\mathbf{F}$ can be written in the form $\mathbf{F} = \mathrm{conv}(\mathbf{S}_1 \cup \mathbf{S}_2$, where $\mathbf{S}_1$ and $\mathbf{S}_2$ are two simplices with $\dim \mathbf{S}_1 + \dim \mathbf{S}_2 = \dim \mathbf{F} = d-1$, and $\mathbf{S}_1 \cap \mathbf{S}_2$ is a singleton $\{\mathbf{x}\}$ in the relative interiors of both $\mathbf{S}_1$ and $\mathbf{S}_2$ (cf. [131]). Let $V(\mathbf{S}_1) = \{\mathbf{p}_i : i = 1, 2, \ldots, m\}$, and $V(\mathbf{S}_2) = \{\mathbf{q}_i : i = 1, 2, \ldots, d+1-m\}$. Then, we may triangulate $\mathbf{F}$ in two different ways:

$$\mathbf{F}_1 = \{\mathrm{conv}(V(\mathbf{F} \setminus \{\mathbf{p}_i\}) : i = 1, 2, \ldots, m\}$$

and

$$\mathbf{F}_2 = \{\mathrm{conv}(V(\mathbf{F} \setminus \{\mathbf{q}_i\}) : i = 1, 2, \ldots, d+1-m\}.$$

Observe that the union of the elements of $\mathbf{F}_1$ containing $\mathbf{p}_1$ is the closure of $\mathbf{F} \setminus \mathrm{conv}\,(V(\mathbf{F}) \setminus \{\mathbf{p}_1\})$, whereas for $\mathbf{F}_2$ it is $\mathbf{F}$. Recall that for any simplex in $\mathbb{E}^s$, with external unit facet normals $\mathbf{m}_1, \mathbf{m}_2, \ldots, \mathbf{m}_{s+1}$ belonging to the facets $\mathbf{F}_1, \mathbf{F}_2, \ldots, \mathbf{F}_{s+1}$, respectively, we have

$$\sum_{i=1}^{s+1} \mathrm{vol}_{s-1}(\mathbf{F}_i)\mathbf{m}_i = \mathbf{o}. \tag{6.13}$$

On the other hand, since the quantity in (i) must be independent from the triangulation, by (6.13) we have reached a contradiction. We remark that if $\mathbf{F}$ is not simplicial, then $\mathbf{F}$ is a $(d-k-1)$-fold pyramid over a k-polytope with $(k+2)$ vertices (cf. [131]), for which a straightforward modification of our argument yields the statement.

Finally, we consider the case that $\mathbf{F}$ has more than $d+1$ vertices. Choose a set S of $d+1$ vertices of $\mathbf{F}$. Note that any triangulation of $\mathrm{conv} S$ for any $S \subset V(\mathbf{F})$ can be extended to a triangulation of $\mathbf{F}$; this can be easily shown by induction. Thus, the assertion follows by applying the argument of the previous paragraph for $\mathrm{conv} S$. □

We state three consequences of Theorem 16, the third of which will play a crucial role in the proofs of Theorems 17 and 18.

Corollary 140 *If $\mathbf{P}$ is simplex satisfying Property Z, then $\mathbf{P}$ is a regular simplex inscribed in $\mathbb{S}^{d-1}$.*

Remark 141 *Assume that $\mathbf{P}$ is a d-dimensional convex polytope satisfying Property Z, and for some $\mathbf{p} \in V(\mathbf{P})$, all the vertices of $\mathbf{P}$ adjacent to $\mathbf{p}$ are contained in a hyperplane H. Then the supporting hyperplane of $\mathbb{S}^{d-1}$ at $\mathbf{p}$ is parallel to H, or in other words, $\mathbf{p}$ is a normal vector to H. Thus, in this case all the edges of $\mathbf{P}$, starting at $\mathbf{p}$, are of equal length.*

Lemma 142 *Let* $\mathbf{P}$ *be a convex d-dimensional polytope satisfying Property Z, and let* $\mathbf{p} \in V(\mathbf{P})$. *Let* $\mathbf{q}_1, \mathbf{q}_2 \in V(\mathbf{P})$ *be adjacent to* $\mathbf{p}$. *Assume that any facet of* $\mathbf{P}$ *containing* $\mathbf{p}$ *contains at least one of* $\mathbf{q}_1$ *and* $\mathbf{q}_2$, *and for any* $S \subset V(\mathbf{P})$ *of cardinality* $d-2$, $\operatorname{conv}(S \cup \{\mathbf{p}, \mathbf{q}_1\})$ *is a facet of* $\mathbf{P}$ *not containing* $\mathbf{q}_2$ *if and only if* $\operatorname{conv}(S \cup \{\mathbf{p}, \mathbf{q}_2\})$ *is a facet of* $\mathbf{P}$ *not containing* $\mathbf{q}_1$. *Then* $||\mathbf{q}_1 - \mathbf{p}|| = ||\mathbf{q}_2 - \mathbf{p}||$.

Proof.
Let

$$U = \{S \subset V(\mathbf{P}) : \operatorname{conv}(S \cup \{\mathbf{p}, \mathbf{q}_1\}) \text{ is a facet of } \mathbf{P}, \text{ but } \mathbf{p}, \mathbf{q}_1, \mathbf{q}_2 \notin S\},$$

and let

$$W = \{S \subset V(\mathbf{P}) : \operatorname{conv}(S \cup \{\mathbf{p}, \mathbf{q}_1, \mathbf{q}_2\}) \text{ is a facet of } \mathbf{P}, \text{ but } \mathbf{p}, \mathbf{q}_1, \mathbf{q}_2 \notin S\}.$$

Note that by our conditions, U is also the family of subsets of $V(\mathbf{P})$, not containing $\mathbf{p}, \mathbf{q}_1, \mathbf{q}_2$, such that $\operatorname{conv}(S \cup \{\mathbf{p}, \mathbf{q}_2\})$ is a facet of $\mathbf{P}$.

Then, using a suitable labelling of the vertices of $\mathbf{P}$, the total volume V of the facial simplices containing $\mathbf{p}$ can be written as

$$V = \frac{1}{d!} \sum_{\{\mathbf{x}_{i_1}, \ldots, \mathbf{x}_{i_{d-2}}\} \in U} \left(|\mathbf{p}, \mathbf{q}_1, \mathbf{x}_{i_1}, \ldots, \mathbf{x}_{i_{d-2}}| - |\mathbf{p}, \mathbf{q}_2, \mathbf{x}_{i_1}, \ldots, \mathbf{x}_{i_{d-2}}| \right)$$

$$+ \frac{1}{d!} \sum_{\{\mathbf{x}_{i_1}, \ldots, \mathbf{x}_{i_{d-3}}\} \in W} |\mathbf{p}, \mathbf{q}_1, \mathbf{q}_2, \mathbf{x}_{i_1}, \ldots, \mathbf{x}_{i_{d-3}}|,$$

where, for simplicity, we denote by $|\mathbf{x}_1, \mathbf{x}_2, \ldots, \mathbf{x}_d|$ the determinant of the $d \times d$ matrix with the vectors $\mathbf{x}_i \in \mathbb{E}^d$ as columns.

Observe that for any $\{\mathbf{x}_{i_1}, \ldots, \mathbf{x}_{i_{d-3}}\} \in W$, we have $|\mathbf{p}, \mathbf{q}_2, \mathbf{q}_2, \mathbf{x}_{i_1}, \ldots, \mathbf{x}_{i_{d-3}}| = 0$, which, for every element of W, we may subtract from V without changing its value. Thus, for some suitable finite set $X \subset (\mathbb{E}^d)^{d-2}$ of $(d-2)$-tuples of points in $\mathbb{E}^d$, we have

$$V = \frac{1}{d!} \sum_{(\mathbf{y}_1, \ldots, \mathbf{y}_{d-2}) \in X} |\mathbf{p}, \mathbf{q}_1 - \mathbf{q}_2, \mathbf{y}_1, \ldots, \mathbf{y}_{d-2}|.$$

Let $f : \mathbb{E}^d \to \mathbb{R}$ be the linear functional $f(\mathbf{x}) = \frac{1}{d!} \sum_{(\mathbf{y}_1, \ldots, \mathbf{y}_{d-2}) \in X} |\mathbf{x}, \mathbf{q}_1 - \mathbf{q}_2, \mathbf{y}_1, \ldots, \mathbf{y}_{d-2}|$. Since $\mathbf{P}$ satisfies Property Z, we have that $\mathbf{p}$ is a normal vector of the hyperplane $\{\mathbf{x} \in \mathbb{E}^d : f(\mathbf{x}) = 0\}$. On the other hand, $f(\mathbf{q}_2 - \mathbf{q}_1) = 0$, due to the properties of determinants. Thus, $\mathbf{q}_2 - \mathbf{q}_1$ and $\mathbf{p}$ are perpendicular, from which it readily follows that $\langle \mathbf{p}, \mathbf{q}_1 \rangle = \langle \mathbf{p}, \mathbf{q}_2 \rangle$, and hence, $||\mathbf{q}_1 - \mathbf{p}|| = ||\mathbf{q}_2 - \mathbf{p}||$. $\square$

6.4.2 Proofs of Theorems 17 and 18

In the proof of our results on d-polytopes with $d+2$ or $d+3$ vertices, we use extensively the properties of the so-called Gale transform of a polytope (cf. [131] or [239]).

Consider a d-polytope $\mathbf{P}$ with vertex set $V(\mathbf{P}) = \{\mathbf{p}_i : i = 1, 2, \dots, n\}$. Regarding $\mathbb{E}^d$ as the hyperplane $\{x_{d+1} = 1\}$ of $\mathbb{E}^{d+1}$, we can represent $V(\mathbf{P})$ as a $(d+1) \times n$ matrix M, in which each column lists the coordinates of a corresponding vertex in the standard basis of $\mathbb{E}^{d+1}$. Clearly, this matrix has rank $d+1$, and thus, it defines a linear mapping $L : \mathbb{E}^n \to \mathbb{E}^{d+1}$, with $\dim \ker L = n-d-1$. Consider a basis $\{\mathbf{w}_1, \mathbf{w}_2, \dots, \mathbf{w}_{n-d-1}\}$ of $\ker L$, and let $\bar{L} : \mathbb{E}^{n-d-1} \to \mathbb{E}^n$ be the linear map mapping the ith vector of the standard basis of $\mathbb{E}^{n-d-1}$ into w_i. Then the matrix $\bar{M}$ of $\bar{L}$ is an $n \times (n-d-1)$ matrix of (maximal) rank $n-d-1$, satisfying the equation $M\bar{M} = O$, where O is the matrix with all entries equal to zero. Note that the rows of $\bar{M}$ can be represented as points of $\mathbb{E}^{n-d-1}$. For any vertex $\mathbf{p}_i \in V(\mathbf{P})$, we call the ith row of $\bar{M}$ the *Gale transform of* $\mathbf{p}_i$, and denote it by $\bar{\mathbf{p}}_i$. Furthermore, the n-element multiset $\{\bar{\mathbf{p}}_i : i = 1, 2, \dots, n\} \subset \mathbb{E}^{n-d-1}$ is called the *Gale transform of* $\mathbf{P}$, and is denoted by $\bar{P}$. If $\operatorname{conv} S$ is a face of $\mathbf{P}$ for some $S \subset V(\mathbf{P})$, then the (multi)set of the Gale transform of the points of S is called a face of $\bar{P}$. If $\bar{S}$ is a face of $\bar{P}$, then $\bar{P} \setminus \bar{S}$ is called a *coface* of $\bar{P}$.

Let $V = \{\mathbf{q}_i : i = 1, 2, \dots, n\} \subset \mathbb{E}^{n-d-1}$ be a (multi)set. We say that V is a *Gale diagram* of $\mathbf{P}$, if for some Gale transform P' the conditions $\mathbf{o} \in \operatorname{relint}\operatorname{conv}\{\mathbf{q}_j : j \in I\}$ and $\mathbf{o} \in \operatorname{relint}\operatorname{conv}\{\bar{\mathbf{p}}_j : j \in I\}$ are satisfied for the same subsets of $\{1, 2, \dots, n\}$. If $V \subset \mathbb{S}^{n-d-2}$, then V is a *normalized Gale diagram* (cf. [169]). A *standard Gale diagram* is a normalized Gale diagram in which the consecutive diameters are equidistant. A *contracted Gale diagram* is a standard Gale diagram which has the least possible number of diameters among all isomorphic diagrams. We note that each d-polytope with at most $d+3$ vertices may be represented by a contracted Gale diagram (cf. [131] or [239]).

In the proofs, we need the following theorem from [131] or also from [239].

Theorem 143 *We have the following.*

(i) A multiset $\bar{P}$ of n points in $\mathbb{E}^{n-d-1}$ is a Gale diagram of a d-polytope $\mathbf{P}$ with n vertices if and only if every open half space in $\mathbb{E}^{n-d-1}$ bounded by a hyperplane through $\mathbf{o}$ contains at least two points of $\bar{P}$ (or, alternatively, all the points of $\bar{P}$ coincide with $\mathbf{o}$ and then $n = d+1$ and $\mathbf{P}$ is a d-simplex).

(ii) If $\mathbf{F}$ is a facet of $\mathbf{P}$, and Z is the corresponding coface, then in any Gale diagram $\bar{P}$ of $\mathbf{P}$, $\bar{Z}$ is the set of vertices of a (non-degenerate) set with $\mathbf{o}$ in the relative interior of its convex hull.

(iii) A polytope $\mathbf{P}$ is simplicial if and only if, for every hyperplane H containing $\mathbf{o} \in \mathbb{E}^{n-d-1}$, we have $\mathbf{o} \notin \operatorname{relint}\operatorname{conv}(\bar{P} \cap H)$.

(iv) *A polytope* $\mathbf{P}$ *is a pyramid if and only if at least one point of* $\bar{P}$ *coincides with the origin* $\mathbf{o} \in \mathbb{E}^{n-d-1}$.

Remark 144 *We note that (ii) can be stated in a more general form:* $\mathbf{F}$ *is a face of* $\mathbf{P}$ *if and only if for the corresponding coface* Z *of* $\mathbf{P}$*, we have* $\mathbf{o} \in \text{relintconv}\bar{Z}$*.*

Proof of Theorem 17. Without loss of generality, we may assume that $d \geq 3$, as otherwise the assertion is trivial. Let $\mathbf{P}$ be a polytope, inscribed in $\mathbb{S}^{d-1}$, with $d+2$ vertices and satisfying $\text{vol}_d(\mathbf{P}) = v_d(d+2)$.

In the proof we use a contracted Gale diagram $\bar{P}$ of $\mathbf{P}$. Since by Theorem 16 $\mathbf{P}$ is simplicial, and since $d+2-d-1=1$, (iii) of Theorem 143 yields that $\bar{P}$ consists of the points -1 and 1 on the real line. We may assume that the multiplicity of -1 is $k+1$ and that of 1 is $d+1-k$. From (i) of Theorem 143, it follows that $2 \leq k \leq d$. Without loss of generality, we may assume that $k+1 \leq d+1-k$, or in other words, that $k \leq \lfloor \frac{d}{2} \rfloor$. By (ii) of Theorem 143, the facets of $\bar{P}$ are the complements of the pairs of the form $\{-1,1\}$.

Let $V_+(\mathbf{P})$ be the set of vertices of $\mathbf{P}$ represented by 1 in $\bar{P}$, and let $V_-(\mathbf{P}) = V(\mathbf{P}) \setminus V_+(\mathbf{P})$. Consider any $\mathbf{p} \in V_+(P)$ and $\mathbf{q}_1, \mathbf{q}_2 \in V_-(P)$. Observe that both $\mathbf{q}_1$ and $\mathbf{q}_2$ are adjacent to $\mathbf{p}$. Furthermore, for $\mathbf{P}$ and these three vertices the conditions of Lemma 142 are satisfied, which yields that $||\mathbf{q}_2 - \mathbf{p}|| = ||\mathbf{q}_1 - \mathbf{p}||$. Hence, there is some $\delta > 0$ such that for any $\mathbf{p} \in V_+(P)$ and $\mathbf{q} \in V_-(P)$, we have $||\mathbf{q} - \mathbf{p}|| = \delta$. Thus, $V_+(\mathbf{P})$ and $V_-(\mathbf{P})$ are contained in orthogonal linear subspaces. Since $\mathbf{P}$ is d-dimensional, it follows that these subspaces are orthogonal complements of each other.

Let $\mathbf{P}_1 = \text{conv}V_+(\mathbf{P})$ and $\mathbf{P}_2 = \text{conv}V_-(\mathbf{P})$. Then $\mathbf{P}_1$ is a k-dimensional, and $\mathbf{P}_2$ is a $(d-k)$-dimensional simplex, and we have

$$\text{vol}_d(\mathbf{P}) = \frac{k!(d-k)!}{d!} \text{vol}_k(\mathbf{P}_1) \text{vol}_{d-k}(\mathbf{P}_2).$$

To find the simplices of maximal volume inscribed in $\mathbb{S}^{k-1}$ and $\mathbb{S}^{d-k-1}$, we may observe that these simplices satisfy Property Z. Thus, we can apply Corollary 140, which yields that $\mathbf{P}_1$ and $\mathbf{P}_2$ are regular.

It is well known (and can be easily computed from its standard representation in $\mathbb{E}^k$) that the volume of a regular k-dimensional simplex inscribed in $\mathbb{S}^{k-1}$ is $\frac{(k+1)^{\frac{k+1}{2}}}{k^{\frac{k}{2}} k!}$. Hence, we have

$$\text{vol}_d(\mathbf{P}) = \frac{k!(d-k)!}{d!} \frac{(k+1)^{\frac{k+1}{2}}}{k^{\frac{k}{2}} k!} \frac{(d-k+1)^{\frac{d-k+1}{2}}}{(d-k)^{\frac{d-k}{2}} (d-k)!},$$

or equivalently,

$$\text{vol}_d(\mathbf{P}) = \frac{1}{d!} \sqrt{\left(1 + \frac{1}{k}\right)^k \left(1 + \frac{1}{d-k}\right)^{d-k}} \sqrt{(k+1)(d-k+1)}.$$

We need to maximize this quantity for $k = 1, 2, \ldots, \lfloor \frac{d}{2} \rfloor$. If d is even, the assertion follows from the inequality for the arithmetic and the geometric means. If d is odd, we may use the strict concavity of the function $x \mapsto x \log\left(1 + \frac{1}{x}\right)$, $x > 0$. □

Remark 145 *The use of Gale diagrams in the proof of Theorem 17 can be avoided if we recall the fact that any simplicial d-polytope is the convex hull of two simplices, having a single point, contained in the relative interiors of both simplices, as their intersection (cf. [239]). The vertex sets of these simplices form the unique Radon partition of the point set.*

Proof of Theorem 18. Assume that $\mathbf{P}$ is a d-polytope, inscribed in $\mathbb{S}^{d-1}$ and satisfying Property Z.

To prove the assertion, we use a contracted Gale diagram $\bar{P}$ of $\mathbf{P}$. Since by Theorem 16 $\mathbf{P}$ is simplicial, $\bar{P}$ is a multiset consisting of the vertices of a regular $(2k+1)$-gon $G(\mathbf{P})$, with $k \geq 1$ and the origin $\mathbf{o} \in \mathbb{E}^2$ as its center, such that the multiplicity of each vertex is at least one, and the sum of their multiplicities is $d + 3$.

Applying Remark 144, we have that $\mathbf{p}, \mathbf{q} \in V(\mathbf{P})$ are not adjacent if and only if there is an open half plane, containing $\mathbf{o}$ in its boundary, that contains only the two points $\bar{\mathbf{p}}, \bar{\mathbf{q}}$ of $\bar{P}$. In this case we have one of the following:

(i) $k = 2$, and the points are two consecutive vertices of the pentagon $G(\mathbf{P})$, with multiplicity one.

(ii) $k = 1$, and the points are either consecutive vertices of $G(\mathbf{P})$ with multiplicity one, or belong to the same vertex of $G(\mathbf{P})$, which has multiplicity exactly two.

Now, consider the case that some point of $\bar{P}$ has multiplicity greater than one, and let $\mathbf{p}_1, \mathbf{p}_2, \ldots, \mathbf{p}_m \in V(\mathbf{P})$ be represented by this point. We set $V_1 = \{\mathbf{p}_1, \mathbf{p}_2, \ldots, \mathbf{p}_m\}$ and $V_2 = V(\mathbf{P}) \setminus V_1$. From the observation in the previous paragraph, it follows that each vertex in V_1 is connected to every vertex in V_2 by an edge, and for any two vertices in V_1, any facet contains at least one of them. Furthermore, if a facet of $\mathbf{P}$ contains exactly $s \geq 1$ elements of V_1, then, replacing them with any other s distinct vertices from V_1 we obtain another facet of $\mathbf{P}$. Thus, we may apply Lemma 142, which, by the simplicity of $\mathbf{P}$, yields that the linear hulls L_1 and L_2 of V_1 and V_2, respectively, are orthogonal. Clearly, we may assume that the sum of the dimensions of these two subspaces is d, as $\mathbf{P}$ is d-dimensional. Hence, we have either $\dim L_1 = m-1$ and $\dim L_2 = d + 1 - m$, or $\dim L_1 = m - 2$ and $\dim L_2 = d + 2 - m$. Note that in the first case $\operatorname{conv} V_1$ is a simplex, and in the second one $\operatorname{conv} V_2$ is a simplex.

Observe that since $\mathbf{P}$ satisfies Property Z, then both $\operatorname{conv} V_1$ and $\operatorname{conv} V_2$ satisfy it in their linear hulls, as otherwise a slight modification of either V_1 or V_2 would yield a polytope $\mathbf{P}' \in$ inscribed in $\mathbb{S}^{d-1}$, having $d+3$ vertices and satisfying $\operatorname{vol}_d(\mathbf{P}) < \operatorname{vol}_d(\mathbf{P}')$, contradicting the definition of Property

Z. Thus, by Corollary 140 and Theorem 17, we have that one of convV_1 and convV_2 is a regular simplex, and the other one is the convex hull of two regular simplices, contained in orthogonal linear subspaces. Hence, **P** is the convex hull of three regular simplices, contained in pairwise orthogonal linear subspaces, and in this case the assertion follows from the argument in the proof of Theorem 17.

Observe that since $\bar{P}$ consists of an odd number of points if we do not count multiplicity, if d is odd, then some vertex of $\bar{P}$ has multiplicity strictly greater than one, and thus, in this case the assertion readily follows. Assume that every vertex of $\bar{P}$ has multiplicity one. Then d is even, and $\bar{P}$ is the vertex set of a regular $d+3$-gon. We need to show only that P is cyclic.

We recall that every cyclic d-polytope is neighborly, that is, the convex hull of any at most $\frac{d}{2}$ vertices is a face of the polytope. Furthermore, every neighborly d-polytope with $n \geq d+3$ vertices is cyclic if d is even (cf. [131]), whereas in odd dimensions there are non-cyclic neighborly polytopes. Thus, we show only that P is neighborly. Indeed, for any $\frac{d}{2}$ vertices of **P**, it is clear that the convex hull of the points of $\bar{P}$ corresponding to the remaining $\frac{d}{2}+3$ vertices of **P** contains **o** in its interior, since every open half plane, containing o in its boundary, contains either $\frac{d}{2}+1$ or $\frac{d}{2}+2$ points of $\bar{P}$.

The rest of the assertion follows from the volume estimates in the proof of Theorem 17. □

6.5 Proof of Theorem 21

As in the formulation of Theorem 21, let $d \geq 2$, and $\mathbf{P} \subset \mathbb{E}^d$ be a d-dimensional convex polytope with vertices $\mathbf{p}_1, \mathbf{p}_2, \ldots, \mathbf{p}_n$, $n \geq 5$, and $n > d$.

First, note that (ii) or (iii) clearly implies (i). We prove that (i) yields (ii).

As a first step, we show that the points $\mathbf{p}_1, \ldots, \mathbf{p}_{d+1}$ are affinely independent; in particular, we show, by induction on s, that for any $2 \leq s \leq d+1$, $\mathrm{aff}\{\mathbf{p}_1, \mathbf{p}_2, \ldots, \mathbf{p}_s\}$ is an $(s-1)$-flat. First, if $\mathbf{p}_1 = \mathbf{p}_2$, then **P** is a single point, a contradiction, and thus, the statement holds for $s = 2$. Now we assume that $\mathrm{aff}\{\mathbf{p}_1, \ldots, \mathbf{p}_s\}$ is an $(s-1)$-flat for some $2 \leq s \leq d$, and show that $\mathrm{aff}\{\mathbf{p}_1, \ldots, \mathbf{p}_{s+1}\}$ is an s-flat. Observe that by (i), for every integer j, $\mathrm{aff}\{\mathbf{p}_{j+1}, \ldots, \mathbf{p}_{j+s}\}$ is also an $(s-1)$-flat. On the other hand, if $\mathrm{aff}\{\mathbf{p}_1, \ldots, \mathbf{p}_{s+1}\}$ is not an s-flat, then $\mathrm{aff}\{\mathbf{p}_1, \ldots, \mathbf{p}_{s+1}\} = \mathrm{aff}\{\mathbf{p}_1, \ldots, \mathbf{p}_s\} = \mathrm{aff}\{\mathbf{p}_2, \ldots, \mathbf{p}_{s+1}\}$, which, by (i) yields that $\mathcal{P} \subset \mathrm{aff}\{\mathbf{p}_1, \ldots, \mathbf{p}_s\}$, a contradiction. Thus, we have that $\mathbf{p}_1, \ldots, \mathbf{p}_{d+1}$ are affinely independent.

Let $\phi : \mathbb{E}^d \to \mathbb{E}^d$ be the affine transformation defined by $\phi(\mathbf{p}_s) = \mathbf{p}_{s+1}$ for $s = 1, 2, \ldots, d+1$. Since $\mathrm{conv}\{\mathbf{p}_1, \ldots, \mathbf{p}_{d+1}\}$ and $\mathrm{conv}\{\mathbf{p}_2, \ldots, \mathbf{p}_{d+2}\}$ are congruent, ϕ is a congruence. Note that as $\mathbf{p}_1, \ldots, \mathbf{p}_{d+1}$ are affinely independent, for any $\mathbf{q} \in \mathbb{E}^d$, the distances of $\mathbf{q}$ from these points determine $\mathbf{q}$. Thus, for any integer j, we have $\phi(\mathbf{p}_j) = \mathbf{p}_{j+1}$, and (ii) holds.

Finally, we prove that (ii) yields (iii). Without loss of generality, let $\mathbf{B}^d$ be the unique smallest ball that contains $\mathbf{P}$. Then $\mathbf{B}^d$ is the smallest ball containing $\phi(\mathbf{P})$ as well. Thus, ϕ is an isometry preserving $\mathbf{B}^d$, from which it follows that $\mathbf{p}_j \in \mathbb{S}^{d-1}$ if and only if $\mathbf{p}_{j+1} \in \mathbb{S}^{d-1}$. This implies that $\mathbf{P}$ is inscribed in $\mathbb{S}^{d-1}$.

We present two different arguments that finish the proof from this point.

First proof of (iii).

Let $\mathbf{E}$ be the unique smallest volume ellipsoid containing $\mathcal{P}$. Since $\mathbf{E}$ is unique, $\mathrm{Sym}(\mathbf{P}) \leq \mathrm{Sym}(\mathbf{E})$. On the other hand, the only ellipsoids whose symmetry groups contain an element ϕ of order $n \geq 5$ such that for some $\mathbf{p} \in \mathbb{E}^d$, the affine hull of the orbit of $\mathbf{p}$ is $\mathbb{E}^d$, are balls. Thus, without loss of generality, we may assume that $\mathbf{E} = \mathbf{B}^d$. We use the following, well-known properties of the smallest volume ellipsoid circumscribed about $\mathbf{P}$ (cf., e.g., [141]).

Theorem 146 *Let $\mathbf{K} \subset \mathbf{B}^d$ be a compact, convex set. Then $\mathbf{B}^d$ is the smallest volume ellipsoid circumscribed about $\mathbf{K}$ if and only if for some $d \leq n \leq \frac{d(d+3)}{2}$ and $k = 1, \ldots, n$, there are $\mathbf{v}_k \in \mathbb{S}^{d-1} \cap \mathrm{bd}\mathbf{K}$ and $\lambda_k > 0$ such that*

$$0 = \sum_{k=1}^{n} \lambda_k \mathbf{v}_k, \quad \mathrm{Id} = \sum_{k=1}^{n} \lambda_k \mathbf{v}_k \otimes \mathbf{v}_k, \tag{6.14}$$

where Id is the d-dimensional identity matrix, and for $\mathbf{x}, \mathbf{y} \in \mathbb{E}^d$, $\mathbf{x} \otimes \mathbf{y}$ denotes the $d \times d$ matrix $\mathbf{x}\mathbf{y}^T$.

Thus, since $\mathbf{P} \cap \mathbf{B}^d$ is the vertex set of $\mathbf{P}$, there are some *non-negative* coefficients $\lambda_1, \lambda_2, \ldots, \lambda_n$ which, together with the points $\mathbf{p}_1, \mathbf{p}_2, \ldots, \mathbf{p}_n$, satisfy the conditions in (6.14). We show that these points, with the coefficients $\lambda_1 = \lambda_2 = \ldots = \lambda_n = \frac{d}{n}$ also satisfy the conditions in (6.14). Indeed, set $0 < \lambda = \frac{\sum_{j=1}^{n} \lambda_j}{n}$. By (ii), we have that $\sum_{j=1}^{n} \lambda_{j+k} \mathbf{p}_j = 0$ for every integer k, and thus, $\sum_{j+1}^{n} \lambda \mathbf{p}_j = 0$. Similarly, since $\mathrm{Id} = \sum_{j=1}^{n} \lambda_{j+k} \mathbf{p}_j \otimes \mathbf{p}_j$ holds for every integer k, it follows that $\mathrm{Id} = \lambda \sum_{j=1}^{n} \mathbf{p}_j \otimes \mathbf{p}_j$. Since $||\mathbf{p}_j|| = 1$ for every value of j, this equality implies that $d = \mathrm{tr}(\mathrm{Id}) = n\lambda$, that is, $\lambda = \frac{d}{n}$.

In the following we set $\bar{\mathbf{p}}_k = \sqrt{\frac{d}{n}} \mathbf{p}_k$ for $k = 1, 2, \ldots, n$, and observe that

$$0 = \sum_{k=1}^{n} \bar{\mathbf{p}}_k, \quad \mathrm{Id} = \sum_{k=1}^{n} \bar{\mathbf{p}}_k \otimes \bar{\mathbf{p}}_k, \tag{6.15}$$

Let G be the Gram matrix of the vectors $\bar{\mathbf{p}}_k$, that is, $G_{jk} = \langle \bar{\mathbf{p}}_j, \bar{\mathbf{p}}_k \rangle$. Then, using an elementary algebraic transformation, from the second equality in (6.15) we obtain that $G^2 = G$, and thus, that G is the matrix of an orthogonal projection in $\mathbb{E}^n$ into a d-dimensional subspace, with rank d. This yields that G has two eigenvalues, 1 and 0, with multiplicities d and $n - d$, respectively, and, furthermore, we can write G in the form AA^T, where A is an $(n \times d)$ matrix, and the columns of A form an orthonormal system in $\mathbb{E}^d$. Equivalently,

G can be written in the form $G = BDB^T$, where D is the diagonal matrix in which the first d diagonal elements are equal to 1, and the last $n-d$ elements are equal to 0, and B is an orthogonal matrix in $\mathbb{E}^n$. Here, A is the matrix composed of the first d columns of B. Observe that $\bar{\mathbf{p}}_j$ is the jth row of A, or equivalently, if we extend A to an orthogonal matrix B, then the coordinates of $\bar{\mathbf{p}}_j$ are the first d coordinates of the jth vector in the orthonormal system formed by the rows of B.

By (ii), G is a circulant matrix. Let $(c_0, c_1, \ldots, c_{n-1})$ be the first row of G. Then the eigenvalues of G are $\mu_k = \sum_{j=0}^{n-1} c_j \epsilon^{jk}$, where $\epsilon = \cos\frac{2\pi}{n} + i \sin\frac{2\pi}{n}$, and the corresponding eigenvectors are $\mathbf{x}_k = \frac{1}{\sqrt{n}}(1, \epsilon^k, \ldots, \epsilon^{(n-1)k})$. Now all eigenvalues are 0 or 1, and hence, $\mu_k = 1$ if and only if $\mu_{n-k} = 1$. Thus, the real 2-flat spanned by $\mathbf{y}_k = \frac{1}{2}(\mathbf{x}_k + \mathbf{x}_{n-k})$ and $\mathbf{y}'_k = \frac{1}{2i}(\mathbf{x}_k - \mathbf{x}_{n-k})$ is contained in one of the two eigenspaces. Let F be the eigenspace associated to 1. Then $F^\perp$ is the eigenspace associated to 0. The definition of G and the fact that $\sum_{k=1}^n \bar{\mathbf{p}}_k = 0$ yield that $\sum_{k=1}^n c_k = 0$. Thus, $\mu_0 = 0$.

If n is even; that is, if $\mu_{n/2}$ exists, then $n-d$ is even if and only if d is even. Hence, $\mu_0 = 0$ implies that if n is even, then $\mu_{n/2} = 0$ if and only if d is even. This yields that if d is even, then F is spanned by pairs of vectors of the form $\mathbf{y}_k, \mathbf{y}'_k$, where $0 < k < \frac{n}{2}$, and if d is odd, then F is spanned by the vector $(1, -1, \ldots, (-1)^n)$, and by pairs of vectors of the form $\mathbf{y}_k, \mathbf{y}'_k$, where $0 < k < \frac{n}{2}$. This shows that for some values $0 < k_1 < k_2 < \ldots < k_{\lfloor d/2 \rfloor} < \frac{n}{2}$, the Gram matrix of the vertices of $\sqrt{\frac{d}{n}}\mathbf{Q}(k_1, k_2, \ldots, k_{\lfloor d/2 \rfloor})$ is equal to G. On the other hand, G determines the pairwise distances between the points $\bar{\mathbf{p}}_1, \bar{\mathbf{p}}_2, \ldots, \bar{\mathbf{p}}_n$. Thus, $\mathbf{P}$ is congruent to $\sqrt{\frac{d}{n}}\mathbf{Q}(k_1, k_2, \ldots, k_{\lfloor d/2 \rfloor})$, and the assertion readily follows.

Second proof of (iii).
Using the fact that $\mathbf{P}$ is inscribed in $\mathbb{S}^{d-1}$ it follows that ϕ in (ii) is an orthogonal linear transformation. Let its matrix be denoted by A. Since ϕ is invertible, the diagonal elements in the Jordan form of A are non-zero, from which it easily follows that A is diagonalizable over $\mathbb{C}$. As ϕ is a real matrix, its complex eigenvalues are either real, or pairs of conjugate non-real complex numbers. Note that if $\mathbf{z} = \mathbf{x} + i\mathbf{y} \in \mathbb{C}^d$ is an eigenvector of A associated to the eigenvalue $\lambda \notin \mathbb{R}$, then the linear subspace in $\mathbb{E}^d$, spanned by $\mathbf{x}$ and $\mathbf{y}$, is invariant under ϕ. Thus, in a suitable orthonormal basis, the matrix of ϕ is a block diagonal matrix, where each block is either 1×1 (belonging to a real eigenvalue), or 2×2 (belonging to a pair of conjugate complex eigenvalues). Let these blocks be $B_1, B_2, \ldots, B_k$.

As ϕ is orthogonal, each 1×1 block is either 1 or -1 (corresponding to the identity, and to the reflection about the origin, respectively), and since 2×2 blocks belong to non-real eigenvalues (that is, they are not diagonalizable as 2×2 matrices), they are 2-dimensional rotation matrices. From the fact that the *affine hull* of the vectors $\mathbf{p}_1, \ldots, \mathbf{p}_n$ is $\mathbb{E}^d$, it follows that there is no block equal to 1, and there is at most one block equal to -1. Thus, if d is even, then each block is a 2-dimensional rotation, and if d is odd, then one block is a reflection about the origin, and every other block is a 2-dimensional rotation. In the latter case, we can assume that the last block belongs to the reflection, that is, $B_{(d+1)/2} = -1$.

Clearly, without loss of generality, we may assume that each angle of rotation is strictly less than π. Thus, and since the order of each rotation is a divisor of n, for $i = 1, 2, \ldots, \lfloor d/2 \rfloor$, we have

$$B_i = \begin{bmatrix} \cos\frac{2k_i\pi}{n} & -\sin\frac{2k_i\pi}{n} \\ \sin\frac{2k_i\pi}{n} & \cos\frac{2k_i\pi}{n} \end{bmatrix}$$

for some $0 < k_i < \frac{n}{2}$. The fact that the affine hull of the points $\mathbf{p}_1, \mathbf{p}_2, \ldots, \mathbf{p}_n$ is $\mathbb{E}^d$ implies that the values k_i are pairwise different. Hence, setting $\mathbf{p}_n = \frac{1}{\sqrt{\lfloor\frac{d+1}{2}\rfloor}}(1, 0, 1, 0, \ldots)$, which we can do without loss of generality, and observing that $\mathbf{p}_i = \phi^i(\mathbf{p}_n)$, the assertion follows.

6.6 Proof of Theorem 22

Consider a 3-connected planar graph $G = (V, E)$ with face set F. A *primal-dual circle representation* of G is a collection of two families of circles $\{C_x : x \in V\}$ and $\{D_y : y \in F\}$ in $\mathbb{E}^2$, called *vertex circles* and *face circles*, respectively, satisfying the following conditions.

(i) The vertex circles C_x have pairwise disjoint interiors.

(ii) All face circles D_y are contained in the circle D_o, corresponding to the outer face $o \in F$, and all other face-circles have pairwise disjoint interiors.

Furthermore, for every edge $xx' \in E$, and the corresponding dual edge yy' in the dual graph of G (that is, the edge connecting the faces in F separated by xx')

(iii) Circles C_x and $C_{x'}$ are tangent at a point $\mathbf{p}$.

(iv) Circles D_y and $D_{y'}$ are tangent at the same point $\mathbf{p}$.

(v) The two tangent lines of the pairs $C_x, C_{x'}$ and $D_y, D_{y'}$ at $\mathbf{p}$ are perpendicular.

Theorem 22 can be formulated in a different form.

Theorem 147 *Every 3-connected plane graph G admits a primal-dual circle representation. Furthermore, this representation is unique up to Möbius transformations.*

First, we show that Theorems 22 and 147 are equivalent. Recall the famous theorem of Steinitz [228] that a graph is the edge graph of a convex polyhedron in $\mathbb{E}^3$ if and only if it is planar and 3-connected. Thus, these graphs correspond to the combinatorial classes of convex polyhedra. Now, if Theorem 147 holds, then projecting a primal-dual representation of such a graph G to $\mathbb{S}^2$ via stereographic projection yields a similar representation of G on $\mathbb{S}^2$. It is easy to see that the planes through the face circles of this representation are the face planes of a convex polyhedron $\mathbf{P}$ whose vertices lie on the rays all starting at $\mathbf{o}$, and each passing through the center of one of the vertex circles. Then the edge graph of $\mathbf{P}$ is G, and $\mathbf{P}$ is midscribed to $\mathbb{S}^2$. On the other hand, if $\mathbf{P}$ is midscribed to $\mathbb{S}^2$, then the tangent points on the edges generate two circle families on $\mathbb{S}^2$, one of them consisting of the inscribed circles of the faces of $\mathbf{P}$, and the other one consisting of all circles which pass through the tangent points of all edges adjacent to a given vertex. Projecting these families to $\mathbb{E}^2$ via stereographic projection, we obtain a primal-dual circle representation of the edge graph of $\mathbf{P}$. On the other hand, any Möbius transformation of $\mathbb{E}^2$ or $\mathbb{S}^2$ can be written as the composition of finitely many stereographic projections from $\mathbb{E}^2$ to $\mathbb{S}^2$, and from $\mathbb{S}^2$ to $\mathbb{E}^2$.

Hence, to prove Theorem 22, it suffices to prove Theorem 147. The proof of Theorem 147 we present here can be found in [112].

Consider any straight-line drawing of G. Using stereographic projection, this drawing can be projected onto $\mathbb{S}^2$. Note that every planar graph, or its dual, has a triangular face. This, since on $\mathbb{S}^2$ the roles of G and its dual can be switched, we may assume that by projecting back the representation from a suitable point, we obtain a representation in which the vertices of the outer face, which we call outer vertices, are the vertices of a regular triangle. In other words, we may restrict the proof to planar graphs whose outer face is a triangle.

Assume, for the moment, that we already found a primal-dual circle representation of G. This representation gives rise to a straight-line representation of G as well as its dual graph G^*. In the representation of G^* we imagine the dual of the outer face as the point at infinity, and edges adjacent to it as rays. The superposition of these two straight-line representations tessellate the convex hull of the outer vertices into axially symmetric quadrangles with two opposite right angles, called *kites*. These kites are in one-to-one correspondence with incident pairs (x, y), where x is a primal vertex, and y is a dual vertex. The symmetry axis of this kite is the line through the centers of C_x and D_y, and the sides are the radius r_x of C_x, and the radius r_y of D_y. Let α_{xy} and α_{yx} denote the angles of this kite at the center of C_x and at the center of D_y, respectively (cf. Figure 6.3). Then,

$$\alpha_{xy} = 2\arctan\frac{r_x}{r_y}, \text{ and } \alpha_{yx} = 2\arctan\frac{r_y}{r_x}, \tag{6.16}$$

and $\alpha_{xy} + \alpha_{yx} = \pi$.

We define the *angle graph* G^a of G as the graph with vertices $V \cup F$, and

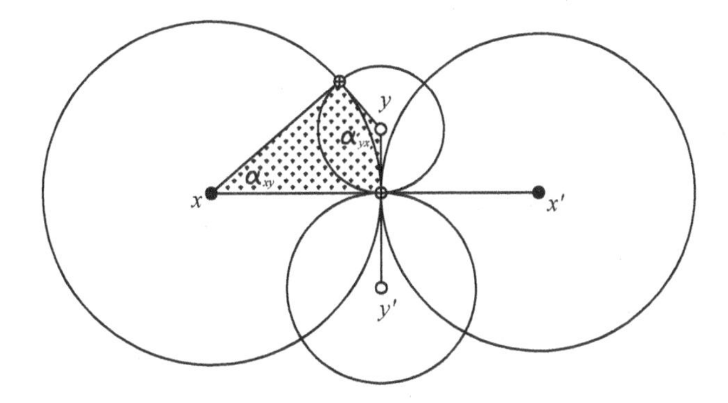

FIGURE 6.3
The kite corresponding to the vertex-face incidence pair (x, y).

whose edges are the incident pairs (x, y) with $x \in V$ and $y \in F$. Note that it is a quadrangulation of the plane, and thus, it is bipartite. The *reduced angle graph* G^a0 of G is defined as the graph obtained from G^a by deleting the vertex corresponding to the outer face of G (Figure 6.4 shows an example of such a graph). Let V_0^a and E_0^a denote the set of the vertices and the edges of G_0^a, respectively.

In the proof we need the next lemma.

Lemma 148 *For any $S \subseteq V_0^a$, the subgraph of G_0^a induced by S has at most $2|S| - 5$ edges.*

Proof. Since G_0^a is bipartite, all faces of G_0^a are adjacent to at least 4 edges. Thus, Euler's formula yields that $E_0^a \leq 2|V_0^a| - 4$, with equality if and only if G_0^a is a quadrangulation. But the outer face of G_0^a is a hexagon, implying that $E_0^a = 2|V_0^a| - 5$. Assume now that $S \subsetneq V_0^a$. Since G is 3-connected, there is no separating 4-cycle in G^a. Thus, the outer face of the subgraph induced by S is not a quadrangle, yet this graph is bipartite. Thus, the subgraph has at most $2|S| - 5$ edges. □

Before proving the theorem, observe that in a primal-dual circle representation of G, for any vertex $x \in V_0^a$, we have

$$\sum_{xy \in E_0^a} \alpha_{xy} = \begin{cases} \frac{\pi}{3} & \text{if } x \text{ is an outer vertex of } G, \\ 2\pi & \text{otherwise.} \end{cases}$$

Thus, we define the *target angles* β for $x \in V_0^a$ such that $\beta(x) = \frac{\pi}{3}$ if x is an outer vertex of G, and $\beta(x) = 2\pi$ otherwise.

First, we assign some radii to the vertices of G_0^a; that is, we consider an

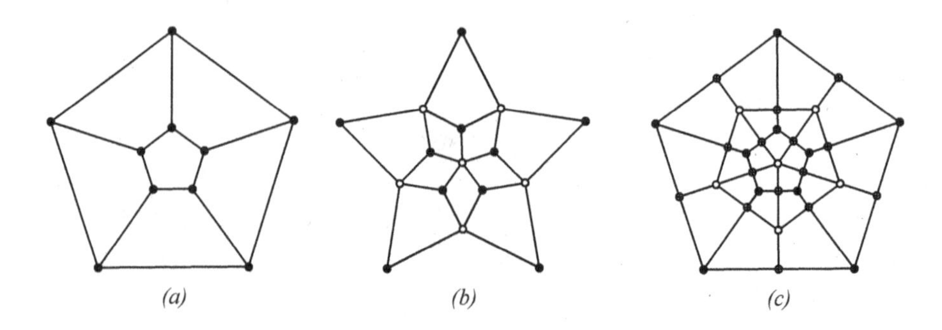

FIGURE 6.4
(a) The edge graph G of a pentagon based prism. (b) The reduced angle graph of G. (c) The primal-dual completion of the graph.

arbitrary positive function $r : V_0^a \to \mathbb{R}$. The first step of the proof is to show that starting with these values and applying an algorithm to modify the radii, for any $x \in V_0^a$, the sum $\sum_{xy \in E_0^a} \alpha_{xy}$ converges to the target value $\beta(x)$. Next, we show that, up to scaling, there are unique values of $r(x)$ satisfying this condition. Finally, we prove that, up to congruence, the kites defined by these values can be uniquely arranged in such a way that preserve the prescribed incidence relations such that their union is the convex hull of the outer vertices of G, and show that this arrangement satisfies the conditions in the definition of a primal-dual circle representation. Note that since any primal-dual circle representation can be transformed into a representation where the outer vertices are vertices of a regular triangle, the second part of Theorem 147 also follows from our argument.
Constructing proper radii r. For any $x \in V_0^a$, let $\alpha(x) = \sum_{xy \in E_0^a} \alpha_{xy}$. First, we observe that 'in average', every choice of the $r(x)$s is suitable. Indeed,

$$\sum_{x \in V_0^a} \alpha(x) = \sum_{xy \in E_0^a} (\alpha_{xy} + \alpha_{yx}) = \sum_{xy \in E_0^a} \pi = |E_0^a|\pi,$$

and on the other hand,

$$\sum_{x \in V_0^a} \beta(x) = (|V_0^a| - 3)2\pi + 3 \cdot \frac{\pi}{3} = (2|V_0^a| - 5)\pi = |E_0^a|\pi.$$

Thus, if $\alpha(x) \neq \beta(x)$ for some $x \in V_0^a$, then both sets

$$V_+ = \{x \in V_0^a : \alpha(x) > \beta(x)\} \quad \text{and} \quad V_- = \{x \in V_0^a : \alpha(x) < \beta(x)\}$$

are non-empty.

If we increase the radius r_x for some $x \in V_+$, and keep the radius of the rest of the vertices, then by (6.16), $\alpha(x)$ decreases, and for any $y \neq x$, $\alpha(y)$ may only increase. Thus, we define the following procedure.

repeat forever:
for all $x \in V_0^a$:
if $x \in V_+$, then increase $r(x)$ to make $\alpha(x) = \beta(x)$.

We show that under this procedure, the radii converge to an assignment satisfying $\alpha(x) = \beta(x)$ for all $x \in V_0^a$. Since increasing $r(x)$ does not decrease the value of $\alpha(y)$ for any $y \neq x$, during this process, a vertex does not move from V_+ to V_-. Thus, if the procedure does not terminate under finitely many steps producing a suitable assignment, then some vertex $x \in V_-$ remains in this set indefinitely. Without loss of generality, omitting the initial finitely many steps in this procedure, we may assume that the set V_- does not change under the process.

It is easy to see that if all radii converge under the process, then the limit radii satisfy the required conditions. Thus, suppose for contradiction that the subset D of V_0^a containing the vertices whose radii do not converge is not empty.

Since radii do not decrease, if $x \in D$, then $r(x) \to \infty$. Furthermore, $D \cap V_- = \emptyset$. Let us denote the set of outer vertices in D by D_o. If $x \in D$ and $y \notin D$, then $\alpha_{xy} \to 0$ by (6.16). Thus, for any given $\varepsilon > 0$, the process leads to an assignment satisfying the inequality $\sum_{y \notin D, xy \in E_0^a} \alpha_{xy} \leq \frac{\varepsilon}{|V_0^a|}$ for all $x \in D$. Then the following holds.

$$\sum_{x \in D} \alpha(x) \leq \varepsilon + \sum_{\text{kite with } x,y \in D} (\alpha_{xy} + \alpha_{yx}) = \varepsilon + \sum_{xy \text{ edge of } G_0^a[D]} \pi$$

$$\leq \varepsilon + (2|D| - 5)\pi. \tag{6.17}$$

$$\sum_{x \in D} \alpha(x) = \sum_{x \in D \setminus V_-} \alpha(x) \geq \sum_{x \in D} \beta(x) = 2\pi|D| - \frac{5\pi}{3}|D_o|. \tag{6.18}$$

Consider the case that $|D| \geq 5$. Comparing these bounds, we obtain that $|D_o| = 3$, and $G_0^a[D]$ has $2|D| - 5$ edges. Thus, $G_0^a[D]$ is an internal quadrangulation. Since G_0^a contains no separating 4-cycle, it follows that $G_0^a[D] = G_0^a$. Thus, $D - V_0^a$, which contradicts our observation that $V_- \neq \emptyset$.

Assume that $3 \leq |D| \leq 4$ and $|D_o| = 3$. Then, by the argument above we have that $G_0^a[D]$ has at most $2|D| - 6$ edges. But then, $\varepsilon + (2|D| - 6)\pi \geq (2|D| - 5)\pi = 2\pi|D| - \frac{5\pi}{3}|D_o|$ leads to a contradiction. If $|D_o| \leq 2$, then $\varepsilon + (2|D| - 5)\pi \geq 2\pi|D| - \frac{5\pi}{3}|D_o|$ is again a contradiction.

If $1 \leq |D| \leq 2$ and $G_0^a[D]$ has no edge, then, $D_o \subseteq D$ yields a contradiction with $\varepsilon \geq 2\pi|D| - \frac{5\pi}{3}|D_o|$. If there is an edge, then $|D| = 2$ and $|D_o| \leq 1$, and $\varepsilon + \pi \geq 4\pi - \frac{5\pi}{3}|D_o|$ is a contradiction.

Uniqueness of radii up to scaling. For contradiction, suppose that there are some vectors of radii r and r' satisfying $\alpha_r(x) = \alpha_{r'}(x) = \beta(x)$ for all $x \in D_0^a$. Without loss of generality, we may assume that $r(z) = r'(z)$ for some $z \in D_0^a$.

Let $S = \{x \in D_0^a : r(x) < r'(x)\}$. Assume that $S \neq \emptyset$, and observe that $z \notin S$. Then

$$0 = \sum_{x \in S} (\alpha_r(x) - \alpha_{r'}(x)) = \sum_{x \in} \sum_{xy \in E_0^a} (\alpha_{xy}(r) - \alpha_{xy}(r'))$$

$$= \sum_{x \in S, y \notin S, xy \in E_0^a} (\alpha_{xy}(r) - \alpha_{xy}(r')) < 0.$$

Here, the second equality holds because if $x, y \in S$, then $\alpha_{xy} + \alpha_{yx} = \pi$, and thus, the contribution of the edge xy cancels out. The last inequality holds since $r(x) > r'(x)$ and $r(y) \leq r'(y)$ imply that $\alpha_{xy}(r) < \alpha_{xy}(r')$, and the connectedness of G_0^a implies that there is an edge xy with $x \in S$ and $y \notin S$. *Arrangement of the kites.* Assume that we manage to arrange the kites defined by the radii in the first step satisfying the intended side-to-side contacts. We show that in this case the circles centered at the vertices of the kites and with the prescribed radii satisfy properties (i) to (v) in the definition of a primal-dual circle representation.

Note that the kites induce a straight-line drawing of G and a straight-line drawing of its dual G^* such that the outer vertex of G^* is at ∞, and the edges ending here are represented by rays. Any two edges $\mathbf{xx'} \in E$ and $\mathbf{yy'} \in E^*$ meet perpendicularly, implying (v).

For a vertex $\mathbf{x} \in V_0^a$, consider the set of kites containing $\mathbf{x}$. These kites are put together in a cyclic order around x, forming a convex polygon $\mathbf{P_x}$. If $\mathbf{x}$ is not an outer vertex of G, then since $\beta(x) = 2\pi$, $\mathbf{P_x}$ surrounds $\mathbf{x}$. All edges of the kites containing $\mathbf{x}$ have the same length $r(x)$, and the circle $C_{\mathbf{x}}$ having radius $r(x)$ is inscribed in $\mathbf{P_x}$, and touches its sides at the right-angled vertices of the kites. If $\mathbf{x}$ is an outer vertex of G, we have a similar configuration, but the union of the kites is a convex polygon whose angle is $\frac{\pi}{3}$ at $\mathbf{x}$. Since the polygons $\mathbf{P_x}$ are mutually non-overlapping, this implies (i) and (iii).

The union of all kites is a triangle $\mathbf{T}$ with all angles equal to $\frac{\pi}{3}$. Thus, $\mathbf{T}$ is regular, and the radii assigned to the vertices of $\mathbf{T}$ are equal. This yields that the three circles assigned to the vertices of $\mathbf{T}$ meet at the midpoints of the sides. Let D_o denote the incircle of $\mathbf{T}$. Since any polygon $\mathbf{P_x}$, associated to a vertex of V_0^a distinct from the outer vertices of G, is contained in the convex hull of the three midpoints of $\mathbf{T}$, D_o contains all circles in the dual family. This implies (ii), and (iv) follows by an argument like the one for (iii).

The unique existence of the layout of the kites follows from the following, more general lemma.

Lemma 149 *Let H be a 3-connected planar graph. For every inner face f of H let $\mathbf{P}_f$ be a simple polygon whose corners are labeled with the vertices of f in the same counterclockwise order. The vertex of $\mathbf{P}_f$ labeled with v is denoted by $\mathbf{p}(f, v)$, and the angle of $\mathbf{P}_f$ at $\mathbf{p}(f, v)$ by $\alpha(f, v)$. Assume that the following conditions are satisfied.*

(i) $\sum_{i=1}^k \alpha(f_i, v) = 2\pi$ *for every inner vertex v of H, with incident faces $f_1, f_2, \ldots, f_k$.*

(ii) $\sum_{i=1}^{k} \alpha(f_i, v) \leq \pi$ *for every outer vertex* v *of* H*, with incident faces* $f_1, f_2, \ldots, f_k$.

(iii) $||\mathbf{p}(f_1, v) - \mathbf{p}(f_1, w)|| = ||\mathbf{p}(f_2, v) - \mathbf{p}(f_2, w)||$ *for every inner edge* vw *of* H*, with incident faces* f_1 *and* f_2.

Then, up to congruence, there is a unique crossing-free straight-line drawing of H *such that the drawing of every inner face* f *can be obtained by* $\mathbf{P}_f$ *by a rigid motion.*

Proof. Let H^* denote the dual graph of H without the vertex associated to the outer face of H, and let S be a spanning tree of H^*. Then, by (iii), we can glue the polygons $\mathbf{P}_f$ of all inner faces of H along the edges of S, in a unique way. This determines the position of every polygon $\mathbf{P}_f$, up to a global motion, which already implies the uniqueness of the drawing.

We prove that the obtained configuration has no overlaps or holes. Since the pairs of faces corresponding to the edges of S do not overlap, we need to show it for pairs corresponding to edges of the complement $\bar{S}$ of S. Note that since H is planar, $\bar{S}$ contains no cycles, and thus, it is a forest. Consider a leaf v of this forest which is an inner vertex of H, and let e be the unique edge in $\bar{S}$ incident to v. By the construction, we have that all other edges $e' \neq e$ in H, incident to v, are attached in a proper way. But then, by (i)-(iii), e is also attached in a proper way. Removing v from $\bar{S}$ again we obtain a forest, and hence, we may continue this process until all inner edges of H are checked. After glueing the polygons $\mathbf{P}_f$, the positions of all vertices are determined, and by (ii), all angles on the boundary are convex. Let V_o be the set of the outer vertices of H.

Note that if $\deg(f) = k$, then the sum of the angles of $\mathbf{P}_f$ is $(k-2)\pi$. Summing up for all polygons $\mathbf{P}_f$ and using Euler's formula, we have $\sum_f (\deg(f) - 2)\pi = (2|E| - 2|F|)\,\pi - (|V_o| - 2)\,\pi = (2|V| - |V_o| - 2)\,\pi$. On the other hand, by (i) and (ii), we have $\sum_f (\deg(f) - 2)\pi = (|V| - |V_o|)\,2\pi + \sum_{v \in V_o} \sum_i \alpha(v, f_i)$. Thus, it follows that $\sum_{v \in V_o} \sum_i \alpha(v, f_i) = (|V_o| - 2)\pi$, which is the sum of the angles of a convex $|V_o|$-gon. Hence, the boundary of the union of the glued polygons is a simple polygonal curve, and therefore this representation contains no holes. $\square$

6.7 Proof of Theorem 27

Imagine $\mathbb{S}^d$ as the 'sphere at infinity' of the hyperbolic space $\mathbb{H}^{d+1}$ in the Poincaré ball model. Then the Möbius transformations of $\mathbb{S}^d$ correspond to the hyperbolic isometries of $\mathbb{H}^{d+1}$. The main idea of the proof is to show that in $\mathbb{H}^{d+1}$, there is a unique point whose sum of 'distances' from the ideal points $\mathbf{v}_1, \mathbf{v}_2, \ldots, \mathbf{v}_n$ is minimal. Since the hyperbolic distance of a point in

$\mathbb{H}^{d+1}$ from any ideal point is infinity, we first 'normalize' these distances via horospheres in the following way.

Let $\mathbf{p} \in \mathbb{H}^{d+1}$ be arbitrary, and let H be a horosphere in $\mathbb{H}^{d+1}$. Note that then H is represented in the model by a sphere touching $\mathbb{S}^d$ from inside at $\mathbf{p}$. We set

$$\delta(\mathbf{p}, H) = \begin{cases} -\mathrm{dist}(\mathbf{p}, H) & \text{if } \mathbf{p} \text{ is inside the sphere representing } H, \\ 0 & \text{if } \mathbf{p} \in H, \\ \mathrm{dist}(\mathbf{p}, H) & \text{if } \mathbf{p} \text{ is outside the sphere representing } H, \end{cases}$$

where $\mathrm{dist}(\mathbf{p}, H)$ denotes the hyperbolic distance between $\mathbf{p}$ and H.

Lemma 150 *Let $\mathbf{v}_1, \mathbf{v}_2, \ldots, \mathbf{v}_n \in \mathbb{S}^d$ be distinct points with $n \geq 3$, and for $i = 1, 2, \ldots, n$, let H_i be a horosphere with ideal point $\mathbf{v}_i$. Then there is a unique point $\mathbf{p} \in \mathbb{H}^{d+1}$ for which the quantity $\sum_{i=1}^n \delta(\mathbf{p}, H_i)$ is minimal. Moreover, this point is independent of the choice of the horospheres.*

Proof. Consider a horosphere H with ideal point $\mathbf{v} \in \mathbb{S}^d$. Then the shortest geodesic segment from $\mathbf{p}$ to H is contained in a geodesic line with ideal point $\mathbf{v}$. Furthermore, if H' is another horosphere with ideal point $\mathbf{v}$, then $\delta(\cdot, H) - \delta(\cdot, H')$ is a constant function on $\mathbb{H}^{d+1}$. Thus, the point $\mathbf{p}$ in Lemma 150 is independent of the choice of the horospheres.

Let $g : \mathbb{R} \to \mathbb{H}^{d+1}$ be a geodesic line parametrized by arclength. It is a straightforward exercise to prove that the function $\delta(g(s), H)$ is a strictly convex function of s unless the ideal points of the geodesic line and H coincide, and in the latter case, it is $\delta(g(s), H) = \pm(s - s_0)$ for some $s_0 \in \mathbb{R}$.

Finally, observe that if H_i, $i = 1, 2, \ldots, n$ are horospheres with pairwise distinct ideal points, and $n \geq 3$, then for any $\mathbf{p} \in \mathbb{H}^{d+1}$ approaching $\mathbb{S}^d$, we have

$$\lim_{\mathbf{p} \to \infty} \sum_{i=1}^n \delta(\mathbf{p}, H_i) = \infty.$$

Thus, the function $\sum_{i=1}^n \delta(\mathbf{p}, H_i)$ is strictly convex, and its limit is ∞ at the boundary of $\mathbb{H}^{d+1}$. This implies the unique existence of the point $\mathbf{p}$ in the lemma. □

We note that our proof yields more than stated, as we also have that point $\mathbf{p}$ in Lemma 150 is the unique *stationary point* of the scalar field $\sum_{i=1}^n \delta(\mathbf{p}, H_i)$ defined on $\mathbb{H}^{d+1}$; that is, $\mathbf{p}$ is the only point where the gradient of this function is zero. We call the point $\mathbf{p}$ defined by Lemma 150 the *point with minimal distance* from the ideal points $\mathbf{v}_1, \mathbf{v}_2, \ldots, \mathbf{v}_n$.

Lemma 151 *Let $\mathbf{v}_1, \mathbf{v}_2, \ldots, \mathbf{v}_n$ be distinct ideal points of $\mathbb{H}^{d+1}$ in the Poincaré ball model, where $n \geq 3$. Then the point of minimal distance from $\mathbf{v}_1, \mathbf{v}_2, \ldots, \mathbf{v}_n$ is the origin if and only if $\sum_{i=1}^n \mathbf{v}_i = \mathbf{o}$.*

Proof. It follows from the metric tensor of the Poincaré ball model that the gradient of $\delta(\mathbf{p}, H_i)$ at $\mathbf{o}$ is $-\frac{1}{2}\mathbf{v}_i$. □

Now we are ready to prove Theorem 27. Let $\mathbf{p} \in \mathbb{H}^{d+1}$ be the point of minimal distance from $\mathbf{v}_1, \mathbf{v}_2, \ldots, \mathbf{v}_n$. Then there is a hyperbolic isometry T moving $\mathbf{p}$ into $\mathbf{o}$. Thus, Lemma 151 implies that $\sum_{i=1}^{n} T(\mathbf{v}_i) = \mathbf{o}$. On the other hand, as the Möbius group acts faithfully on $\mathbb{S}^d$ as well as on $\mathbb{H}^{d+1}$, we have that if $\tilde{T}$ is a hyperbolic isometry moving $\mathbf{p}$ to $\mathbf{o}$, then $\tilde{T} = RT$, where R is an isometry.

6.8 Proofs of Theorems 28 and 29

First, in Subsection 6.8.1, we present some preliminary observations and notations, and introduce the main idea of the proofs. In Subsection 6.8.2 we prove the main lemma of the proof of Theorem 28. In Subsections 6.8.3 and 6.8.4, we prove Theorems 28 and 29, respectively.

6.8.1 Preliminaries and the main idea of the proofs

In the following, let $\mathbf{P}$ be a Koebe polyhedron. The centers and the spherical radii of the vertex circles of $\mathbf{P}$ are denoted by $\mathbf{v}_i \in \mathbb{S}^2$ and α_i, respectively, where $i = 1, 2, \ldots, n$, and the centers and the radii of its face circles by $\mathbf{f}_j \in \mathbb{S}^2$ and β_j, respectively, where $j = 1, 2, \ldots, m$. We note that by Theorem 27 (cf. [222]), we may assume that the barycenter of the tangency points of $\mathbf{P}$ is the origin $\mathbf{o}$, implying that $\mathbf{o}$ is contained both in $\mathbf{P}$ and in its polar, or in other words, the radii of all vertex and face circles of $\mathbf{P}$ are less than $\frac{\pi}{2}$. Thus, for any vertex or face circle there is an associated spherical cap, obtained as the union of the circle and its interior.

In the proof, we regard the sphere $\mathbb{S}^2$ (or, in the proof of Theorem 29, $\mathbb{S}^d$) as the set of the ideal points in the Poincaré ball model of the hyperbolic space $\mathbb{H}^3$ (or $\mathbb{H}^{d+1}$). Thus, every circle on $\mathbb{S}^2$ is associated to a hyperbolic plane, and every spherical cap is associated to a closed hyperbolic half space. We note that since the Poincaré ball model is conformal, the dihedral angle between two circles on $\mathbb{S}^2$ is equal to the dihedral angle between the two corresponding hyperbolic planes (cf. [147, Observation 0.1]).

For the vertex circle with center $\mathbf{v}_i$ we denote the corresponding hyperbolic plane by V_i and the associated closed half space by $\mathbf{V}_i$. Similarly, the hyperbolic plane corresponding to the face circle with center f_j is denoted by F_j, and the associated closed half space by $\mathbf{F}_j$. We set $\mathbf{D} = \mathbb{H}^3 \setminus \left((\bigcup_{i=1}^{n} \mathbf{V}_i) \cup \left(\bigcup_{j=1}^{m} \mathbf{F}_j \right) \right)$. Observe that as the radii of all vertex and face circles of $\mathbf{P}$ are less than $\frac{\pi}{2}$, we have $\mathbf{o} \in \mathbf{D}$, and thus, $\mathbf{D}$ is a non-empty, open convex set in $\mathbb{H}^3$.

Let $\mathbf{p} \in \mathbf{D} \subset \mathbb{H}^3$ be a point. For any plane V_i, consider the geodesic line through $\mathbf{p}$ and perpendicular to V_i. Let $\mathbf{v}_i(p) \in T_p\mathbb{H}^3$ denote the unit tangent vector of this line at $\mathbf{p}$, pointing towards V_i, and let $d_i^v(\mathbf{p})$ denote the hyperbolic distance of $\mathbf{p}$ from V_i. We define $\mathbf{f}_j(p)$ and $d_j^f(\mathbf{p})$ similarly for the plane F_j.

An important point of the proof is the following simple observation. Recall that the *angle of parallelism* of a point $\mathbf{p}$ not lying on a hyperbolic hyperplane H is the hyperbolic half angle of the hyperbolic cone with apex $\mathbf{p}$ formed by the half lines starting at $\mathbf{p}$ and parallel to H. Thus, Remark 152 is a consequence of the fact that the Poincaré ball model is conformal, and of a well-known hyperbolic formula [151]. Even though we state it for the 3-dimensional space $\mathbb{H}^3$, it also holds in any dimensions.

Remark 152 *Let H be a hyperbolic plane in $\mathbb{H}^3$ whose set of ideal points is a circle C on $\mathbb{S}^2$ with spherical radius α. Then α is the angle of parallelism of H from the origin $\mathbf{o}$ (cf. Figure 6.5). In particular, $\cos\alpha = \tanh d$, where d is the hyperbolic distance between H and $\mathbf{o}$.*

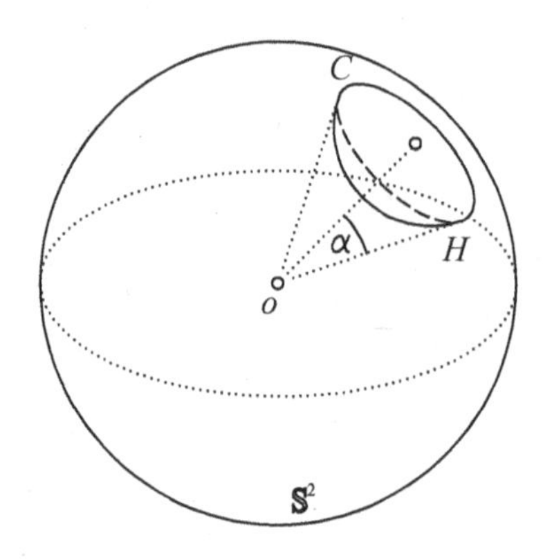

FIGURE 6.5
The angle of parallelism from the origin of the model is the spherical radius of the circle C.

Among other things, it follows by Remark 152 that

$$\tanh d_i^v(\mathbf{o}) = \cos\alpha_i \text{ and } \tanh d_j^f(\mathbf{o}) = \cos\beta_j \text{ for all values of } i, j. \quad (6.19)$$

Furthermore, the metric tensor of the Poincaré ball model yields (cf. [222]) that

$$\mathbf{v}_i(\mathbf{o}) = \frac{1}{2}\mathbf{v}_i \text{ and } \mathbf{f}_j(\mathbf{o}) = \frac{1}{2}\mathbf{f}_j \text{ for all values of } i, j. \quad (6.20)$$

The idea of the proof of Theorem 28 in most cases is as follows. Let $g(\cdot)$ be one of the points in Theorem 28. First, we compute $g(\mathbf{P})$ in terms of the radii and the centers of its vertex and face circles; that is, in a form $g(\mathbf{P}) = \sum_{i=1}^n w_i\mathbf{v}_i + \sum_{j=1}^m W_j\mathbf{f}_j$, where the coefficients w_i and W_j are smooth functions depending on the values $0 < \alpha_i, \beta_j < \frac{\pi}{2}$. Applying the formulas in (6.19) to the coefficients w_i and W_j, we obtain a smooth hyperbolic vector field $h : \mathbf{D} \to T\mathbf{D}$. Since in this model Möbius transformations on $\mathbb{S}^2$ are associated to hyperbolic isometries of $\mathbb{H}^3$, this function has the property that if T corresponds to a hyperbolic isometry that maps $\mathbf{p}$ into $\mathbf{o}$, then $h(\mathbf{p}) = g(T(\mathbf{P}))$ for all $\mathbf{p} \in \mathbf{D}$. It is well known that hyperbolic isometries act transitively on $\mathbb{H}^3$. Thus, to prove the existence of a suitable Möbius transformation, it suffices to prove that $h(\mathbf{p}) = \mathbf{o}_\mathbf{p}$ for some $\mathbf{p} \in \mathbf{D}$. In the cases of $\mathrm{cc}(\cdot)$ and $\mathrm{IC}(\cdot)$ the function h is not C^∞-class; here we use similar, geometric arguments. In the remaining cases h is smooth; here we examine the properties of the integral curves of h. To prove Theorem 29, we use an analogous consideration.

In the proof we often use the following geometric observation.

Remark 153 *For $i = 1, 2, \ldots, n$ and $j = 1, 2, \ldots, m$, the ith vertex of $\mathbf{P}$ is $\frac{\mathbf{v}_i}{\cos \alpha_i}$, and the incenter of the jth face of $\mathbf{P}$ is $\cos \beta_j \mathbf{f}_j$.*

Most of the computations will be carried out in the Poincaré half space model.

In this model, we regard $\mathbb{H}^3$ embedded in $\mathbb{E}^3$ as the open half space $\{z > 0\}$. Hyperbolic planes having the 'point at infinity' as an ideal point are represented in this model by the intersections of the Euclidean half space $\{z > 0\}$ with Euclidean planes parallel to the z-axis, we call these hyperbolic planes *vertical.* Hyperbolic planes not having the 'point at infinity' as an ideal point are represented by open hemispheres in the Euclidean half space $\{z > 0\}$, with their centers on the Euclidean plane $\{z = 0\}$, we call these planes *spherical.* For any plane H in this model, we denote the set of its ideal points, different from the point at infinity, by H^*. We use the same terminology and notation for this model in any dimension.

The last remark in this section is the result of elementary computations using distance formulas from the Poincaré half plane model.

Remark 154 *Let $\mathbf{p} = (a, t)$, $a, t > 0$ be a point in the Poincaré half plane model, and let $\mathbf{m} \in T_\mathbf{p}\mathbb{H}^2$ denote the tangent unit vector of the geodesic line through $\mathbf{p}$ and perpendicular to the y-axis, pointing towards the axis. Furthermore, let C be the hyperbolic line represented by the circle centered at the origin $\mathbf{o}$ and Euclidean radius r, and let $\mathbf{v} \in T_\mathbf{p}\mathbb{H}^2$ denote the tangent unit vector of the geodesic line through $\mathbf{p}$ and perpendicular to C, pointing towards C. Assume that $r < \sqrt{a^2 + t^2}$. Then the hyperbolic distance of $\mathbf{p}$ from the y-axis and from C are $\operatorname{arsinh} \frac{a}{t}$ and $\operatorname{arsinh} \frac{t^2+a^2-r^2}{2rt}$, respectively. In addition, the y-coordinates of $\mathbf{m}$ and $\mathbf{v}$ are $\frac{a}{\sqrt{a^2+t^2}}$ and $-\frac{t^2+r^2-a^2}{\sqrt{(r^2+a^2+t^2)^2-4r^2a^2}}$, respectively.*

6.8.2 The main lemma of the proofs

The main goal of this section is to prove Lemma 155. In its formulation and proof we use the notations introduced in Subsection 6.8.1. We note that two hyperbolic planes V_i and F_j intersect if and only if the ith vertex of $\mathbf{P}$ lies on the jth face of $\mathbf{P}$. In this case, the two planes have a common ideal point, coinciding with a tangency point of $\mathbf{P}$. This point is the ideal point of one pair of V_is and of one pair of F_js, and these two pairs are orthogonal.

If $\mathbf{q}$ is a boundary point of $\mathbf{D}$ in the Euclidean topology, by a *neighborhood* of $\mathbf{q}$ we mean the intersection of a neighborhood of $\mathbf{q}$ with $\mathbf{D}$, induced by the Euclidean topology of $\mathbb{E}^3$. Before stating our main lemma, we note that if $h : \mathbf{D} \to T\mathbf{D}$ is a smooth vector field, then by the Picard-Lindelöf Theorem for any $\mathbf{p} \in \mathbf{D}$ with $h(\mathbf{p}) \neq \mathbf{o}$ there is a unique integral curve of h passing through $\mathbf{p}$. These integral curves are either closed, or start and end at boundary points of $\mathbf{D}$ or at points $\mathbf{q}$ with $h(\mathbf{q}) = \mathbf{o}$.

Lemma 155 *Let*

$$h : \mathbf{D} \to T\mathbf{D},\ h(\mathbf{p}) = \sum_{i=1}^{n} w_i \mathbf{v}_i(\mathbf{p}) + \sum_{j=1}^{m} W_j \mathbf{f}_j(\mathbf{p}),$$

be a vector field where the coefficients w_i and W_j are positive smooth functions of $n+m$ variables, depending on $d_i^v(\mathbf{p})$, $i = 1, 2, \ldots, n$ and $d_j^f(\mathbf{p})$, $j = 1, 2, \ldots, m$. Assume that for any boundary point $\mathbf{q}$ of $\mathbf{D}$,

(i) *$\mathbf{q}$ has a neighborhood disjoint from any closed integral curve of h.*

(ii) *If $\mathbf{q} \in F_j$ for some value of j, then there is no integral curve of h ending at $\mathbf{q}$.*

(iii) *If $\mathbf{q} \in V_i$ for some value of i and $\mathbf{q} \notin F_j$ for all values of j, then $\mathbf{q}$ has a neighborhood in which the integral curve through any point ends at a point of V_i.*

(iv) *If $\mathbf{q} \in \mathbb{S}^2$ is a tangency point of $\mathbf{P}$, then there is a codimension 1 foliation of a neighborhood of $\mathbf{q}$ in $\mathbf{D}$ such that $\mathbf{q}$ is not an ideal point of any leaf, and for any point $\mathbf{p}$ on any leaf $h(\mathbf{p}) \neq \mathbf{o}$, the integral curve through $\mathbf{p}$ crosses the leaf, either in the direction of $\mathbf{q}$ or from this direction, independently of the choice of $\mathbf{p}$, the leaf and $\mathbf{q}$.*

Then $h(\mathbf{p}) = \mathbf{o}_{\mathbf{p}}$ for some $\mathbf{p} \in \mathbf{D}$.

First, we prove Lemma 156.

Lemma 156 *Let $\mathbf{X} = (\mathrm{int}\mathbf{B}^{d+1}) \setminus (1-\varepsilon)\mathbf{B}^{d+1}$, where $0 < \varepsilon < 1$, and $d \geq 2$. Let $\mathbf{Z}_1, \ldots, \mathbf{Z}_k$ be pairwise disjoint closed sets in $\mathbf{X}$, where $k \geq 1$. If $\mathbf{X} \setminus \mathbf{Z}_i$ is connected for all values of i, then $\mathbf{X} \setminus \left(\bigcap_{i=1}^{k} \mathbf{Z}_i\right)$ is connected.*

Proof of Lemma 156. We prove the assertion by induction for k. If $k = 1$, then the statement is trivial. Assume that Lemma 156 holds for any $k-1$ closed sets. Let $\mathbf{Z}' = \bigcup_{i=1}^{k-1} \mathbf{Z}_i$. Then $\overline{\mathbf{Z}}_k = \mathbf{X} \setminus \mathbf{Z}_k$ and $\overline{\mathbf{Z}'} = \mathbf{Z} \setminus \mathbf{Z}'$ are open sets whose union is $\mathbf{X}$. Consider the Mayer-Vietoris exact sequence [180] of these subspaces:

$$H_1(\mathbf{X}) \to H_0(\overline{\mathbf{Z}_k} \cap \overline{\mathbf{Z}'}) \to H_0(\overline{\mathbf{Z}_k} \oplus \overline{\mathbf{Z}'}) \to H_0(\mathbf{X}) \to 0.$$

Note that by the induction hypothesis, $\overline{\mathbf{Z}'}$ is connected. On the other hand, since $\mathbb{S}^d$ is a deformation retract of $\mathbf{X}$, their homology groups coincide, implying that rank$H_1(\mathbf{X}) = 0$, rank$H_0(\mathbf{X}) = 1$. Since $\mathbf{X}$ is locally path-connected, any connected subset of $\mathbf{X}$ is path-connected, and thus, rank$H_0(\mathbf{X})$ is the number of connected components of $\mathbf{X}$, implying that rank$(H_0(\overline{\mathbf{Z}_k} \oplus \overline{\mathbf{Z}'})) = 2$, and rank$(H_0(\overline{\mathbf{Z}_k} \cap \overline{\mathbf{Z}'})) = t$, where t is the number of the connected components of $\overline{\mathbf{Z}_k} \cap \overline{\mathbf{Z}'}$. The exactness of the Mayer-Vietoris sequence yields that $1 - 2 + t = 0$, that is, $t = 1$. □

Proof of Lemma 155. We prove the lemma by contradiction. Assume that $h(\mathbf{p}) \neq \mathbf{o}$ for any $\mathbf{p} \in \mathbf{D}$, and let S denote the set of tangency points of $\mathbf{P}$. Furthermore, let $\mathbf{Z}$ denote the set of the points of $\mathbf{D}$ belonging to a closed integral curve. For $i = 1, 2, \ldots, n$, let $\mathbf{Y}_i$ denote the set of points whose integral curve terminates at a point of V_i, and let $\mathbf{W}_\mathbf{s}$ be the set of the points with their integral curves ending at $\mathbf{s} \in S$. By (iii), every set $\mathbf{Y}_i$ is open, and it is easy to see that every set $\mathbf{W}_\mathbf{s}$ is closed.

First, assume that for any $\mathbf{s} \in S$, the integral curve through any point $\mathbf{p}$ on a leaf of the codimension 1 foliation in a neighborhood of $\mathbf{s}$ points away from the direction of $\mathbf{s}$. This implies, in particular, that $\mathbf{W}_\mathbf{s} = \emptyset$ for all $\mathbf{s} \in S$. For all $\mathbf{q} \in$ bd$\mathbf{D}$, let $\mathbf{U}_\mathbf{q}$ denote a neighborhood of q satisfying the conditions of the lemma. By the definition of induced topology, $\mathbf{U}_\mathbf{q} = \mathbf{U}^*_\mathbf{q}$ for some neighborhood of $\mathbf{q}$ in $\mathbb{E}^3$. We may assume that $\mathbf{U}^*_\mathbf{q}$ is open for all $\mathbf{q} \in$ bd$D$. Since the sets $\mathbf{U}^*_\mathbf{q}$ cover the compact set bd$\mathbf{D}$, we may choose a finite subfamily that covers bd$\mathbf{D}$. By finiteness, it follows that there is some $\varepsilon > 0$ such that the set $\mathbf{D}_\varepsilon$ of points at Euclidean distance less than ε from bd$\mathbf{D}$ is disjoint from $\mathbf{Z}$. On the other hand, $\mathbf{D}_\varepsilon$ is connected, yet it is the disjoint union of the finitely many open sets $\mathbf{Y}_i \cap \mathbf{D}_\varepsilon$, a contradiction.

Assume now that for any $\mathbf{s} \in S$, the integral curve through any point $\mathbf{p}$ on a leaf of the codimension 1 foliation in a neighborhood $\mathbf{U}_\mathbf{s}$ of $\mathbf{s}$ points towards $\mathbf{s}$. By this, if $\mathbf{s} \in S$ is the tangency point connecting the ith and jth vertices, then $\mathbf{U}_\mathbf{s} \subseteq \mathbf{W}_\mathbf{s} \cup \mathbf{Y}_i \cup \mathbf{Y}_j$. On the other hand, by (iii), all $\mathbf{Y}_i$s are connected. Thus, for any walk on the edge graph of $\mathbf{P}$ starting at the kth and ending at the lth vertex, there is a continuous curve in $\mathbf{D}$ starting at a point of $\mathbf{Y}_k$ and ending at a point of $\mathbf{Y}_l$, and passing through points of only those $\mathbf{Y}_i$s and $\mathbf{W}_\mathbf{s}$s for which the associated vertices and edges of $\mathbf{P}$ are involved in the walk. In addition, the curve may pass arbitrarily close to bd$\mathbf{D}$, measured in Euclidean metric.

We choose the set $\mathbf{D}_\varepsilon$ as in the previous case. Note that $\mathbf{D}_\varepsilon$ is homeomorphic to $(\text{int}\mathbf{B}^3) \setminus (1-\varepsilon)\mathbf{B}^3$, and thus, we may apply Lemma 156 with the $\mathbf{W}_\mathbf{s}$s playing the roles of the $\mathbf{Z}_j$s. Then it follows that for some $\mathbf{s} \in S$, $\mathbf{D}_\varepsilon \setminus \mathbf{W}_\mathbf{s}$ is disconnected. Since the union of finitely many closed sets is closed, there are some $\mathbf{Y}_k$ and $\mathbf{Y}_l$ in different components. By Steinitz's theorem [228, 230], there is a path in the edge graph of $\mathbf{P}$ that connects the kth and lth vertices and avoids the edge associated to $\mathbf{s}$. Hence, there is a continuous curve in $\mathbf{D}$, starting at a point of $\mathbf{Y}_k$ and ending at a point of $\mathbf{Y}_l$ that avoids $\mathbf{W}_\mathbf{s}$; a contradiction. $\square$

6.8.3 Proof of Theorem 28

Proof of Theorem 28 for $\mathrm{cc}(\cdot)$ *and* $\mathrm{IC}\cdot$.
First, we prove the statement for $\mathrm{cc}(\cdot)$. During the proof, we set $\mathbf{D}^v = \mathbb{H}^3 \setminus (\bigcup_{i=1}^n \mathbf{V}_i)$. Observe that a ball $\mathbf{B}$ is the smallest ball containing $\mathbf{P}$ if and only if it contains $\mathbf{P}$, and its center belongs to the convex hull of the vertices of $\mathbf{P}$ lying on the boundary of the ball.

Let I be the set of indices such that $\frac{1}{\cos\alpha_i} = \max\left\{\frac{1}{\cos\alpha_j} : j = 1, 2, \ldots, n\right\}$. Thus, by Remark 153, $\mathbf{o} = \mathrm{cc}(\mathbf{P})$ if and only if $\mathbf{o} \in \mathrm{conv}\left\{\frac{1}{\cos\beta_i}\mathbf{v}_i : i \in I\right\}$, which is equivalent to $\mathbf{o} \in \mathrm{conv}\{\mathbf{v}_i : i \in I\}$. Furthermore, I is the set of indices with the property that $d_i^v(\mathbf{o}) = \min\{d_j^v(\mathbf{o}) : j = 1, 2, \ldots, n\}$. We may extend this definition for any $\mathbf{p} \in \mathbf{D}^v$, and let $I(\mathbf{p})$ denote the set of indices with the property that $d_i^v(\mathbf{p}) = \min\{d_j^v(\mathbf{p}) : j = 1, 2, \ldots, n\}$. Since Möbius transformations act transitively on $\mathbb{H}^3$, we need only to show the existence of a point $\mathbf{p} \in \mathbf{D}^v$ such that $\mathbf{o}_\mathbf{p} \in \mathrm{conv}\{\mathbf{v}_i(\mathbf{p}) \subset T_\mathbf{p}\mathbb{H}^3 : i \in I(\mathbf{p})\}$.

For any plane V_i and $\tau > 0$, consider the set $\mathbf{V}_i(\tau)$ of points in $\mathbf{D}^v$ at distance at most τ from $\mathbf{V}_i$. This set is bounded by $\mathbf{V}_i$ and a hypersphere, which, in the model, is represented by the intersection of a sphere with the interior of $\mathbb{S}^2$, and having the same ideal points as V_i. Hence, if τ is sufficiently small, then the sets $\mathbf{V}_i(\tau)$ and $\mathbf{V}_j(\tau)$, where $i \neq j$, intersect if and only if the ith and the jth vertices of $\mathbf{P}$ are connected by an edge. On the other hand, if τ is sufficiently large, then all $\mathbf{V}_i(\tau)$s intersect. Let τ_0 be the smallest value such that some $\mathbf{V}_i(\tau_0)$ and $\mathbf{V}_j(\tau_0)$ intersect, where $i \neq j$ and the ith and jth vertices are not neighbors, and let $\mathbf{p} \in \mathbf{V}_i(\tau_0) \cap \mathbf{V}_j(\tau_0)$. Note that $\mathbf{v}_i(\mathbf{p})$ is an inner surface normal of the boundary of $\mathbf{V}_i(\tau_0)$ at $\mathbf{p}$. Thus, the definition of τ_0 yields that the system of inequalities $\langle \mathbf{x}, \mathbf{v}_i(\mathbf{p})\rangle > 0$, $i \in I(\mathbf{p})$ has no solution for $\mathbf{x}$, from which it follows that there is no plane in $T_\mathbf{p}\mathbb{H}^3$ that strictly separates $\mathbf{o}_\mathbf{p}$ from the $\mathbf{v}_i(\mathbf{p})$s, implying that $\mathbf{o}_\mathbf{p} \in \mathrm{conv}\{\mathbf{v}_i(\mathbf{p}) : i \in I(\mathbf{p})\}$. This proves the statement for $\mathrm{cc}(\cdot)$. To prove it for $\mathrm{IC}(\cdot)$, we may apply the same argument for the face circles of P.

Proof of Theorem 28 for $\mathrm{cm}_0(\cdot)$.
We show that this case of Theorem 28 is an immediate consequence of Theorem 29.

By Remark 153, we have $\mathrm{cm}_0(\mathbf{P}) = \frac{1}{n}\sum_{i=1}^n \frac{1}{\cos\alpha_i}\mathbf{v}_i$. Thus, it is sufficient to show that the conditions of Theorem 29 are satisfied for the family of vertex circles of $\mathbf{P}$ with the weight functions $w_i(t) = \frac{1}{\cos t}$ for all is.

First, observe that if $n = 4$ (i.e., if $\mathbf{P}$ is a tetrahedron), then $\mathrm{cm}_0(\mathbf{P}) = \mathbf{o}$ if $\mathbf{P}$ is regular. Thus, we may assume that $n \geq 5$. Note that the weight functions $w_i(t) = \frac{1}{\cos t}$ are positive smooth functions on $\left(0, \frac{\pi}{2}\right)$ and satisfy

$\lim_{t \to \frac{\pi}{2}} w_i(t) = \infty$. Furthermore, since $|I(\mathbf{q})| \leq 2$ for all points $\mathbf{q} \in \mathbb{S}^2$, the inequality in (1.3) holds, and Theorem 29 implies Theorem 28 for $\mathrm{cm}_0(\cdot)$.

Proof of Theorem 28 for $\mathrm{cm}_1(\cdot)$.
Let E denote the set of edges of the edge graph of $\mathbf{P}$; that is, $\{i,j\} \in E$ if and only if the ith and jth vertices are connected by an edge. An elementary computation yields that if $\{i,j\} \in E$, the length of the corresponding edge of $\mathbf{P}$ is $\tan\alpha_i + \tan\alpha_j$, and its center of mass is $\frac{1}{2}\left(\frac{\mathbf{v}_i}{\cos\alpha_i} + \frac{\mathbf{v}_j}{\cos\alpha_j}\right)$. Thus, letting $A = \sum_{\{i,j\}\in E} (\tan\alpha_i + \tan\alpha_j)$, we have

$$\mathrm{cm}_1(P) = \frac{1}{2A} \sum_{\{i,j\} in E} (\tan\alpha_i + \tan\alpha_j)\left(\frac{\mathbf{v}_i}{\cos\alpha_i} + \frac{\mathbf{v}_j}{\cos\alpha_j}\right). \tag{6.21}$$

Set $\mathbf{D}^v = \mathbb{H}^3 \setminus (\bigcup_{i=1}^n \mathbf{V}_i)$, and define the functions $h_{i,j}^v : \mathbf{D}^v \to T\mathbf{D}^v$ and $h^v : \mathbf{D}^v \to T\mathbf{D}^v$ as

$$h_{i,j}^v(\mathbf{p}) = \left(\frac{1}{\sinh d_i^v(\mathbf{p})} + \frac{1}{\sinh d_j^v(\mathbf{p})}\right)(\coth(d_i^v(\mathbf{p}))\mathbf{v}_i(\mathbf{p}) + \coth(d_j(\mathbf{p}))\mathbf{v}_j(\mathbf{p}))$$

and

$$h^v(\mathbf{p}) = \sum_{\{i,j\}\in E} h_{i,j}^v(\mathbf{p}). \tag{6.22}$$

Then h^v is a smooth function on $\mathbf{D}^v$ and the coefficient of each vector $\mathbf{v}_i(\mathbf{p})$ is positive. By Remark 152, it follows that if there is a point $\mathbf{p} \in \mathbf{D}^v$ such that $h^v(\mathbf{p}) = \mathbf{o}$, then, choosing a Möbius transformation T that maps $\mathbf{p}$ into $\mathbf{o}$, we have $\mathrm{cm}_1(T(\mathbf{P})) = \mathbf{o}$. We denote the restriction of h^v to $\mathbf{D}$ by h, and show that h satisfies the conditions in Lemma 155.

Let $\mathbf{q}$ be a boundary point of $\mathbf{D}$ in some plane F_j associated to a face circle of $\mathbf{P}$. Assume that $\mathbf{q}$ is not contained in V_i for any value of i. Observe that if the ith vertex lies on the jth face, then $\mathbf{v}_i(\mathbf{q})$ and $\mathbf{f}_j(\mathbf{q})$ are orthogonal, and otherwise $\mathbf{v}_i(\mathbf{q})$ points inward to $\mathbf{D}$. Thus, by the continuity of h^v, there is no integral curve of h that ends at $\mathbf{q}$, and $\mathbf{q}$ has a neighborhood disjoint from the set $\mathbf{Z}$ of the points of the closed integral curves of h. If $\mathbf{q}$ is contained in $\mathbf{V}_i$ for some i, then a slight modification of this argument can be applied. This proves (ii) in Lemma 155.

Let $\mathbf{q}$ be a point of some V_i not contained in any of the F_js. Then, denoting the coefficient of $\mathbf{v}_j(\mathbf{p})$ by $\mu_j(\mathbf{p})$ for any j, we have that $\frac{\mu_i(\mathbf{p})}{\mu_j(\mathbf{p})} \to \infty$ for all $j \neq i$, as $\mathbf{p} \to \infty$, which shows that if $\mathbf{p}$ is 'close' to $\mathbf{q}$, then $h(\mathbf{p})$ is 'almost orthogonal' to F_j. This shows (iii), and the fact that a neighborhood of $\mathbf{q}$ is disjoint from $\mathbf{Z}$.

Finally, let $\mathbf{q}$ be a tangency point of P. Without loss of generality, we may assume that $\mathbf{q}$ is the ideal point of V_1, V_2, F_1 and F_2. To prove (iv), we imagine the configuration in the Poincaré half space model, with $\mathbf{q}$ as the 'point at infinity'; geometrically, it means that we apply an inversion to $\mathbb{E}^3$ about a sphere centered at $\mathbf{q}$. Then $\mathbf{D}$ is contained in the half-infinite cylinder bounded by the four vertical planes V_1, V_2, F_1 and F_2 (for the definition of vertical and spherical planes, see Subsection 6.8.1). Note that the cross-section of this cyclinder is a rectangle, and that all other V_is and F_js are spherical planes centered at ideal points of $\mathbf{D}$ in the Euclidean plane $\{z = 0\}$.

For any $t > 0$, let D_t denote the intersection of the set $\{z = t\}$ with $\mathbf{D}$. We remark that $\{z = t\}$ is a horosphere whose only ideal point is $\mathbf{q}$, and thus, the sets D_t, where t is sufficiently large, form a codimension 1 foliation of a neighborhood of $\mathbf{q}$ in $\mathbf{D}$. Hence, to show that the conditions of Lemma 155 are satisfied, it is sufficient to show that if t is sufficiently large, then $h(\mathbf{p})$ has a positive z-coordinate for any $\mathbf{p} \in D_t$.

For any $\{i, j\} \in E$, denote the term in $h(\mathbf{p})$ belonging to $\{i, j\}$ by $h_{i,j}(\mathbf{p})$, and the z-coordinate of $h_{i,j}(\mathbf{p})$ by $z_{i,j}(\mathbf{p})$. Let $\{i, j\}$ and $\{1, 2\}$ be disjoint. Note that the closure of D_t is compact. Thus, by Remark 154, if $t \to \infty$, then $h(\mathbf{p})$ uniformly converges to 0. Assume that $\{i, j\} \cap \{1, 2\}$ is a singleton, say $i = 1$ and $j \neq 2$. Then, by Remark 154, the z-coordinate of $\coth d_1^v(\mathbf{p})\mathbf{v}_1(\mathbf{p})$ is 1, and that of $\coth d_j^v(\mathbf{p})\mathbf{v}_j(\mathbf{p})$ is less than 1. Thus, $z_{1,j}(\mathbf{p}) > 0$ in this case. Finally, $z_{1,2}(\mathbf{p}) > Ct$ for any $\mathbf{p} \in D_t$ for some universal constant $C > 0$. Thus $h(\mathbf{p})$ has a positive z-coordinate for large values of t, and Lemma 155 implies Theorem 28 for the case of $\mathrm{cm}_1(\cdot)$.

Proof of Theorem 28 for $\mathrm{cm}_2(\cdot)$.
Let I denote the edge set of the vertex-face incidence graph of $\mathbf{P}$; that is, $(i, j) \in I$ if and only if the ith vertex lies on the jth face. Consider some $(i, j) \in I$. Then there are exactly two edges of $\mathbf{P}$ adjacent to both the vertex and the face. Let the tangency points on these two edges be denoted by $\mathbf{e}_{i,j}^1$ and $\mathbf{e}_{i,j}^2$. Then, by Remark 153, the points $\frac{\mathbf{v}_i}{\cos\alpha_i}$, $\mathbf{e}_{i,j}^1$, $\cos\beta_j \mathbf{f}_j$ and $\mathbf{e}_{i,j}^2$ are coplanar, and they are the vertices of a symmetric right trapezoid $Q_{i,j}$ (cf. Figure 6.6). Note that bd$\mathbf{P}$ can be decomposed into the mutually non-overlapping trapezoids $Q_{i,j}$, $(i, j) \in I$. An elementary computation yields that the center of mass of $Q_{i,j}$ is

$$\mathbf{x}_{i,j} = \frac{1}{3}\left(\frac{2\tan^2\alpha_i + \sin^2\beta_j}{\tan^2\alpha_i + \sin^2\beta_j}\cos\beta_j \mathbf{f}_j + \frac{\tan^2\alpha_i + 2\sin^2\beta_j}{\tan^2\alpha_i + \sin^2\beta_j}\frac{1}{\cos\alpha_i}\mathbf{v}_i\right).$$

The area of $Q_{i,j}$ is $\tan\alpha_i \sin\beta_j$. Thus, letting $A = \sum_{(i,j)\in I} \tan\alpha_i \sin\beta_j$, we have

$$\mathrm{cm}_2(\mathbf{P}) = \frac{1}{3A}\sum_{(i,j)\in I} \tan\alpha_i \sin\beta_j \mathbf{x}_{i,j}. \tag{6.23}$$

Let us define the smooth vector field $h : \mathbf{D} \to T\mathbf{D}$ as

$$h(\mathbf{p}) = \sum_{(i,j)\in I} h_{i,j}(\mathbf{p}), \tag{6.24}$$

where

$$h_{i,j}(\mathbf{p}) = \frac{1}{\sinh d_i^v \cosh d_j^f}\left(\frac{2\cosh^2 d_j^f + \sinh^2 d_i^v}{\cosh^2 d_j^f + \sinh^2 d_i^v}\tanh d_j^f \mathbf{f}_j(p)\right.$$

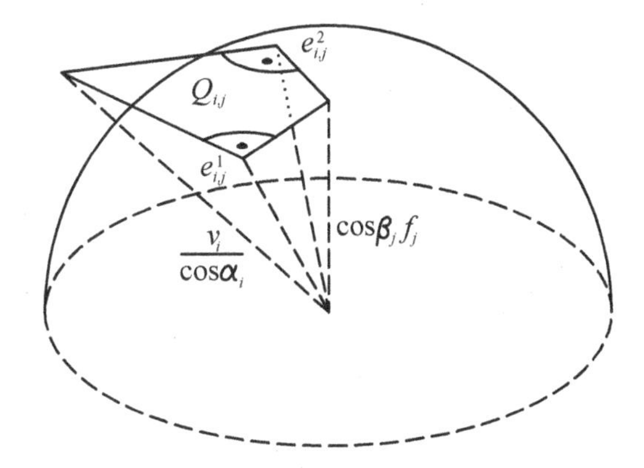

FIGURE 6.6
The right trapezoid $Q_{i,j}$.

$$+\frac{\cosh^2 d_j^f + 2\sinh^2 d_i^v}{\cosh^2 d_j^f + \sinh^2 d_i^v} \coth d_i^v \mathbf{v}_i(p) \Bigg).$$

Here, for brevity, we set $d_i^v = d_i^v(\mathbf{p})$ and $d_j^f = d_j^f(\mathbf{p})$. The function h is a smooth function on $\mathbf{D}$ with positive coefficients. Furthermore, by Remark 152, if $h(\mathbf{p}) = \mathbf{o}$ for some $\mathbf{p} \in \mathbf{D}$ and T is a Möbius transformation mapping $\mathbf{p}$ into $\mathbf{o}$, then $\mathrm{cm}_2(T(\mathbf{P})) = \mathbf{o}$. Similarly like in case of $\mathrm{cm}_1(\cdot)$, we show that the conditions of Lemma 155 are satisfied for h.

To prove (ii) and (iii), we apply the same argument as in case of $\mathrm{cm}_1(\cdot)$. To prove (iv), we follow the line of the same proof, and imagine the configuration in the half space model. Let $\mathbf{q}$ be the ideal point of V_1, V_2, F_1 and F_2. Then $\mathbf{D}$ is bounded by the vertical planes V_1, V_2, F_1 and F_2 which form a rectangle-based half-infinite cylinder. We adapt the notations from the previous subsection, and set $D_t = \mathbf{D} \cap \{z = t\}$ for all $t > 0$. We denote the z-coordinate of $h_{i,j}(\mathbf{p})$ by $z_{i,j}(\mathbf{p})$, and show that their sum is positive if t is sufficiently large.

By Remark 154 and an elementary computation, if $i \notin \{1,2\}$, then $z_{i,j}(\mathbf{p})$ uniformly tends to zero for all $\mathbf{p} \in D_t$ as $t \to \infty$. To examine the remaining cases, for $i = 1,2$, let $x_i(\mathbf{p})$ denote the Euclidean distance of the point $\mathbf{p}$ from V_i. Then $x_1(\mathbf{p}) + x_2(\mathbf{p}) = x$ is the Euclidean distance of V_1 and V_2. By Remark 154, there is some constant $C_1 > 0$ independent of $\mathbf{p}$, t, i and j such that for all $\mathbf{p} \in D_t$, $j \geq 3$ and $i \in \{1,2\}$, we have $z_{i,j}(\mathbf{p}) \geq -\frac{C_1}{x_i}$. Similarly, there is some constant $C_2 > 0$ independent of $\mathbf{p}, t, i, j$ such that for all $\mathbf{p} \in D_t$, $i, j \in \{1,2\}$, we have $z_{i,j}(\mathbf{p}) \geq \frac{C_2 t^2}{x_i}$. This implies that if t is sufficiently large (and in particular, if $t > \sqrt{\frac{C_1 k}{C_2}}$, where k is the maximal degree of a vertex of $\mathbf{P}$), then the z-coordinate of $h(\mathbf{p})$ is positive for all $\mathbf{p} \in D_t$. From this, Theorem 28 readily follows for $\mathrm{cm}_2(\cdot)$.

Proof of Theorem 28 for ccm(·).
During this proof, we assume that **P** is simplicial.

Like in the proof for $\mathrm{cm}_2(\cdot)$, we denote by I the set of edges of the vertex-face incidence graph of **P**, and by $\mathcal{V}_j = \{a_j, b_j, c_j\}$ the set of the indices of the vertices adjacent to the jth face of **P**.

Let the convex hull of the jh face of **P** and **o** be denoted by $\mathbf{S}_j$. To compute ccm(**P**), we need to compute the volume and the circumcenter of $\mathbf{S}_j$, which we denote by m_j and $\mathbf{p}_j$. To do this, in the next lemma for simplicity we omit the index j, and in addition denote $\tan \alpha_{x_j}$ by t_x for $x \in \{a, b, c\}$.

Lemma 157 *The volume of* $\mathbf{S}_j$ *is*

$$m_j = \frac{1}{3}\sqrt{t_a t_b t_c \left(t_a + t_b + t_c - t_a t_b t_c\right)}. \tag{6.25}$$

The circumcenter of $\mathbf{S}_j$ *is*

$$\mathbf{p}_j = \sum_{s \in \{a,b,c\}} N_s \mathbf{v}_s, \tag{6.26}$$

where

$$N_a = \frac{(t_b + t_c)\left((t_b + t_c)t_a^2 + (2t_b^2 t_c^2 + t_b^2 + t_c^2)t_a - t_b t_c (t_b + t_c)\right)}{4 t_a t_b t_c \left(t_a + t_b + t_c - t_a t_b t_c\right)}, \tag{6.27}$$

and N_b *and* N_c *are defined analogously.*

Proof. Note that the three edges of $\mathbf{S}_j$ starting at **o** are of length $\frac{1}{\cos \alpha_x}$ with $x \in \{a, b, c\}$. Furthermore, the edge opposite of the one with length $\frac{1}{\cos \alpha_x}$ is $t_y + t_z$, where $\{x, y, z\} = \{a, b, c\}$. Thus, the volume of $\mathbf{S}_j$ can be computed from its edge lengths using a Cayley-Menger determinant. It is worth noting that since the projection of the jth face onto $\mathbb{S}^2$ is a spherical triangle of edge lengths $\alpha_a + \alpha_b$, $\alpha_a + \alpha_c$ and $\alpha_b + \alpha_c$, and such a triangle is spherically convex, its perimeter is $\alpha_a + \alpha_b + \alpha_c < \pi$. From this, an elementary computation yields that $t_i + t_j + t_k - t_i t_j t_k > 0$, and the formula in (6.25) is valid.

We compute $\mathbf{p}_j$. Since the vectors $\mathbf{v}_a, \mathbf{v}_b$ and $\mathbf{v}_c$ are linearly independent, we may write this point in the form $\mathbf{p}_j = \sum_{s \in \{i,j,k\}} N_s \mathbf{v}_s$ for some coefficients N_a, N_b, N_c. We multiply both sides of this equation by $\mathbf{v}_r$ with some $r \in \{a, b, c\}$. Since all $\mathbf{v}_i$s are unit vectors, we have that $\langle \mathbf{v}_s, \mathbf{v}_r \rangle = \cos(\alpha_s + \alpha_r)$ if $s \neq r$, and $\langle \mathbf{v}_r, \mathbf{v}_r \rangle = 1$. On the other hand, for any value of r, $\mathbf{p}_j$ is contained in the plane with normal vector $\mathbf{v}_r$ passing through the point $\frac{\mathbf{v}_r}{2 \cos \alpha_r}$. Hence, it follows that $[N_a, N_b, N_c]^T$ is the solution of the system of linear equations with coefficient matrix

$$\begin{bmatrix} 1 & \cos(\alpha_a + \alpha_b) & \cos(\alpha_a + \alpha_c) \\ \cos(\alpha_a + \alpha_b) & 1 & \cos(\alpha_b + \alpha_c) \\ \cos(\alpha_a + \alpha_c) & \cos(\alpha_b + \alpha_c) & 1 \end{bmatrix}$$

and with constants $\frac{1}{2\cos\alpha_r}$, where $r = a, b, c$. The determinant of the coefficient matrix is $36(m_j)^2(1+t_a^2)(1+t_b^2)(1+t_c^2) > 0$. Thus, this system has a unique solution, which can be computed by Cramer's rule, yielding the formula in (6.27).

Now we prove Theorem 28 for $\mathrm{ccm}(\cdot)$. For $s = 1, 2, \ldots, n$, let us denote the value $\operatorname{csch} d_s^v(\mathbf{p}) = \frac{1}{\operatorname{sh} d_s^v(\mathbf{p})}$ by $\tau_s(\mathbf{p})$. Observe that Remark 152 implies that $\operatorname{csch} d_s^v(\mathbf{o}) = \tan\alpha_s$. For any $\mathbf{p} \in \mathbf{D}$, let us define the vector field

$$h(\mathbf{p}) = \sum_{j=1}^{m} \sum_{s \in \mathcal{V}_j} B_s(p)\mathbf{v}_s(p), \tag{6.28}$$

where, using the notation $\mathcal{V}_j = \{a, b, c\}$ and for brevity omitting the variable $\mathbf{p}$, we have

$$B_a(\mathbf{p}) = \frac{\tanh d_a\,(\tau_b+\tau_c)}{\sqrt{\tau_a\tau_b\tau_c}}$$
$$\cdot \frac{\tau_a^2\,(\tau_b+\tau_c) + \tau_a\left(2\tau_b^2\tau_c^2 + \tau_b^2 + \tau_c^2\right) - \tau_b\tau_c\,(\tau_b+\tau_c)}{\sqrt{(\tau_a+\tau_b+\tau_c-\tau_a\tau_b\tau_c)}}, \tag{6.29}$$

and $B_b(\mathbf{p})$ and $B_c(\mathbf{p})$ are defined similarly. If $h(\mathbf{p}) = \mathbf{o}_\mathbf{p}$ and T is a Möbius transformation that maps $\mathbf{p}$ into $\mathbf{o}$, then $\mathrm{ccm}(T(\mathbf{P})) = \mathbf{o}$. Thus, to prove the statement it is sufficient to prove that for some $\mathbf{p} \in \mathbf{D}$, $h(\mathbf{p}) = \mathbf{o}_\mathbf{p}$. To do this, we check that the conditions of Lemma 155 are satisfied.

Let $\mathbf{Z}$ denote the set of points of $\mathbf{D}$ whose integral curve is closed. Since for any value of j, F_j is perpendicular to any V_i with $(i, j) \in I$ and does not intersect any other V_i, like in the proof for $\mathrm{cm}_1(\cdot)$, it follows that if $\mathbf{q} \in F_j$ for some plane F_j associated to a face circle of $\mathbf{P}$, then $\mathbf{q}$ has a neighborhood disjoint from $\mathbf{Z}$, and no integral curve ends at $\mathbf{q}$.

Let $\mathbf{q} \in V_i$ for some value of i. It is an elementary computation to check that if $\alpha + \beta + \gamma = \pi$, and $0 < \alpha, \beta, \gamma < \frac{\pi}{2}$, then $\tan\alpha + \tan\beta + \tan\gamma = \tan\alpha\tan\beta\tan\gamma$. This and Remark 152 imply that if $\mathbf{p} \to \mathbf{q}$ and $i \in \{a, b, c\}$, then the denominator of $B_a(\mathbf{p})$ tends to zero. Since the numerator tends to a positive number if $a = i$, and to zero if $i = b$ or $i = c$, it follows that if $i \in \mathcal{V}_j$, then the length of $\sum_{s \in \mathcal{V}_j} B_s(\mathbf{p})\mathbf{v}_s(\mathbf{p})$ tends to ∞, and its direction tends to that of $\mathbf{v}_i(\mathbf{p})$. Since $i \notin \mathcal{V}_j$ implies that $\sum_{s \in \mathbf{V}_j} B_s(p)\mathbf{v}_s(p)$ can be continuously extended to $\mathbf{q}$, it follows that the angle of $h(\mathbf{p})$ and the external normal vector of V_i at $\mathbf{q}$ is 'almost' zero in a suitable neighborhood of $\mathbf{q}$. This yields (iii).

We prove (iv) in the Poincaré half space model with $\mathbf{q}$ being the 'point at infinity'. Without loss of generality, we may assume that $\mathbf{q}$ is the ideal point of V_1, V_2, F_1 and F_2. Then these two pairs of hyperbolic planes are represented by two perpendicular pairs of vertical hyperbolic planes. As before, let D_t denote the set of points in $\mathbf{D}$ with z-coordinates equal to t. We show that the z-coordinate of $h(\mathbf{p})$ is positive for any $\mathbf{p} \in D_t$, if t is sufficiently large. For any j and any $i \in \mathcal{V}_j$, let us denote the z-coordinate of $B_i(\mathbf{p})\mathbf{v}_i(\mathbf{p})$ by $z_i^j(\mathbf{p})$.

Let $\mathbf{p} \in D_t$, and denote by x_1 and x_2 the Euclidean distance of $\mathbf{p}$ from V_1 and V_2, respectively. Consider some value of j. If $\mathcal{V}_j$ is disjoint from $\{1,2\}$, then Remark 154 and (6.29) show that there is some $C_1 > 0$ independent of $\mathbf{p}$ such that $|z_i^j(\mathbf{p})| \leq \frac{C_1}{t^2}$ if t is sufficiently large. Assume that $\mathcal{V}_j$ contains exactly one of $1,2$, say 1. Then, an elementary computation and Remark 154 yield the existence of some $C_2, C_3 > 0$ independent of $\mathbf{p}$, such that $|z_1^j(\mathbf{p})| \leq \frac{C_2}{t^2}$, and for $1 \neq i \in \mathcal{V}_j$, $|z_i^j(\mathbf{p})| \leq C_3 \frac{t^2}{x_1^2}$.

Finally, let $\mathcal{V}_j = \{1,2,i\}$. Note that in this case $j = 1$ or $j = 2$. Furthermore, since $\mathbf{P}$ is simplicial, we have that the Euclidean radius of the hemisphere representing V_i is $\frac{x_1+x_2}{2}$, and the Euclidean distance of the center of this hemisphere from the projection of $\mathbf{p}$ onto the $\{z = 0\}$ plane is $\sqrt{\left(\frac{x_1-x_2}{2}\right)^2 + y_j^2}$, where y_j is the Euclidean distance of $\mathbf{p}$ from F_j (cf. Figure 6.7). An elementary computation yields that by this and Remark 154, the denominator in (6.29) is $\frac{t^3(x_1+x_2)y_j}{(t^2+y_j^2-x_1x_2)x_1x_2}$. Using this, we have $|z_1^j(\mathbf{p})| \leq \frac{2x_1^2}{x_2 y_j} t$, $|z_2^j(\mathbf{p})| \leq \frac{2x_2^2}{x_1 y_j} t$ and $z_i^j(\mathbf{p}) \geq \frac{x_1+x_2}{2x_1^2x_2^2y_j} t^3$ if t is sufficiently large. Using these estimates, we have $z_1^j(\mathbf{p}) + z_2^j(\mathbf{p}) + z_i^j(\mathbf{p}) \geq \frac{C_4 t^3}{x_1^2 x_2^2}$ for some $C_4 > 0$ independently of t and $\mathbf{p}$. Thus, there is some $C > 0$ such that if t is sufficiently large, $\sum_{j=1}^n \sum_{i \in \mathcal{V}_j} z_i^j(\mathbf{p}) \geq Ct^3$, and, in particular, this expression is positive. The regions D_t form a codimension 1 foliation of a neighborhood of $\mathbf{q}$, and thus Theorem 28 for $\mathrm{ccm}(\cdot)$ follows from Lemma 155.

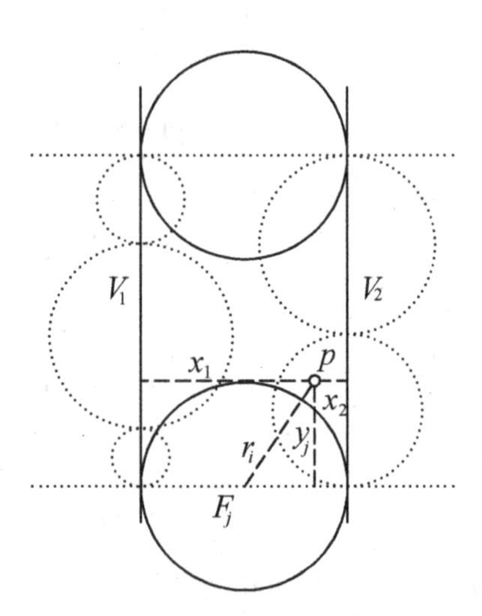

FIGURE 6.7
The ideal points of hyperbolic planes associated to a simplicial polyhedron in the Euclidean plane $\{z = 0\}$. Continuous lines represent planes associated to vertex circles. Dotted lines represent planes associated to face circles.

Proof of Theorem 29 for points of the Euler line.
Again, we assume that $\mathbf{P}$ is simplicial. Using the calculations in the proof for $\mathrm{cm}_2(\mathbf{P})$, we have that the center of mass of $\mathbf{P}$ is

$$\mathrm{cm}_3(\mathbf{P}) = \frac{1}{4A} \sum_{(i,j)\in I} \tan\alpha_i \sin\beta_j \cos\beta_j \left(\frac{2\tan^2\alpha_i + \sin^2\beta_j}{\tan^2\alpha_i + \sin^2\beta_j} \cos\beta_j \mathbf{f}_j \right.$$

$$\left. + \frac{\tan^2\alpha_i + 2\sin^2\beta_j}{\tan^2\alpha_i + \sin^2\beta_j} \frac{1}{\cos\alpha_i} \mathbf{v}_i \right),$$

where $A = \sum_{(i,j)\in I} \tan\alpha_i \sin\beta_j \cos\beta_j$.

By Remark 152, we define the smooth vector field $h_{cm} : \mathbf{D} \to T\mathbf{D}$ as

$$h_{cm}(p) = \sum_{(i,j)\in I} h_{i,j}(\mathbf{p}), \tag{6.30}$$

where

$$h_{i,j}(\mathbf{p}) = \frac{\sinh d_j^f}{\sinh d_i^v \cosh^2 d_j^f} \left(\frac{2\cosh^2 d_j^f + \sinh^2 d_i^v}{\cosh^2 d_j^f + \sinh^2 d_i^v} \tanh d_j^f \mathbf{f}_j(\mathbf{p}) \right.$$

$$\left. + \frac{\cosh^2 d_j^f + 2\sinh^2 d_i^v}{\cosh^2 d_j^f + \sinh^2 d_i^v} \coth d_i^v \mathbf{v}_i(\mathbf{p}) \right).$$

Furthermore, for any $\lambda \in (0,1)$, we set $h_\lambda(\mathbf{p}) = \lambda h_{cm}(\mathbf{p}) + (1-\lambda) h_{ccm}(\mathbf{p})$, where $h_{ccm} : \mathbf{D} \to T\mathbf{D}$ is the vector field defined in (6.28). We observe that if there is some $\mathbf{p} \in \mathbf{D}$ such that $h_\lambda(\mathbf{p}) = \mathbf{o}$, and T is a Möbius transformation moving $\mathbf{p}$ to $\mathbf{o}$, then $\mathbf{o} = \lambda\, \mathrm{cm}_3(T(\mathbf{P})) + (1-\lambda)\, \mathrm{ccm}(T(\mathbf{P}))$.

We show that the conditions of Lemma 155 are satisfied for h_λ. Note that since $\lambda \in (0,1)$, all coefficients in the definition of h_λ are positive. To check (i), (ii) and (iii), we may apply an argument similarly as before. To prove (iv), again we represent the configuration in the half space model. Let D_t be the intersection of D with the horosphere $\{z = t\}$, and $z_{cm}(\mathbf{p})$ and $z_\lambda(\mathbf{p})$ denote the z-coordinate of $h_{cm}(\mathbf{p})$ and $h_\lambda(\mathbf{p})$, respectively. Then an elementary computation yields by Remark 154 that there is some $\bar{C} > 0$ such that $|z_{cm}(\mathbf{p})| \leq \bar{C}$ for all $\mathbf{p} \in D_t$, if t is sufficiently large. Thus, by the estimates in the proof for $\mathrm{ccm}(\cdot)$ and since $\lambda < 1$ it follows that if t is sufficiently large, then $z_\lambda(\mathbf{p}) > 0$ for all $\mathbf{p} \in D_t$. Consequently, Lemma 155 can be applied, and Theorem 28 holds for the considered point of the Euler line.

6.8.4 Proof of Theorem 29

To prove Theorem 29, we follow the line of the proof of Theorem 28. To do this, we need a lemma for polyhedral regions in Euclidean space.

Lemma 158 *Let* $\mathbf{S}_1, \ldots, \mathbf{S}_k$ *be closed half spaces in* $\mathbb{E}^d$*, with outer normal vectors* $\mathbf{m}_1, \ldots, \mathbf{m}_k$*. Then there are unit normal vectors* $\mathbf{v}_1, \ldots, \mathbf{v}_m$ *such that* $\langle \mathbf{m}_i, \mathbf{v}_j \rangle \leq 0$*, for all* $1 \leq i \leq l$ *and* $1 \leq j \leq m$*, and for arbitrary closed half spaces* $\mathbf{S}'_1, \ldots, \mathbf{S}'_m$ *with outer unit normal vectors* $\mathbf{v}_1, \ldots, \mathbf{v}_m$*, respectively, the set* $\mathbf{Q} = \left(\bigcap_{i=1}^k \mathbf{S}_i \right) \cap \left(\bigcap_{j=1}^m \mathbf{S}'_j \right)$ *is bounded.*

Proof. First, observe that the property that $\mathbf{Q}$ is bounded is equivalent to the property that there is no unit vector $\mathbf{v} \in \mathbb{S}^{d-1}$ such that $\langle \mathbf{v}, \mathbf{m}_i \rangle \leq 0$ and $\langle \mathbf{v}, \mathbf{v}_j \rangle \leq 0$ holds for all $1 \leq i \leq k$ and $1 \leq j \leq m$. In other words, $\mathbf{Q}$ is bounded if and only if the open hemispheres of $\mathbb{S}^{d-1}$, centered at the $\mathbf{m}_i$s and the $\mathbf{v}_j$s, cover $\mathbb{S}^{d-1}$. If $\bigcap_{i=1}^k \mathbf{S}_i$ is bounded, there is nothing to prove, and thus, we may consider the set $\mathbf{Z}$ of vectors in $\mathbb{S}^{d-1}$ not covered by any open hemisphere centered at some $\mathbf{m}_i$. Note that since $\mathbf{Z}$ is the intersection of finitely many closed hemispheres, it is compact. Let $\mathbf{F}(\mathbf{v})$ denote the open hemisphere centered at $\mathbf{v}$. Then the family $\{\mathbf{F}(\mathbf{v}) : \mathbf{v} \in \mathbf{Z}\}$ is an open cover of $\mathbf{Z}$, and thus it has a finite subcover $\{\mathbf{F}(\mathbf{v}_j) : i = 1, \ldots, m\}$. By its construction, the vectors $\mathbf{v}_1, \ldots, \mathbf{v}_m$ satisfy the required conditions. □

Now we prove Theorem 29, and for any $i = 1, 2, \ldots, n$, we let ρ_i denote the spherical radius of $\mathbf{C}_i$. We imagine $\mathbb{S}^d$ as the set of ideal points of the Poincaré ball model of $\mathbb{H}^{d+1}$. Then each spherical cap is associated to a closed hyperbolic half space. We denote the half space associated to $\mathbf{C}_i$ by $\mathbf{H}_i$, and the hyperplane bounding $\mathbf{H}_i$ by H_i. Let $\mathbf{D} = \mathbb{H}^{d+1} \setminus (\bigcup_{i=1}^n \mathbf{H}_i)$, and note that as $\rho_i < \frac{\pi}{2}$ for all indices, $\mathbf{D}$ is an open, convex set in $\mathbb{H}^{d+1}$ containing the origin $\mathbf{o}$.

For any value of i, let us define the function $f_i(d) = w_i(\arccos \tanh d)$. Then $f_i : (0, \infty) \to (0, \infty)$ is a positive smooth function on its domain satisfying $\lim_{d \to 0+0} f_i(d) = \infty$. Let $\mathbf{v}_i(\mathbf{p}) \in T_{\mathbf{p}}\mathbb{H}^{d+1}$ denote the unit tangent vector of the geodesic half line starting at $\mathbf{p}$ and perpendicular to H_i, and let $d_i(\mathbf{p})$ denote the hyperbolic distance of $\mathbf{p}$ from H_i. Finally, let the smooth vector field $f : \mathbf{D} \to T\mathbf{D}$ be defined as

$$f(\mathbf{p}) = \sum_{i=1}^{n} f_i(d_i(\mathbf{p}))\mathbf{v}_i(\mathbf{p}).$$

By (6.19) and (6.20), if T is a Möbius transformation mapping $\mathbf{p}$ into $\mathbf{o}$, then $f(\mathbf{p}) = \sum_{i=1^n} w_i(\rho_T(\mathbf{C}_i))\mathbf{c}_T(\mathbf{C}_i)$. Since hyperbolic isometries act transitively on $\mathbb{H}^{d+1}$, it is sufficient to show that $f(\mathbf{p}) = \mathbf{o}_{\mathbf{p}}$ for some $\mathbf{p} \in \mathbf{D}$.

We prove it by contradiction, and assume that $f(\mathbf{p}) \neq \mathbf{o}_{\mathbf{p}}$ for any $\mathbf{p} \in \mathbf{D}$. Consider the integral curves of this vector field. Then, by the Picard-Lindelöf Theorem, they are either closed, or start and terminate at boundary points of $\mathbf{D}$. On the other hand, since f_i is smooth for all values of i, f_i has an antiderivative function F_i on its domain. It is easy to check that $\mathrm{grad}(-\sum_{i=1}^n F_i(d_i(\mathbf{p}))) = f(\mathbf{p})$, implying that f is a gradient field, and thus it has no closed integral curves.

Our main tool is the next lemma. To state it, we define a *neighborhood* of a point $\mathbf{q}$ in the boundary of $\mathbf{D}$ as the intersection of $\mathbf{D}$ with a neighborhood of $\mathbf{q}$ in $\mathbb{E}^{d+1}$ induced by the Euclidean topology (cf. Subsection 6.8.2). Recall from Theorem 29 that if $\mathbf{q} \in \mathbb{S}^d$, then $I(\mathbf{q})$ denotes the set of indices of the spherical caps $\mathbf{C}_i$ that contain $\mathbf{q}$ in their boundaries.

Lemma 159 *Let $\mathbf{q}$ be a boundary point of $\mathbf{D}$, and if $\mathbf{q} \notin \mathbb{S}^d$, then let $I(\mathbf{q})$ denote the set of indices such that $\mathbf{q} \in H_i$.*

(i) If $\mathbf{q} \notin \mathbb{S}^d$, then $\mathbf{q}$ has a neighborhood $\mathbf{V}$ such that any integral curve intersecting $\mathbf{V}$ terminates at a point of H_j for some $j \in I(\mathbf{q})$.

(ii) If $\mathbf{q} \in \mathbb{S}^d$, then there is no integral curve terminating at $\mathbf{q}$.

Proof. First, we prove (i) for the case that $I(\mathbf{q}) = \{i\}$ is a singleton. Let $\mathbf{v}$ be the external unit normal vector of bd$\mathbf{D}$ at $\mathbf{q}$. For any $\mathbf{p} \in \mathbf{D}$, if $\mathbf{p} \to \mathbf{q}$, then $f_i(d_i(\mathbf{p})) \to \infty$, and $\mathbf{v}_i(\mathbf{p})$ tends to a vector of unit hyperbolic length, perpendicular to H_i at $\mathbf{q}$ and pointing outward. On the other hand, $\sum_{j \neq i} f_j(d_j(\mathbf{p}))\mathbf{v}_j(\mathbf{p})$ is continuous at $\mathbf{q}$ and hence it tends to a vector of fixed hyperbolic length. Thus, for every $\varepsilon > 0$ there is a neighborhood $\mathbf{V}$ of q such that the angle between $\mathbf{v}$ and $f(\mathbf{p})$ is at most ε, for any $\mathbf{p} \in \mathbf{V}$. This implies (i) in this case. If $I(\mathbf{q}) = \{j_1, \ldots, j_k\}$ is not a singleton and the inner unit normal vectors of $H_{j_1}, \ldots, H_{j_k}$ are denoted by $\mathbf{v}_{j_1}, \ldots, \mathbf{v}_{j_k}$, respectively, then a similar argument shows that if $\mathbf{p}$ is 'close to $\mathbf{q}$', then $f(\mathbf{p})$ is 'close' to the conic hull of these vectors.

Now we prove (ii). Our method is to show that $\mathbf{q}$ has a basis of closed neighborhoods with the property that no integral curve enters any of them, which clearly implies (ii). For computational reasons, we imagine the configuration in the Poincaré half space model, with $\mathbf{q}$ as the 'point at infinity'. The region $\mathbf{D}$ in this model is the intersection of finitely many open hyperbolic half spaces with vertical and spherical bounding hyperplanes, where H_i is vertical if and only if $i \in I(\mathbf{q})$ (cf. Figure 6.8).

Consider a neighborhood $\mathbf{U}$ of $\mathbf{q}$. Then $\mathbf{U}$ is the complement of a set which is bounded in $\mathbb{E}^{d+1}$. Thus, without loss of generality, we may assume that $\mathbf{U}$ is disjoint from all spherical H_is, and it is bounded by a spherical hyperbolic hyperplane H. For any $i \in I(\mathbf{q})$, let $\mathbf{y}_i \in \mathbf{S}$ be the outer unit normal vector of H_i in $\mathbb{E}^{d+1}$, where we set $\mathbf{S} = \mathbb{S}^d \cap \{x_{d+1} = 0\}$.

Note that as $\mathbf{q}$ is an ideal point of $\mathbf{D}$, $\mathbf{D}$ is not bounded in this model. Let $\mathbf{D}^*$ denote the set of ideal points of $\mathbf{D}$ on the Euclidean hyperplane $\{x_{d+1} = 0\}$. This set is the intersection of the closed half spaces $\mathbf{H}_i$, $i \in I(\mathbf{q})$ in the Euclidean d-space $\{x_{d+1} = 0\}$ (for the definition of $\mathbf{H}_i$, see Subsection 6.8.1). Thus, if $\mathbf{D}^*$ is not bounded, Lemma 158 implies that there are some closed vertical half spaces in $\mathbb{H}^{d+1}$ whose intersection contains H, and whose outer unit normal vectors $\mathbf{y}'_1, \mathbf{y}'_2, \ldots, \mathbf{y}'_m$ satisfy $\langle \mathbf{y}'_j, \mathbf{y}_i \rangle < 0$ for any $\mathbf{y}_i$ and $\mathbf{y}'_j$. Let the intersection of these half spaces with $\mathbf{D}$ be $\mathbf{D}'$, and their bounding hyperbolic hyperplanes be $H'_1, H'_2, \ldots, H'_m$, where $\mathbf{y}'_j$ is the outer unit normal vector of H'_j for all values of j.

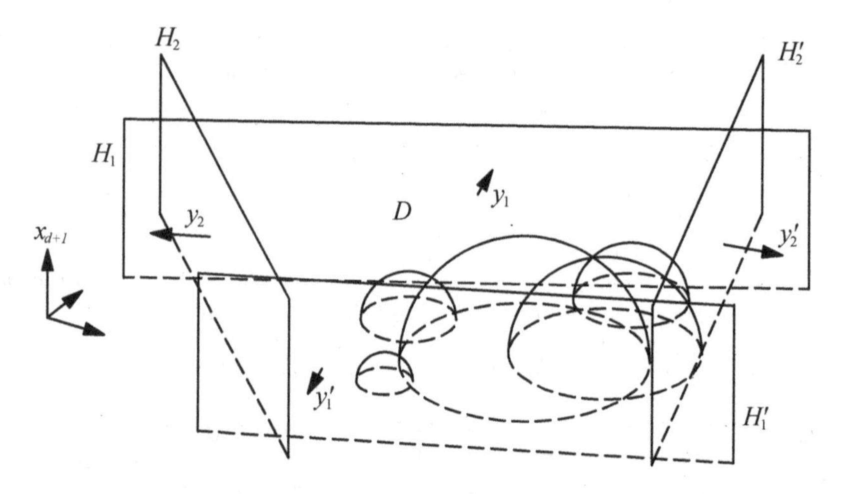

FIGURE 6.8
The configuration in the Poincaré half space model.

Let $\mathbf{p}$ be a boundary point of $\mathbf{D}'$ in $\mathbb{H}^{d+1}$. Then $\mathbf{p} \in H_j' \cap \mathbf{D}'$ for some js. Observe that if $i \in I(\mathbf{q})$, then the geodesic line through $\mathbf{p}$ and perpendicular to H_i, which in the model is a circle arc perpendicular to the hyperplane $\{x_{d+1} = 0\}$, is contained in the vertical plane through $\mathbf{p}$ and perpendicular to H_i. Thus, $\mathbf{v}_i(\mathbf{p})$ points strictly inward into $\mathbf{D}'$ at every boundary point of $\mathbf{D}'$. A similar argument shows the same statement for any $i \notin I(\mathbf{q})$ as well. As a result, we have that the integral curve through any point $\mathbf{p} \in \mathrm{bd}\mathbf{D}'$ enters $\mathbf{D}'$ at $\mathbf{p}$.

Let X_t denote the set $\{x_{d+1} = t\}$ for any $t > 0$, and note that this is a horosphere in $\mathbb{H}^{d+1}$ with $\mathbf{q}$ as its unique ideal point. Set $D_t = X_t \cap \mathbf{D}'$. We show that if t is sufficiently large, then $f(\mathbf{p})$ has a negative x_{d+1}-coordinate. We denote this coordinate by $z(\mathbf{p})$.

Let $\mathbf{p} \in D_t$. It follows from Remark 154 and an elementary computation that if $i \in I(\mathbf{q})$, then the x_{d+1}-coordinate of $\mathbf{v}_i(\mathbf{p})$ is $\tanh d_i(\mathbf{p})$, and if $i \notin I(\mathbf{q})$, then it tends to -1 as $d_i(\mathbf{p}) \to \infty$. On the other hand, for any $\varepsilon, K > 0$ there is some value t_0 such that if $t > t_0$, then $d_i(\mathbf{p}) < \varepsilon$ for all $i \in I(\mathbf{q})$, and $d_i(\mathbf{p}) > K$ for all $i \notin I(\mathbf{q})$ and for all $\mathbf{p} \in D_t$. This implies that

$$\lim_{t\to\infty} \sup_{\mathbf{p}\in D_t} z(\mathbf{p}) = \lim_{d\to 0+0} \sum_{i\in I(\mathbf{q})} f_i(d) \tanh d - \lim_{d\to\infty} \sum_{i\notin I(\mathbf{q})} f_i(d).$$

By the condition (1.3) and the relation (6.19), we have that this quantity is negative, implying that $z(\mathbf{p})$ is negative for all $\mathbf{p} \in D_t$ if t is sufficiently large. Let t' be chosen to satisfy this property. Without loss of generality, we may also assume that $X_{t'}$ does not intersect the hyperplane H. Let $\bar{\mathbf{V}}$ denote the set of points in $\mathbf{D}'$ with x_{d+1}-coordinates less than t', and let $\mathbf{V} = \mathbb{H}^{d+1} \setminus \bar{\mathbf{V}}$.

Then $\mathbf{V}$ is a neighborhood of $\mathbf{q}$ in $\mathbb{H}^{d+1}$, contained in $\mathbf{U}$, and $\mathbf{V}$ has the property that the integral curve through any boundary point $\mathbf{p}$ of $\mathbf{V}$ leaves $\mathbf{V}$ at $\mathbf{p}$. This proves (ii). □

Now we finish the proof of Theorem 29. By the conditions in the formulation of the theorem, the set $\bigcup_{i=1}^{n} \mathrm{int}\mathbf{C}_i \subset \mathbb{S}^d$ is disconnected. Let the components of this set be $X_1, X_2, \ldots, X_r$. By Lemma 159, the integral curve of every point $\mathbf{p} \in \mathbf{D}$ terminates at some point of these sets. Let $\mathbf{Y}_j$ denote the points of $\mathbf{D}$ whose integral curve ends at a point of X_j. By Lemma 159, no $\mathbf{Y}_j$ is empty, and it also implies that $\mathbf{Y}_j$ is open in $\mathbf{D}$ for all js. Thus, $\mathbf{D}$ is the disjoint union of the r open sets $\mathbf{Y}_1, \mathbf{Y}_2, \ldots, \mathbf{Y}_r$, where $r > 1$. On the other hand, $\mathbf{D}$ is an open convex set, and thus, it is connected; a contradiction.

7

Proofs on the Volume of the Convex Hull of a Pair of Convex Bodies

Summary. In this chapter we collect selected proofs of some theorems from Chapter 2. In Section 7.1 we investigate the properties of the volume of a linear parameter system, and the Steiner symmetrization of a convex body. We use these theorems to prove sharp estimates about the volume of translation, reflection, and associated $(d+1)$-dimensional body of a d-dimensional convex body in Section 7.2. In Section 7.3 we find similar estimates for $c_i(\mathbf{K})$ for certain values of $\mathbf{K}$, and characterize the plane convex bodies satisfying the translative constant volume property. In Section 7.4 we prove the Blaschke-Santaló inequality about the maximum of the Holmes-Thompson volume of the unit ball of a normed space. In Section 7.5 we examine the dual problem in the plane, and determine the minimum of the Holmes-Thompson volume of the unit disks of 2-dimensional norms. In Section 7.6 we prove sharp estimates for the Gromov's mass and mass* of the unit disk of a normed plane, and also for the normed volumes of unit disks of Radon norms. Finally, in Section 7.7 we determine the minimum and the maximum values of the normed variants of the quantity $c^{tr}(\mathbf{K})$ both if $\mathbf{K}$ is an arbitrary plane convex body, and if it is assumed to be centrally symmetric.

7.1 Proofs of Theorems 32 and 33

7.1.1 Proof of Theorem 32

Observe that since the sets $\mathcal{P}$ and L are bounded, the volume function

$$V(t) = \mathrm{vol}_d(\mathbf{K}(t)) = \mathrm{vol}_d\left(\mathrm{conv}\{\mathbf{p}_i + t\lambda_i \mathbf{e} : i \in I\}\right)$$

is a continuous function of t. Thus, it is sufficient to prove that for arbitrary values $t_1, t_2 \in \mathbb{R}$, we have

$$V\left(\frac{t_1+t_2}{2}\right) \leq \frac{1}{2}\left(V(t_1)+V(t_2)\right).$$

Furthermore, as the property in Theorem 32 is invariant under linear transformations of t, it is sufficient to show that

$$V(0) \leq \frac{1}{2}\left(V(-1) + V(1)\right). \tag{7.1}$$

Without loss of generality, let $\mathbf{e} = (1, 0, 0, \ldots, 0) \in \mathbb{E}^d$, and let H denote the hyperplane $\{x_1 = 0\}$. For any $\mathbf{x} \in H$, we denote by $l(\mathbf{x}, t)$ the length of the intersection of the line $\{\mathbf{x} + \tau\mathbf{e} : \tau \in \mathbb{R}\}$ with $K(t)$. Note that it may happen that $l(\mathbf{x}, t) = 0$ for some values of $\mathbf{x}$ and t. Since $V(t) = \int\limits_{\mathbf{x}\in H} l(\mathbf{x}, t)\, d\mathbf{x}$, to prove (7.1), it is sufficient to prove that

$$l(\mathbf{x}, 0) \leq \frac{1}{2}\left(l(\mathbf{x}, -1) + l(\mathbf{x}, 1)\right) \tag{7.2}$$

for all $\mathbf{x} \in H$.

If $l(\mathbf{x}, 0) = 0$, the inequality in (7.2) trivially holds. Thus, assume that $l(\mathbf{x}, 0) > 0$. Let $\varepsilon > 0$ be fixed. Then, by the definition of a linear parameter system, there are some $\alpha, \beta \in \mathbb{R}$ such that $\mathbf{x} + \beta\mathbf{e}, \mathbf{x} + \alpha\mathbf{e} \in \mathbf{K}(0)$ and $\beta - \alpha \geq l(\mathbf{x}, 0) - \varepsilon$.

Since $\mathbf{K}(0) = \operatorname{conv}\mathcal{P} = \operatorname{conv}\{\mathbf{p}_i : i \in I\}$, by Carathéodory's theorem there are some indices $\mathbf{p}_{i_1}, \ldots, \mathbf{p}_{i_{d+1}}, \mathbf{p}_{j_1}, \ldots, \mathbf{p}_{j_{d+1}} \in I$ and non-negative real numbers $\alpha_1, \ldots, \alpha_{d+1}, \beta_1, \ldots, \beta_{d+1} \in \mathbb{R}$ satisfying

$$\mathbf{x} + \alpha\mathbf{e} = \sum_{s=1}^{d+1} \alpha_s \mathbf{p}_{i_s}, \quad \sum_{s=1}^{d+1} \alpha_s = 1,$$

and

$$\mathbf{x} + \beta\mathbf{e} = \sum_{s=1}^{d+1} \beta_s \mathbf{p}_{j_s}, \quad \sum_{s=1}^{d+1} \beta_s = 1.$$

Now, since the points $\mathbf{p}_{i_s} \pm \lambda_{i_s}\mathbf{e}$ and $\mathbf{p}_{j_s} \pm \lambda_{j_s}\mathbf{e}$ lie in the sets $\mathbf{K}(\pm 1)$, it follows that

$$l(\mathbf{x}, 1) \geq \left(\beta + \sum_{s=1}^{d+1} \beta_s \lambda_{j_s}\right) - \left(\alpha + \sum_{s=1}^{d+1} \alpha_s \lambda_{j_s}\right)$$

and

$$l(\mathbf{x}, -1) \geq \left(\beta - \sum_{s=1}^{d+1} \beta_s \lambda_{j_s}\right) - \left(\alpha - \sum_{s=1}^{d+1} \alpha_s \lambda_{j_s}\right),$$

implying that

$$\frac{1}{2} l(\mathbf{x}, -1) + \frac{1}{2} l(\mathbf{x}, 1) \geq \beta - \alpha \leq l(\mathbf{x}, 0) - \varepsilon.$$

Since $\varepsilon > 0$ is arbitrary, this yields $\frac{1}{2} l(\mathbf{x}, -1) + \frac{1}{2} l(\mathbf{x}, 1) \geq l(\mathbf{x}, 0)$. From this, the assertion in Theorem 32 readily follows.

7.1.2 Proof of Theorem 33

By Theorem 10.3.2 in [217], for any convex bodies $\mathbf{K}, \mathbf{L} \subset \mathbb{E}^d$, there is a sequence of hyperplanes such that under subsequent Steiner symmetrizations with respect to these hyperplanes, the images of the convex bodies $\mathbf{K}$ and $\mathbf{L}$ converge to Euclidean balls of volume $\mathrm{vol}_d(\mathbf{K})$ and $\mathrm{vol}_d(\mathbf{L})$, respectively. Thus, it is sufficient to prove (i) of Theorem 33.

To do this, without loss of generality, we may assume that H is the hyperplane $\{x_1 = 0\}$. Since during the proof we are going to apply Theorem 32, we assume also that $\mathbf{K}$ and $\mathbf{L}$ are disjoint, as then we can regard the index set I in Definition 6 as $\mathbf{K} \cup \mathbf{L}$.

Consider some point $\mathbf{p}_H \in \mathbb{E}^d$ such that $S_H(\mathbf{K}) \cap (\mathbf{p}_H + S_H(\mathbf{L})) \neq \emptyset$, and $\mathrm{vol}_d(\mathrm{conv}(S_H(\mathbf{K}) \cup (\mathbf{p}_H + S_H(\mathbf{L})))) = V^*(\mathbf{K}, \mathbf{L})$. Let $\mathbf{x}_H \in S_H(\mathbf{K}) \cap (\mathbf{p}_H + S_H(\mathbf{L}))$. Then $\mathbf{x}_H \in S_H(\mathbf{K})$, and $\mathbf{y}_H = \mathbf{x}_H - \mathbf{p}_H \in S_H(\mathbf{L})$.

As in the proof of Theorem 32, let $\mathbf{e} = (1, 0, \ldots, 0)$, and consider some point $\mathbf{q}_\mathbf{K} \in \mathbf{K}$. Then, during the symmetrization process, the image of any point $\mathbf{q}_\mathbf{K} + t\mathbf{e}$ takes the form $\mathbf{q}_\mathbf{K} + (t - \sigma(\mathbf{q}_\mathbf{K}))\mathbf{e}$, where $\sigma(\mathbf{q}_\mathbf{K}) \in \mathbb{R}$ depends only on $\mathbf{q}_\mathbf{K}$ and not on t. Similarly, for any point $\mathbf{q}_\mathbf{L} \in \mathbf{L}$, the image of every point $\mathbf{q}_\mathbf{L} + t\mathbf{e}$ takes the form $\mathbf{q}_\mathbf{L} + (t - \sigma(\mathbf{q}_\mathbf{L}))\mathbf{e}$. In particular, there are points $\mathbf{x} \in \mathbf{K}$ and $\mathbf{y} \in \mathbf{L}$ such that $\mathbf{x}_H = \mathbf{x} - \sigma(\mathbf{x})\mathbf{e}$ and $\mathbf{y}_H = \mathbf{y} - \sigma(\mathbf{y})\mathbf{e}$.

Let $I = \mathbf{K} \cup \mathbf{L}$. For any point $\mathbf{q}_\mathbf{K} \in \mathbf{K}$, let $\mathbf{p}_{\mathbf{q}_\mathbf{K}} = \mathbf{q}_\mathbf{K} - \sigma(\mathbf{q}_\mathbf{K})\mathbf{e}$, and $\lambda_{\mathbf{q}_\mathbf{K}} = \sigma_{\mathbf{q}_\mathbf{K}} - \sigma(\mathbf{x})$, and similarly, for any $\mathbf{q}_\mathbf{L} \in \mathbf{L}$, let $\mathbf{p}_{\mathbf{q}_\mathbf{L}} = \mathbf{q}_\mathbf{L} + \mathbf{p}_H - \sigma(\mathbf{q}_\mathbf{L})\mathbf{e}$, and $\lambda_{\mathbf{q}_\mathbf{L}} = \sigma_{\mathbf{q}_\mathbf{L}} - \sigma(\mathbf{y})$. Clearly, the sets $\mathcal{P} = \{\mathbf{p}_\mathbf{q} : \mathbf{q} \in \mathbf{K} \cup \mathbf{L}\}$, and $L = \{\lambda_\mathbf{q} : \mathbf{q} \in \mathbf{K} \cup \mathbf{L}\}$ are bounded, and hence, they define a linear parameter system. Using the notation in the proof of Theorem 32, we consider the quantities $V(1)$, $V(0)$ and $V(-1)$.

The set $\mathbf{K}(1)$ is the convex hull of the points which are either of the form $\mathbf{q} - \sigma(\mathbf{x})\mathbf{e}$, where $\mathbf{q} \in \mathbf{K}$, or of the form $\mathbf{q} + \mathbf{p}_H - \sigma(\mathbf{y})\mathbf{e}$, where $\mathbf{q} \in \mathbf{L}$. Thus,

$$V(1) = \mathrm{vol}_d(\mathrm{conv}((\mathbf{K} - \sigma(\mathbf{x})\mathbf{e}) \cup (\mathbf{L} + \mathbf{p}_H - \sigma(\mathbf{y})\mathbf{e}))).$$

On the other hand, $\mathbf{x} - \sigma(\mathbf{x})\mathbf{e} = \mathbf{x}_H = \mathbf{y}_H + \mathbf{p}_H = \mathbf{y} + \mathbf{p}_H - \sigma(\mathbf{y})\mathbf{e} \in (\mathbf{K} - \sigma(\mathbf{x})\mathbf{e}) \cap (\mathbf{L} + \mathbf{p}_H - \sigma(\mathbf{y})\mathbf{e})$. This and the definition of $V^*(\cdot, \cdot)$ in Definition 8 imply that

$$V(1) \leq V^*(\mathbf{K}, \mathbf{L}).$$

Similarly, $\mathbf{K}(-1)$ is the convex hull of the points which are either of the form $\mathbf{q} - 2\sigma(\mathbf{q})\mathbf{e} + \sigma(\mathbf{x})\mathbf{e}$, where $\mathbf{q} \in \mathbf{K}$, or of the form $\mathbf{q} + \mathbf{p}_H - 2\sigma(\mathbf{q})\mathbf{e} + \sigma(\mathbf{y})\mathbf{e}$, where $\mathbf{q} \in \mathbf{L}$. On the other hand, the sets $\mathbf{K}' = \{\mathbf{q} - 2\sigma(\mathbf{q})\mathbf{e} : \mathbf{q} \in \mathbf{K}$ and $\mathbf{L}' = \{\mathbf{q} - 2\sigma(\mathbf{q})\mathbf{e} : \mathbf{q} \in \mathbf{L}$ are the reflections of $\mathbf{K}$ and $\mathbf{L}$, respectively, about the hyperplane H of the symmetrization. Thus,

$$V(-1) = \mathrm{vol}_d(\mathrm{conv}((\mathbf{K}' + \sigma(\mathbf{x})\mathbf{e}) \cup (\mathbf{L}' + \mathbf{p}_H + \sigma(\mathbf{y})\mathbf{e}))).$$

Since $\mathbf{x} - 2\sigma(\mathbf{x})\mathbf{e} \in \mathbf{K}'$ and $\mathbf{y} - 2\sigma(\mathbf{y})\mathbf{e} \in \mathbf{L}'$, it follows that

$$\mathbf{x} - 2\sigma(\mathbf{x})\mathbf{e} + \sigma(\mathbf{x})\mathbf{e} = \mathbf{x}_H = \mathbf{y}_H + \mathbf{p}_H = \mathbf{y} - 2\sigma(\mathbf{y})\mathbf{e} + \mathbf{p}_H + \sigma(\mathbf{y})\mathbf{e}$$

is a common point of $(\mathbf{K}' + \sigma(\mathbf{x})\mathbf{e}) \cap (\mathbf{L}' + \mathbf{p}_H - \sigma(\mathbf{y})\mathbf{e})$, implying that

$$V(-1) \leq V^*(\mathbf{K}', \mathbf{L}') = V^*(\mathbf{K}, \mathbf{L}).$$

Finally, $\mathbf{K}(0)$ is the convex hull of the sets $S_H(\mathbf{K})$ and $\mathbf{p}_H + S_H(\mathbf{L})$. Hence, by our choice of $\mathbf{p}_H$, we have

$$V(0) = \mathrm{vol}_d\,(\mathrm{conv}\,(S_H(\mathbf{K}) \cup (\mathbf{p}_H + S_H(\mathbf{L})))) = V^*(S_H(\mathbf{K}), S_H(\mathbf{L})).$$

Thus, Theorem 33 follows from Theorem 32.

7.2 Proofs of Theorems 34, 36, 37 and 40

7.2.1 Preliminaries

In this subsection we prove two lemmas that permit us to prove the theorems.

Lemma 160 *Let L_1 and L_2 be linear subspaces in $\mathbb{E}^d$ that are orthogonal complements of each other. For $i = 1, 2$, let d_i denote the dimension of L_i, and let $\mathbf{K}_i$ be a convex body in L_i. Then*

$$\mathrm{vol}_d\,(\mathrm{conv}\,(\mathbf{K}_1 \cup \mathbf{K}_2)) = \frac{d_1! \cdot d_2!}{(d_1 + d_2)!} \mathrm{vol}_{d_1}(\mathbf{K}_1)\mathrm{vol}_{d_2}(\mathbf{K}_2). \qquad (7.3)$$

Proof. For simplicity, we regard the points of $\mathbb{E}^d$ as the pairs $(\mathbf{x}_1, \mathbf{x}_2)$, where $\mathbf{x}_1 \in L_1$ and $\mathbf{x}_2 \in L_2$. Then,

$$\mathbf{K} = \mathrm{conv}\,(\mathbf{K}_1 \cup \mathbf{K}_2) = \{(t\mathbf{x}_1, (1-t)\mathbf{x}_2) : \mathbf{x}_1 \in \mathbf{K}_1, \mathbf{x}_2 \in \mathbf{K}_2, 0 \leq t \leq 1\}.$$

For any point $\mathbf{p} \in L_1$, consider the intersection of $\mathbf{K}$ with the affine subspace through $\mathbf{p}$ and parallel to L_2. This intersection contains the points of the form

$$(\mathbf{p}, \mathbf{q}) = (t\mathbf{x}_1, (1-t)\mathbf{x}_2),$$

for some $0 \leq t \leq 1$, $\mathbf{x}_1 \in \mathbf{K}_1$ and $\mathbf{x}_2 \in \mathbf{K}_2$. In particular, the intersection is not empty if and only if $\mathbf{p} \in \mathbf{K}_1$.

Let $t(\mathbf{p})$ denote the Minkowski functional of $\mathbf{K}_1$; that is, $t(\mathbf{p}) = \min\{t : \mathbf{p} \in t\mathbf{K}_1\}$. Then, for any $\mathbf{p} \in \mathbf{K}_1$, the point $(\mathbf{p}, \mathbf{q})$ is contained in $\mathbf{K}$ if and only if $\mathbf{q} = (1 - t(\mathbf{p}))\mathbf{x}_2$ for some $\mathbf{x}_2 \in \mathbf{K}_2$, or equivalently, $\mathbf{q} \in (1 - t(\mathbf{p}))\mathbf{K}_2$. Thus, the volume of this intersection is $(1 - t(\mathbf{p}))^{d_2}\mathrm{vol}_{d_2}(\mathbf{K}_2)$.

By definition, the value of $t(\mathbf{p})$ is the constant t on the boundary of any homothetic copy $t\mathbf{K}_1$ of $\mathbf{K}_1$. This yields that

$$\mathrm{vol}_d(\mathbf{K}) = \int_{\mathbf{x}_1 \in \mathbf{K}_1} (1 - t(\mathbf{p}))^{d_2}\mathrm{vol}_{d_2}(\mathbf{K}_2)\, d\mathbf{x}_1$$

$$= \mathrm{vol}_{d_2}(\mathbf{K}_2) \int_0^1 (1-t)^{d_2} \left(t^{d_1} \mathrm{vol}_{d_1}(\mathbf{K}_1)\right)' \, dt$$

$$= \mathrm{vol}_{d_1}(\mathbf{K}_1)\mathrm{vol}_{d_2}(\mathbf{K}_2) \int_0^1 t^{d_1}(1-t)^{d_2} \, dt = \frac{d_1! \cdot d_2!}{(d_1+d_2)!} \mathrm{vol}_{d_1}(\mathbf{K}_1)\mathrm{vol}_{d_2}(\mathbf{K}_2).$$

□

Lemma 161 *Let L_1 and L_2 be linear subspaces in $\mathbb{E}^d$ that are orthogonal complements of each other. For $i = 1, 2$, let d_i denote the dimension of L_i, and let $\mathbf{K}_i$ be a convex body in L_i. Let $\mathbf{K}$ be a convex body in $\mathbb{E}^d$ such that its intersection with L_1 is $\mathbf{K}_1$, and its orthogonal projection onto L_2 is $\mathbf{K}_2$, then*

$$\mathrm{vol}_d(\mathbf{K}) \geq \frac{d_1! \cdot d_2!}{(d_1+d_2)!} \mathrm{vol}_{d_1}(\mathbf{K}_1)\mathrm{vol}_{d_2}(\mathbf{K}_2).$$

Furthermore, in case of equality, every d_1-dimensional section of $\mathbf{K}$ parallel to L_2 is a homothetic copy of $\mathbf{K}_2$.

Proof. For any $\mathbf{q} \in \mathbf{K}_2$, let $V_{d_1}(\mathbf{q})$ denote the d_1-dimensional volume of the section of $\mathbf{K}$ with the affine subspace $\mathbf{q} + L_1$. Furthermore, let $\mathbf{B}(\mathbf{q})$ denote the d_1-dimensional ball in $\mathbf{q} + L_1$, with center $\mathbf{q}$ and volume $V_{d_1}(\mathbf{q})$. Finally, let $\mathbf{K}^* = \bigcup_{\mathbf{q} \in \mathbf{K}_2} \mathbf{B}(\mathbf{q})$. Note that $\mathrm{vol}_d(\mathbf{K}^*) = \mathrm{vol}_d(\mathbf{K})$, and that starting with $\mathbf{K}$, $\mathbf{K}^*$ can be approached by subsequent Steiner symmetrizations. Thus, in particular, $\mathbf{K}^*$ is convex.

The section of $\mathbf{K}^*$ with L_1 is $\mathbf{B}(\mathbf{o})$, and with L_2 it is $\mathbf{K}_2$. Thus, we have $\mathrm{conv}(\mathbf{B}(\mathbf{o}) \cup \mathbf{K}_2) \subseteq \mathbf{K}^*$, and, by $\mathrm{vol}_{d_1}(\mathbf{K}_1) = \mathrm{vol}_{d_1}(\mathbf{B}(\mathbf{o}))$, Lemma 160 implies that

$$\mathrm{vol}_d(\mathbf{K}) = \mathrm{vol}_d(\mathbf{K}^*) \geq \frac{d_1! \cdot d_2!}{(d_1+d_2)!} \mathrm{vol}_{d_1}(\mathbf{K}_1)\mathrm{vol}_{d_2}(\mathbf{K}_2). \tag{7.4}$$

Finally, assume that equality holds in (7.4). Then $\mathrm{conv}(\mathbf{B}(\mathbf{o}) \cup \mathbf{K}_2) = \mathbf{K}^*$. Since for any $\mathbf{q} \in \mathrm{bd}\mathbf{K}_2$ and $0 \leq t \leq 1$, the volume of the section of $\mathbf{K}^*$ with $\mathbf{q} + L_1$ is $V_{d_1}(t\mathbf{q})$, and the volume of the section of $\mathrm{conv}(\mathbf{B}(\mathbf{o}) \cup \mathbf{K}_2)$ with $\mathbf{q} + L_1$ is $\mathrm{vol}_{d_1}((1-t)\mathbf{B}(\mathbf{o})) = (1-t)^{d_1} V_{d_1}(\mathbf{o}) = (1-t)^{d_1} \mathrm{vol}_{d_1}(\mathbf{K}_1)$, we have

$$V_{d_1}(t\mathbf{q}) = (1-t)^{d_1} \mathrm{vol}_{d_1}(\mathbf{K}_1).$$

Observe that for any fixed point $\mathbf{q} \in \mathrm{bd}\mathbf{K}_2$, the volume in the formula above is the volume of the section of the (d_1+1)-dimensional cone with apex $\mathbf{q}$ and base $\mathbf{K}_1$. Thus, by the conditions of equality in the Brunn-Minkowski Inequality, it follows that equality occurs in (7.4) if and only if $\mathbf{K} \cap (\mathbf{q} + L_1)$ is a homothetic copy of $\mathbf{K} \cap L_1$ for all $\mathbf{q} \in \mathbf{K}_2$. □

7.2.2 Proofs of the Theorems

First, note that the lower bounds in Theorems 36, 37 and 40 follow immediately from (ii) of Theorem 33. To prove the upper bounds, we apply Lemma 161 for suitably chosen convex bodies $\mathbf{K}_1$ and $\mathbf{K}_2$.

To show how our method works, first we prove the inequality in (2.1). Let $\mathbf{K}$ be an arbitrary convex body in $\mathbb{E}^d$. We define the $(2d)$-dimensional convex body $\mathbf{L}$ as the set of points $(\mathbf{x}_1, \mathbf{x}_2) \in \mathbb{E}^{2d}$ satisfying

$$\mathbf{x}_1 \in \mathbf{K}, \quad \mathbf{x}_1 + \mathbf{x}_2 \in \mathbf{K}.$$

Let L_1 and L_2 be d-dimensional linear subspaces of $\mathbb{E}^{2d}$ spanned by the first, respectively last, n unit vectors of the standard orthonormal basis.

Note that for any $\mathbf{x} = (\mathbf{x}_1, \mathbf{o}) \in L_1$, the intersection $(\mathbf{x} + L_2) \cap \mathbf{L}$ is non-empty if and only if $\mathbf{x}_1 \in \mathbf{K}$, and in this case the section is congruent to $\mathbf{K}$. Thus,

$$\mathrm{vol}_{2d}(\mathbf{L}) = \int_{\mathbf{x} \in \mathbf{K}} \mathrm{vol}_d(\mathbf{K}) \, d\mathbf{x} = (\mathrm{vol}_d(\mathbf{K}))^2 .$$

On the other hand, $\mathbf{x} = (\mathbf{o}, \mathbf{x}_2) \in L_2$ is on the projection of $\mathbf{L}$ onto L_2 if and only if there is some $\mathbf{x}_1 \in \mathbf{K}$ such that $\mathbf{x}_1 + \mathbf{x}_2 \in \mathbf{K}$. Thus, setting $\mathbf{y} = \mathbf{x}_1 + \mathbf{x}_2 \in \mathbf{K}$, $\mathbf{x}$ is on the projection if and only if $\mathbf{x}_2$ can be written in the form $\mathbf{x}_2 = \mathbf{y} - \mathbf{x}_1$ for some $\mathbf{y}, \mathbf{x}_1 \in \mathbf{K}$. This occurs if and only if $\mathbf{x}_2 \in \mathbf{K} - \mathbf{K}$. As the intersection of $\mathbf{L}$ with L_1 is the set of points $(\mathbf{x}_1, \mathbf{o})$, where $\mathbf{x}_1 \in \mathbf{K}$, applying Lemma 161 we obtain that

$$(\mathrm{vol}_d(\mathbf{K}))^2 = \mathrm{vol}_{2d}(\mathbf{L}) \geq \frac{(d!)^2}{(2d)!} \mathrm{vol}_d(\mathbf{K}) \mathrm{vol}_d(\mathbf{K} - \mathbf{K}),$$

from which the inequality in (2.1) readily follows.

Proof of Theorem 40. Consider the $(2d+1)$-dimensional convex body $\mathbf{L}$ defined as

$$\mathbf{L} = \{(\mathbf{x}_1, \mathbf{x}_2, t) : \mathbf{x}_1 \in t\mathbf{K}, \mathbf{x}_1 + \mathbf{x}_2 \in (1-t)\mathbf{K}\} .$$

Fix some $0 \leq t \leq 1$ and $\mathbf{x}_1 \in t\mathbf{K}$. Then the set of points of $\mathbf{L}$ of the form $(\mathbf{x}_1, \cdot, t)$ is congruent to $-\mathbf{x}_1 + (1-t)\mathbf{K}$, and hence, its volume is $(1-t)^d \mathrm{vol}_d(\mathbf{K})$. Thus, we have

$$\mathrm{vol}_{2d+1}(\mathbf{L}) = \int_0^1 \left(\int_{\mathbf{x}_1 \in t\mathbf{K}} (1-t)^d \mathrm{vol}_d(\mathbf{K}) \, d\mathbf{x}_1 \right) dtf$$

$$= \int_0^1 t^d (1-t)^d \, dt \, (\mathrm{vol}_d(\mathbf{K}))^2 = \frac{(d!)^2}{(2d+1)!} (\mathrm{vol}_d(\mathbf{K}))^2 .$$

The section of $\mathbf{L}$ with the subspace $\left\{(\mathbf{x}_1, \mathbf{x}_2, t) : t = \frac{1}{2}, \mathbf{x}_2 = \mathbf{o}\right\}$ is

$$\left\{ \left(\mathbf{x}, \mathbf{o}, \frac{1}{2} \right) : \mathbf{x} \in \frac{1}{2}\mathbf{K} \right\},$$

and its volume is $\frac{1}{2^d} \mathrm{vol}_d(\mathbf{K})$.

The projection of $\mathbf{L}$ onto the subspace $\left\{(\mathbf{o}, \mathbf{x}_2, t) : \mathbf{x}_2 \in \mathbb{E}^d, t \in \mathbb{R}\right\}$ is the set of points $(\mathbf{o}, \mathbf{x}_2, t)$ such that there is some $\mathbf{x}_1 \in t\mathbf{K}$ satisfying $\mathbf{x}_1 + \mathbf{x}_2 \in (1-t)\mathbf{K}$. Set $\mathbf{y} = \mathbf{x}_1 + \mathbf{x}_2$. Then $(\mathbf{o}, \mathbf{x}_2, t)$ belongs to this projection if and only if

$0 \leq t \leq 1$, and $\mathbf{x}_2 = \mathbf{y} - \mathbf{x}_1$ for some $\mathbf{x}_1 \in t\mathbf{K}$ and $\mathbf{y} \in (1-t)\mathbf{K}$. The latter condition is equivalent to saying that $\mathbf{x}_2 \in (1-t)\mathbf{K} + t(-\mathbf{K})$. Thus, this projection is congruent to $C(\mathbf{K})$.

Based on these observations, Lemma 161 yields that

$$\frac{(d!)^2}{(2d+1)!}\left(\mathrm{vol}_d(\mathbf{K})\right)^2 = \mathrm{vol}_{2d+1}(\mathbf{L}) \geq \frac{d!(d+1)!}{(2d+1)!}\frac{1}{2^d}\mathrm{vol}_d(\mathbf{K})\mathrm{vol}_d(C(\mathbf{K})),$$

which implies

$$\mathrm{vol}_d(C(\mathbf{K})) \leq \frac{2^d}{d+1}\mathrm{vol}_d(\mathbf{K}).$$

□

Proof of Theorem 34. Without loss of generality, we may assume that $\mathbf{p} = \mathbf{o} \in \mathbf{K}$. Consider the body $C(\mathbf{K})$ in $\mathbb{E}^{d+1}$. We have

$$C(\mathbf{K}) = \{(\mathbf{x}, t) : \mathbf{x} = t\mathbf{x}_1 + (1-t)(-\mathbf{x}_2) \text{ for some } 0 \leq t \leq 1, \mathbf{x}_1, \mathbf{x}_2 \in \mathbf{K}\}.$$

Since $\mathbf{o} \in \mathbf{K}$, the section of $C(\mathbf{K})$ with the line $\{(\mathbf{o}, t) : t \in \mathbb{R}\}$ is the segment $\{(\mathbf{o}, t) : t \in [0, 1]\}$. Furthermore, the projection of $C(\mathbf{K})$ onto the hyperplane $\{(\mathbf{x}, 0) : \mathbf{x} \in \mathbb{E}^d\}$ is $\mathrm{conv}(\mathbf{K} \cup (-\mathbf{K})) = R_{\mathbf{o}}(\mathbf{K})$.

Now, Lemma 161 implies that

$$\mathrm{vol}_{d+1}(C(\mathbf{K})) \geq \frac{d! \cdot 1!}{(d+1)!}\mathrm{vol}_d(R_{\mathbf{o}}(\mathbf{K})),$$

or equivalently, that

$$\mathrm{vol}_d(R_{\mathbf{o}}(\mathbf{K})) \leq (d+1)\mathrm{vol}_{d+1}(C(\mathbf{K})).$$

By Theorem 40, we have $\mathrm{vol}_{d+1}(C(\mathbf{K})) \leq \frac{2^d}{d+1}\mathrm{vol}_d(\mathbf{K})$, which readily implies

$$\mathrm{vol}_d(R_{\mathbf{o}}(\mathbf{K})) \leq 2^d \mathrm{vol}_d(\mathbf{K}).$$

□

Proof of Theorem 37. Without loss of generality, we assume that the translation vector $\mathbf{x}$ is not zero.

First, note that for any $\mathbf{x} \in \mathbb{E}^d$, the property that $\mathbf{K} \cap (\mathbf{x} + \mathbf{K}) \neq \emptyset$ is equivalent to saying that $\mathbf{x}$ is not longer than a longest chord of $\mathbf{K}$ parallel to $\mathbf{x}$. Let $d_{\mathbf{K}}(\mathbf{x})$ denote the length of such a chord, $\mathbf{x}^{\perp}$ the hyperplane through $\mathbf{o}$ and perpendicular to $\mathbf{x}$, and $\mathbf{K}|\mathbf{x}^{\perp}$ the orthogonal projection of $\mathbf{K}$ onto $\mathbf{x}^{\perp}$. Furthermore, for any $\mathbf{y} \in \mathbf{K}|\mathbf{x}^{\perp}$, let $l(\mathbf{y})$ denote the length of the intersection of $\mathbf{K}$ with the line through $\mathbf{y}$ and parallel to $\mathbf{x}$. Then, for any $\mathbf{x} \neq \mathbf{o}$ satisfying $\mathbf{K} \cap (\mathbf{x} + \mathbf{K}) \neq \emptyset$, we have

$$\mathrm{vol}_d\left(\mathrm{conv}(\mathbf{K} \cup (\mathbf{x} + \mathbf{K}))\right) = \int_{\mathbf{y} \in \mathbf{K}|\mathbf{x}^{\perp}} ||\mathbf{x}|| + l(\mathbf{y})\, d\mathbf{y}$$

$$= \mathrm{vol}_d(\mathbf{K}) + ||\mathbf{x}||\mathrm{vol}_{d-1}(\mathbf{K}|\mathbf{x}^{\perp}) \leq \mathrm{vol}_d(\mathbf{K}) + d_{\mathbf{K}}(\mathbf{x}) \cdot \mathrm{vol}_{d-1}(\mathbf{K}|\mathbf{x}^{\perp}).$$

Let L be a line parallel to $\mathbf{x}$ such that its intersection with $\mathbf{K}$ is a segment of length $d_{\mathbf{K}}(\mathbf{x})$; such a line exists since $\mathbf{K}$ has chords of maximal length parallel to $\mathbf{x}$. Then, since $\mathbf{K}|\mathbf{x}^{\perp}$ is perpendicular to L, Lemma 161 yields that

$$\mathrm{vol}_d(\mathbf{K}) \geq \frac{(d-1)! \cdot 1!}{d!} d_{\mathbf{K}}(\mathbf{x}) \cdot \mathrm{vol}_{d-1}(\mathbf{K}|\mathbf{x}^{\perp}),$$

implying that

$$\mathrm{vol}_d\left(\mathrm{conv}(\mathbf{K} \cup (\mathbf{x}+\mathbf{K}))\right) \leq \mathrm{vol}_d(\mathbf{K}) + d\mathrm{vol}_d(\mathbf{K}) = (d+1)\mathrm{vol}_d(\mathbf{K}).$$

□

Proof of Theorem 36. The upper bound in (2.2) readily follows from Theorem 34, whereas the one in (2.3) is a consequence of Theorem 37. □

7.3 Proofs of Theorems 41 and 46

In this section, for any $\mathbf{m} \in \mathbb{S}^{d-1}$ and any convex body $\mathbf{K}$ in $\mathbb{E}^d$, we denote by $d_{\mathbf{K}}(\mathbf{m})$ the length of a longest chord of $\mathbf{K}$ parallel to $\mathbf{m}$ and by $w_{\mathbf{K}}(\mathbf{m})$ the width of $\mathbf{K}$ in the direction of $\mathbf{m}$. Furthermore, by $\mathbf{m}^{\perp}$ we denote the hyperplane through the origin and perpendicular to $\mathbf{m}$, and by $\mathbf{K}|\mathbf{m}^{\perp}$ the orthogonal projection of $\mathbf{K}$ onto $\mathbf{m}^{\perp}$ (cf. also Section 7.2).

7.3.1 Proof of Theorem 41

To prove the theorem, we need the next lemma, which appeared as Lemma 10 in [177].

Lemma 162 *If* $\mathbf{K}$ *is any not centrally symmetric convex body in* $\mathbb{E}^d$*, where* $d \geq 2$*, then there exist* d *pairwise orthogonal vectors* $\mathbf{m}_1, \mathbf{m}_2, \dots, \mathbf{m}_d \in \mathbb{S}^{d-1}$ *such that applying subsequent Steiner symmetrizations through* $\mathbf{m}_1^{\perp}, \mathbf{m}_2^{\perp}, \dots, \mathbf{m}_d^{\perp}$ *to* $\mathbf{K}$ *leads to an* $\mathbf{o}$*-symmetric convex body* $S_{\mathbf{m}_d^{\perp}}(S_{\mathbf{m}_{d-1}^{\perp}}(\dots S_{\mathbf{m}_1^{\perp}}(\mathbf{K})\dots))$ *that is not an ellipsoid.*

Proof. First, let $d = 2$. We prove that there is some $\mathbf{m}_1 \in \mathbb{S}^1$ such that

$$\frac{\mathrm{vol}_1(K|\mathbf{m}_1^{\perp}) d_{\mathbf{K}}(\mathbf{m}_1)}{\mathrm{area}(\mathbf{K})} = \frac{w_{\mathbf{K}}(\mathbf{m}_1) d_{\mathbf{K}}(\mathbf{m}_1)}{\mathrm{area}(\mathbf{K})} > \frac{4}{\pi}. \tag{7.5}$$

Let $\mathbf{K}_0 = \frac{1}{2}(\mathbf{K} - \mathbf{K})$. Then, since central symmetrization does not change the width and the length of a longest chord in any direction, we have $w_{\mathbf{K}}(\mathbf{m}) d_{\mathbf{K}}(\mathbf{m}) = w_{\mathbf{K}_0}(\mathbf{m}) d_{\mathbf{K}_0}(\mathbf{m})$ for any $\mathbf{m} \in \mathbb{S}^1$. On the other hand, by the Brunn-Minkowski Inequality, we have $\text{area}(\mathbf{K}) < \text{area}(\mathbf{K}_0)$. Thus, replacing $\mathbf{K}$ by $\mathbf{K}_0$, the left-hand side quantity strictly decreases for any $\mathbf{m}_1 \in \mathbb{S}^1$. On the other hand, Theorem 39 implies that for some $\mathbf{m}_1 \in \mathbb{S}^1$, we have

$$\frac{w_{\mathbf{K}_0}(\mathbf{m}_1)d_{\mathbf{K}_0}(\mathbf{m}_1)}{\text{area}(\mathbf{K}_0)} \geq \frac{2\kappa_1}{\kappa_2} = \frac{4}{\pi}.$$

This implies (7.5).

Note that for any $\mathbf{m} \in \mathbb{S}^{d-1}$ and convex body $\mathbf{L} \subset \mathbb{E}^d$, Steiner symmetrization through $\mathbf{m}^\perp$ does not change $\text{vol}_{d-1}(\mathbf{L}|\mathbf{m}^\perp)$, $d_{\mathbf{L}}(\mathbf{m})$ and $\text{vol}_d(\mathbf{L})$. Thus, if we let $\mathbf{m}_2 \in \mathbb{S}^1$ be orthogonal to $\mathbf{m}_1$, then, applying Steiner symmetrization to $\mathbf{K}$ first through $\mathbf{m}_1^\perp$ and then through $\mathbf{m}_2^\perp$, we obtain an $\mathbf{o}$-symmetric convex body $\mathbf{K}'$ satisfying

$$\frac{w_{\mathbf{K}'}(\mathbf{m}_1)d_{\mathbf{K}'}(\mathbf{m}_1)}{\text{area}(\mathbf{K}')} > \frac{4}{\pi}.$$

Since for ellipses, the quantity on the left-hand side is equal to $\frac{4}{\pi}$ for any direction $\mathbf{m}_1 \in \mathbb{S}^1$, it follows that $\mathbf{K}'$ is not an ellipse.

Now, let $d \geq 3$. We show that there is a 2-dimensional linear subspace P of $\mathbb{E}^d$ such that the orthogonal projection $\mathbf{K}|P$ of $\mathbf{K}$ onto P is not centrally symmetric. Suppose for contradiction that $\mathbf{K}|P$ is centrally symmetric for any plane P. Without loss of generality, we may assume that the origin $\mathbf{o}$ is the midpoint of a diameter $[\mathbf{p}, \mathbf{q}]$ of $\mathbf{K}$. Then $\mathbf{K}|P$ is $\mathbf{o}$-symmetric for any plane P satisfying $[\mathbf{p}, \mathbf{q}] \subset P$. Thus, $\mathbf{K}^\circ \cap P$ is $\mathbf{o}$-symmetric for every plane P containing $[\mathbf{p}, \mathbf{q}]$, where $\mathbf{K}^\circ$ denotes the polar of $\mathbf{K}$. This implies that $\mathbf{K}^\circ$, and consequently also $\mathbf{K}$, are $\mathbf{o}$-symmetric, a contradiction.

Let P be a 2-dimensional linear subspace such that $\mathbf{K}|P$ is not centrally symmetric, and set $\mathbf{L} = \mathbf{K}|P$. Let $\mathbf{m}_1, \mathbf{m}_2, \ldots, \mathbf{m}_{d-1}$ be an orthonormal basis in the orthogonal complement of P. Let $\bar{\mathbf{K}}$ be the convex body obtained from $\mathbf{K}$ by subsequent Steiner symmetrizations through $\mathbf{m}_1^\perp, \mathbf{m}_2^\perp, \ldots, \mathbf{m}_{d-2}^\perp$. Then $\mathbf{L} = \bar{\mathbf{K}}|P$. Applying the result of the $d = 2$ case, there are orthogonal unit vectors $\mathbf{m}_{d-1}, \mathbf{m}_d$ in P such that subsequent Steiner symmetrizations of $\mathbf{L}$ in P through $\mathbf{m}_{d-1}^\perp$ and $\mathbf{m}_d^\perp$ result in an o-symmetric plane convex body $\mathbf{L}'$ in P which is not an ellipse. Now, applying Steiner symmetrizations to $\bar{\mathbf{K}}$ through $\mathbf{m}_{d-1}^\perp$ and $\mathbf{m}_d^\perp$, we obtain an $\mathbf{o}$-symmetric convex body $\mathbf{K}'$ such that $\mathbf{K}'|P = \mathbf{L}'$. Thus, the fact that $\mathbf{L}'$ is not an ellipse implies that $\mathbf{K}'$ is not an ellipsoid. □

Now we prove (i) of Theorem 41.

Since for ellipsoids $c^{tr}(\mathbf{K}) = 1 + \frac{2\kappa_{d-1}}{\kappa_n}$, it suffices to show that if $c^{tr}(\mathbf{K}) \leq 1 + \frac{2v_{n-1}}{v_n}$, then $\mathbf{K}$ is an ellipsoid.

Let $\mathbf{K} \in$ be a convex body in $\mathbb{E}^d$ such that $c^{tr}(\mathbf{K}) \leq 1 + \frac{2\kappa_{d-1}}{\kappa_d}$. Consider the case that $\mathbf{K}$ is not centrally symmetric. Let $H \subset \mathbb{E}^d$ be any hyperplane. Then (i) of Theorem 33 yields that $c^{tr}(\mathbf{K}) \geq c^{tr}(S_H(\mathbf{K}))$. On the other hand, Lemma 162 states that, for any not centrally symmetric convex body, there is an orthonormal basis such that subsequent Steiner symmetrizations, through hyperplanes perpendicular to its vectors, yields a centrally symmetric convex body, different from ellipsoids. Combining these statements, we obtain that there is an $\mathbf{o}$-symmetric convex body $\mathbf{K}'$ in $\mathbb{E}^d$ that is not an ellipsoid and satisfies $c^{tr}(\mathbf{K}) \geq c^{tr}(\mathbf{K}')$. Thus, it suffices to prove the assertion in the case that $\mathbf{K}$ is centrally symmetric.

Assume that $\mathbf{K}$ is $\mathbf{o}$-symmetric, and that $c^{tr}(\mathbf{K}) \leq 1 + \frac{2\kappa_{d-1}}{\kappa_s}$. Observe that for any $\mathbf{m} \in \mathbb{S}^{d-1}$, $\mathbf{K}$ and $d_{\mathbf{K}}(\mathbf{m})\mathbf{m} + \mathbf{K}$ touch each other and

$$\frac{\operatorname{vol}_d(\operatorname{conv}(\mathbf{K} \cup (d_{\mathbf{K}}(\mathbf{m})\mathbf{m} + \mathbf{K})))}{\operatorname{vol}_d(\mathbf{K})} = 1 + \frac{d_{\mathbf{K}}(\mathbf{m})\operatorname{vol}_{d-1}(\mathbf{K}|\mathbf{m}^{\perp})}{\operatorname{vol}_d(\mathbf{K})}. \tag{7.6}$$

Clearly, $c^{tr}(\mathbf{K})$ is the maximum of this quantity over $\mathbf{m} \in \mathbb{S}^{d-1}$.

Let $\mathbf{m} \mapsto r_{\mathbf{K}}(\mathbf{m}) = \frac{d_{\mathbf{K}}(\mathbf{m})}{2}$ be the radial function of $\mathbf{K}$. From (7.6) and the inequality $c^{tr}(\mathbf{K}) \leq 1 + \frac{2\kappa_{d-1}}{\kappa_d}$, we obtain that for any $\mathbf{m} \in \mathbb{S}^{d-1}$

$$\frac{\kappa_{d-1}\operatorname{vol}_d(\mathbf{K})}{\kappa_d \operatorname{vol}_{d-1}(\mathbf{K}|\mathbf{m}^{\perp})} \geq r_{\mathbf{K}}(\mathbf{m}). \tag{7.7}$$

Applying this for the polar form of the volume of K, we obtain

$$\operatorname{vol}_d(\mathbf{K}) = \frac{1}{d} \int_{\mathbb{S}^{d-1}} (r_{\mathbf{K}}(\mathbf{m}))^d \, d\mathbf{m} \leq \frac{1}{d} \frac{\kappa_{d-1}^d}{\kappa_d^d} (\operatorname{vol}_d(\mathbf{K}))^d \int_{\mathbb{S}^{d-1}} \frac{1}{(\operatorname{vol}_{d-1}(\mathbf{K}|\mathbf{m}^{\perp}))^d} \, d\mathbf{m},$$

which yields

$$\frac{\kappa_d^d d}{\kappa_{d-1}^d (\operatorname{vol}_d(\mathbf{K}))^{d-1}} \leq \int_{\mathbb{S}^{d-1}} \frac{1}{(\operatorname{vol}_{d-1}(\mathbf{K}|\mathbf{m}^{\perp}))^d} \, d\mathbf{m}. \tag{7.8}$$

On the other hand, combining Cauchy's surface area formula with Petty's projection inequality, we obtain that for every $p \geq -d$,

$$\kappa_d^{1/d} (\operatorname{vol}_d(\mathbf{K}))^{\frac{d-1}{d}} \leq \kappa_d \left(\frac{1}{d\kappa_d} \int_{\mathbb{S}^{d-1}} \left(\frac{\operatorname{vol}_{d-1}(\mathbf{K}|\mathbf{m}^{\perp})}{\kappa_{d-1}} \right)^p d\mathbf{m} \right)^{\frac{1}{p}},$$

with equality only for Euclidean balls if $p > -d$, and for ellipsoids if $p = -d$ (cf., e.g., Theorems 9.3.1 and 9.3.2 in [119]).

This inequality, with $p = -d$ and after some algebraic transformations, implies that

$$\int_{\mathbb{S}^{n-1}} \frac{1}{(\operatorname{vol}_{d-1}(\mathbf{K}|\mathbf{m}^{\perp}))^d} \, d\mathbf{m} \leq \frac{\kappa_d^d d}{\kappa_{d-1}^d (\operatorname{vol}_d(\mathbf{K}))^{d-1}} \tag{7.9}$$

with equality if and only if $\mathbf{K}$ is an ellipsoid. Combining (7.8) and (7.9), we can immediately see that if $c^{tr}(\mathbf{K})$ is minimal, then $\mathbf{K}$ is an ellipsoid, and in this case $c^{tr}(\mathbf{K}) = 1 + \frac{2\kappa_{d-1}}{\kappa_d}$, implying (i) of Theorem 41.

To prove (ii) of Theorem 41, it is sufficient to observe that, like in the proof of (i), by Lemma 162 we may assume that $\mathbf{K}$ is $\mathbf{o}$-symmetric, and apply (i).

Finally, we prove (iii).

For a hyperplane $H \subset \mathbb{E}^d$, let $\mathbf{K}_H$ denote the reflected copy of $\mathbf{K}$ about H. Furthermore, if H is a supporting hyperplane of $\mathbf{K}$, let $\mathbf{K}_{-H}$ be the reflected copy of $\mathbf{K}$ about the other supporting hyperplane of $\mathbf{K}$ parallel to H. Clearly,

$$c_{d-1}(\mathbf{K}) = \frac{1}{\mathrm{vol}_d(\mathbf{K})} \max\{\mathrm{vol}_d(\mathrm{conv}(\mathbf{K} \cup \mathbf{K}_H)) : H \text{ supports } \mathbf{K}\}.$$

For any direction $\mathbf{m} \in \mathbb{S}^{d-1}$, let $\mathbf{F_K}(\mathbf{m})$ be the right cylinder circumscribed about $\mathbf{K}$ and with generators parallel to $\mathbf{m}$. Observe that for any $\mathbf{m} \in \mathbb{S}^{d-1}$ and supporting hyperplane H perpendicular to $\mathbf{m}$, we have $\mathrm{vol}_d(\mathrm{conv}(\mathbf{K} \cup K_H)) + \mathrm{vol}_d(\mathrm{conv}(\mathbf{K} \cup \mathbf{K}_{-H}) = 2\mathrm{vol}_d(\mathbf{K}) + 2\mathrm{vol}_d(\mathbf{F_K}(\mathbf{m})) = 2\mathrm{vol}_d(\mathbf{K}) + 2w_{\mathbf{K}}(u)\mathrm{vol}_{d-1}(\mathbf{K}|\mathbf{m}^{\perp})$. Thus, for any convex body $\mathbf{K}$ in $\mathbb{E}_d$,

$$c_{d-1}(\mathbf{K}) \geq 1 + \frac{\max\{w_{\mathbf{K}}(\mathbf{m})\mathrm{vol}_{d-1}(\mathbf{K}|\mathbf{m}^{\perp}) : \mathbf{m} \in \mathbb{S}^{d-1}\}}{\mathrm{vol}_d(\mathbf{K})}. \tag{7.10}$$

Recall that $d_{\mathbf{K}}(\mathbf{m})$ is the length of a longest chord of $\mathbf{K}$ parallel to $\mathbf{m} \in \mathbb{S}^{d-1}$. Observe that for any $\mathbf{m} \in \mathbb{S}^{d-1}$, $d_{\mathbf{K}}(\mathbf{m}) \leq w_{\mathbf{K}}(\mathbf{m})$, and thus for any convex body $\mathbf{K}$,

$$c_{d-1}(\mathbf{K}) \geq c^{tr}(\mathbf{K}).$$

This readily implies that $c_{d-1}(\mathbf{K}) \geq 1 + \frac{2\kappa_{d-1}}{\kappa_d}$, and if here there is equality for some convex body $\mathbf{K}$, then $\mathbf{K}$ is an ellipsoid. On the other hand, in case of equality, for any $\mathbf{m} \in \mathbb{S}^{d-1}$ we have $d_{\mathbf{K}}(\mathbf{m}) = w_{\mathbf{K}}(\mathbf{m})$, which yields that $\mathbf{K}$ is a Euclidean ball. This finishes the proof of the theorem.

7.3.2 Proof of Theorem 46

Recall that a convex body $\mathbf{K}$ is a body of constant width in a normed space with unit ball $\mathbf{M}$ if and only if its central symmetrization $\frac{1}{2}(\mathbf{K} - \mathbf{K})$ is a homothetic copy of $\mathbf{M}$. Thus, (ii) and (iii) are clearly equivalent, and we need to show only that (i) and (ii) are equivalent.

Let $\mathbf{K}$ be a plane convex body. Like in Subsection 7.3.1, note that, using the notation $\mathbf{m} = \frac{\mathbf{w}-\mathbf{v}}{||\mathbf{w}-\mathbf{v}||}$, for any touching pair of translates $\mathbf{v}+\mathbf{K}$ and $\mathbf{w}+\mathbf{K}$, we have

$$\mathrm{area}(\mathrm{conv}((\mathbf{v} + \mathbf{K}) \cup (\mathbf{w} + \mathbf{K}))) = \mathrm{area}(\mathbf{K}) + d_{\mathbf{K}}(\mathbf{m})w_{\mathbf{K}}(\mathbf{m}^{\perp}). \tag{7.11}$$

Since for any direction $\mathbf{m} \in \mathbb{S}^1$, we have $d_{\mathbf{K}}(\mathbf{m}) = d_{\frac{1}{2}(\mathbf{K}-\mathbf{K})}(\mathbf{m})$ and $w_{\mathbf{K}}(\mathbf{m}) = w_{\frac{1}{2}(\mathbf{K}-\mathbf{K})}(\mathbf{m})$, $\mathbf{K}$ satisfies the translative constant volume property if and only if its central symmetrization does. Thus, we may assume that $\mathbf{K}$ is $\mathbf{o}$-symmetric. Now let $\mathbf{x} \in \mathrm{bd}\mathbf{K}$. Then the boundary of $\mathrm{conv}(\mathbf{K} \cup (2\mathbf{x} + \mathbf{K}))$ consists of an arc of $\mathrm{bd}\mathbf{K}$, its reflection about $\mathbf{x}$, and two parallel segments, each contained in one of the two common supporting lines of $\mathbf{K}$ and

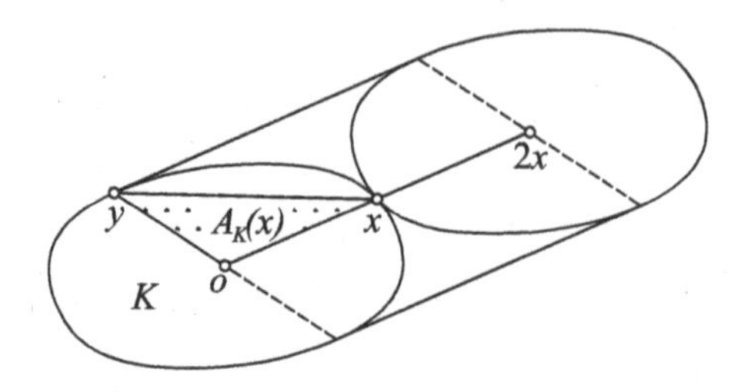

FIGURE 7.1
Touching translates of an **o**-symmetric plane convex body.

$2\mathbf{x} + \mathbf{K}$, which are parallel to $\mathbf{x}$. For some point $\mathbf{y}$ on one of these two segments, set $A_{\mathbf{K}}(\mathbf{x}) = \text{area}(\text{conv}\{\mathbf{o}, \mathbf{x}, \mathbf{y}\})$ (cf. Figure 7.1). Clearly, $A_{\mathbf{K}}(\mathbf{x})$ is independent of the choice of $\mathbf{y}$. Then we have for every $\mathbf{x} \in \text{bd}\mathbf{K}$, that $d_{\mathbf{K}}(\mathbf{x}) w_{\mathbf{K}}(\mathbf{x}^{\perp}) = 8 A_{\mathbf{K}}(\mathbf{x})$.

Assume that $A_{\mathbf{K}}(\mathbf{x})$ is independent of $\mathbf{x}$. We need to show that in this case bd$\mathbf{K}$ is a Radon curve. It is known (cf. [179]) that bd$\mathbf{K}$ is a Radon curve if and only if in the norm of $\mathbf{K}$, *Birkhoff orthogonality* is a symmetric relation. Recall that in a normed plane with unit disk $\mathbf{K}$, a vector $\mathbf{x}$ is called *Birkhoff orthogonal* to a vector $\mathbf{y}$, denoted by $\mathbf{x} \dashv_{\mathbf{K}} \mathbf{y}$, if $\mathbf{y}$ is parallel to a line supporting $||\mathbf{x}||_{\mathbf{K}} \text{bd}\mathbf{K}$ at $\mathbf{x}$ (cf. Definition 23 or [7] for an equivalent formulation).

Observe that for any $\mathbf{x}, \mathbf{y} \in \text{bd}\mathbf{K}$, $\mathbf{y} \dashv_{\mathbf{K}} \mathbf{x}$ if and only if $A_{\mathbf{K}}(\mathbf{x}) = \text{area}(\text{conv}\{\mathbf{o}, \mathbf{x}, \mathbf{y}\})$, or in other words, if $\text{area}(\text{conv}\{\mathbf{o}, \mathbf{x}, \mathbf{y}\})$ is maximal over $\mathbf{x} \in \mathbf{K}$. Clearly, it suffices to prove the symmetry of Birkhoff orthogonality for $\mathbf{x}, \mathbf{y} \in \text{bd}\mathbf{K}$. Consider a sequence $\mathbf{z} \dashv_{\mathbf{K}} \mathbf{y} \dashv_{\mathbf{K}} \mathbf{x}$ for some $\mathbf{x}, \mathbf{y}, \mathbf{z} \in \text{bd}\mathbf{K}$. Then we have $A_{\mathbf{K}}(\mathbf{x}) = \text{area}(\text{conv}\{\mathbf{o}, \mathbf{x}, \mathbf{y}\})$ and $A_{\mathbf{K}}(\mathbf{y}) = \text{area}(\text{conv}\{\mathbf{o}, \mathbf{y}, \mathbf{z}\})$. By the maximality of $\text{area}(\text{conv}\{\mathbf{o}, \mathbf{y}, \mathbf{z}\})$, we have $A_{\mathbf{K}}(\mathbf{x}) \leq A_{\mathbf{K}}(\mathbf{y})$ with equality if and only if $\mathbf{x} \dashv_{\mathbf{K}} \mathbf{y}$. This readily implies that Birkhoff orthogonality is symmetric, and thus, that bd$\mathbf{K}$ is a Radon curve. The opposite direction follows from the definition of Radon curves and polar sets.

7.4 Proof of Theorem 53

In this section we prove the Blaschke-Santaló Inequality, apart from the case of equality. The proof presented here follows the proof given by Meyer and Pajor [182] (cf. also [142]), and is based on Steiner symmetrization, which we introduced in Definition 7.

We intend to examine the properties of Steiner symmetrization. To do this, recall that the Hausdorff distance of the convex bodies $\mathbf{K}, \mathbf{L} \subset \mathbb{E}^d$ (cf.

Definition 28) is defined as

$$h(\mathbf{K}, \mathbf{L}) = \min\left\{\varepsilon : \mathbf{K} \subseteq \mathbf{L} + \varepsilon\mathbf{B}^d, \mathbf{L} \subseteq \mathbf{K} + \varepsilon\mathbf{B}^d\right\}.$$

Remark 163 *Let $\mathbf{K} \subset \mathbb{E}^d$ be a convex body, and let H be a hyperplane. Then*

(i) $\mathrm{vol}_d(S_H(\mathbf{K})) = \mathrm{vol}_d(\mathbf{K})$, *and*

(ii) $S_H(\mathbf{K})$ is a continuous function of $\mathbf{K}$, with respect to the topology on the family of convex bodies in $\mathbb{E}^d$, induced by Hausdorff distance,

where $S_H(\mathbf{K})$ is the Steiner symmetrization of $\mathbf{K}$ to H.

Lemma 164 *Let $\mathbf{K}$ be a convex body in $\mathbb{E}^d$, and let $\mathcal{K}$ denote the family of convex bodies in $\mathbb{E}^d$ that can be obtained from $\mathbf{K}$ by taking finitely many subsequent Steiner symmetrizations with respect to hyperplanes containing $\mathbf{o}$. Let $\mathbf{B}(\mathbf{K})$ denote the d-dimensional Euclidean ball centered at $\mathbf{o}$ and volume equal to* $\mathrm{vol}_d(\mathbf{K})$. *Then $\mathcal{K}$ contains a sequence of convex bodies which converges to $\mathbf{B}(\mathbf{K})$ with respect to Hausdorff distance.*

Proof. Clearly, without loss of generality, we may assume that $\mathbf{o} \in \mathrm{int}\mathbf{K}$, and observe that in this case the origin is contained in the interior of every convex body in $\mathcal{K}$.

For any convex body $\mathbf{L}$ with $\mathbf{o} \in \mathbf{L}$, let $\rho(\mathbf{L})$ denote the smallest positive number r satisfying $\mathbf{L} \subseteq r\mathbf{B}^d$. Let $\rho = \inf\{\rho(\mathbf{L}) : \mathbf{L} \in \mathcal{K}\}$. For simplicity and without loss of generality, we assume that $\rho = 1$. Then there is a sequence $\{\mathbf{K}_n\}$ of convex bodies in $\mathcal{K}$ such that $\rho(\mathbf{K}_n) \to 1$. Let $\mathcal{K}'$ be the set of the elements of this sequence.

Observe that if H is any hyperplane through $\mathbf{o}$, then for every convex body $\mathbf{L}$ with $\mathbf{o} \in \mathbf{L}$, we have $\rho(S_H(\mathbf{L})) \leq \rho(\mathbf{L})$. Applying this observation for $\mathcal{K}$, it follows that for any $\mathbf{L} \in \mathcal{K}$ or in particular for any $\mathbf{L} \in \mathcal{K}'$, we have $\rho(\mathbf{L}) \leq \rho(\mathbf{K})$, or in other words, $\mathbf{L} \subseteq \rho(\mathbf{K})\mathbf{B}^d$. Thus, $\mathcal{K}'$ is uniformly bounded, implying by Blaschke's selection theorem (cf. [217]) that $\mathcal{K}'$ contains a convergent sequence with respect to Hausdorff distance.

With a little abuse of notations we assume that $\{\mathbf{K}_n\}$ converges to a convex body $\mathbf{M}$ in $\mathbb{E}^d$. Since the functional $\rho(\cdot)$ is continuous on the family of convex bodies containing $\mathbf{o}$, we have $\rho(\mathbf{M}) = 1$, implying $\mathbf{M} \subseteq \mathbf{B}^d$.

We show that $\mathbf{M} = \mathbf{B}^d$. Suppose for contradiction that there is some point $\mathbf{p} \in \mathbb{S}^{d-1} \setminus \mathbf{M}$. Then there is an open spherical cap $C(\mathbf{p}) \subset \mathbb{S}^{d-1}$ centered at $\mathbf{p}$ such that $C(\mathbf{p}) \cap \mathbf{M} = \emptyset$. For any $\mathbf{q} \in \mathbb{S}^{d-1}$, let $C(\mathbf{q})$ be the spherical cap congruent to $C(\mathbf{p})$ and centered at $\mathbf{q}$. Then, since $\mathbb{S}^{d-1}$ is compact and the family $\{C(\mathbf{q}) : \mathbf{q} \in \mathbb{S}^{d-1}\}$ is an open cover of $\mathbb{S}^{d-1}$, there are finitely many points $\mathbf{q}_1, \ldots, \mathbf{q}_k \in \mathbb{S}^{d-1}$ such that $\{C(\mathbf{q}_i) : i = 1, 2, \ldots, k\}$ cover $\mathbb{S}^{d-1}$.

Note that for any two congruent open spherical caps $C_1, C_2 \subset \mathbb{S}^{d-1}$ there is a unique hyperplane H through o such that C_2 is the reflected copy of C_1 with respect to H. Furthermore, if $\mathbf{L} \subset \mathbf{B}^d$ is a convex body disjoint from C_1 *or* C_2, then its symmetrization $S_H(\mathbf{L})$ is disjoint from *both* C_1 and C_2.

For any $1 \leq i \leq k$, let H_i be the hyperplane such that $C(\mathbf{q}_i)$ is the reflected copy of $C(\mathbf{p})$. Then, by the previous observation, $\mathbf{M}' = S_{H_1}(S_{H_2}(\dots S_{H_k}(\mathbf{M})\dots)) \subset \mathbf{B}^d$ is a convex body disjoint from $\mathbb{S}^{d-1}$. Clearly, $\mathbf{M}'$ is the limit of a sequence of convex bodies in $\mathbf{K}$, and by the compactness of $\mathbf{M}'$, $\rho(\mathbf{M}') < 1$, a contradiction. Thus, $\mathbf{M} = \mathbf{B}^d$. □

Lemma 165 *Let* $\mathbf{K}$ *be an o-symmetric convex body in* $\mathbb{E}^d$*, and let* H *be a hyperplane through* $\mathbf{o}$*. Then*

$$\mathrm{vol}_d(\mathbf{K}^\circ) \leq \mathrm{vol}_d((S_H(\mathbf{K}))^\circ).$$

Proof. Without loss of generality, we may assume that H is the hyperplane $\{x_d = 0\}$. Then each point $\mathbf{p} \in \mathbb{E}^d$ can be uniquely represented in the form $(\mathbf{x}, t)$ for some $\mathbf{x} \in H$ and $t \in \mathbb{R}$. Let $\mathbf{K}|H$ denote the orthogonal projection of $\mathbf{K}$ onto H, and for any $\mathbf{x} \in \mathbf{K}|H$, let $(\mathbf{x}, s_1(\mathbf{x}))$ and $(\mathbf{x}, s_2(\mathbf{x}))$ denote the endpoints of the closed segment in $\mathbf{K}$ whose orthogonal projection onto H is $\mathbf{x}$. We use the labeling in such a way that $s_1(\mathbf{x}) \leq s_2(\mathbf{x})$ for all $\mathbf{x} \in \mathbf{K}|H$. Then the Steiner symmetrization of $\mathbf{K}$ can be written as

$$S_H(\mathbf{K}) = \left\{ (\mathbf{x}, s) | \, \mathbf{x} \in \mathbf{K}|H, |s| \leq \frac{1}{2} \left(s_2(\mathbf{x}) - s_1(\mathbf{x}) \right) \right\},$$

and the polar of $\mathbf{K}$ is

$$\mathbf{K}^\circ = \{ (\mathbf{y}, t) | \, \langle \mathbf{x}, \mathbf{y} \rangle + st \leq 1 \text{ for all } \mathbf{x} \in \mathbf{K}|H, s_1(\mathbf{x}) \leq s \leq s_2(\mathbf{x}) \}.$$

From these representations, we obtain that

$$(S_H(\mathbf{K}))^\circ = \left\{ (\mathbf{y}, t) | \, \langle \mathbf{x}, \mathbf{y} \rangle + st \leq 1 \forall \mathbf{x} \in \mathbf{K}|H, |s| \leq \frac{1}{2} \left(s_2(\mathbf{x}) - s_1(\mathbf{x}) \right) \right\}.$$

For any $t \in \mathbb{R}$ and convex body $\mathbf{L}$ in $\mathbb{E}^d$, we denote by $\mathbf{L}(t) = \{\mathbf{x} \in \mathbf{L}|H : (\mathbf{x}, t) \in \mathbf{L}\}$ the orthogonal projection of the intersection of $\mathbf{L}$ with the hyperplane $\{x_d = t\}$, and show that for all $t \in \mathbb{R}$, we have $\frac{1}{2} \left(\mathbf{K}^\circ(t) + \mathbf{K}^\circ(-t) \right) \subseteq S_H(\mathbf{K})^\circ(t)$.

To do it, consider some $\mathbf{y}_1 \in \mathbf{K}^\circ(t)$ and $\mathbf{y}_2 \in \mathbf{K}^\circ(-t)$. Then, for any $\mathbf{x} \in \mathbf{K}|H$ and $s_1(\mathbf{x}) \leq s_1, s_2 \leq s_2(\mathbf{x})$, we have

$$\langle \mathbf{x}, \mathbf{y}_1 \rangle + s_1 t \leq 1, \quad \langle \mathbf{x}, \mathbf{y}_2 \rangle - s_2 t \leq 1,$$

implying that

$$\langle \mathbf{x}, \frac{1}{2}(\mathbf{y}_1 + \mathbf{y}_2) \rangle + \frac{1}{2}(s_1 - s_2) \leq 1,$$

holds for any $\mathbf{x} \in \mathbf{K}|H$ and $\left| \frac{1}{2}(s_1 - s_2) \right| \leq \frac{1}{2}(s_2(\mathbf{x}) - s_1(\mathbf{x}))$. Thus, by the formula for $(S_H(\mathbf{K}))^\circ$, it follows that $\frac{1}{2}(\mathbf{y}_1 + \mathbf{y}_2) \in S_H(\mathbf{K})^\circ(t)$.

Now we have that for all $t \in \mathbb{R}$, we have $\frac{1}{2} \left(\mathbf{K}^\circ(t) + \mathbf{K}^\circ(-t) \right) \subseteq S_H(\mathbf{K})^\circ(t)$.

On the other hand, since $\mathbf{K}$ is $\mathbf{o}$-symmetric, so is $\mathbf{K}^\circ$, implying that $\mathbf{K}^\circ(-t) = -\mathbf{K}^\circ(-t)$. Thus, the Brunn-Minkowski Inequality yields that

$$\mathrm{vol}_{d-1}(\mathbf{K}^\circ) \leq \mathrm{vol}_{d-1}\left(\frac{1}{2}\left(\mathbf{K}^\circ(t) + \mathbf{K}^\circ(-t)\right)\right)$$

$$= \mathrm{vol}_{d-1}\left(\frac{1}{2}\left(\mathbf{K}^\circ(t) + \mathbf{K}^\circ(-t)\right)\right) \leq \mathrm{vol}_{d-1}\left(S_H(\mathbf{K})^\circ(t)\right).$$

Using this inequality and by Cavalieri's principle, we have

$$\mathrm{vol}_d(\mathbf{K}) = \int_{t\in\mathbb{R}} \mathrm{vol}_{d-1}(\mathbf{K}^\circ)\, dt \leq \int_{t\in\mathbb{R}} \mathrm{vol}_{d-1}\left(S_H(\mathbf{K})^\circ(t)\right) dt = \mathrm{vol}_d(S_H(\mathbf{K})),$$

and the assertion follows. □

Now we prove Theorem 53.

Let $\mathbf{M}$ be an $\mathbf{o}$-symmetric convex body in $\mathbb{E}^d$. By Lemma 164, there is a sequence of convex bodies $\{\mathbf{M}_n\}$, each obtained from $\mathbf{M}$ by finitely many Steiner symmetrizations, such that $\mathbf{M}_n$ converges to the ball $\left(\frac{\mathrm{vol}_d(\mathbf{M})}{\kappa_d}\right)^d \mathbf{B}^d$ measured in Hausdorff distance. Then the sequence $\{\mathbf{M}_n^\circ\}$ converges to the ball $\left(\frac{\kappa_d}{\mathrm{vol}_d(\mathbf{M})}\right)^d \mathbf{B}^d$. Since both $\mathrm{vol}_d(\mathbf{L})$ and $\mathrm{vol}_d(\mathbf{L}^\circ)$ are continuous functions on the family of convex bodies containing $\mathbf{o}$ in their interiors, it follows that

$$\mathrm{vol}_d(\mathbf{M}_n)\mathrm{vol}_d(\mathbf{M}_n^\circ) \to \kappa_d^2.$$

On the other hand, by Remark 163 and Lemma 165, we have $\mathrm{vol}_d(\mathbf{M}_n) = \mathrm{vol}_d(\mathbf{M})$ and $\mathrm{vol}_d(\mathbf{M}_n^\circ) \geq \mathrm{vol}_d(\mathbf{M}^\circ)$ for all values of n. Thus, $\mathrm{vol}_d(\mathbf{M})\mathrm{vol}_d(\mathbf{M}^\circ) \leq \lim \mathrm{vol}_d(\mathbf{M}_n)\mathrm{vol}_d(\mathbf{M}_n^\circ) = \kappa_d^2$. Finally, by the formula for Holmes-Thompson volume in (2.7), it follows that

$$\mathrm{vol}_{\mathbf{M}}^{HT}(\mathbf{M}) = \frac{\mathrm{vol}_d(\mathbf{M}^\circ)}{\kappa_d}\mathrm{vol}_d(\mathbf{M}) \leq \kappa_d,$$

which proves the assertion.

7.5 Proof of Theorem 54

We prove the following.

Theorem 166 *For any* $\mathbf{o}$*-symmetric convex polygon* $\mathbf{P}$ *with* $2k$ *sides, where* $k \geq 3$*, we have*

$$\mathrm{area}(\mathbf{P})\mathrm{area}(\mathbf{P}^\circ) > 8.$$

Note that since for parallelograms $\text{area}(\mathbf{P})\text{area}(\mathbf{P}^\circ) = 8$, by the continuity of the quantities $\text{area}(\mathbf{P})$ and $\text{area}(\mathbf{P}^\circ)$, Theorem 166 implies that for any $\mathbf{o}$-symmetric plane convex body $\mathbf{M}$, $\text{area}(\mathbf{M})\text{area}(\mathbf{M}^\circ) \geq 8$. This inequality, by the formula for Holmes-Thompson volume in (2.7), proves the inequality in Theorem 54. Here, by Theorem 166, equality holds only for parallelograms within the family of $\mathbf{o}$-symmetric convex polygons. Nevertheless, the continuity argument cannot be used to investigate equality in the larger family of $\mathbf{o}$-symmetric plane convex bodies.

To prove Theorem 166, we momentarily use a modified notion of polarity for convex polygons. For any point $\mathbf{x} = (x_1, x_2) \in \mathbb{E}^2 \setminus \{\mathbf{o}\}$, the *polar line* $L_{\mathbf{x}}$ of $\mathbf{x}$ is defined by the equation $t_1x_1 + t_2x_2 = 1$, and vice versa. We note that this correspondence is a bijection between the points in $\mathbb{E}^2 \setminus \{\mathbf{o}\}$ and the lines not containing $\mathbf{o}$. Let $\mathbf{P}$ be a convex polygon with vertices $\mathbf{x}_1, \mathbf{x}_2, \ldots, \mathbf{x}_k$ such that no sideline of $\mathcal{P}$ contains $\mathbf{o}$. For any edge $[\mathbf{x}_i, \mathbf{x}_j]$ of $\mathbf{P}$, let $\mathbf{v}_{i,j}$ denote the polar point of line through $[\mathbf{x}_i, \mathbf{x}_j]$. Then the *polar* of the polygon $\mathbf{P}$ is defined as

$$\mathbf{P}^* = \text{conv}\{\mathbf{v}_{i,j} : [\mathbf{x}_i, \mathbf{x}_j] \text{ is an edge of } \mathbf{P}\}.$$

Observe that if $\mathbf{P}$ contains the origin in its interior, then $\mathbf{P}^* = \mathbf{P}^\circ$. On the other hand, if $\mathbf{o} \notin \mathbf{P}$, then $\mathbf{P}^*$ is bounded, whereas $\mathbf{P}^\circ$ is not.

In the following, for any $\mathbf{x} = (x_1, x_2), \mathbf{y} = (y_1, y_2) \in \mathbb{E}^2$ we set

$$D_{\mathbf{x},\mathbf{y}} = \det \begin{bmatrix} x_1 & y_1 \\ x_2 & y_2 \end{bmatrix} = x_1y_2 - x_2y_1.$$

Lemma 167 *Let* $\mathbf{x} = (x_1, x_2), \mathbf{y} = (y_1, y_2), \mathbf{z} = (z_1, z_2) \in \mathbb{E}^2$ *be points such that no two of them are linearly dependent. Let* $\mathbf{T}^* = \text{conv}\{\mathbf{x}', \mathbf{y}', \mathbf{z}'\}$ *be the polar of the triangle* $\mathbf{T} = \text{conv}\{\mathbf{x}, \mathbf{y}, \mathbf{z}\}$. *Then the (signed) area of* $\mathbf{T}^*$ *is*

$$\text{area}(\mathbf{T}^*) = \frac{2(\text{area}(T))^2}{|D_{\mathbf{x},\mathbf{y}}D_{\mathbf{y},\mathbf{z}}D_{\mathbf{z},\mathbf{x}}|}. \tag{7.12}$$

Proof. First, note that the property that no two of $\mathbf{x}, \mathbf{y}, \mathbf{z}$ are linearly dependent is equivalent to the property that no sideline of $\mathbf{T}$ contains $\mathbf{o}$, implying that $\mathbf{T}^*$ exists. For any $\mathbf{w} \in \{\mathbf{x}, \mathbf{y}, \mathbf{z}\}$, let $\mathbf{w}'$ denote the polar of the sideline of $\mathbf{T}$ not containing $\mathbf{w}$. The coordinates of $\mathbf{x}', \mathbf{y}', \mathbf{z}'$ can be computed by solving systems of linear equations. Hence, by Kramer's rule we obtain

$$\begin{aligned} \mathbf{x}' &= \frac{1}{D_{\mathbf{x},\mathbf{y}}}(y_2 - x_2, x_1 - y_1), \\ \mathbf{y}' &= \frac{1}{D_{\mathbf{y},\mathbf{z}}}(z_2 - y_2, y_1 - z_1), \\ \mathbf{z}' &= \frac{1}{D_{\mathbf{z},\mathbf{x}}}(x_2 - z_2, z_1 - x_1). \end{aligned}$$

After computing $\text{area}(\mathbf{T}^*)$ via determinants, using these formulas, and simplification, we have

$$\text{area}(\mathbf{T}^*) = \frac{1}{2}|\det[\mathbf{y}' - \mathbf{x}', \mathbf{z}' - \mathbf{x}']| = \frac{1}{2}|D_{\mathbf{x}',\mathbf{y}'} + D_{\mathbf{y}',\mathbf{z}'} + D_{\mathbf{z}',\mathbf{x}'}|$$

$$= \frac{D_{\mathbf{x},\mathbf{y}}^2 + D_{\mathbf{y},\mathbf{z}}^2 + D_{\mathbf{z},\mathbf{x}}^2 + 2D_{\mathbf{x},\mathbf{y}}D_{\mathbf{y},\mathbf{z}} + 2D_{\mathbf{x},\mathbf{y}}D_{\mathbf{z},\mathbf{x}} + 2D_{\mathbf{y},\mathbf{z}}D_{\mathbf{z},\mathbf{x}}}{|D_{\mathbf{x},\mathbf{y}}D_{\mathbf{y},\mathbf{z}}D_{\mathbf{z},\mathbf{x}}|}$$

$$= \frac{2(\text{area}(\mathbf{T}))^2}{|D_{\mathbf{x},\mathbf{y}}D_{\mathbf{y},\mathbf{z}}D_{\mathbf{z},\mathbf{x}}|},$$

finishing the proof. □

Lemma 168 *Let* $\mathbf{P}$ *be an* $\mathbf{o}$*-symmetric polygon with* $2k$ *vertices, where* $k \geq 3$*. Then there is a polygon* $\mathbf{Q}$ *with* $2k-2$ *vertices such that*

$$\text{area}(\mathbf{P})\text{area}(\mathbf{P}^\circ) > \text{area}(\mathbf{Q})\text{area}(\mathbf{Q}^\circ).$$

Proof. Let $\mathbf{x}, \mathbf{y}$ and $\mathbf{z}$ be three consecutive vertices of $\mathbf{P}$ in counterclockwise order. Since any line through $\mathbf{o}$ and not containing any vertex of $\mathbf{P}$ strictly separates exactly k vertices of $\mathbf{P}$ from their reflections about $\mathbf{o}$, the triangle $\mathbf{T} = \text{conv}\{\mathbf{x}, \mathbf{y}, \mathbf{z}\}$ is disjoint from $\mathbf{o}$, implying, in particular, that no two of $\mathbf{x}, \mathbf{y}, \mathbf{z}$ are linearly dependent. Let $\mathbf{P}_0$ be the $\mathbf{o}$-symmetric $(2k-2)$-gon whose vertices are all the vertices of $\mathbf{P}$ but $\mathbf{y}$ and $-\mathbf{y}$. Then $\mathbf{o} \in \text{int}\mathbf{P}_0$, implying that $\mathbf{P}_0^* = \mathbf{P}_0^\circ$ is bounded.

The lines polar to $\mathbf{x}$ and $\mathbf{z}$ are sidelines of $\mathbf{P}_0^*$, whose intersections with the line polar to $\mathbf{y}$ are $\mathbf{z}'$ and $\mathbf{x}'$, respectively. The intersection point of these sidelines is the polar $\mathbf{y}'$ of the line through $\mathbf{x}$ and $\mathbf{z}$. Since $\mathbf{y} \notin \mathbf{P}_0$ but $[\mathbf{x}, \mathbf{y}]$ and $[\mathbf{y}, \mathbf{z}]$ are sides of $\text{conv}(\mathbf{P}_0 \setminus \{\mathbf{y}\})$, the line polar to $\mathbf{y}$ cuts off a triangle from $\mathbf{P}_0^*$, or in other words, $\mathbf{x}'$ and $\mathbf{z}'$ are relative interior points of two consecutive sides of $\mathbf{P}_0^*$. The same argument can be repeated for the polars of the vertices $-\mathbf{x}$, $-\mathbf{y}$ and $-\mathbf{z}$, implying that $\mathbf{P}_0^* = \mathbf{P}^* \cup \mathbf{T}^* \cup (-\mathbf{T}^*)$, and the three polygons in this union are pairwise non-overlapping.

Now we move $\mathbf{y}$ parallel to the segment $[\mathbf{x}, \mathbf{z}]$. This modification does not change the value of $\text{area}(\mathbf{P})$. We show that $\text{area}(\mathbf{T}^*)$ is maximal, and thus $\text{area}(\mathbf{P}^*)$ is minimal, if $\mathbf{y}$ is moved to a sideline of $\mathbf{P}_0$. Replacing $\mathbf{y}$ with this point and $-\mathbf{y}$ with its reflection about $\mathbf{o}$ yields an $\mathbf{o}$-symmetric $(2k-2)$-gon $\mathbf{Q}$ such that $\text{area}(\mathbf{P})\text{area}(\mathbf{P}^\circ) > \text{area}(\mathbf{Q})\text{area}(\mathbf{Q}^\circ)$, implying the statement.

We show the maximality of $\text{area}(\mathbf{T}^*)$ by two different, one algebraic and one geometric, arguments.

We first present an algebraic proof. Note that by our labeling of the points $\mathbf{x}, \mathbf{y}, \mathbf{z}$, $D_{\mathbf{x},\mathbf{y}}, D_{\mathbf{y},\mathbf{z}} > 0$ and $D_{\mathbf{z},\mathbf{x}} < 0$. Thus, by Lemma 167,

$$\text{area}(\mathbf{T}^*) = -\frac{2(\text{area}(\mathbf{T}))^2}{D_{\mathbf{x},\mathbf{y}}D_{\mathbf{y},\mathbf{z}}D_{\mathbf{z},\mathbf{x}}}.$$

Consider the point $\mathbf{y}(t) = \mathbf{y} + s(\mathbf{x} - \mathbf{z})$, and let $\mathbf{T}(s) = \text{conv}\{\mathbf{x}, \mathbf{y}(s), \mathbf{z}\}$. Then, as $\text{area}(\mathbf{T}(s)) = \text{area}(\mathbf{T})$ for all $s \in \mathbb{R}$, we have

$$\text{area}((\mathbf{T}(s))^*) = -\frac{2(\text{area}(\mathbf{T}))^2}{D_{\mathbf{x},\mathbf{y}s(\mathbf{x}-\mathbf{z})} + D_{\mathbf{y}+s(\mathbf{x}-\mathbf{z}),\mathbf{z}}D_{\mathbf{z},\mathbf{x}}}.$$

Using the properties of determinants, we have

$$D_{\mathbf{x},\mathbf{y}s(\mathbf{x}-\mathbf{z})} D_{\mathbf{y}+s(\mathbf{x}-\mathbf{z},\mathbf{z})} = (D_{\mathbf{x},\mathbf{y}} + sD_{\mathbf{x},\mathbf{x}} - sD_{\mathbf{x},\mathbf{z}})(D_{\mathbf{y},\mathbf{z}} + sD_{\mathbf{x},\mathbf{z}} - sD_{\mathbf{z},\mathbf{z}})$$

$$= D_{\mathbf{x},\mathbf{y}} D_{\mathbf{y},\mathbf{z}} - D_{\mathbf{z},\mathbf{x}} (D_{\mathbf{x},\mathbf{y}} + D_{\mathbf{y},\mathbf{z}}) s + D_{\mathbf{z},\mathbf{x}}^2 s^2.$$

This function is strictly convex on its domain, and thus, it is maximal at an extremal value of s; that is, if $\mathbf{y}(s)$ is on a sideline of $\mathbf{P}_0$. Thus, the inequality $D_{\mathbf{z},\mathbf{x}} < 0$ implies that $\text{area}((\mathbf{T}(s))^*)$ is minimal if $\mathbf{y}(s)$ is on a sideline of $\mathbf{P}_0$, which readily yields the assertion.

Next, we show the same statement using a geometric argument. Let X, Y and Z denote the lines polar to $\mathbf{x}$, $\mathbf{y}$ and $\mathbf{z}$. Let L be the line through $\mathbf{y}$ and parallel to $[\mathbf{x}, \mathbf{z}]$. Then the point $\mathbf{l}$ polar to L lies on the line polar to $\mathbf{y}$, and on the ray starting at $\mathbf{o}$ and perpendicular to the line of $[\mathbf{x}, \mathbf{z}]$. Note that this ray passes through $\mathbf{y}'$. Furthermore, since L is a supporting line of $\mathbf{P}$, it follows that $\mathbf{l} \in \mathbf{P}^*$. In particular, if $\mathbf{R}$ is the convex angular region whose boundary consists of a half line of X and a half line of Z starting at $\mathbf{y}'$ such that $\mathbf{o} \in \text{int}\mathbf{R}$, then $\mathbf{l} \in \mathbf{R}$.

The polar of any point of L is a line through $\mathbf{l}$. Thus, to maximize $\text{area}(T^*)$ we need to find a line L' through $\mathbf{l}$ which cuts off a triangle of maximal area from $\mathbf{R}$. It is an elementary computation to show that a minimal area triangle is attained if $\mathbf{l}$ is the midpoint of the segment $L' \cap \mathbf{R}$, and a maximal area triangle is attained in an 'extremal' position, i.e., if the polar of L' is on a sideline of $\mathbf{P}_0$. This finishes the proof. □

Finally, to prove Theorem 166, consider any $\mathbf{o}$-symmetric polygon $\mathbf{P}$ with $2k$ sides. Then repeated applications of Lemma 168 yield a sequence of $\mathbf{o}$-symmetric polygons $\mathbf{P}_0, \mathbf{P}_1, \mathbf{P}_2, \mathbf{P}_{k-2}$ such that $\mathbf{P}_0 = \mathbf{P}$, and for any $i = 1, \ldots, k-2$, $\mathbf{P}_i$ is a polygon with $2(k-i)$ sides satisfying

$$\text{area}(\mathbf{P}_i)\text{area}(\mathbf{P}_i^\circ) < \text{area}(\mathbf{P}_{i-1})\text{area}(\mathbf{P}_{i-1}^\circ).$$

Since any $\mathbf{o}$-symmetric quadrilateral is a parallelogram, this proves Theorem 166.

7.6 Proofs of Theorems 57 and 58

7.6.1 Proof of Theorem 57

First, observe that if $\mathbf{M}$ is an $\mathbf{o}$-symmetric plane convex body, then for any planar set $\mathbf{S}$, $\text{vol}_{\mathbf{M}}^m(\mathbf{S}) = \frac{2}{\text{area}(\mathbf{P}_i)}\text{area}(\mathbf{S})$, where $\mathbf{P}_i$ is a largest area parallelogram inscribed in $\mathbf{M}$. Since $\text{area}(\mathbf{M}) \geq \text{area}(\mathbf{P}_i)$, with equality if and only if $\mathbf{M}$ is a parallelogram, it follows that

$$\text{vol}_{\mathbf{M}}^m(\mathbf{M}) \geq 2,$$

with equality if and only if $\mathbf{M}$ is a parallelogram.

We obtain by a similar argument that

$$\mathrm{vol}_{\mathbf{M}}^{m*}(\mathbf{M}) \leq 4,$$

with equality if and only if $\mathbf{M}$ is a parallelogram.

To show that the remaining inequalities hold, first we prove the next lemma.

Lemma 169 *For any* $\mathbf{o}$*-symmetric plane convex body* $\mathbf{M}$*, we have*

$$2\mathrm{vol}_{\mathbf{M}}^{m}(\mathbf{M}) \leq M_{\mathbf{M}}(\mathrm{bd}\mathbf{M}) \leq 2\mathrm{vol}_{\mathbf{M}}^{m*}(\mathbf{M}). \tag{7.13}$$

Furthermore, if $\mathrm{bd}\mathbf{M}$ *is a regular curve, then equality holds in any of the two inequalities if and only if* $\mathrm{bd}\mathbf{M}$ *is a Radon curve.*

Proof. By the continuity of the quantities in (7.13), without loss of generality we may assume that $\mathrm{bd}\mathbf{M}$ is a regular curve.

Let $c = M_{\mathbf{M}}(\mathrm{bd}\mathbf{M})$, and let $\gamma : [0, c] \to \mathbb{E}^2$ be an arclength parametrization of $\mathrm{bd}\mathbf{M}$ with respect to Minkowski arclength, in counterclockwise direction. In other words, assume that for any $0 \leq s \leq c$, $\dot{\gamma}(s)$ is a vector of unit normed length, and $\gamma(s), \dot{\gamma}(s)$ is a positively oriented basis of $\mathbb{E}^2$. Then, by Green's theorem, we have

$$\mathrm{area}(\mathbf{M}) = \int_{\mathbf{M}} dx\, dy = \frac{1}{2}\int_{\gamma} x\, dy - y\, dx = \frac{1}{2}\int_0^c \det(\gamma(s), \dot{\gamma}(s))\, ds.$$

Let $\mathbf{P}_s$ denote the parallelogram with vertices $\pm\gamma(s), \pm\dot{\gamma}(s)$. Since $\dot{\gamma}(s)$ is a unit vector in the norm of $\mathbf{M}$, $\mathbf{P}_s$ is inscribed in $\mathbf{M}$. Thus, $\mathrm{area}(\mathbf{P}_s) \leq \mathrm{area}(\mathbf{P}_i)$, where $\mathbf{P}_i$ is a maximal area parallelogram inscribed in $\mathbf{M}$. This implies that

$$\mathrm{area}(\mathbf{M}) = \frac{1}{4}\int_0^c \mathrm{area}(\mathbf{P}_s)\, ds \leq \frac{1}{4}\int_0^c \mathrm{area}(\mathbf{P}_i)\, ds = \frac{1}{4}\mathrm{area}(\mathbf{P}_i) \cdot M_{\mathbf{M}}(\mathrm{bd}\mathbf{M}),$$

and hence,

$$2\mathrm{vol}_{\mathbf{M}}^{m}(\mathbf{M}) = \frac{4\mathrm{area}(\mathbf{M})}{\mathrm{area}(\mathbf{P}_i)} \leq M_{\mathbf{M}}(\mathrm{bd}\mathbf{M}). \tag{7.14}$$

To prove the other inequality, observe that since $\pm\gamma(s), \pm\dot{\gamma}(s)$ are two pairs of antipodal points of $\mathrm{bd}\mathbf{M}$, the tangent lines through these pairs form a circumscribed parallelogram of $\mathbf{M}$. The area of this parallelogram is $2\mathrm{area}(\mathbf{P}_s) = 4\det(\gamma(s), \dot{\gamma}(s))$, and clearly, $2\mathrm{area}(\mathbf{P}_s) \geq \mathrm{area}(\mathbf{P}_c)$, where $\mathbf{P}_c$ is a minimal area parallelogram circumscribed about $\mathbf{M}$. This implies that

$$\mathrm{area}(\mathbf{M}) = \frac{1}{8}\int_0^c 2\mathrm{area}(\mathbf{P}_s)\, ds \geq \frac{1}{8}\int_0^c \mathrm{area}(\mathbf{P}_c)\, ds = \frac{1}{8}\mathrm{area}(\mathbf{P}_c) M_{\mathbf{M}}(\mathrm{bd}\mathbf{M}),$$

from which we have

$$2\mathrm{vol}_{\mathbf{M}}^{m*}(\mathbf{M}) = \frac{8\mathrm{area}(\mathbf{M})}{\mathrm{area}(\mathbf{P}_c)} \geq M_{\mathbf{M}}(\mathrm{bd}\mathbf{M}). \tag{7.15}$$

To examine the case of equality, note that in any of (7.14) or (7.15), $\mathrm{area}(\mathbf{P}_s) = \mathrm{area}(\mathbf{P}_i)$ independently of the value of s. Recall the notation $A_{\mathbf{M}}(\mathbf{x})$ from Subsection 7.3.2, which is one fourth of the area of a largest area parallelogram, inscribed in $\mathbf{M}$ and having a vertex at $\mathbf{x} \in \mathrm{bd}\mathbf{M}$. By the definition of $\mathbf{P}_s$, now we have $\mathrm{area}(\mathbf{P}_s) = 4A_{\mathbf{M}}(\dot{\gamma}(s))$. Since $\dot{\gamma}(s)$ runs over all points of $\mathrm{bd}\mathbf{M}$ as s takes all values $s \in [0, c]$, the property that $\mathrm{area}(\mathbf{P}_s)$ is independent of s is equivalent to saying that $A_{\mathbf{M}}(\mathbf{x})$ is independent of $\mathbf{x} \in \mathrm{bd}\mathbf{M}$. As we have seen in Subsection 7.3.2, this property holds if and only if $\mathrm{bd}\mathbf{M}$ is a Radon curve.
□

For the proof, we also need a theorem of Mustafaev [190], which we do not prove here.

Theorem 170 *For any* $\mathbf{o}$*-symmetric plane convex body,* $2\mathrm{vol}_{\mathbf{M}}^{HT}(\mathbf{M}) \leq M_{\mathbf{M}}(\mathrm{bd}\mathbf{M})$*, with equality if and only if* $\mathbf{M}$ *is an ellipse.*

Now we prove the remaining inequalities.

By Lemma 169, $M_{\mathbf{M}}(\mathrm{bd}\mathbf{M}) \leq 2\mathrm{vol}_{\mathbf{M}}^{m*}(\mathbf{M})$ with equality if and only if $\mathrm{bd}\mathbf{M}$ is a Radon curve. By Theorem 49, $6 \leq M_{\mathbf{M}}(\mathrm{bd}\mathbf{M})$ with equality if and only if $\mathbf{M}$ is an affinely regular hexagon. Since affinely regular hexagons are Radon curves, we have

$$3 \leq \mathrm{vol}_{\mathbf{M}}^{m*}(\mathbf{M}),$$

with equality if and only if $\mathbf{M}$ is an affinely regular hexagon.

By Theorem 170 and Lemma 169, it follows that $\mathrm{vol}^{HT}(\mathbf{M}) \leq \mathrm{vol}^{m*}(\mathbf{M})$ for all o-symmetric plane convex body $\mathbf{M}$. Observe that for any parallelogram $\mathbf{P}$, $\mathbf{P}$ is a largest area parallelogram inscribed in $\mathbf{M}$ if and only if its polar $\mathbf{P}^\circ$ is a smallest area parallelogram circumscribed about $\mathbf{M}^\circ$. Thus, by the definition of Holmes-Thompson volume and Gromov's mass*, the above inequality yields that

$$\frac{\mathrm{area}(\mathbf{M})\mathrm{area}(\mathbf{M}^\circ)}{\pi} \leq \frac{4}{\mathrm{area}(\mathbf{P}_i^\circ)}\mathrm{area}(\mathbf{M}^\circ),$$

where $\mathbf{P}_i$ is a largest area parallelogram inscribed in $\mathbf{M}$. On the other hand, since the volume product of an o-symmetric parallelogram is 8, we have

$$\frac{\mathrm{area}(\mathbf{M})\mathrm{area}(\mathbf{M}^\circ)}{\pi} \leq \frac{\mathrm{area}(\mathbf{P}_i)}{2}\mathrm{area}(\mathbf{M}^\circ),$$

from which

$$\mathrm{vol}_{\mathbf{M}}^{m}(\mathbf{M}) = \frac{2}{\mathrm{area}(\mathbf{P}_i)}\mathrm{area}(\mathbf{M}) \leq \pi.$$

Here, by Theorem 170, equality holds if and only if $\mathbf{M}$ is an ellipse.

7.6.2 Proof of Theorem 58

Let $\mathbf{M}$ be an $\mathbf{o}$-symmetric plane convex body such that $\mathrm{bd}\mathbf{M}$ is a Radon curve.

Lemma 171 *We have* $4\mathrm{area}(\mathbf{M})\mathrm{area}(\mathbf{M}^\circ) = (M_{\mathbf{M}}(\mathrm{bd}\mathbf{M}))^2$.

Proof. Note that

$$\mathrm{vol}_{\mathbf{M}}^{m}(\mathbf{M})\mathrm{vol}_{\mathbf{M}^\circ}^{m*}(\mathbf{M}^\circ) = \mathrm{area}(\mathbf{M})\mathrm{area}(\mathbf{M}^\circ).$$

On the other hand, by Lemma 169, $2\mathrm{vol}_{\mathbf{M}}^{m}(\mathbf{M}) = M_{\mathbf{M}}(\mathrm{bd}\mathbf{M})$, and $2\mathrm{vol}_{\mathbf{M}^\circ}^{m*}(\mathbf{M}^\circ) = M_{\mathbf{M}^\circ}(\mathrm{bd}\mathbf{M}^\circ)$, which, by Theorem 52, is equal to $M_{\mathbf{M}}(\mathrm{bd}\mathbf{M})$. This implies the statement. □

Lemma 171, Theorems 49 and 53 immediately imply the following for the unit disk $\mathbf{M}$ of any Radon norm:

$$6 \leq M_{\mathbf{M}}(\mathbf{M}) \leq 2\pi, \tag{7.16}$$

with equality on the left if and only if $\mathbf{M}$ is an affinely regular hexagon, and on the right if and only if $\mathbf{M}$ is an ellipse. This proves (i) of Theorem 58.

Since for the unit disk $\mathbf{M}$ of any Radon norm we have

$$2\mathrm{vol}_{\mathbf{M}}^{m}(\mathbf{M}) = M_{\mathbf{M}}(\mathrm{bd}\mathbf{M}) = 2\mathrm{vol}_{\mathbf{M}}^{m*}(\mathbf{M})$$

by Lemma 169, (iii) and (iv) of Theorem 58 readily follows.

We prove the inequalities in (iii). By Lemma 171, we have

$$\mathrm{vol}_{\mathbf{M}}^{HT}(\mathbf{M}) = \frac{\mathrm{area}(\mathbf{M})\mathrm{area}(\mathbf{M}^\circ)}{\pi} = \frac{(M_{\mathbf{M}}(\mathbf{M}))^2}{4\pi}.$$

Thus, (7.16) implies (ii).

7.7 Proofs of Theorems 59 and 60

Since the proof of Theorem 60 is a straightforward modification of that of Theorem 59, we present only the proof of Theorem 59.

7.7.1 The proof of the left-hand side inequality in (ii)

Let $\mathbf{K}$ be a plane convex body, and $\mathbf{M} = \frac{1}{2}(\mathbf{K} - \mathbf{K})$. From (2.7) and (2.8), one can deduce that

$$c_{tr}^{HT}(\mathbf{K}) = \frac{\mathrm{area}(\mathbf{M}^\circ)}{\pi}\left(\mathrm{area}(\mathbf{K}) + \max\{d_{\mathbf{K}}(\mathbf{m})w_{\mathbf{K}}(\mathbf{m}^\perp) : \mathbf{m} \in \mathbb{S}^1\}\right), \tag{7.17}$$

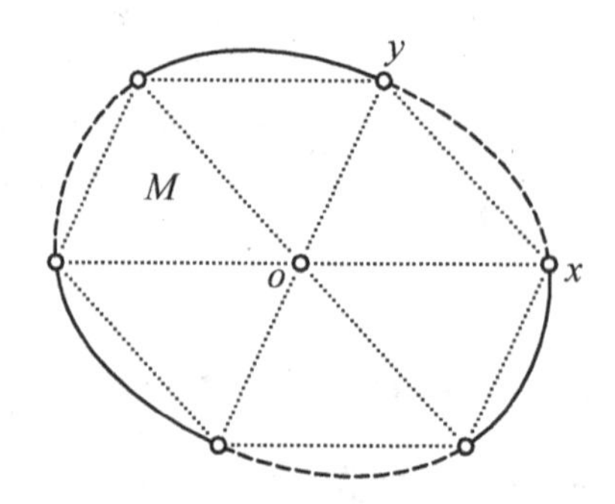

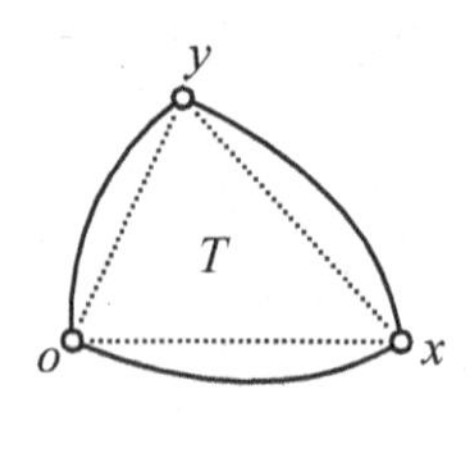

FIGURE 7.2
The construction of Reuleaux triangles in a normed plane.

where $d_{\mathbf{K}}(\mathbf{m})$ is the length of a longest chord of $\mathbf{K}$ in the direction of $\mathbf{m}$, and $w_{\mathbf{K}}(\mathbf{m}^{\perp})$ is the width of $\mathbf{K}$ in the direction perpendicular to $\mathbf{m}$ (cf. also Section 7.3).

Observe that for any direction $\mathbf{m}$, we have $d_{\mathbf{K}}(\mathbf{m}) = d_{\mathbf{M}}(\mathbf{m})$ and $w_{\mathbf{K}}(\mathbf{m}) = w_{\mathbf{M}}(\mathbf{m})$, which yields that minimizing $c_{tr}^{Bus}(\mathbf{K})$, over the class of convex disks with a given central symmetrization, is equivalent to minimizing area($\mathbf{K}$) within this class. For the special case that $\mathbf{M}$ is a Euclidean unit ball, this problem is solved by a theorem of Blaschke [53] and Lebesgue [168], which states that the smallest area convex disks of constant width two are the Reuleaux triangles of width two. This result was generalized by Chakerian [76] for normed planes in the following way.

Let $\mathbf{M}$ be an o-symmetric plane convex body. Then, for every $\mathbf{x} \in \mathrm{bd}\mathbf{M}$, there is an affine-regular hexagon, inscribed in $\mathbf{M}$, with $\mathbf{x}$ as a vertex. Let $\mathbf{y}$ be a consecutive vertex of this hexagon. By joining the points $\mathbf{o}$, $\mathbf{x}$ and $\mathbf{y}$ with the corresponding arcs in bd$\mathbf{M}$, we obtain a 'triangle' $\mathbf{T}$ with three arcs from bd$\mathbf{M}$ as its 'sides' (cf. Figure 7.2). These 'triangles', and their homothetic copies, are called the *Reuleaux triangles in the norm of* $\mathbf{M}$. Chakerian proved that, given a normed plane with unit disk $\mathbf{M}$, the area of any convex disk $\mathbf{K}$ of constant width two in the norm of $\mathbf{M}$ is minimal for some Reuleaux triangle in the norm. It is not too difficult to see, and was also proven by Chakerian, that the area of such a triangle is equal to $\mathrm{area}(\mathbf{K}) = 2\mathrm{area}(\mathbf{M}) - \frac{4}{3}\mathrm{area}(\mathbf{H})$, where $\mathbf{H}$ is a largest area affinely regular hexagon inscribed in the unit disk $\mathbf{M}$.

Now, assume that a plane convex body $\mathbf{K}$ is a minimizer of $c_{tr}^{HT}(\mathbf{K})$; by compactness arguments, such a minimizer exists. Then, from Chakerian's result, we obtain that $\mathbf{K}$ is a Reuleaux triangle in its relative norm, and that its area is $\mathrm{area}(\mathbf{K}) = 2\mathrm{area}(\mathbf{M}) - \frac{4}{3}\mathrm{area}(\mathbf{H})$, where $\mathbf{H}$ is a largest area affinely regular hexagon inscribed in $\mathbf{M}$. Now let $\mathbf{P}$ be a largest area parallelogram inscribed in $\mathbf{M}$. Then, by (7.17) and the equality

$$\max\{d_{\mathbf{K}}(\mathbf{m})w_{\mathbf{K}}(\mathbf{m}^{\perp}) : \mathbf{m} \in \mathbb{S}^1\} = 2\text{area}(\mathbf{P}),$$

we have

$$c_{tr}^{HT}(\mathbf{K}) = \frac{\text{area}(\mathbf{M}^{\circ})}{\pi}\left(2\text{area}(\mathbf{M}) - \frac{4}{3}\text{area}(\mathbf{H}) + 2\text{area}(\mathbf{P})\right). \tag{7.18}$$

It is easy to see that if $\mathbf{K}$ is a triangle, then $\mathbf{M}$ is an affinely regular hexagon, and vice versa, if $\mathbf{M}$ is an affinely regular hexagon, then the smallest area Reuleaux triangles in its norm are (Euclidean) triangles. Thus, we only need to show that the quantity in (7.18) is minimal if and only if $\mathbf{M} = \mathbf{H}$. Observe that $\text{area}(\mathbf{H}) \leq \text{area}(\mathbf{M})$, and hence, it suffices to prove that

$$f(\mathbf{M}) = \frac{\text{area}(\mathbf{M}^{\circ})\left(\frac{2}{3}\text{area}(\mathbf{M}) + 2\text{area}(\mathbf{P})\right)}{\pi} \tag{7.19}$$

is minimal if and only if $\mathbf{M}$ is an affine-regular hexagon.

Now we show that if $f(\mathbf{M})$ is minimal for $\mathbf{M}$, then its norm is a *Radon norm* (cf. [179] or [10]). Recall that a norm is Radon if, for some affine image $\mathbf{C}$ of its unit disk, the polar $\mathbf{C}^{\circ}$ is a rotated copy of $\mathbf{C}$ by $\frac{\pi}{2}$; in this case, the boundary of the unit disk is called a *Radon curve*.

Since $f(\mathbf{M})$ is an affine invariant quantity, we may assume that $\mathbf{P}$ is a square, with vertices $(\pm 1, 0)$ and $(0, \pm 1)$ in a Cartesian coordinate system. Note that as $\mathbf{P}$ is a largest area inscribed parallelogram, the lines $x = \pm 1$ and $y = \pm 1$ support $\mathbf{M}$. Thus, the arc of $\text{bd}\mathbf{M}$ in the first quadrant determines the corresponding part of $\text{bd}\mathbf{M}^{\circ}$. On the other hand, the maximality of the area of $\mathbf{P}$ yields that for any point $\mathbf{p} \in \text{bd}\mathbf{M}$, the two lines, parallel to the segment $[\mathbf{o}, \mathbf{p}]$ and at the distance $\frac{1}{\|\mathbf{p}\|}$ from the origin, are either disjoint from $\mathbf{M}$ or support it. Thus, the rotated copy of $\mathbf{M}^{\circ}$ by $\frac{\pi}{2}$ contains $\mathbf{M}$, and the two bodies coincide if and only if $\text{bd}\mathbf{M}$ is a Radon curve.

Let $\mathbf{Q}_1$ and $\mathbf{Q}_2$ denote the parts of $\mathbf{M}$ in the first and the second quadrants, respectively. We define $\mathbf{Q}_1^{\circ}$ and $\mathbf{Q}_2^{\circ}$ similarly for $\mathbf{M}^{\circ}$. Then $\text{area}(\mathbf{Q}_2^{\circ}) = \text{area}(\mathbf{Q}_1) + x_1$ and $\text{area}(\mathbf{Q}_1^{\circ}) = \text{area}(\mathbf{Q}_2) + x_2$ for some $0 \leq x_1, x_2 \leq \frac{1}{2}$. Using this notation, we have $f(\mathbf{M}) = \frac{1}{\pi}\left(\text{area}(\mathbf{M}) + 2x_1 + 2x_2\right)\left(\frac{2}{3}\text{area}(\mathbf{M}) + 4\right)$. Let $\mathbf{M}_1$ denote the convex disk obtained by replacing the part of $\text{bd}\mathbf{M}$ in the second and fourth quadrants by the rotated copy of the arc of $\text{bd}\mathbf{M}^{\circ}$ in the first quadrant (cf. Figure 7.3). Similarly, let $\mathbf{M}_2$ be the disk obtained by replacing the part of $\text{bd}\mathbf{M}$ in the other two quadrants by the rotated copy of the arc of $\text{bd}\mathbf{M}^{\circ}$ in the second quadrant.

By our previous observations, we have that $\mathbf{M}_1$ and $\mathbf{M}_2$ are unit disks of Radon norms, and $\mathbf{M} \subset \mathbf{M}_1$ and $\mathbf{M} \subset \mathbf{M}_2$. On the other hand, the area of a largest area parallelogram inscribed in $\mathbf{M}_1$ or $\mathbf{M}_2$ is equal to $\text{area}(\mathbf{P}) = 2$. Now an elementary computation shows that

$$f(\mathbf{M}_i) = \frac{1}{\pi}\left(\text{area}(\mathbf{M}) + 2x_{i+1}\right)\left(\frac{2}{3}(\text{area}(\mathbf{M}) + 2x_{i+1}) + 4\right) \quad \text{for } i = 1, 2,$$

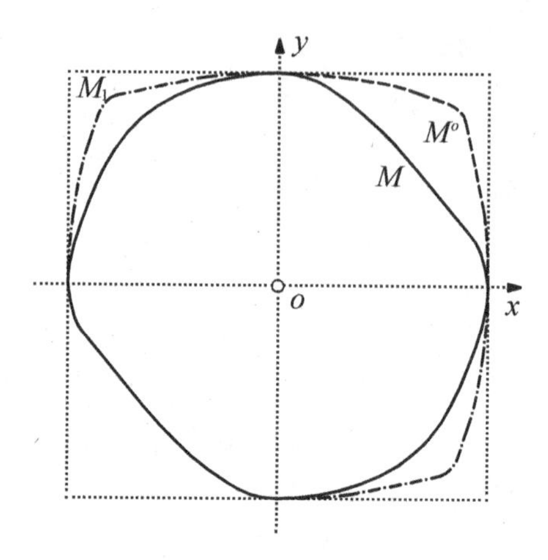

FIGURE 7.3
The extension of $\mathbf{M}$ to the unit disk of a Radon norm.

which, since $0 \le x_1, x_2 \le \frac{1}{2}$, yields that

$$2f(\mathbf{M}) - f(\mathbf{M}_1) - f(\mathbf{M}_2) = \frac{1}{\pi}\left(8x_1 + 8x_2 - \frac{8}{3}x_1^2 - \frac{8}{3}x_2^2\right) \ge 0,$$

with equality if and only if $x_1 = x_2 = 0$. From this, it follows that $f(\mathbf{M}) \ge \min\{f(\mathbf{M}_1), f(\mathbf{M}_2)\}$, with equality if and only if $x_1 = x_2 = 0$ and $\mathbf{M}_1 = \mathbf{M}_2 = \mathbf{M}$. This readily implies that if $f(\mathbf{M})$ is minimal for $\mathbf{M}$, then $\mathbf{M}$ is the unit disk of a Radon norm.

In the following, we assume that the norm of $\mathbf{M}$ is Radon. Observe that, under our assumption about $\mathbf{P}$, we have $\text{area}(\mathbf{M}) = \text{area}(\mathbf{M}^\circ)$, since $\mathbf{M}^\circ$ is a rotated copy of $\mathbf{M}$. On the other hand, since the *volume product* $\text{area}(\mathbf{M})\text{area}(\mathbf{M}^\circ)$ of $\mathbf{M}$ (cf., e.g., [61]) does not change under affine transformations, the definition of Radon norm implies that, in general,

$$\text{area}(\mathbf{M}^\circ) = \frac{4\text{area}(\mathbf{M})}{(\text{area}(\mathbf{P}))^2}.$$

Since $\text{vol}_{\mathbf{M}}^m(\mathbf{M}) = \frac{2}{\text{area}(\mathbf{P})}\text{area}(\mathbf{M})$ (cf. the definition in Section 2.2, this yields that

$$\begin{aligned} f(\mathbf{M}) &= \frac{4\text{area}(\mathbf{M})}{(\pi\text{area}(\mathbf{P}))^2}\left(\frac{2}{3}\text{area}(\mathbf{M}) + 2\text{area}(\mathbf{P})\right) \\ &= \frac{2}{3\pi}\left(\text{vol}_{\mathbf{M}}^m(\mathbf{M})\right)^2 + \frac{2}{\pi}\text{vol}_{\mathbf{M}}^m(\mathbf{M}). \end{aligned}$$

Hence, we need to find the minimum of $\text{vol}_{\mathbf{M}}^m(\mathbf{M})$ under the condition that $\mathbf{M}$ defines a Radon norm. This problem is examined in Theorem 58, which states, among other things, that for any Radon norm with unit disk

$\mathbf{M}$, $\mathrm{vol}^m_{\mathbf{M}}(\mathbf{M})$ is at least 3, with equality if and only if $\mathbf{M}$ is an affinely regular hexagon. Thus, the left-hand side of (ii) immediately follows.

7.7.2 The proof of the right-hand side inequality in (ii)

Assume that $c^{HT}_{tr}(\mathbf{K})$ is maximal for some plane convex body $\mathbf{K}$ and let $\mathbf{M} = \frac{1}{2}(\mathbf{K} - \mathbf{K})$. Note that by the Brunn-Minkowski Inequality, we have $\mathrm{area}(\mathbf{K}) \leq \mathrm{area}(\mathbf{M})$, with equality if and only if $\mathbf{K}$ is centrally symmetric. Thus, (7.17) implies that $\mathbf{K}$ is centrally symmetric and, without loss of generality, we may assume that $\mathbf{K} = \mathbf{M}$.

Let $\mathbf{P}$ be a largest area parallelogram inscribed in $\mathbf{M}$. Since $c^{HT}_{tr}(\mathbf{M})$ is affine invariant, we may assume that $\mathbf{P}$ is the square with vertices $(\pm 1, 0)$ and $(0, \pm 1)$ in a Cartesian coordinate system. Then the lines $x = \pm 1$ and $y = \pm 1$ support $\mathbf{M}$. Let $\mathbf{M}^* = S_H(\mathbf{M})$ be the Steiner symmetrization of $\mathbf{M}$ to a symmetry axis of $\mathbf{P}$. Then, clearly, $\mathrm{area}(\mathbf{M}^*) = \mathrm{area}(\mathbf{M})$. Observe that $\mathbf{P}$ is inscribed in $\mathbf{M}^*$ as well, which yields that if $\mathbf{P}^*$ is a maximal area parallelogram inscribed in $\mathbf{M}^*$, then $\mathrm{area}(\mathbf{P}^*) \geq \mathrm{area}(\mathbf{P})$. For the Euclidean version of the problem (cf. Theorem 41), we have

$$c_{tr}(\mathbf{M}) = 1 + \frac{2\mathrm{area}(\mathbf{P})}{\mathrm{area}(\mathbf{M})}. \tag{7.20}$$

Then, (i) of Theorem 33 yields that $c_{tr}(\mathbf{M})$ does not increase under Steiner symmetrization, which implies that $\mathrm{area}(\mathbf{P}^*) \leq \mathrm{area}(\mathbf{P})$. Thus, we have $\mathrm{area}(\mathbf{P}^*) = \mathrm{area}(\mathbf{P})$.

Now we apply a result of Meyer and Pajor [182] about the Blaschke-Santaló Inequality, who proved that volume product does not decrease under Steiner symmetrizations, which yields that $\mathrm{area}((\mathbf{M}^*)^\circ) \geq \mathrm{area}(\mathbf{M}^\circ)$. Thus, since $\mathbf{M}$ maximizes $c^{HT}_{tr}(\mathbf{M})$, (7.17) implies that $\mathrm{area}((\mathbf{M}^*)^\circ) = \mathrm{area}(\mathbf{M}^\circ)$. Unfortunately, no geometric condition is known that characterizes the equality case for Steiner symmetrization. Nevertheless, we may apply another method, used by Saint-Raymond [212], which he used to characterize the equality case of the Blaschke-Santaló Inequality. This method, described also in [236], is as follows.

Let $\mathbf{C}$ be an $\mathbf{o}$-symmetric convex body in $\mathbb{E}^d$, and let H be the hyperplane with the equation $x_d = 0$. For any $t \in \mathbb{R}$, let C_t be the section of $\mathbf{C}$ with the hyperplane $\{x_d = t\}$. Define $\bar{\mathbf{C}}$ as the union of the $(d-1)$-dimensional convex bodies $t\mathbf{e}_d + \frac{1}{2}(C_t - C_t)$, where $\mathbf{e}_d$ is the dth coordinate unit vector. Then we have the following (cf. Lemma 5.3.1 and the proof of Theorem 5.3.2 of [236]).

(i) $\bar{\mathbf{C}}$ is an $\mathbf{o}$-symmetric convex body.

(ii) $\mathrm{vol}_d(\bar{\mathbf{C}}) \geq \mathrm{vol}_d(\mathbf{C})$, with equality if and only if every t-section C_t has a center of symmetry.

(iii) $\mathrm{vol}_d(\bar{\mathbf{C}}^\circ) \geq \mathrm{vol}_d(\mathbf{C}^\circ)$.

(iv) If $\text{vol}_d(\bar{\mathbf{C}}^\circ)\text{vol}_d(\bar{\mathbf{C}}) = \text{vol}_d(\mathbf{C}^\circ)\text{vol}_d(\mathbf{C})$, then the centers of symmetry of the sets C_t lie on a straight line segment.

We note that this symmetrization procedure in the plane coincides with the Steiner symmetrization with respect to the second coordinate axis.

Since in our case $\text{area}(\mathbf{M}^*) = \text{area}(\mathbf{M})$ and $\text{area}((\mathbf{M}^*)^\circ) = \text{area}(\mathbf{M}^\circ)$, it follows from the theorem of Saint-Raymond that the midpoints of the chords of $\mathbf{M}$, perpendicular to H, lie on a straight line segment. On the other hand, as $S_H(\mathbf{P}) = \mathbf{P}$, we have that this segment is contained in H. Thus, $\mathbf{M}$ is symmetric to H. Since H was an arbitrary symmetry axis of $\mathbf{P}$, we obtain that the symmetry group of $\mathbf{M}$ contains that of $\mathbf{P}$, and, in particular, $\mathbf{M}$ has a 4-fold rotational symmetry.

Observe that in this case $\mathbf{M} \subseteq \mathbf{B}^2$. Indeed, if for some $\mathbf{p} \in \mathbf{M}$ we have $||\mathbf{p}|| > 1$, then, by the 4-fold rotational symmetry of $\mathbf{M}$, it follows that $\mathbf{M}$ contains a square of area greater than $\text{area}(\mathbf{P}) = 2$, which contradicts our assumption that $\mathbf{P}$ is a largest area parallelogram inscribed in $\mathbf{M}$. Since it is easy to check that $c_{tr}^{HT}(\mathbf{M})$ is not maximal if $\mathbf{M} = \mathbf{B}^2$, this implies, in particular, that $\text{area}(\mathbf{M}) < \pi$. Note that in our case the area of the part of $\mathbf{M}$ in each quadrant is equal.

In the next step, we use the following Proposition from [59].

Proposition 172 *Let $\mathbf{Q} = \text{conv}\{\mathbf{o}, \mathbf{x}, \mathbf{y}, \mathbf{z}\}$ be a convex deltoid symmetric about the line containing the diagonal $[\mathbf{o}, \mathbf{y}]$. Assume that $\mathbf{x}, \mathbf{y} \in \mathbb{S}^1$ and that the lines containing $[\mathbf{x}, \mathbf{z}]$ and $[\mathbf{y}, \mathbf{z}]$ support $\mathbf{B}^2$. Let $\mathbf{C}$ be any $\mathbf{o}$-symmetric plane convex body such that $\mathbf{x}, \mathbf{y} \in \text{bd}\mathbf{C}$ and the lines containing $[\mathbf{x}, \mathbf{z}]$ and $[\mathbf{y}, \mathbf{z}]$ support $\mathbf{C}$, and set $\mathbf{K} = \mathbf{C} \cap \mathbf{Q}$ and $\mathbf{K}^\circ = \mathbf{C}^\circ \cap \mathbf{Q}$. Let $\text{area}(\mathbf{K}) = \alpha \leq \text{area}(\mathbf{Q} \cap \mathbf{B}^2)$ be fixed. Then $\lambda(\mathbf{K}^\circ)$ is maximal, e.g., if $\mathbf{C}$ is an $\mathbf{o}$-symmetric ellipse $\mathbf{E}$ satisfying $\text{area}(\mathbf{E} \cap \mathbf{Q}) = \alpha$.*

Applying this theorem for the part of $\mathbf{M}$, say, in the first quadrant, we have that, under our assumption about $\mathbf{P}$, $\mathbf{M}$ is a convex body bounded by four congruent elliptic arcs, having centers at $\mathbf{o}$. Then it is a matter of computation to verify that $f(\mathbf{M})$ is maximal for a rotated copy of the body $\mathbf{M}_0$ described in Section 2.2.

7.7.3 The proofs of (i), (iii) and (iv)

First, we prove (i). Observe that for any plane convex body $\mathbf{K}$,

$$c_{tr}^{Bus}(\mathbf{K}) = \frac{\pi}{\text{area}(\mathbf{M})} \left(\text{area}(\mathbf{K}) + 2\text{area}(\mathbf{P})\right),$$

where $\mathbf{M} = \frac{1}{2}(\mathbf{K} - \mathbf{K})$, and $\mathbf{P}$ is a largest area parallelogram inscribed in $\mathbf{M}$. By the result of Chakerian [76] described in Subsection 7.7.1, we have that if $\mathbf{K}$ minimizes $c_{tr}^{Bus}(\mathbf{K})$ over the family of plane convex bodies, then $\mathbf{K}$ is a minimal area Reuleaux triangle in the norm of $\mathbf{M}$, and its area is

$$\text{area}(\mathbf{K}) = 2\text{area}(\mathbf{M}) - \frac{4}{3}\text{area}(\mathbf{H}), \tag{7.21}$$

where $\mathbf{H}$ is a largest area affinely regular hexagon inscribed in $\mathbf{M}$. Thus, we may assume, without loss of generality, that

$$c_{tr}^{Bus}(\mathbf{K}) = \frac{\pi}{\text{area}(\mathbf{M})}\left(2\text{area}(\mathbf{M}) + 2\text{area}(\mathbf{P}) - \frac{4}{3}\text{area}(\mathbf{H})\right). \tag{7.22}$$

Note that in this case $\mathbf{K}$ is a (Euclidean) triangle if and only if $\mathbf{M} = \mathbf{H}$.

From (7.22), it readily follows that

$$c_{tr}^{Bus}(\mathbf{K}) = 2\pi + 2\pi\frac{3\text{area}(\mathbf{P}) - 2\text{area}(\mathbf{H})}{3\text{area}(\mathbf{M})}.$$

Observe that $\mathbf{H}$ contains a parallelogram whose area is equal to $\text{area}(\bar{\mathbf{P}}) = \frac{2}{3}\text{area}(\mathbf{H})$. Since $\mathbf{H} \subseteq \mathbf{M}$, this yields that $\text{area}(\mathbf{P}) \geq \frac{2}{3}\text{area}(\mathbf{H})$, with equality if and only if $\mathbf{M} = \mathbf{H}$. This means that $c_{tr}^{Bus}(\mathbf{K}) \geq 2\pi$, with equality if and only if $\mathbf{M} = \mathbf{H}$, which proves the left-hand side inequality about Busemann area.

Now we prove the right-hand side inequality. The formula in (7.21) and the Brunn-Minkowski Inequality shows, like in Subsection 7.7.2, that if $c_{tr}^{Bus}(\mathbf{K})$ is maximal over the family of plane convex bodies, then $\mathbf{K}$ is centrally symmetric. Thus we may apply Theorem 34 about the maximum of $c_{tr}(\mathbf{K})$, which yields the assertion.

Next, we prove (iii). Let $\mathbf{P}$ be a largest area parallelogram inscribed in $\mathbf{M} = \frac{1}{2}(\mathbf{K} - \mathbf{K})$. Then we have

$$c_{tr}^{m}(\mathbf{K}) = \frac{2\left(\text{area}(\mathbf{K}) + 2\text{area}(\mathbf{P})\right)}{\text{area}(\mathbf{P})} = 4 + \frac{2\text{area}(\mathbf{K})}{\text{area}(\mathbf{P})}. \tag{7.23}$$

Observe that for any plane convex body $\mathbf{K}$, we have

$$c_{tr}(\mathbf{K}) = \frac{\text{area}(\mathbf{K}) + 2\text{area}(\mathbf{P})}{\text{area}(\mathbf{K})} = 1 + \frac{2\text{area}(\mathbf{P})}{\text{area}(\mathbf{K})}.$$

By Theorem 34, the latter expression is maximal if and only if $\mathbf{K}$ is a convex quadrilateral, and by Theorem 41, it is minimal if and only if $\mathbf{K}$ is an ellipse. Thus the assertion readily follows.

Our next case is the left-hand side inequality of (iv). Observe that

$$c_{tr}^{m^*}(K) = \frac{4\left(\text{area}(\mathbf{K}) + 2\text{area}(\mathbf{P})\right)}{\text{area}(\mathbf{P}')}, \tag{7.24}$$

where $\mathbf{P}$ is a largest area inscribed, and $\mathbf{P}'$ is a smallest area circumscribed parallelogram in $\mathbf{M} = \frac{1}{2}(\mathbf{K} - \mathbf{K})$.

As in the previous sections, if $c_{tr}^{m^*}(\mathbf{K})$ is minimal for some plane convex body $\mathbf{K}$, then, by [76], we may assume that $\mathbf{K}$ is a Reuleaux triangle in its relative norm, and its area is $\text{area}(\mathbf{K}) = 2\text{area}(\mathbf{M}) - \frac{4}{3}\text{area}(\mathbf{H})$, where $\mathbf{H}$

is a largest area affinely regular hexagon inscribed in $\mathbf{M}$. Thus, area($\mathbf{M}$) $\geq$ area($\mathbf{H}$) implies that

$$c_{tr}^{m^*}(\mathbf{K}) \geq \frac{8\,(\text{area}(\mathbf{M}) + 3\text{area}(\mathbf{P}))}{3\text{area}(\mathbf{P}')}. \tag{7.25}$$

On the other hand, we clearly have area($\mathbf{P}$) $\geq \frac{1}{2}$area($\mathbf{P}'$), where we have equality, for example, if $\mathbf{M}$ is an affinely regular hexagon. Furthermore, Theorem 57 states that Gromov's mass* of any $\mathbf{o}$-symmetric convex disk is at least three, with equality if and only if $\mathbf{M}$ is an affinely regular hexagon. This implies that area($\mathbf{M}$) $\geq \frac{3}{4}$area($\mathbf{P}'$), and thus, we obtain $c_{tr}^{m^*}(\mathbf{K}) \geq 6$. Here, we have equality if and only if $\mathbf{M}$ is an affinely regular hexagon, which immediately implies that $\mathbf{K}$ is a triangle.

Finally, we prove the right-hand side of (iv). Like in the previous cases, we may assume that $\mathbf{K} = \mathbf{M}$. But then, clearly, area($\mathbf{M}$) $\leq$ area($\mathbf{P}'$), area($\mathbf{P}$) $\leq$ area($\mathbf{P}'$) and (7.24) yields that $c_{tr}^{m^*}(\mathbf{K}) \leq 12$. Since in both inequalities equality is possible only if $\mathbf{M}$ is a parallelogram; the assertion follows.

8

Proofs on the Kneser-Poulsen Conjecture

Summary. In this chapter we present selected proofs of some theorems from Chapter 3 about the Kneser-Poulsen Conjecture. In Sections 8.1-8.3 we prove this conjecture, both for unions and for intersections of Euclidean disks in $\mathbb{E}^2$. In Section 8.4 we prove a Blaschke-Santaló-type inequality for r-ball bodies in spaces of constant curvature. In Section 8.5 we prove that the volume of the intersection of equal balls in a space of constant curvature increases under uniform contractions. In Section 8.6, we drop the condition of uniform contractions, and prove the Kneser-Poulsen Conjecture for unions and intersections of closed hemispheres in $\mathbb{S}^d$ under arbitrary contractions. In Section 8.7 we extend the result of Section 8.4 to all intrinsic volumes of r-ball bodies in $\mathbb{E}^d$. In Section 8.8 we apply this result to prove the Kneser-Poulsen Conjecture for intersections of sufficiently many equal balls in $\mathbb{E}^d$ under uniform contractions. In Section 8.9 we prove the dual of this result for unions of sufficiently many equal balls in $\mathbb{E}^d$ under uniform contractions. Finally, in Section 8.10 we present the proof of the Kneser-Poulsen Conjecture for unions (resp., intersections) of disks in $\mathbb{S}^2$ or $\mathbb{H}^2$, under the condition that the first set has a simply connected interior.

8.1 Proof of Theorem 67

As a first step in our proof of Theorem 67, we recall the following underlying system of (truncated) Voronoi cells. For a given point configuration $\mathbf{p} = (\mathbf{p}_1, \mathbf{p}_2, \ldots, \mathbf{p}_N)$ in $\mathbb{E}^d$ and radii $r_1, r_2, \ldots, r_N$ consider the following sets,

$$\mathbf{V}_i = \{\mathbf{x} \in \mathbb{E}^d \mid \text{for all } j,\ \|\mathbf{x} - \mathbf{p}_i\|^2 - r_i^2 \leq \|\mathbf{x} - \mathbf{p}_j\|^2 - r_j^2\},$$

$$\mathbf{V}^i = \{\mathbf{x} \in \mathbb{E}^d \mid \text{for all } j,\ \|\mathbf{x} - \mathbf{p}_i\|^2 - r_i^2 \geq \|\mathbf{x} - \mathbf{p}_j\|^2 - r_j^2\}.$$

The set $\mathbf{V}_i$ (resp., $\mathbf{V}^i$) is called the nearest (resp., farthest) point Voronoi cell of the point $\mathbf{p}_i$. We now restrict each of these sets as follows,

$$\mathbf{V}_i(r_i) = \mathbf{V}_i \cap \mathbf{B}^d[\mathbf{p}_i, r_i],$$

$$\mathbf{V}^i(r_i) = \mathbf{V}^i \cap \mathbf{B}^d[\mathbf{p}_i, r_i].$$

We call the set $\mathbf{V}_i(r_i)$ (resp., $\mathbf{V}^i(r_i)$) the nearest (resp., farthest) point truncated Voronoi cell of the point $\mathbf{p}_i$. For each $i \neq j$ let $W_{ij} = \mathbf{V}_i \cap \mathbf{V}_j$ and $W^{ij} = \mathbf{V}^i \cap \mathbf{V}^j$. The sets W_{ij} and W^{ij} are the walls between the nearest point and farthest point Voronoi cells. Finally, it is natural to define the relevant truncated walls as follows.

$$W_{ij}(\mathbf{p}_i, r_i) = W_{ij} \cap \mathbf{B}^d[\mathbf{p}_i, r_i]$$
$$= W_{ij}(\mathbf{p}_j, r_j) = W_{ij} \cap \mathbf{B}^d[\mathbf{p}_j, r_j],$$

$$W^{ij}(\mathbf{p}_i, r_i) = W^{ij} \cap \mathbf{B}^d[\mathbf{p}_i, r_i]$$
$$= W^{ij}(\mathbf{p}_j, r_j) = W^{ij} \cap \mathbf{B}^d[\mathbf{p}_j, r_j].$$

Second, for each $i = 1, 2, \ldots, N$ and $0 \leq s$, define $r_i(s) = \sqrt{r_i^2 + s}$. Clearly,

$$\frac{d}{ds} r_i(s) = \frac{1}{2r_i(s)}. \tag{8.1}$$

Now, define $\mathbf{r}(s) = (r_1(s), \ldots, r_N(s))$, and introduce

$$V_d(t, s) := \mathrm{vol}_d \left(\mathbf{B}^d_{\bigcup}[\mathbf{p}(t), \mathbf{r}(s)] \right),$$

and

$$V^d(t, s) := \mathrm{vol}_d \left(\mathbf{B}^d_{\bigcap}[\mathbf{p}(t), \mathbf{r}(s)] \right)$$

as functions of the variables t and s, where

$$\mathbf{B}^d_{\bigcup}[\mathbf{p}(t), \mathbf{r}(s)] := \bigcup_{i=1}^{N} \mathbf{B}^d[\mathbf{p}_i(t), r_i(s)],$$

and

$$\mathbf{B}^d_{\bigcap}[\mathbf{p}(t), \mathbf{r}(s)] := \bigcap_{i=1}^{N} \mathbf{B}^d[\mathbf{p}_i(t), r_i(s)].$$

Throughout we assume that all $r_i > 0$.

Lemma 173 *Let $d \geq 2$ and let $\mathbf{p}(t), 0 \leq t \leq 1$ be a smooth motion of a point configuration in $\mathbb{E}^d$ such that for each t, the points of the configuration are pairwise distinct. Then the volume functions $V_d(t, s)$ and $V^d(t, s)$ are continuously differentiable in t and s simultaneously, and for any fixed t, the nearest point and farthest point Voronoi cells are constant.*

Proof. Let $t = t_0$ be fixed. Then recall that the point $\mathbf{x}$ belongs to the Voronoi cell $\mathbf{V}_i(t_0, s)$ (resp., $\mathbf{V}^i(t_0, s)$), when for all j, $\|\mathbf{x} - \mathbf{p}_i(t_0)\|^2 - \|\mathbf{x} - \mathbf{p}_j(t_0)\|^2 - r_i(s)^2 + r_j(s)^2$ is non-positive (resp., non-negative). But $r_i(s)^2 - r_j(s)^2 = r_i^2 - r_j^2$ is constant. So each $\mathbf{V}_i(t_0, s)$ and $\mathbf{V}^i(t_0, s)$ is a constant function of s.

As $\mathbf{p}(t)$ is continuously differentiable, therefore the partial derivatives of $V_d(t, s)$ and $V^d(t, s)$ with respect to t exist and are continuous by Theorem 65. Each ball $\mathbf{B}^d[\mathbf{p}_i(t), r_i(s)], d \geq 2$ is strictly convex. Hence, the $(d-1)$-dimensional surface volume of the boundaries of $\mathbf{B}^d_{\bigcup}[\mathbf{p}(t), \mathbf{r}(s)]$ and $\mathbf{B}^d_{\bigcap}[\mathbf{p}(t), \mathbf{r}(s)]$ are continuous functions of s, and the partial derivatives of $V_d(t, s)$ and $V^d(t, s)$ with respect to s exist and are continuous. Thus, $V_d(t, s)$ and $V^d(t, s)$ are both continuously differentiable with respect to t and s simultaneously. □

Lemma 174 *Let $\mathbf{p}(t), 0 \leq t \leq 1$ be an analytic motion of a point configuration in $\mathbb{E}^d, d \geq 2$. Then there exists an open dense set U in $[0,1] \times (0, \infty)$ such that for any $(t, s) \in U$ the following hold.*

$$\frac{\partial^2}{\partial t \partial s} V_d(t, s) = \sum_{1 \leq i < j \leq N} \left(\frac{d}{dt} d_{ij}(t) \right) \cdot \frac{\partial}{\partial s} \mathrm{vol}_{d-1} \left[W_{ij}(\mathbf{p}_i(t), r_i(s)) \right],$$

and

$$\frac{\partial^2}{\partial t \partial s} V^d(t, s) = \sum_{1 \leq i < j \leq N} - \left(\frac{d}{dt} d_{ij}(t) \right) \cdot \frac{\partial}{\partial s} \mathrm{vol}_{d-1} \left[W^{ij}(\mathbf{p}_i(t), r_i(s)) \right].$$

Hence, if $\mathbf{p}(t)$ is contracting, then $\frac{\partial}{\partial s} V_d(t, s)$ is monotone decreasing in t, and $\frac{\partial}{\partial s} V^d(t, s)$ is monotone increasing in t.

Proof. Given that $\mathbf{p}(t), 0 \leq t \leq 1$ is an analytic function of t, we wish to define an open dense set U in $[0,1] \times (0, \infty)$, where the volume functions $V_d(t, s)$ and $V^d(t, s)$ are analytic in t and s simultaneously. Lemma 173 implies that the Voronoi cells $\mathbf{V}_i$ and $\mathbf{V}^i$ are functions of t alone. Moreover, clearly there are only a finite number of values of t in the interval $[0, 1]$, where the combinatorial type of the above Voronoi cells changes. The volume of the truncated Voronoi cells $\mathbf{V}_i\,(r_i(s))$ and $\mathbf{V}^i\,(r_i(s))$ are obtained from the volume of the d-dimensional Euclidean ball of radius $r_i(s)$ by removing or adding the volumes of the regions obtained by conning over the walls $W_{ij}(\mathbf{p}_i(t), r_i(s))$ or $W^{ij}(\mathbf{p}_i(t), r_i(s))$ from the point $\mathbf{p}_i(t)$. By induction on d, starting at $d = 1$, each W_{ij} and W^{ij} is an analytic function of t and s, when the ball of radius $r_i(s)$ is not tangent to any of the faces of $\mathbf{V}_i$ or $\mathbf{V}^i$. So, for any fixed t the ball of radius $r_i(s)$ will not be tangent to any of the faces $\mathbf{V}_i$ or $\mathbf{V}^i$ for all but a finite number of values of s. Thus, we define U to be the set of those (t, s), where for some open interval about t in $[0, 1]$, the combinatorial type of the Voronoi cells is constant and for all i, the ball of radius $r_i(s)$ is not tangent to any of the faces of $\mathbf{V}_i$ or $\mathbf{V}^i$. We also assume that the points of the configuration $\mathbf{p}(t)$ are distinct for any $(t, s) \in U$. If, for $i \neq j$ and for infinitely many values of t in the interval $[0, 1]$, $\mathbf{p}_i(t) = \mathbf{p}_j(t)$, then they are the same point for all t, and those points may be identified. Then the set U is open and dense in $[0, 1] \times (0, \infty)$ and the volume functions $V_d(t, s)$ and $V^d(t, s)$ are analytic in t and s. Thus, the formulas for the mixed partial derivatives in Lemma 174 follow from the definition of U and from Theorem 65. (Note also that here we could interchange the order of partial differentiation with respect to the variables t and s.)

To show that $\frac{\partial}{\partial s}V_d(t, s)$ and $\frac{\partial}{\partial s}V^d(t, s)$ are monotone, suppose they are not. We show a contradiction. If we perturb s slightly to s_0, say, then using the formulas for the mixed partial derivatives in Lemma 174 we get that the partial derivative of $\frac{\partial}{\partial s}V_d(t, s)$ and $\frac{\partial}{\partial s}V^d(t, s)$ with respect to t exists and has the appropriate sign, except for a finite number of values of t for $s = s_0$. (Here we have also used the following rather obvious monotonicity property of the walls: $W_{ij}(\mathbf{p}_i(t), r_i(s)) \subset W_{ij}(\mathbf{p}_i(t), r_i(s^*))$ and $W^{ij}(\mathbf{p}_i(t), r_i(s)) \subset W^{ij}(\mathbf{p}_i(t), r_i(s^*))$ for any $s \leq s^*$.) Since $\frac{\partial}{\partial s}V_d(t, s)$ and $\frac{\partial}{\partial s}V^d(t, s)$ are continuous as a function of t at $s = s_0$ by Lemma 173, they are monotone. But the functions at s_0 approximate the functions at s (again by Lemma 173) providing the contradiction. So, $\frac{\partial}{\partial s}V_d(t, s)$ and $\frac{\partial}{\partial s}V^d(t, s)$ are indeed monotone. This completes the proof of Lemma 174. □

First, note that

$$F_i(t) = \mathbf{V}_i(t, 0) \cap \mathrm{bd}\left(\mathbf{B}^d_{\bigcup}[\mathbf{p}(t), \mathbf{r}(0)]\right) \tag{8.2}$$

and

$$\left(\text{resp., } F_i(t) = \mathbf{V}^i(t, 0) \cap \mathrm{bd}\left(\mathbf{B}^d_{\bigcap}[\mathbf{p}(t), \mathbf{r}(0)]\right)\right). \tag{8.3}$$

Second, (8.1), (8.2) and (8.3) imply in a straightforward way that

$$\frac{\partial}{\partial s}V_d(t, s)\bigg|_{s=0} = \frac{1}{2}\sum_{i=1}^{N}\frac{1}{r_i}\mathrm{svol}_{d-1}(F_i(t)) = \lim_{s_0 \to 0+}\frac{\partial}{\partial s}V_d(t, s)\bigg|_{s=s_0} \tag{8.4}$$

$$\left(\text{resp., } \frac{\partial}{\partial s}V^d(t, s)\bigg|_{s=0} = \frac{1}{2}\sum_{i=1}^{N}\frac{1}{r_i}\mathrm{svol}_{d-1}(F_i(t)) = \lim_{s_0 \to 0+}\frac{\partial}{\partial s}V_d(t, s)\bigg|_{s=s_0}\right). \tag{8.5}$$

Thus, (8.4) and (8.5) together with Lemma 174 finish the proof of Theorem 67.

8.2 Proof of Theorem 68

We start our proof with the following volume formula from calculus, which is based on cylindrical shells.

Lemma 175 *Let* $\mathbf{X}$ *be a compact measurable set in* $\mathbb{E}^d, d \geq 3$ *that is a solid of revolution about* $\mathbb{E}^{d-2}$*. In other words, the orthogonal projection of* $\mathbf{X} \cap \{\mathbb{E}^{d-2} \times (s\cos\theta, s\sin\theta)\}$ *onto* $\mathbb{E}^{d-2}$ *is a measurable set* $X(s)$ *independent of* θ*. Then*

$$\mathrm{vol}_d(\mathbf{X}) = \int_0^{\infty} (2\pi s)\mathrm{vol}_{d-2}(X(s))\, ds.$$

By assumption the centers of the closed d-dimensional balls $\mathbf{B}^d[\mathbf{p}_i, r_i]$, $1 \leq i \leq N$ lie in the $(d-2)$-dimensional affine subspace L of $\mathbb{E}^d$. Now, recall the construction of the following (truncated) Voronoi cells.

$$\mathbf{V}_i(d) = \{\mathbf{x} \in \mathbb{E}^d \mid \text{for all } j,\ \|\mathbf{x}-\mathbf{p}_i\|^2 - r_i^2 \leq \|\mathbf{x}-\mathbf{p}_j\|^2 - r_j^2\},$$

$$\mathbf{V}^i(d) = \{\mathbf{x} \in \mathbb{E}^d \mid \text{for all } j,\ \|\mathbf{x}-\mathbf{p}_i\|^2 - r_i^2 \geq \|\mathbf{x}-\mathbf{p}_j\|^2 - r_j^2\}.$$

The set $\mathbf{V}_i(d)$ (resp., $\mathbf{V}^i(d)$) is called the nearest (resp., farthest) point Voronoi cell of the point $\mathbf{p}_i$ in $\mathbb{E}^d$. Then we restrict each of these sets as follows:

$$\mathbf{V}_i(r_i, d) = \mathbf{V}_i \cap \mathbf{B}^d[\mathbf{p}_i, r_i],$$

$$\mathbf{V}^i(r_i, d) = \mathbf{V}^i \cap \mathbf{B}^d[\mathbf{p}_i, r_i].$$

We call the set $\mathbf{V}_i(r_i, d)$ (resp., $\mathbf{V}^i(r_i, d)$) the nearest (resp., farthest) point truncated Voronoi cell of the point $\mathbf{p}_i$ in $\mathbb{E}^d$. As the point configuration $\mathbf{p} = (\mathbf{p}_1, \mathbf{p}_2, \ldots, \mathbf{p}_N)$ lies in the $(d-2)$-dimensional affine subspace $L \subset \mathbb{E}^d$ and as without loss of generality we may assume that $L = \mathbb{E}^{d-2}$; therefore, one can introduce the relevant $(d-2)$-dimensional truncated Voronoi cells $\mathbf{V}_i(r_i, d-2)$ and $\mathbf{V}^i(r_i, d-2)$ in a straightforward way. We are especially interested in the relation of the volume of $\mathbf{V}_i(r_i, d-2)$ and $\mathbf{V}^i(r_i, d-2)$ in $\mathbb{E}^{d-2}$ to the volume of the corresponding truncated Voronoi cells $\mathbf{V}_i(r_i, d)$ and $\mathbf{V}^i(r_i, d)$ in $\mathbb{E}^d$.

Lemma 176 *We have that*

$$\mathrm{vol}_d\left(\mathbf{V}_i(r_i, d)\right) = \int_0^{r_i} (2\pi s)\mathrm{vol}_{d-2}\left(\mathbf{V}_i(s, d-2)\right) ds,$$

and

$$\mathrm{vol}_d\left(\mathbf{V}^i(r_i, d)\right) = \int_0^{r_i} (2\pi s)\mathrm{vol}_{d-2}\left(\mathbf{V}^i(s, d-2)\right) ds.$$

Proof. It is clear, in both cases, that $\mathbf{V}_i(r_i, d)$ and $\mathbf{V}^i(r_i, d)$ are compact measurable sets of revolution (about $\mathbb{E}^{d-2}$). Note that the orthogonal projection of $\mathbf{B}^d[\mathbf{p}_i, r_i] \cap \left\{\mathbb{E}^{d-2} \times (s\cos\theta, s\sin\theta)\right\}$ onto $\mathbb{E}^{d-2}$ is the $(d-2)$-dimensional ball of radius $\sqrt{r_i^2 - s^2}$ centered at $\mathbf{p}_i$. Thus, by Lemma 175 we have that

$$\mathrm{vol}_d\left(\mathbf{V}_i(r_i, d)\right) = \int_0^{r_i} (2\pi s)\mathrm{vol}_{d-2}\left(\mathbf{V}_i\left(\sqrt{r_i^2 - s^2}, d-2\right)\right) ds\,.$$

But if we make the change of variable $u = \sqrt{r_i^2 - s^2}$, we get the desired integral. A similar calculation works for $\mathrm{vol}_d\left(\mathbf{V}^i(r_i, d)\right)$. □

The following is an immediate corollary of Lemma 176.

Corollary 177 *We have that*

$$\left.\frac{d}{dr}\mathrm{vol}_d\left(\mathbf{V}_i(r, d)\right)\right|_{r=r_i} = 2\pi r_i \mathrm{vol}_{d-2}\left(\mathbf{V}_i(r_i, d-2)\right),$$

and

$$\left.\frac{d}{dr}\mathrm{vol}_d\left(\mathbf{V}^i(r, d)\right)\right|_{r=r_i} = 2\pi r_i \mathrm{vol}_{d-2}\left(\mathbf{V}^i(r_i, d-2)\right).$$

Moreover, it is clear that if F_i stands for the contribution of the ith ball to the boundary of the union $\bigcup_{i=1}^N \mathbf{B}^d[\mathbf{p}_i, r_i]$, then

$$\mathrm{svol}_{d-1}(F_i) = \left.\frac{d}{dr}\mathrm{vol}_d\left(\mathbf{V}_i(r, d)\right)\right|_{r=r_i}. \tag{8.6}$$

Similarly, if F_i denotes the contribution of the ith ball to the boundary of the intersection $\bigcap_{i=1}^N \mathbf{B}^d[\mathbf{p}_i, r_i]$, then

$$\mathrm{svol}_{d-1}(F_i) = \left.\frac{d}{dr}\mathrm{vol}_d\left(\mathbf{V}^i(r, d)\right)\right|_{r=r_i}. \tag{8.7}$$

Finally, it is obvious that

$$\mathrm{vol}_{d-2}\left(\bigcup_{i=1}^N \mathbf{B}^{d-2}[\mathbf{p}_i, r_i]\right) = \sum_{i=1}^N \mathrm{vol}_{d-2}\left(\mathbf{V}_i(r_i, d-2)\right), \tag{8.8}$$

and

$$\mathrm{vol}_{d-2}\left(\bigcap_{i=1}^N \mathbf{B}^{d-2}[\mathbf{p}_i, r_i]\right) = \sum_{i=1}^N \mathrm{vol}_{d-2}\left(\mathbf{V}^i(r_i, d-2)\right). \tag{8.9}$$

Thus, Corollary 177 and (8.6), (8.8) (resp., (8.7), (8.9)) finish the proof of Theorem 68.

8.3 Proof of Theorem 69

Actually, we are going to prove the following even stronger statement. For more information on the background of this theorem, we refer the interested reader to [38].

Theorem 178 *Suppose that* $\mathbf{p}$ *and* $\mathbf{q}$ *are two configurations in* $\mathbb{E}^d, d \geq 1$. *Then the following is a continuous motion* $\mathbf{p}(t) = (\mathbf{p}_1(t), \ldots, \mathbf{p}_N(t))$ *in* $\mathbb{E}^{2d}$, *that is analytic in* t, *such that* $\mathbf{p}(0) = \mathbf{p}$, $\mathbf{p}(1) = \mathbf{q}$ *and for* $0 \leq t \leq 1$, $\|\mathbf{p}_i(t) - \mathbf{p}_j(t)\|$ *is monotone:*

$$\mathbf{p}_i(t) = \left(\frac{\mathbf{p}_i + \mathbf{q}_i}{2} + (\cos \pi t) \frac{\mathbf{p}_i - \mathbf{q}_i}{2}, (\sin \pi t) \frac{\mathbf{p}_i - \mathbf{q}_i}{2} \right), \quad 1 \leq i < j \leq N.$$

Proof. We calculate:

$$4\|\mathbf{p}_i(t) - \mathbf{p}_j(t)\|^2 = \|(\mathbf{p}_i - \mathbf{p}_j) - (\mathbf{q}_i - \mathbf{q}_j)\|^2$$

$$+\|(\mathbf{p}_i - \mathbf{p}_j) + (\mathbf{q}_i - \mathbf{q}_j)\|^2 + 2(\cos \pi t)(\|\mathbf{p}_i - \mathbf{p}_j\|^2 - \|\mathbf{q}_i - \mathbf{q}_j\|^2) \,.$$

This function is monotone, as required. □

8.4 Proof of Theorem 72

The following proof from [36] adapts the two-point symmetrization method of the proof of the Gao-Hug-Schneider theorem from [117]. For this we need to recall the definition of two-point symmetrization, which is also known under the names "two-point rearrangement", "compression", or "polarization". (For more details on two-point symmetrization, we refer the interested reader to the relevant section in [117] and the references mentioned there.)

Definition 32 *Let* H *be a hyperplane in* $\mathbb{M}^d$ *with an orientation, which determines* $\mathbf{H}^+$ *and* $\mathbf{H}^-$ *the two closed half spaces bounded by* H *in* $\mathbb{M}^d$, $d > 1$. *Let* σ_H *denote the reflection about* H *in* $\mathbb{M}^d$. *If* $K \subseteq \mathbb{M}^d$, *then the two-point symmetrization* τ_H *with respect to* H *transforms* K *into the set*

$$\tau_H K := (K \cap \sigma_H K) \cup \left((K \cup \sigma_H K) \cap \mathbf{H}^+ \right).$$

If $K_H := K \cap \sigma_H K$ *stands for the* H*-symmetric core of* K, *then we call*

$$\tau_H K = K_H \cup \left((K \cap H^+) \setminus K_H \right) \cup \sigma_H \left((K \cap \mathbf{H}^-) \setminus K_H \right) \tag{8.10}$$

the canonical decomposition of $\tau_H K$.

Remark 179 *The canonical decomposition of $\tau_H K$ is a disjoint decomposition of $\tau_H K$, which easily implies that two-point symmetrization preserves volume.*

Definition 33 *Let $K \subset \mathbb{M}^d$, $d > 1$ and $r \in \mathbf{R}_+$. Then the r-convex hull $\mathrm{conv}_r K$ of K is defined by*

$$\mathrm{conv}_r K := \bigcap \{\mathbf{B}_{\mathbb{M}^d}[\mathbf{x}, r] \mid K \subseteq \mathbf{B}_{\mathbb{M}^d}[\mathbf{x}, r]\}.$$

Moreover, let the r-convex hull of $\mathbb{M}^d$ be $\mathbb{M}^d$. Furthermore, we say that $K \subseteq \mathbb{M}^d$ is an r-convex set if $K = \mathrm{conv}_r K$.

Remark 180 *We note that clearly, $\mathrm{conv}_r K = \emptyset$ if and only if $K^r = \emptyset$.*

Lemma 181 *If $K \subseteq \mathbb{M}^d$, $d > 1$ and $r \in \mathbf{R}_+$, then*

$$K^r = (\mathrm{conv}_r K)^r. \tag{8.11}$$

Proof. Based on Remark 180, the claim holds for $\mathrm{conv}_r K = \emptyset$. Thus, in what follows, we assume that $\mathrm{conv}_r K \neq \emptyset$, that is, $K^r \neq \emptyset$. Then $K \subseteq \mathrm{conv}_r K$ and therefore $(\mathrm{conv}_r K)^r \subseteq K^r$. On the other hand, we show that $K^r \subseteq (\mathrm{conv}_r K)^r$. So let $\mathbf{y} \in K^r$. Then clearly, $K \subseteq \mathbf{B}_{\mathbb{M}^d}[\mathbf{y}, r]$ and so, $\mathrm{conv}_r K \subseteq \mathbf{B}_{\mathbb{M}^d}[\mathbf{y}, r]$ implying that $\mathbf{y} \in (\mathrm{conv}_r K)^r$. Thus, (8.11) follows. □

The core part of our proof of Theorem 72 is

Lemma 182 *If $K \subseteq \mathbb{M}^d$, $d > 1$ and $r \in \mathbf{R}_+$, then*

$$\tau_H(K^r) \subseteq (\mathrm{conv}_r(\tau_H K))^r.$$

Proof. Lemma 181 implies that $(\mathrm{conv}_r(\tau_H K))^r = (\tau_H K)^r$ and so, it is sufficient to prove that $\tau_H(K^r) \subseteq (\tau_H K)^r$. For this, we need to show that if $\mathbf{x} \in \tau_H(K^r)$, then $\mathbf{x} \in (\tau_H K)^r$, i.e.,

$$\tau_H K \subseteq \mathbf{B}_{\mathbb{M}^d}[\mathbf{x}, r]. \tag{8.12}$$

Remark 179 implies that

$$\tau_H(K^r) = (K^r)_H \cup ((K^r \cap H^+) \setminus (K^r)_H) \cup \sigma_H ((K^r \cap H^-) \setminus (K^r)_H)$$

is a disjoint decomposition of $\tau_H(K^r)$ with $(K^r)_H = K^r \cap \sigma_H(K^r)$. Thus, either $\mathbf{x} \in (K^r)_H$ (Case 1), or $\mathbf{x} \in (K^r \cap H^+) \setminus (K^r)_H$ (Case 2), or $\mathbf{x} \in \sigma_H ((K^r \cap H^-) \setminus (K^r)_H)$ (Case 3). In all three cases, we use (8.10) for the proof of (8.12).

Case 1: As $(K^r)_H = K^r \cap \sigma_H(K^r)$ therefore $\mathbf{x}, \sigma_H\mathbf{x} \in (K^r)_H$. As $\mathbf{x} \in (K^r)_H \subseteq K^r$ therefore $K_H \cup ((K \cap H^+) \setminus K_H) \subseteq K \subseteq \mathbf{B}_{\mathbb{M}^d}[\mathbf{x}, r]$. On the other hand, as $\sigma_H\mathbf{x} \in (K^r)_H \subseteq K^r$ therefore $(K \cap H^-) \setminus K_H \subseteq K \subseteq \mathbf{B}_{\mathbb{M}^d}[\sigma_H\mathbf{x}, r]$ and so, $\sigma_H ((K \cap H^-) \setminus K_H) \subseteq \mathbf{B}_{\mathbb{M}^d}[\mathbf{x}, r]$, finishing the proof of (8.12).

Case 2: As $\mathbf{x} \in (K^r \cap H^+) \setminus (K^r)_H \subseteq K^r$ therefore $K_H \cup ((K \cap H^+) \setminus K_H) \subseteq K \subseteq \mathbf{B}_{\mathbb{M}^d}[\mathbf{x}, r]$. So, we are left to show that

$$\sigma_H \left((K \cap H^-) \setminus K_H\right) \subseteq \mathbf{B}_{\mathbb{M}^d}[\mathbf{x}, r]. \tag{8.13}$$

On the one hand, $\mathbf{x} \in (K^r \cap H^+) \setminus (K^r)_H \subseteq K^r$ implies that $(K \cap H^-) \setminus K_H \subseteq K \subseteq \mathbf{B}_{\mathbb{M}^d}[\mathbf{x}, r]$. On the other hand, for any $\mathbf{y} \in (K \cap H^-) \setminus K_H$ we have $\sigma_H \mathbf{y} \in \sigma_H ((K \cap H^-) \setminus K_H)$. As $\mathbf{x}, \sigma_H \mathbf{y} \in H^+$ and $\mathbf{y} \in H^-$ therefore $\mathrm{dist}_{\mathbb{M}^d}(\sigma_H \mathbf{y}, \mathbf{x}) \leq \mathrm{dist}_{\mathbb{M}^d}(\mathbf{y}, \mathbf{x}) \leq r$. Thus, (8.13) follows.
Case 3: It follows from the assumption that $\sigma_H \mathbf{x} \in (K^r \cap H^-) \setminus (K^r)_H \subseteq K^r$ and therefore $(K \cap H^-) \setminus K_H \subseteq K \subseteq \mathbf{B}_{\mathbb{M}^d}[\sigma_H \mathbf{x}, r]$ implying that $\sigma_H ((K \cap H^-) \setminus K_H) \subseteq \mathbf{B}_{\mathbb{M}^d}[\mathbf{x}, r]$. So, we are left to show that

$$K_H \cup \left((K \cap H^+) \setminus K_H\right) \subseteq \mathbf{B}_{\mathbb{M}^d}[\mathbf{x}, r]. \tag{8.14}$$

As $\sigma_H \mathbf{x} \in (K^r \cap H^-) \setminus (K^r)_H \subseteq K^r$ therefore $K_H \cup ((K \cap H^+) \setminus K_H) \subseteq K \subseteq \mathbf{B}_{\mathbb{M}^d}[\sigma_H \mathbf{x}, r]$. Moreover, as $\sigma_H \mathbf{x} \in H^-$ and $\mathbf{x} \in H^+$ therefore for all $\mathbf{y} \in (K \cap H^+) \setminus K_H \subseteq H^+$ (resp., $\mathbf{y} \in K_H \cap H^+ \subseteq H^+$) we have $\mathrm{dist}_{\mathbb{M}^d}(\mathbf{x}, \mathbf{y}) \leq \mathrm{dist}_{\mathbb{M}^d}(\sigma_H \mathbf{x}, \mathbf{y}) \leq r$ implying that $(K_H \cap H^+) \cup ((K \cap H^+) \setminus K_H) \subseteq \mathbf{B}_{\mathbb{M}^d}[\mathbf{x}, r]$. Finally, for any $\mathbf{y} \in K_H \cap H^-$ we have $\sigma_H \mathbf{y} \in K_H \cap H^+ \subseteq K_H$ with $\mathrm{dist}_{\mathbb{M}^d}(\mathbf{x}, \mathbf{y}) = \mathrm{dist}_{\mathbb{M}^d}(\sigma_H \mathbf{x}, \sigma_H \mathbf{y}) \leq r$ implying that $K_H \cap H^- \subseteq \mathbf{B}_{\mathbb{M}^d}[\mathbf{x}, r]$. This completes the proof of (8.14). □

Now, we are ready to prove Theorem 72. To avoid any trivial case, we may assume that $V_{\mathbb{M}^d}(\mathbf{A}^r) > 0$ for $\mathbf{A} \subseteq \mathbb{M}^d$ with $a := V_{\mathbb{M}^d}(\mathbf{A}) > 0$. In fact, our goal is to maximize the volume $V_{\mathbb{M}^d}(\mathbf{A}^r)$ for compact sets $\mathbf{A} \subseteq \mathbb{M}^d$ of given volume $V_{\mathbb{M}^d}(\mathbf{A}) = a > 0$ and for given $d > 1$ and $r \in \mathbf{R}_+$. As according to Lemma 181 we have $\mathbf{A}^r = (\mathrm{conv}_r \mathbf{A})^r$ with $\mathbf{A} \subseteq \mathrm{conv}_r \mathbf{A}$, it follows from the monotonicity of $V_{\mathbb{M}^d}((\cdot)^r)$ in a straightforward way that for the proof of Theorem 72 it is sufficient to maximize the volume $V_{\mathbb{M}^d}(\mathbf{A}^r)$ for r-convex sets $\mathbf{A} \subseteq \mathbb{M}^d$ of given volume $V_{\mathbb{M}^d}(\mathbf{A}) = a$ with given d and r. Next, consider the extremal family $\mathcal{E}_{a,r,d}$ of r-convex sets $\mathbf{A} \subseteq \mathbb{M}^d$ with $V_{\mathbb{M}^d}(\mathbf{A}) = a$ and maximal $V_{\mathbb{M}^d}(\mathbf{A}^r)$ for given a, d and r. As the Blaschke selection theorem ([217]) for non-empty, compact, convex subsets of $\mathbb{M}^d$ (using the convenient Hausdorff metric) extends to non-empty, r-convex subsets of $\mathbb{M}^d$ in a rather straightforward way; therefore, one obtains by standard arguments that $\mathcal{E}_{a,r,d} \neq \emptyset$.

Lemma 183 *The extremal family $\mathcal{E}_{a,r,d}$ is closed under two-point symmetrization.*

Proof. Let $\mathbf{A} \in \mathcal{E}_{a,r,d}$ be an arbitrary extremal set and consider $\tau_H \mathbf{A}$ for an arbitrary hyperplane H in $\mathbb{M}^d$. Lemmas 181 and 182 imply that $\tau_H(\mathbf{A}^r) \subseteq (\mathrm{conv}_r(\tau_H \mathbf{A}))^r = (\tau_H \mathbf{A})^r$ and therefore

$$V_{\mathbb{M}^d}(\mathbf{A}^r) = V_{\mathbb{M}^d}\left(\tau_H(\mathbf{A}^r)\right) \leq V_{\mathbb{M}^d}\left((\mathrm{conv}_r(\tau_H \mathbf{A}))^r\right) = V_{\mathbb{M}^d}((\tau_H \mathbf{A})^r). \tag{8.15}$$

Here $\tau_H \mathbf{A} \subseteq \mathrm{conv}_r(\tau_H \mathbf{A})$ implying that

$$a = V_{\mathbb{M}^d}(\mathbf{A}) = V_{\mathbb{M}^d}(\tau_H \mathbf{A}) \leq V_{\mathbb{M}^d}\left(\mathrm{conv}_r(\tau_H \mathbf{A})\right). \tag{8.16}$$

We are left to show that $\tau_H\mathbf{A} \in \mathcal{E}_{a,r,d}$. Based on (8.15) and (8.16) we need to prove only that $\tau_H\mathbf{A}$ is r-convex, i.e., $\tau_H\mathbf{A} = \mathrm{conv}_r(\tau_H\mathbf{A})$. We prove this by contradiction, i.e., assume that $\tau_H\mathbf{A} \neq \mathrm{conv}_r(\tau_H\mathbf{A})$. As $\tau_H\mathbf{A} \subseteq \mathrm{conv}_r(\tau_H\mathbf{A})$, this means that $\tau_H\mathbf{A} \subset \mathrm{conv}_r(\tau_H\mathbf{A})$. Then there exists an r-convex set $\mathbf{A}' \subset \mathrm{conv}_r(\tau_H\mathbf{A})$ with $V_{\mathbb{M}^d}(\mathbf{A}') = a$. Thus, $(\mathrm{conv}_r(\tau_H\mathbf{A}))^r \subset (\mathbf{A}')^r$ implying that $V_{\mathbb{M}^d}\left((\mathrm{conv}_r(\tau_H\mathbf{A}))^r\right) < V_{\mathbb{M}^d}\left((\mathbf{A}')^r\right)$, a contradiction via (8.15). $\square$

We finish the proof of Theorem 72 by adapting an argument from [117]. Namely, we are going to show that $\mathbf{B} \in \mathcal{E}_{a,r,d}$, where $\mathbf{B} \subseteq \mathbb{M}^d$ is a ball with $a = V_{\mathbb{M}^d}(\mathbf{A}) = V_{\mathbb{M}^d}(\mathbf{B})$. By a standard argument there exists an r-convex set $\mathbf{C} \in \mathcal{E}_{a,r,d}$ for which $V_{\mathbb{M}^d}(\mathbf{B} \cap \mathbf{C})$ is maximal. Suppose that $\mathbf{B} \neq \mathbf{C}$. As $a = V_{\mathbb{M}^d}(\mathbf{B}) = V_{\mathbb{M}^d}(\mathbf{C})$ therefore there are congruent balls $\mathbf{C}_1 \subseteq \mathbf{C} \setminus \mathbf{B}$ and $\mathbf{C}_2 \subseteq \mathbf{B} \setminus \mathbf{C}$. Let H be the hyperplane in $\mathbb{M}^d$ with an orientation, which determines $\mathbf{H}^+$ and $\mathbf{H}^-$ the two closed half spaces bounded by H in $\mathbb{M}^d$, $d > 1$ such that $\sigma_H\mathbf{C}_1 = \mathbf{C}_2$ with $\mathbf{C}_1 \subset H^-$. Clearly, $V_{\mathbb{M}^d}(\mathbf{B} \cap \tau_H\mathbf{C}) > V_{\mathbb{M}^d}(\mathbf{B} \cap \mathbf{C})$; moreover, Lemma 183 implies that $\tau_H\mathbf{C} \in \mathcal{E}_{a,r,d}$, a contradiction. Thus, $\mathbf{B} = \mathbf{C} \in \mathcal{E}_{a,r,d}$, finishing the proof of Theorem 72.

8.5 Proof of Theorem 73

Following [36], our proof is based on estimates of the following functionals.

Definition 34 *Let*

$$f_{\mathbb{M}^d}(N, \lambda, \delta) := \min\{V_{\mathbb{M}^d}(Q^\delta) \mid Q := \{\mathbf{q}_1, \ldots, \mathbf{q}_N\} \subset \mathbb{M}^d,$$
$$\mathrm{dist}_{\mathbb{M}^d}(\mathbf{q}_i, \mathbf{q}_j) \leq \lambda \text{ for all } 1 \leq i < j \leq N\} \tag{8.17}$$

and

$$g_{\mathbb{M}^d}(N, \lambda, \delta) := \max\{V_{\mathbb{M}^d}(P^\delta) \mid P := \{\mathbf{p}_1, \ldots, \mathbf{p}_N\} \subset \mathbb{M}^d,$$
$$\lambda \leq \mathrm{dist}_{\mathbb{M}^d}(\mathbf{p}_i, \mathbf{p}_j) \text{ for all } 1 \leq i < j \leq N\} \tag{8.18}$$

(We note that in this proof the maximum of the empty set is zero.) We need also

Definition 35 *The circumradius* $\mathrm{cr}X$ *of the set* $X \subseteq \mathbb{M}^d$, $d > 1$ *is defined by*

$$\mathrm{cr}X := \inf\{r \mid X \subseteq \mathbf{B}_{\mathbb{M}^d}[\mathbf{x}, r]\}.$$

For the proof that follows, we need the following straightforward extension of the rather obvious but very useful Euclidean identity (9) of [48]: for any $X = \{\mathbf{x}_1, \ldots, \mathbf{x}_n\} \subset \mathbb{M}^d, n > 1, d > 1, r \in \mathbf{R}_+, r^* \in \mathbf{R}_+$ with $r + r^* \in \mathbf{R}_+$ one has

$$X^r = \left(\bigcup_{i=1}^{n} \mathbf{B}_{\mathbb{M}^d}[\mathbf{x}_i, r^*]\right)^{r+r^*}. \tag{8.19}$$

8.5.1 Proof of (i) in Theorem 73

First, we give a lower bound for (8.17). Jung's theorem ([87]) implies in a straightforward way that $\mathrm{cr}Q \leq \sqrt{\frac{2d}{d+1}}\frac{\lambda}{2} < \frac{1}{\sqrt{2}}\lambda$ and so, $\mathbf{B}_{\mathbb{E}^d}\left[\mathbf{x}, \delta - \frac{1}{\sqrt{2}}\lambda\right] \subset Q^{\delta}$ for some $\mathbf{x} \in \mathbb{E}^d$. (We note that by assumption $\delta - \frac{1}{\sqrt{2}}\lambda \geq 0$.) As a result, we get that

$$f_{\mathbb{E}^d}(N, \lambda, \delta) > V_{\mathbb{E}^d}\left(\mathbf{B}_{\mathbb{E}^d}\left[\mathbf{x}, \delta - \frac{1}{\sqrt{2}}\lambda\right]\right). \tag{8.20}$$

Second, we give an upper bound for (8.18). (8.19) implies that

$$P^{\delta} = \left(\bigcup_{i=1}^{N} \mathbf{B}_{\mathbb{E}^d}\left[\mathbf{p}_i, \frac{\lambda}{2}\right]\right)^{\delta+\frac{\lambda}{2}}, \tag{8.21}$$

where the balls $\mathbf{B}_{\mathbb{E}^d}[\mathbf{p}_1, \frac{\lambda}{2}], \ldots, \mathbf{B}_{\mathbb{E}^d}[\mathbf{p}_N, \frac{\lambda}{2}]$ are pairwise non-overlapping in $\mathbb{E}^d$. Thus,

$$V_{\mathbb{E}^d}\left(\bigcup_{i=1}^{N} \mathbf{B}_{\mathbb{E}^d}\left[\mathbf{p}_i, \frac{\lambda}{2}\right]\right) = N V_{\mathbb{E}^d}\left(\mathbf{B}_{\mathbb{E}^d}\left[\mathbf{p}_1, \frac{\lambda}{2}\right]\right). \tag{8.22}$$

Let $\mu > 0$ be chosen such that $N V_{\mathbb{E}^d}\left(\mathbf{B}_{\mathbb{E}^d}\left[\mathbf{p}_1, \frac{\lambda}{2}\right]\right) = V_{\mathbb{E}^d}\left(\mathbf{B}_{\mathbb{E}^d}\left[\mathbf{p}_1, \mu\right]\right)$. Clearly,

$$\mu = \frac{1}{2} N^{\frac{1}{d}} \lambda \tag{8.23}$$

Now Theorem 72, (8.21), (8.22), and (8.23) imply in a straightforward way that

$$V_{\mathbb{E}^d}\left(P^{\delta}\right) = V_{\mathbb{E}^d}\left(\left(\bigcup_{i=1}^{N} \mathbf{B}_{\mathbb{E}^d}\left[\mathbf{p}_i, \frac{\lambda}{2}\right]\right)^{\delta+\frac{\lambda}{2}}\right) \leq V_{\mathbb{E}^d}\left(\left(\mathbf{B}_{\mathbb{E}^d}\left[\mathbf{p}_1, \frac{1}{2} N^{\frac{1}{d}} \lambda\right]\right)^{\delta+\frac{\lambda}{2}}\right) \tag{8.24}$$

Clearly, $\left(\mathbf{B}_{\mathbb{E}^d}\left[\mathbf{p}_1, \frac{1}{2}N^{\frac{1}{d}}\lambda\right]\right)^{\delta+\frac{\lambda}{2}} = \mathbf{B}_{\mathbb{E}^d}\left[\mathbf{p}_1, \delta - \frac{N^{\frac{1}{d}}-1}{2}\lambda\right]$ with the convention that if $\delta - \frac{N^{\frac{1}{d}}-1}{2}\lambda < 0$, then $\mathbf{B}_{\mathbb{E}^d}\left[\mathbf{p}_1, \delta - \frac{N^{\frac{1}{d}}-1}{2}\lambda\right] = \emptyset$. Hence (8.24) yields

$$g_{\mathbb{E}^d}(N, \lambda, \delta) \leq V_{\mathbb{E}^d}\left(\mathbf{B}_{\mathbb{E}^d}\left[\mathbf{p}_1, \delta - \frac{N^{\frac{1}{d}} - 1}{2}\lambda\right]\right) \tag{8.25}$$

(with $V_{\mathbb{E}^d}(\emptyset) = 0$). Finally, as $N \geq (1+\sqrt{2})^d$ therefore $\frac{N^{\frac{1}{d}}-1}{2}\lambda \geq \frac{1}{\sqrt{2}}\lambda$ and so, (8.20) and (8.25) yield $g_{\mathbb{E}^d}(N, \lambda, \delta) < f_{\mathbb{E}^d}(N, \lambda, \delta)$, finishing the proof of (i) in Theorem 73.

8.5.2 Proof of (ii) in Theorem 73

First, we give a lower bound for (8.17). Let $R := \mathrm{cr}Q$. Then Jung's theorem ([87]) yields $\sin R \leq \sqrt{\frac{2d}{d+1}} \sin\frac{\lambda}{2}$. By assumption $0 < \lambda < \frac{\pi}{2}$ and so,

$$0 \leq \frac{2}{\pi}R < \sin R \leq \sqrt{\frac{2d}{d+1}} \sin\frac{\lambda}{2} < \sqrt{\frac{2d}{d+1}} \frac{\lambda}{2} < \frac{1}{\sqrt{2}}\lambda$$

implying that $0 \leq R < \frac{\pi}{2\sqrt{2}}\lambda$. Thus, $\mathbf{B}_{\mathbb{S}^d}\left[\mathbf{x}, \delta - \frac{\pi}{2\sqrt{2}}\lambda\right] \subset Q^\delta$ for some $\mathbf{x} \in \mathbb{S}^d$. (We note that by assumption $\delta - \frac{\pi}{2\sqrt{2}}\lambda > 0$.) As a result we get that

$$f_{\mathbb{S}^d}(N, \lambda, \delta) > V_{\mathbb{S}^d}\left(\mathbf{B}_{\mathbb{S}^d}\left[\mathbf{x}, \delta - \frac{\pi}{2\sqrt{2}}\lambda\right]\right). \tag{8.26}$$

Second, we give an upper bound for (8.18). By assumption $0 < \delta + \frac{\lambda}{2} < \frac{\pi}{2}$ and therefore (8.19) implies that

$$P^\delta = \left(\bigcup_{i=1}^{N} \mathbf{B}_{\mathbb{S}^d}\left[\mathbf{p}_i, \frac{\lambda}{2}\right]\right)^{\delta + \frac{\lambda}{2}}, \tag{8.27}$$

where the balls $\mathbf{B}_{\mathbb{S}^d}[\mathbf{p}_1, \frac{\lambda}{2}], \ldots, \mathbf{B}_{\mathbb{S}^d}[\mathbf{p}_N, \frac{\lambda}{2}]$ are pairwise non-overlapping in $\mathbb{S}^d$. Thus,

$$V_{\mathbb{S}^d}\left(\bigcup_{i=1}^{N} \mathbf{B}_{\mathbb{S}^d}\left[\mathbf{p}_i, \frac{\lambda}{2}\right]\right) = N V_{\mathbb{S}^d}\left(\mathbf{B}_{\mathbb{S}^d}\left[\mathbf{p}_1, \frac{\lambda}{2}\right]\right). \tag{8.28}$$

Let $\mu > 0$ be chosen such that

$$N V_{\mathbb{S}^d}\left(\mathbf{B}_{\mathbb{S}^d}\left[\mathbf{p}_1, \frac{\lambda}{2}\right]\right) = V_{\mathbb{S}^d}\left(\mathbf{B}_{\mathbb{S}^d}\left[\mathbf{p}_1, \mu\right]\right). \tag{8.29}$$

Proposition 184 *If* $0 < \mu < \frac{\pi}{2}$, *then* $\left(\frac{1}{2ed\pi^{d-1}}\right)^{\frac{1}{d}} N^{\frac{1}{d}}\lambda < \mu$.

Proof. One can rewrite (8.29) using the integral representation of volume of balls in $\mathbb{S}^d$ ([78]) as follows:

$$N d\kappa_d \int_{\frac{\pi}{2}-\frac{\lambda}{2}}^{\frac{\pi}{2}} (\cos t)^{d-1} dt = d\kappa_d \int_{\frac{\pi}{2}-\mu}^{\frac{\pi}{2}} (\cos t)^{d-1} dt,$$

where $\kappa_d := V_{\mathbb{E}^d}(\mathbf{B}_{\mathbb{E}^d}[\mathbf{x}, 1])$, $\mathbf{x} \in \mathbb{E}^d$. Then Lemma 4.7 of [47] yields the following chain of inequalities in a rather straightforward way:

$$\frac{N}{2ed\pi^{d-1}}\lambda^d < \frac{N}{ed}\frac{\lambda}{2}\left(\sin\frac{\lambda}{2}\right)^{d-1} \leq N \int_{\frac{\pi}{2}-\frac{\lambda}{2}}^{\frac{\pi}{2}} (\cos t)^{d-1} dt$$

$$= \int_{\frac{\pi}{2}-\mu}^{\frac{\pi}{2}} (\cos t)^{d-1} dt \leq \mu(\sin\mu)^{d-1} \leq \mu^d.$$

From this, the claim follows. □

Now Theorem 72, (8.27), (8.28), and (8.29) imply in a straightforward way that

$$V_{\mathbb{S}^d}\left(P^\delta\right) = V_{\mathbb{S}^d}\left(\left(\bigcup_{i=1}^{N} \mathbf{B}_{\mathbb{S}^d}\left[\mathbf{p}_i, \frac{\lambda}{2}\right]\right)^{\delta+\frac{\lambda}{2}}\right) \leq V_{\mathbb{S}^d}\left(\left(\mathbf{B}_{\mathbb{S}^d}\left[\mathbf{p}_1, \mu\right]\right)^{\delta+\frac{\lambda}{2}}\right) \quad (8.30)$$

Clearly, $\left(\mathbf{B}_{\mathbb{S}^d}\left[\mathbf{p}_1, \mu\right]\right)^{\delta+\frac{\lambda}{2}} = \mathbf{B}_{\mathbb{S}^d}\left[\mathbf{p}_1, \delta + \frac{\lambda}{2} - \mu\right]$ (with the usual convention that if $\delta + \frac{\lambda}{2} - \mu < 0$, then $\mathbf{B}_{\mathbb{S}^d}\left[\mathbf{p}_1, \delta + \frac{\lambda}{2} - \mu\right] = \emptyset$). By assumption $0 < \delta + \frac{\lambda}{2} < \frac{\pi}{2}$ and so, if $\delta + \frac{\lambda}{2} - \mu \geq 0$, then necessarily $0 < \mu < \frac{\pi}{2}$. Thus, Proposition 184 and (8.30) yield

$$g_{\mathbb{S}^d}(N, \lambda, \delta) \leq V_{\mathbb{S}^d}\left(\mathbf{B}_{\mathbb{S}^d}\left[\mathbf{p}_1, \delta - \left(\left(\frac{1}{2ed\pi^{d-1}}\right)^{\frac{1}{d}} N^{\frac{1}{d}} - \frac{1}{2}\right)\lambda\right]\right) \quad (8.31)$$

(with $V_{\mathbb{S}^d}(\emptyset) = 0$). As $N \geq 2ed\pi^{d-1}\left(\frac{1}{2} + \frac{\pi}{2\sqrt{2}}\right)^d$ therefore

$$\left(\left(\frac{1}{2ed\pi^{d-1}}\right)^{\frac{1}{d}} N^{\frac{1}{d}} - \frac{1}{2}\right)\lambda \geq \frac{\pi}{2\sqrt{2}}\lambda$$

and so, (8.26) and (8.31) yield $g_{\mathbb{S}^d}(N, \lambda, \delta) < f_{\mathbb{S}^d}(N, \lambda, \delta)$, finishing the proof of (ii) in Theorem 73.

8.5.3 Proof of (iii) in Theorem 73

Let us give a lower bound for (8.17) in a way similar to the previous cases. Let $R := \mathrm{cr}Q$. Then Jung's theorem ([87]) yields $\sinh R \leq \sqrt{\frac{2d}{d+1}} \sinh \frac{\lambda}{2}$. By assumption we have $0 < \frac{1}{2}\lambda < \frac{\sinh k}{\sqrt{2}k}\lambda \leq \delta < k$ and so,

$$0 \leq R \leq \sinh R \leq \sqrt{\frac{2d}{d+1}} \sinh \frac{\lambda}{2} < \sqrt{2}\frac{\sinh k}{k}\frac{\lambda}{2}, \quad (8.32)$$

where for the last inequality we have used $0 < x < \sinh x < \frac{\sinh k}{k} x$ that holds for all $0 < x < k$. From (8.32) it follows that $0 \leq R < \frac{\sinh k}{\sqrt{2}k}\lambda$. Thus, $\mathbf{B}_{\mathbb{H}^d}\left[\mathbf{x}, \delta - \frac{\sinh k}{\sqrt{2}k}\lambda\right] \subset Q^\delta$ for some $\mathbf{x} \in \mathbb{H}^d$. (We note that by assumption $\delta - \frac{\sinh k}{\sqrt{2}k}\lambda \geq 0$.) As a result we get that

$$f_{\mathbb{H}^d}(N, \lambda, \delta) > V_{\mathbb{H}^d}\left(\mathbf{B}_{\mathbb{H}^d}\left[\mathbf{x}, \delta - \frac{\sinh k}{\sqrt{2}k}\lambda\right]\right). \quad (8.33)$$

Next, we give an upper bound for (8.18). (8.19) implies that

$$P^\delta = \left(\bigcup_{i=1}^{N} \mathbf{B}_{\mathbb{H}^d}\left[\mathbf{p}_i, \frac{\lambda}{2}\right]\right)^{\delta+\frac{\lambda}{2}}, \quad (8.34)$$

where the balls $\mathbf{B}_{\mathbb{H}^d}[\mathbf{p}_1, \frac{\lambda}{2}], \dots, \mathbf{B}_{\mathbb{H}^d}[\mathbf{p}_N, \frac{\lambda}{2}]$ are pairwise non-overlapping in $\mathbb{H}^d$. Thus,

$$V_{\mathbb{H}^d}\left(\bigcup_{i=1}^{N} \mathbf{B}_{\mathbb{H}^d}\left[\mathbf{p}_i, \frac{\lambda}{2}\right]\right) = N V_{\mathbb{H}^d}\left(\mathbf{B}_{\mathbb{H}^d}\left[\mathbf{p}_1, \frac{\lambda}{2}\right]\right). \tag{8.35}$$

Let $\mu > 0$ be chosen such that

$$N V_{\mathbb{H}^d}\left(\mathbf{B}_{\mathbb{H}^d}\left[\mathbf{p}_1, \frac{\lambda}{2}\right]\right) = V_{\mathbb{H}^d}\left(\mathbf{B}_{\mathbb{H}^d}\left[\mathbf{p}_1, \mu\right]\right). \tag{8.36}$$

Now Theorem 72, (8.34), (8.35), and (8.36) imply in a straightforward way that

$$V_{\mathbb{H}^d}\left(P^{\delta}\right) = V_{\mathbb{H}^d}\left(\left(\bigcup_{i=1}^{N} \mathbf{B}_{\mathbb{H}^d}\left[\mathbf{p}_i, \frac{\lambda}{2}\right]\right)^{\delta+\frac{\lambda}{2}}\right)$$

$$\leq V_{\mathbb{H}^d}\left(\left(\mathbf{B}_{\mathbb{H}^d}\left[\mathbf{p}_1, \mu\right]\right)^{\delta+\frac{\lambda}{2}}\right) = V_{\mathbb{H}^d}\left(\mathbf{B}_{\mathbb{H}^d}\left[\mathbf{p}_1, \delta + \frac{\lambda}{2} - \mu\right]\right) \tag{8.37}$$

with the usual convention that if $\delta + \frac{\lambda}{2} - \mu < 0$, then $\mathbf{B}_{\mathbb{H}^d}\left[\mathbf{p}_1, \delta + \frac{\lambda}{2} - \mu\right] = \emptyset$.

Proposition 185 *If* $0 < \mu \leq \delta + \frac{\lambda}{2}$, *then* $\left(\frac{2k}{\sinh 2k}\right)^{\frac{d-1}{d}} N^{\frac{1}{d}} \frac{\lambda}{2} < \mu$.

Proof. One can rewrite (8.36) using the integral representation of volume of balls in $\mathbb{H}^d$ ([78]) as follows:

$$N d\kappa_d \int_0^{\frac{\lambda}{2}} (\sinh t)^{d-1} dt = d\kappa_d \int_0^{\mu} (\sinh t)^{d-1} dt.$$

As $0 < \mu \leq \delta + \frac{\lambda}{2}$ therefore by assumption also the inequalities $0 < \mu \leq \delta + \frac{\lambda}{2} < 2\delta < 2k$ hold. Hence the following inequalities follow in a rather straightforward way:

$$\frac{N}{d}\left(\frac{\lambda}{2}\right)^d = N \int_0^{\frac{\lambda}{2}} t^{d-1} dt < N \int_0^{\frac{\lambda}{2}} (\sinh t)^{d-1} dt = \int_0^{\mu} (\sinh t)^{d-1} dt$$

$$< \int_0^{\mu} \left(\frac{\sinh 2k}{2k} t\right)^{d-1} dt = \left(\frac{\sinh 2k}{2k}\right)^{d-1} \frac{\mu^d}{d},$$

where for the last inequality we have used $0 < x < \sinh x < \frac{\sinh 2k}{2k} x$ that holds for all $0 < x < 2k$. From this, the claim follows. □

Thus, Proposition 185 and (8.37) yield

$$g_{\mathbb{H}^d}(N, \lambda, \delta) \leq V_{\mathbb{H}^d}\left(\mathbf{B}_{\mathbb{H}^d}\left[\mathbf{p}_1, \delta - \left(\left(\frac{2k}{\sinh 2k}\right)^{\frac{d-1}{d}} N^{\frac{1}{d}} - 1\right)\frac{\lambda}{2}\right]\right) \tag{8.38}$$

(with $V_{\mathbb{H}^d}(\emptyset) = 0$). As $N \geq \left(\frac{\sinh 2k}{2k}\right)^{d-1}\left(\frac{\sqrt{2}\sinh k}{k}+1\right)^d$ therefore

$$\left(\left(\frac{2k}{\sinh 2k}\right)^{\frac{d-1}{d}} N^{\frac{1}{d}} - 1\right)\frac{\lambda}{2} \geq \frac{\sinh k}{\sqrt{2}k}\lambda$$

and so, (8.33) and (8.38) yield $g_{\mathbb{H}^d}(N, \lambda, \delta) < f_{\mathbb{H}^d}(N, \lambda, \delta)$, finishing the proof of (iii) in Theorem 73.

8.6 Proof of Theorem 74

8.6.1 The spherical leapfrog lemma

As usual, let $\mathbb{S}^d, d \geq 2$ denote the unit sphere centered at the origin $\mathbf{o}$ in $\mathbb{E}^{d+1}$, and let $\mathbf{X}(\mathbf{p})$ be a finite intersection of closed balls of radius $\frac{\pi}{2}$ (i.e., of closed hemispheres) in $\mathbb{S}^d$ whose configuration of centers is $\mathbf{p} = (\mathbf{p}_1, \ldots, \mathbf{p}_N)$. We say that another configuration $\mathbf{q} = (\mathbf{q}_1, \ldots, \mathbf{q}_N)$ is a contraction of $\mathbf{p}$ if, for all $1 \leq i < j \leq N$, the spherical distance between $\mathbf{p}_i$ and $\mathbf{p}_j$ is not less than the spherical distance between $\mathbf{q}_i$ and $\mathbf{q}_j$. We denote the d-dimensional spherical volume measure by $\mathrm{Svol}_d(\cdot)$. Thus, Theorem 74, which we need to prove, can be phrased as follows: if $\mathbf{q}$ is a configuration in $\mathbb{S}^d$ that is a contraction of the configuration $\mathbf{p}$, then

$$\mathrm{Svol}_d\left(\mathbf{X}(\mathbf{p})\right) \leq \mathrm{Svol}_d\left(\mathbf{X}(\mathbf{q})\right). \tag{8.39}$$

We note that the part of Theorem 74 on the union of closed hemispheres is a simple set-theoretic consequence of (8.39).

Next, we recall Theorem 178, which is called the Euclidean Leapfrog Lemma ([38]).

Theorem 186 *Suppose that* $\mathbf{p}$ *and* $\mathbf{q}$ *are two configurations in* $\mathbb{E}^d, d \geq 1$. *Then the following is a continuous motion* $\mathbf{p}(t) = (\mathbf{p}_1(t), \ldots, \mathbf{p}_N(t))$ *in* $\mathbb{E}^{2d}$, *that is analytic in* t, *such that* $\mathbf{p}(0) = \mathbf{p}$, $\mathbf{p}(1) = \mathbf{q}$ *and for* $0 \leq t \leq 1$, $\|\mathbf{p}_i(t) - \mathbf{p}_j(t)\|$ *is monotone:*

$$\mathbf{p}_i(t) = \left(\frac{\mathbf{p}_i + \mathbf{q}_i}{2} + (\cos \pi t)\frac{\mathbf{p}_i - \mathbf{q}_i}{2}, (\sin \pi t)\frac{\mathbf{p}_i - \mathbf{q}_i}{2}\right), \quad 1 \leq i < j \leq N.$$

We need to apply this to a sphere. Here we consider the unit spheres $\mathbb{S}^d \subset \mathbb{S}^{d+1} \subset \mathbb{S}^{d+2} \ldots$ in such a way that each $\mathbb{S}^d$ is the set of points that are a unit distance from the origin $\mathbf{o}$ in $\mathbb{E}^{d+1}$. The following statement is called the Spherical Leapfrog Lemma.

Corollary 187 *Suppose that* $\mathbf{p}$ *and* $\mathbf{q}$ *are two configurations in* $\mathbb{S}^d$. *Then there is a monotone analytic motion from* $\mathbf{p}$ *to* $\mathbf{q}$ *in* $\mathbb{S}^{2d+1}$.

Proof. Apply Theorem 186 to each configuration $\mathbf{p}$ and $\mathbf{q}$ with $\mathbf{o}$ as an additional configuration point for each. So for each t, the configuration $\mathbf{p}(t) = (\mathbf{p}_1(t), \ldots, \mathbf{p}_N(t))$ lies at a unit distance from $\mathbf{o}$ in $\mathbb{E}^{2d+2}$, which is just $\mathbb{S}^{2d+1}$. □

8.6.2 Smooth contractions via Schläfli's differential formula

We look at the case when there is a smooth motion $\mathbf{p}(t)$ of the configuration $\mathbf{p}$ in $\mathbb{S}^d$. More precisely, we consider the family $\mathbf{X}(t) = \mathbf{X}(\mathbf{p}(t))$ of convex spherical d-polytopes in $\mathbb{S}^d$ having the same combinatorial face structure with facet hyperplanes being differentiable in the parameter t. The following classical theorem of Schläfli describes how the volume of $\mathbf{X}(t)$ changes as a function of its dihedral angles and the volume of its $(d-2)$-dimensional faces.

Lemma 188 *For each $(d-2)$-face $F_{ij}(t)$ of the convex spherical d-polytope $\mathbf{X}(t)$ in $\mathbb{S}^d$ let $\alpha_{ij}(t)$ represent the (inner) dihedral angle between the two facets $F_i(t)$ and $F_j(t)$ meeting at $F_{ij}(t)$. Then the following holds.*

$$\frac{d}{dt}\mathrm{Svol}_d\left(\mathbf{X}(t)\right) = \frac{1}{d-1}\sum_{F_{ij}} \mathrm{Svol}_{d-2}\left(F_{ij}(t)\right) \cdot \frac{d}{dt}\alpha_{ij}(t),$$

to be summed over all $(d-2)$-faces.

Corollary 189 *Let $\mathbf{q}$ be a configuration in $\mathbb{S}^d$ with a differentiable contraction $\mathbf{p}(t)$ in t of the configuration $\mathbf{p}$ in $\mathbb{S}^d$ and assume that the convex spherical d-polytopes $\mathbf{X}(t) = \mathbf{X}(\mathbf{p}(t))$ of $\mathbb{S}^d$ have the same combinatorial face structure. Then*

$$\frac{d}{dt}\mathrm{Svol}_d\left(\mathbf{X}(t)\right) \geq 0 \ .$$

Proof. As the spherical distance between $\mathbf{p}_i(t)$ and $\mathbf{p}_j(t)$ is decreasing, the derivative of the dihedral angle $\frac{d}{dt}\alpha_{ij}(t) \geq 0$. The result then follows from Lemma 188. □

8.6.3 From higher- to lower-dimensional spherical volume

The last piece of information that we need before we get to the proof of Theorem 74 is a way of relating higher-dimensional spherical volumes to lower-dimensional ones. Let X be any integrable set in $\mathbb{S}^n$. Recall that we regard

$$X \subset \mathbb{S}^n = \mathbb{S}^n \times \{\mathbf{o}\} \subset \mathbb{E}^{n+1} \times \mathbb{E}^{k+1} \ .$$

Regard

$$\{\mathbf{o}\} \times \mathbb{S}^k \subset \mathbb{E}^{n+1} \times \mathbb{E}^{k+1} \ .$$

Let $X * \mathbb{S}^k$ be the subset of $\mathbb{S}^{n+k+1}$ consisting of the union of the geodesic arcs from each point of X to each point of $\{\mathbf{o}\} \times \mathbb{S}^k$. (So, in particular, $\mathbb{S}^n * \mathbb{S}^k = \mathbb{S}^{n+k+1}$).

Lemma 190 *For any integrable subset X of $\mathbb{S}^n$,*

$$\mathrm{Svol}_{n+k+1}\left(X * \mathbb{S}^k\right) = \frac{\omega_{n+k+1}}{\omega_n}\mathrm{Svol}_n(X) ,$$

where $\omega_n = \mathrm{Svol}_n\left(\mathbb{S}^n\right)$, $\omega_{n+k+1} = \mathrm{Svol}_{n+k+1}\left(\mathbb{S}^n * \mathbb{S}^k\right) = \mathrm{Svol}_{n+k+1}\left(\mathbb{S}^{n+k+1}\right)$.

Proof. Since the $*$ operation (a kind of spherical join) is associative, we only need to consider the case when $k = 0$. Regard $\{\mathbf{o}\} \times \mathbb{S}^0 = \mathbb{S}^0 = \{\mathbf{n}, \mathbf{s}\}$, the north pole and the south pole of $\mathbb{S}^{n+1}$. We use polar coordinates centered at $\mathbf{n}$ to calculate the $(n+1)$-dimensional volume of $X * \mathbb{S}^0$. Let $X(z) = (X * \mathbb{S}^0) \cap \left(\mathbb{E}^{n+1} \times \{z\}\right)$, and let θ be the angle that a point in $\mathbb{S}^{n+1}$ makes with $\mathbf{n}$, the north pole in $\mathbb{S}^{n+1}$. So $z = z(\theta) = \cos\theta$. Then the spherical volume element for $\mathbb{S}^n(z) = \mathbb{S}^{n+1} \cap \left(\mathbb{E}^{n+1} \times \{z\}\right)$ is $dV_n(z) = (\sin^n\theta)dV_n(0)$ because $\mathbb{S}^n(z)$ is obtained from $\mathbb{S}^n(0)$ by a dilation by $\sin\theta$. Then

$$\mathrm{Svol}_{n+1}\left(X * \mathbb{S}^0\right) = \int_{X*\mathbb{S}^0} dV_n(z)d\theta \tag{8.40}$$

$$= \int_0^{\pi}\int_{X(z(\theta))} dV_n(z)d\theta = \int_0^{\pi}(\sin^n\theta)V_n(X)d\theta \tag{8.41}$$

$$= \mathrm{Svol}_n(X)\int_0^{\pi}(\sin^n\theta)d\theta = \mathrm{Svol}_n(X)\frac{\omega_{n+1}}{\omega_n} , \tag{8.42}$$

where (8.42) can be seen by taking $X = \mathbb{S}^n$, or by performing the integral explicitly. □

8.6.4 Putting pieces together

Now, we are ready for the proof of Theorem 74.

Let the configuration $\mathbf{q} = (\mathbf{q}_1, \ldots, \mathbf{q}_N)$ be a contraction of the configuration $\mathbf{p} = (\mathbf{p}_1, \ldots, \mathbf{p}_N)$ in $\mathbb{S}^d$. By Corollary 187, there is an analytic motion $\mathbf{p}(t)$, in $\mathbb{S}^{2d+1}$ for $0 \leq t \leq 1$, where $\mathbf{p}(0) = \mathbf{p}$, and $\mathbf{p}(1) = \mathbf{q}$, and all the pairwise distances between the points of $\mathbf{p}(t)$ decrease in t.

Without loss of generality we may assume that $\mathbf{X}^d(\mathbf{p}(0)) := \mathbf{X}(\mathbf{p}(0))$ is a convex spherical d-polytope in $\mathbb{S}^d$. Since $\mathbf{p}(t)$ is analytic in t, the intersection $\mathbf{X}^{2d+1}(\mathbf{p}(t))$ of the (closed) hemispheres centered at the points of the configuration $\mathbf{p}(t)$ in $\mathbb{S}^{2d+1}$ is a convex spherical $(2d+1)$-polytope with a constant combinatorial structure, except for a finite number of points in the interval $[0, 1]$. By Corollary 189, $\mathrm{Svol}_{2d+1}\left(\mathbf{X}^{2d+1}(\mathbf{p}(t))\right)$ is monotone increasing in t.

Recall that $\mathbf{X}^d(\mathbf{p})$ and $\mathbf{X}^d(\mathbf{q})$ are the intersections of the (closed) hemispheres centered at the points of $\mathbf{p}$ and $\mathbf{q}$ in $\mathbb{S}^d$. From the definition of the spherical join $*$,

$$\mathbf{X}^d(\mathbf{p}) * \mathbb{S}^d = \mathbf{X}^{2d+1}(\mathbf{p}) = \mathbf{X}^{2d+1}(\mathbf{p}(0))$$

$$\mathbf{X}^d(\mathbf{q}) * \mathbb{S}^d = \mathbf{X}^{2d+1}(\mathbf{q}) = \mathbf{X}^{2d+1}(\mathbf{p}(1)).$$

Hence, by Lemma 190,

$$\mathrm{Svol}_d\left(\mathbf{X}^d(\mathbf{p})\right) = \frac{\kappa_d}{\kappa_{2d+1}}\mathrm{Svol}_{2d+1}\left(\mathbf{X}^{2d+1}(\mathbf{p}(0))\right)$$

$$\leq \frac{\kappa_d}{\kappa_{2d+1}}\mathrm{Svol}_{2d+1}\left(\mathbf{X}^{2d+1}(\mathbf{p}(1))\right) = \mathrm{Svol}_d\left(\mathbf{X}^d(\mathbf{q})\right) .$$

This finishes the proof of Theorem 74.

8.7 Proof of Theorem 75

Let $\mathbf{A} \subset \mathbb{E}^d$, $d > 1$ be a compact set of volume $V_d(\mathbf{A}) > 0$ and $\mathbf{B} \subset \mathbb{E}^d$ is a ball with $V_d(\mathbf{A}) = V_d(\mathbf{B})$ and $r > 0$. Clearly, if $\mathbf{B}^r = \emptyset$, then $\mathbf{A}^r = \emptyset$ and (3.1) follows. Similarly, it is easy to see that if $\mathbf{B}^r$ is a point in $\mathbb{E}^d$, then (3.1) follows. Hence, we may assume that $\mathbf{B}^r = \mathbf{B}^d[\mathbf{o}, R]$ and $\mathbf{B} = \mathbf{B}^d[\mathbf{o}, r-R]$ with $0 < R < r$. Next recall that a special case of the Alexandrov-Fenchel inequality yields the following statement ([217]): if $\mathbf{K}$ is a convex body in $\mathbb{E}^d$ satisfying $V_i(\mathbf{K}) \leq V_i(\mathbf{B}^d[\mathbf{o}, R])$ for given $1 \leq i < d$ and $R > 0$, then

$$V_j(\mathbf{K}) \leq V_j(\mathbf{B}^d[\mathbf{o}, R]) \tag{8.43}$$

holds for all j with $i < j \leq d$. Thus, it is sufficient to prove (3.1) for $k = 1$ and $\mathbf{B}^r = \mathbf{B}^d[\mathbf{o}, R]$ with $0 < R < r$. We need the following special cases of the corresponding definition, remark, and lemma from Section 8.4.

Definition 36 *Let $\emptyset \neq K \subset \mathbb{E}^d$, $d > 1$ and $r > 0$. Then the* r-ball convex hull $\mathrm{conv}_r K$ *of K is defined by*

$$\mathrm{conv}_r K := \bigcap\{\mathbf{B}^d[\mathbf{x}, r] \mid K \subseteq \mathbf{B}^d[\mathbf{x}, r]\}.$$

Moreover, let the r-ball convex hull of $\mathbb{E}^d$ be $\mathbb{E}^d$. Furthermore, we say that $K \subseteq \mathbb{E}^d$ is r-ball convex *if $K = \mathrm{conv}_r K$.*

Remark 191 *We note that clearly, $\mathrm{conv}_r K = \emptyset$ if and only if $K^r = \emptyset$. Moreover, $\emptyset \neq K \subset \mathbb{E}^d$ is r-ball convex if and only if K is an r-ball body.*

We need the Euclidean version of Lemma 181 (cf. also [36]).

Lemma 192 *If $K \subseteq \mathbb{E}^d$, $d > 1$ and $r > 0$, then $K^r = (\mathrm{conv}_r K)^r$.*

Hence, using Lemma 192 we may assume that $\mathbf{A} \subset \mathbb{E}^d$ is an r-ball body of volume $V_d(\mathbf{A}) > 0$ and $\mathbf{B} = \mathbf{B}^d[\mathbf{o}, r-R]$ with $0 < R < r$ such that $V_d(\mathbf{A}) = V_d(\mathbf{B})$. Our goal is to prove that

$$V_1(\mathbf{A}^r) \leq V_1(\mathbf{B}^r) = V_1(\mathbf{B}^d[\mathbf{o}, R]). \tag{8.44}$$

Next recall Theorem 1 of [73], which we state as follows.

Lemma 193 *If $\mathbf{A} \subset \mathbb{E}^d$ is an r-ball body (for $r > 0$), then $\mathbf{A}+\mathbf{A}^r$ is a convex body of constant width $2r$, where $+$ denotes the Minkowski sum.*

Thus, we have:

Corollary 194 *If $\mathbf{A} \subset \mathbb{E}^d$ is an r-ball body (for $r > 0$), then*

$$V_1(\mathbf{A}) + V_1(\mathbf{A}^r) = \frac{d\kappa_d}{\kappa_{d-1}} r = V_1(\mathbf{B}) + V_1(\mathbf{B}^r),$$

where $\mathbf{B} = \mathbf{B}^d[\mathbf{o}, r - R]$ and $\mathbf{B}^r = \mathbf{B}^d[\mathbf{o}, R]$ with $0 < R < r$ and $V_d(\mathbf{A}) = V_d(\mathbf{B})$.

Finally, recall ([217]) that (8.43) for $j = d$ can be restated as follows: if $\mathbf{A}$ is a convex body in $\mathbb{E}^d$ satisfying $V_d(\mathbf{A}) = V_d(\mathbf{B}^d[\mathbf{o}, r - R])$ for given $d > 1$ and $0 < R < r$, then

$$V_i(\mathbf{A}) \geq V_i(\mathbf{B}^d[\mathbf{o}, r - R]) \tag{8.45}$$

holds for all i with $1 \leq i < d$.

Hence, Corollary 194 and (8.45) for $i = 1$ imply (8.44) in a straightforward way. This completes the proof of Theorem 75.

8.8 Proof of Theorem 76

8.8.1 Proof of Part (i) of Theorem 76

Let $d > 1$, $\lambda > 0$, $r > 0$, and $k \in [d] := \{1, \ldots, d\}$ be given. If $\lambda > 2r$, then $V_k(P^r) = V_k(\emptyset) = 0$ and (3.2) follows. Thus, we may assume that $0 < \lambda \leq 2r$, and as in [48], we proceed by proving the following theorem, which implies part (i) of Theorem 76 in a straightforward way.

Theorem 195 *Let $d > 1$, $\lambda > 0$, $r > 0$, and $k \in [d]$ be given such that $0 < \lambda \leq 2r$. Let $Q := \{\mathbf{q}_1, \ldots, \mathbf{q}_N\} \subset \mathbb{E}^d$ be a uniform contraction of $P := \{\mathbf{p}_1, \ldots, \mathbf{p}_N\} \subset \mathbb{E}^d$ with separating value λ in $\mathbb{E}^d$. If*

(i) $N \geq \left(1 + \frac{2r}{\lambda}\right)^d$, or

(ii) $0 < \lambda \leq \sqrt{2} r$ and $N \geq \left(1 + \sqrt{\frac{2d}{d+1}}\right)^d$,

then (3.2) holds.

Proof. Following [48], our proof is based on proper estimates of the following functionals.

Definition 37 *Let*

$$f_{k,d}(N,\lambda,r) := \min\{V_k(Q^r) \mid \|\mathbf{q}_i - \mathbf{q}_j\| \leq \lambda \text{ for all } 1 \leq i < j \leq N\} \quad (8.46)$$

and

$$g_{k,d}(N,\lambda,r) := \max\{V_k(P^r) \mid \lambda \leq \|\mathbf{p}_i - \mathbf{p}_j\| \text{ for all } 1 \leq i < j \leq N\} \quad (8.47)$$

We note that in this paper the maximum of the empty set is zero. We need also

Definition 38 *The circumradius* cr$X$ *of the set* $X \subseteq \mathbb{E}^d$, $d > 1$ *is defined by*

$$\mathrm{cr}X := \inf\{r \mid X \subseteq \mathbf{B}^d[\mathbf{x}, r]\}.$$

Part (i): By assumption $N \geq \left(1 + \frac{2r}{\lambda}\right)^d$ and so

$$N\left(\frac{\lambda}{2}\right)^d \kappa_d \geq \left(\frac{\lambda}{2} + r\right)^d \kappa_d, \quad (8.48)$$

where κ_d denotes the volume of a d-dimensional unit ball in $\mathbb{E}^d$. As the balls $\mathbf{p}_1 + \mathbf{B}^d[\mathbf{o}, \frac{\lambda}{2}], \dots, \mathbf{p}_N + \mathbf{B}^d[\mathbf{o}, \frac{\lambda}{2}]$ are pairwise non-overlapping in $\mathbb{E}^d$ therefore (8.48) yields in a straightforward way that cr$P > r$. Thus, $P^r = \emptyset$ and therefore clearly, $g_{k,d}(N,\lambda,r) = 0 \leq f_{k,d}(N,\lambda,r)$ holds, finishing the proof of Theorem 195, part (i).

Part (ii): For the proof that follows, we need the following straightforward extension of the rather obvious but very useful Euclidean identity (9) of [48]: for any $X = \{\mathbf{x}_1, \dots, \mathbf{x}_n\} \subset \mathbb{E}^d, n > 1, d > 1, r > 0$, and $r^* > 0$ one has

$$X^r = \left(\bigcup_{i=1}^{n} \mathbf{B}^d[\mathbf{x}_i, r^*]\right)^{r+r^*}. \quad (8.49)$$

First, we give a lower bound for (8.46). Jung's theorem ([87]) implies in a straightforward way that cr$Q \leq \sqrt{\frac{2d}{d+1}}\frac{\lambda}{2}$ and so, $\mathbf{B}^d\left[\mathbf{x}, r - \sqrt{\frac{2d}{d+1}}\frac{\lambda}{2}\right] \subset Q^r$ for some $\mathbf{x} \in \mathbb{E}^d$. (We note that by assumption $N \geq \left(1 + \sqrt{\frac{2d}{d+1}}\right)^d > 1$ and $r - \sqrt{\frac{2d}{d+1}}\frac{\lambda}{2} > r - \frac{1}{\sqrt{2}}\lambda \geq 0$.) As a result, we get that

$$f_{k,d}(N,\lambda,r) > V_k\left(\mathbf{B}^d\left[\mathbf{x}, r - \sqrt{\frac{2d}{d+1}}\frac{\lambda}{2}\right]\right). \quad (8.50)$$

Second, we give an upper bound for (8.47). (8.49) implies that

$$P^r = \left(\bigcup_{i=1}^{N} \mathbf{B}^d\left[\mathbf{p}_i, \frac{\lambda}{2}\right]\right)^{r+\frac{\lambda}{2}}, \quad (8.51)$$

where the balls $\mathbf{B}^d[\mathbf{p}_1, \frac{\lambda}{2}], \ldots, \mathbf{B}^d[\mathbf{p}_N, \frac{\lambda}{2}]$ are pairwise non-overlapping in $\mathbb{E}^d$. Thus,

$$V_d\left(\bigcup_{i=1}^{N} \mathbf{B}^d\left[\mathbf{p}_i, \frac{\lambda}{2}\right]\right) = NV_d\left(\mathbf{B}^d\left[\mathbf{p}_1, \frac{\lambda}{2}\right]\right). \tag{8.52}$$

Let $\mu > 0$ be chosen such that $NV_d\left(\mathbf{B}^d\left[\mathbf{p}_1, \frac{\lambda}{2}\right]\right) = V_d\left(\mathbf{B}^d\left[\mathbf{p}_1, \mu\right]\right)$. Clearly,

$$\mu = \frac{1}{2}N^{\frac{1}{d}}\lambda \tag{8.53}$$

Now Theorem 75, (8.51), (8.52), and (8.53) imply in a straightforward way that

$$V_k\left(P^r\right) = V_k\left(\left(\bigcup_{i=1}^{N} \mathbf{B}^d\left[\mathbf{p}_i, \frac{\lambda}{2}\right]\right)^{r+\frac{\lambda}{2}}\right) \leq V_k\left(\left(\mathbf{B}^d\left[\mathbf{p}_1, \frac{1}{2}N^{\frac{1}{d}}\lambda\right]\right)^{r+\frac{\lambda}{2}}\right). \tag{8.54}$$

Clearly, $\left(\mathbf{B}^d\left[\mathbf{p}_1, \frac{1}{2}N^{\frac{1}{d}}\lambda\right]\right)^{r+\frac{\lambda}{2}} = \mathbf{B}^d\left[\mathbf{p}_1, r - \frac{N^{\frac{1}{d}}-1}{2}\lambda\right]$ with the convention that if $r - \frac{N^{\frac{1}{d}}-1}{2}\lambda < 0$, then $\mathbf{B}^d\left[\mathbf{p}_1, r - \frac{N^{\frac{1}{d}}-1}{2}\lambda\right] = \emptyset$. Hence (8.54) yields

$$g_{k,d}(N, \lambda, r) \leq V_k\left(\mathbf{B}^d\left[\mathbf{p}_1, r - \frac{N^{\frac{1}{d}}-1}{2}\lambda\right]\right) \tag{8.55}$$

(with $V_k(\emptyset) = 0$). Finally, as $N \geq \left(1 + \sqrt{\frac{2d}{d+1}}\right)^d$ therefore $\frac{N^{\frac{1}{d}}-1}{2}\lambda \geq \sqrt{\frac{2d}{d+1}}\frac{\lambda}{2}$ and so, (8.50) and (8.55) yield $g_{k,d}(N, \lambda, r) < f_{k,d}(N, \lambda, r)$, finishing the proof of Theorem 195, part (ii). □

8.8.2 Proof of Part (ii) of Theorem 76

The following strengthening of Theorem 195 implies Part (ii) of Theorem 76 in a straightforward way. Thus, we are left to prove:

Theorem 196 *Let $d \geq 42$, $\lambda > 0$, $r > 0$, and $k \in \{1, 2, \ldots, d\}$ be given such that $0 < \lambda \leq 2r$. Let $Q := \{\mathbf{q}_1, \ldots, \mathbf{q}_N\} \subset \mathbb{E}^d$ be a uniform contraction of $P := \{\mathbf{p}_1, \ldots, \mathbf{p}_N\} \subset \mathbb{E}^d$ with separating value λ in $\mathbb{E}^d$. If*

(i) $N \geq \sqrt{\frac{\pi}{2d}}\left(1 + \frac{2r}{\lambda}\right)^d + 1$, or

(ii) $0 < \lambda \leq \sqrt{2}r$ and $N \geq \sqrt{\frac{\pi}{2d}}\left(1 + \sqrt{\frac{2d}{d+1}}\right)^d + 1$,

then (3.2) holds.

Proof. We use the notations and methods of the proof of Theorem 195. Furthermore, we need the following well-known result of U. Betke and M. Henk

[24], which proves the sausage conjecture of L. Fejes Tóth in $\mathbb{E}^d$ for $d \geq 42$: whenever the balls $\mathbf{B}^d[\mathbf{p}_1, \frac{\lambda}{2}], \ldots, \mathbf{B}^d[\mathbf{p}_N, \frac{\lambda}{2}]$ are pairwise non-overlapping in $\mathbb{E}^d$ then

$$V_d\left(\operatorname{conv}\left(\bigcup_{i=1}^{N} \mathbf{B}^d\left[\mathbf{p}_i, \frac{\lambda}{2}\right]\right)\right) \geq (N-1)\lambda\left(\frac{\lambda}{2}\right)^{d-1}\kappa_{d-1} + \left(\frac{\lambda}{2}\right)^d \kappa_d, \quad (8.56)$$

where conv$(\cdot)$ denotes the convex hull of the given set. Using the inequality $\frac{\kappa_{d-1}}{\kappa_d} > \sqrt{\frac{d}{2\pi}}$ for $d \geq 1$ (see Lemma 1 in [23]) we get in a straightforward way from (8.56) that

$$V_d\left(\operatorname{conv}\left(\bigcup_{i=1}^{N} \mathbf{B}^d\left[\mathbf{p}_i, \frac{\lambda}{2}\right]\right)\right) > \left((N-1)\sqrt{\frac{2d}{\pi}} + 1\right)\left(\frac{\lambda}{2}\right)^d \kappa_d. \quad (8.57)$$

Part (i): By assumption $N \geq \sqrt{\frac{\pi}{2d}}\left(1 + \frac{2r}{\lambda}\right)^d + 1 > \sqrt{\frac{\pi}{2d}}\left[\left(1 + \frac{2r}{\lambda}\right)^d - 1\right] + 1$

and so

$$\left((N-1)\sqrt{\frac{2d}{\pi}} + 1\right)\left(\frac{\lambda}{2}\right)^d \kappa_d > \left(\frac{\lambda}{2} + r\right)^d \kappa_d, \quad (8.58)$$

As the balls $\mathbf{p}_1 + \mathbf{B}^d[\mathbf{o}, \frac{\lambda}{2}], \ldots, \mathbf{p}_N + \mathbf{B}^d[\mathbf{o}, \frac{\lambda}{2}]$ are pairwise non-overlapping in $\mathbb{E}^d$; therefore, (8.57) and (8.58) yield in a straightforward way that cr$P > r$. Thus, $P^r = \emptyset$ and therefore clearly, $g_{k,d}(N, \lambda, r) = 0 \leq f_{k,d}(N, \lambda, r)$ holds, finishing the proof of Theorem 196, part (i).

Part (ii): In the same way as in the proof of Part (ii) of Theorem 195, one can derive that

$$f_{k,d}(N, \lambda, r) > V_k\left(\mathbf{B}^d\left[\mathbf{x}, r - \sqrt{\frac{2d}{d+1}}\frac{\lambda}{2}\right]\right). \quad (8.59)$$

Next, recall that

$$P^r = \left(\bigcup_{i=1}^{N} \mathbf{B}^d\left[\mathbf{p}_i, \frac{\lambda}{2}\right]\right)^{r+\frac{\lambda}{2}}, \quad (8.60)$$

where the balls $\mathbf{B}^d[\mathbf{p}_1, \frac{\lambda}{2}], \ldots, \mathbf{B}^d[\mathbf{p}_N, \frac{\lambda}{2}]$ are pairwise non-overlapping in $\mathbb{E}^d$. Lemma 192 applied to (8.60) yields

$$P^r = \left(\bigcup_{i=1}^{N} \mathbf{B}^d\left[\mathbf{p}_i, \frac{\lambda}{2}\right]\right)^{r+\frac{\lambda}{2}} = \left(\operatorname{conv}_{r+\frac{\lambda}{2}}\left(\bigcup_{i=1}^{N} \mathbf{B}^d\left[\mathbf{p}_i, \frac{\lambda}{2}\right]\right)\right)^{r+\frac{\lambda}{2}}$$

$$\subset \left(\operatorname{conv}\left(\bigcup_{i=1}^{N} \mathbf{B}^d\left[\mathbf{p}_i, \frac{\lambda}{2}\right]\right)\right)^{r+\frac{\lambda}{2}}. \tag{8.61}$$

Hence, Theorem 75, (8.57), and (8.61) imply in a straightforward way that

$$V_k(P^r) < V_k\left(\left(\mathbf{B}^d[\mathbf{o}, \mu]\right)^{r+\frac{\lambda}{2}}\right), \tag{8.62}$$

where $V_d\left(\mathbf{B}^d[\mathbf{o}, \mu]\right) = \left((N-1)\sqrt{\frac{2d}{\pi}} + 1\right)\left(\frac{\lambda}{2}\right)^d \kappa_d$. Thus, (8.62) yields

$$g_{k,d}(N, \lambda, r) < V_k\left(\mathbf{B}^d\left[\mathbf{o}, r - \left(\left((N-1)\sqrt{\frac{2d}{\pi}} + 1\right)^{\frac{1}{d}} - 1\right)\frac{\lambda}{2}\right]\right) \tag{8.63}$$

(with $V_k(\emptyset) = 0$). Finally, as $N \geq \sqrt{\frac{\pi}{2d}}\left(1 + \sqrt{\frac{2d}{d+1}}\right)^d + 1 > \sqrt{\frac{\pi}{2d}}\left(1 + \sqrt{\frac{2d}{d+1}}\right)^d + (1 - \sqrt{\frac{\pi}{2d}})$ therefore $\left(\left((N-1)\sqrt{\frac{2d}{\pi}} + 1\right)^{\frac{1}{d}} - 1\right)\frac{\lambda}{2} > \sqrt{\frac{2d}{d+1}}\frac{\lambda}{2}$ and so, (8.59) and (8.63) yield $g_{k,d}(N, \lambda, r) < f_{k,d}(N, \lambda, r)$, finishing the proof of Theorem 196, part (ii). □

8.9 Proof of Theorem 78

As the claim holds trivially for $0 < r \leq \frac{\lambda}{2}$, therefore we may assume that $0 < \frac{\lambda}{2} < r$. The diameter of $\bigcup_{i=1}^{N} \mathbf{B}^d[\mathbf{q}_i, r]$ is at most $2r + \lambda$. Thus, the isodiametric inequality ([217]) implies that

$$\mathrm{V}_d\left(\bigcup_{i=1}^{N} \mathbf{B}^d[\mathbf{q}_i, r]\right) \leq \left(r + \frac{\lambda}{2}\right)^d \kappa_d. \tag{8.64}$$

On the other hand, $\left\{\mathbf{B}^d[\mathbf{p}_i, \lambda/2] \,:\, i = 1, \ldots, N\right\}$ is a packing of balls. Next, we are going to use the following form of the isoperimetric inequality. Namely, the Brunn-Minkowski inequality ([118]) shows that if $\mathbf{B} \subset \mathbb{E}^d$ is a ball of the same volume as $\mathbf{A} \subset \mathbb{E}^d$, then

$$\mathrm{V}_d\left(\mathbf{A}_\varepsilon\right) \geq \mathrm{V}_d\left(\mathbf{B}_\varepsilon\right) \tag{8.65}$$

holds for any $\varepsilon > 0$, where $X_\varepsilon := X + \mathbf{B}^d[\mathbf{o}, \varepsilon]$ is called the *ε-neighborhood* of $X \subseteq \mathbb{E}^d$. Thus, (8.65) implies in a straightforward way that

$$\mathrm{V}_d\left(\bigcup_{i=1}^{N} \mathbf{B}^d[\mathbf{p}_i, r]\right) = \mathrm{V}_d\left(\left(\bigcup_{i=1}^{N} \mathbf{B}^d\left[\mathbf{p}_i, \frac{\lambda}{2}\right]\right)_{r-\frac{\lambda}{2}}\right)$$

$$\geq V_d\left(\left(\mathbf{B}^d\left[\mathbf{o}, N^{\frac{1}{d}}\frac{\lambda}{2}\right]\right)_{r-\frac{\lambda}{2}}\right) = \left(r + \left(N^{\frac{1}{d}} - 1\right)\frac{\lambda}{2}\right)^d \kappa_d \tag{8.66}$$

By assumption $N \geq 2^d$ and therefore $\left(r + \left(N^{\frac{1}{d}} - 1\right)\frac{\lambda}{2}\right)^d \kappa_d \geq \left(r + \frac{\lambda}{2}\right)^d \kappa_d$. This inequality, (8.64), and (8.66) complete the proof of Theorem 78.

8.10 Proof of Theorem 79

8.10.1 Basic results on central sets of ball-polytopes

Recall that $\mathbb{M}^d$ denotes the d-dimensional Euclidean, hyperbolic, or spherical space. In this subsection, we collect the definitions and the relevant basic statements that are needed for the proof of Theorem 79 and hold in $\mathbb{M}^d$ for all $d \geq 2$. The proofs of these statements are not difficult but technical and lengthy and so, for details on them, we refer the interested reader to Section 4 in [124] as well as the references mentioned there.

Definition 39 *Let $U \subset \mathbb{M}^d$ be a compact set. We say that the closed ball $\mathbf{B} \subset U$ (possibly with zero radius) is* maximal *in U, if it is not a proper subset of any other ball $\mathbf{B}' \subset U$. The set $C_U \subset U$ that consists of the centers of all maximal balls in U, is called* the central set *of U.*

Definition 40 *For a continuous map $f : C_U \to \mathbb{M}^d$ that rearranges the centers of the maximal balls in U, we denote by $U_f \subset \mathbb{M}^d$ the union of these maximal balls after their rearrangement by the map f.*

The following statement has been explained in [124] by saying that "if there is a counterexample to the Kneser-Poulsen conjecture, then that counterexample can be chosen so that the centers of the balls before the rearrangement lie in the central set of their union".

Lemma 197 *Let $\mathbf{U} \subset \mathbb{M}^d$ be the union of finitely many closed balls and let $\mathbf{V} \subset \mathbb{M}^d$ be the union of the same balls after a contractive rearrangement. Let $f : U \to \mathbb{M}^d$ be an arbitrary contraction (1-Lipschitz map), such that its restriction to the centers of the balls provides the considered contractive rearrangement. Then the set $\mathbf{U}_f$ is compact and $\mathbf{V} \subseteq \mathbf{U}_f$. In particular, if $\mathrm{vol}_{\mathbb{M}^d}(\mathbf{U}_f) \leq \mathrm{vol}_{\mathbb{M}^d}(\mathbf{U})$, then $\mathrm{vol}_{\mathbb{M}^d}(\mathbf{V}) \leq \mathrm{vol}_{\mathbb{M}^d}(\mathbf{U})$, where $\mathrm{vol}_{\mathbb{M}^d}(\cdot)$ denotes the Lebesgue measure in $\mathbb{M}^d$.*

According to Lemma 197, in order to prove the Kneser-Poulsen conjecture, it is enough to rule out the counterexample with (possibly infinitely many) balls, whose centers in the initial configuration lie in the central set of their union. Thus, the following combinatorial and topological description of the structure of the central set will be helpful for the proof of Theorem 79.

Definition 41 *We say that a set* $\mathbf{U} \subset \mathbb{M}^d$ *is a d-dimensional ball-polytope, if it can be represented as a union of finitely many closed balls of positive radii, and its boundary* $\mathrm{bd}\mathbf{U}$ *is a codimension one topological submanifold of* $\mathbb{M}^d$.

Lemma 198 *Let* $\mathbf{U} \subset \mathbb{M}^d$ *be a ball-polytope. Then the following holds:*

(i) The central set $C_{\mathbf{U}}$ *has a structure of a finite* $(d-1)$*-dimensional cell complex whose k-dimensional cells are k-dimensional convex polytopes;*

(ii) The set $\mathbf{U}$ *can be represented as the union of finitely many balls with centers at the 0-dimensional cells of the cell complex* $C_{\mathbf{U}}$*;*

(iii) The sets $C_{\mathbf{U}}$ *and* $\mathbf{U}$ *are homotopy equivalent.*

The following statement is called the Splitting Lemma in [124] because it provides the basis for applying the "divide and conquer" principle and constructing the inductive argument.

Definition 42 *Let* $X \subseteq C_{\mathbf{U}}$ *be a closed non-empty subset. We denote by* $\mathbf{U}_X$ *the union of all maximal balls in* $\mathbf{U}$ *that are centered at the points of* X.

Lemma 199 *Let* $\mathbf{U} \subset \mathbb{M}^d$ *be a ball-polytope and let* $X, Y \subseteq C_U$ *be closed non-empty sets such that* $X \cup Y = C_{\mathbf{U}}$*. Then*

$$\mathbf{U}_X \cap \mathbf{U}_Y = \mathbf{U}_{X \cap Y}.$$

The following statement formulates a version of the "divide and conquer" principle, which is sufficient for the proof of Theorem 79.

Definition 43 *For a closed set* $X \subseteq \mathbf{U}$ *and a contraction* $f : C_{\mathbf{U}} \to \mathbb{M}^d$ *that rearranges the centers of the maximal balls in* $\mathbf{U}$*, let* $\mathbf{U}_{X,f} \subseteq \mathbf{U}_f$ *represent the union of the balls that are obtained from all maximal balls in* $\mathbf{U}$ *with centers in* X *after these balls are rearranged by the map* f.

Proposition 200 *Let* $\mathbf{U} \subset \mathbb{M}^d$ *be a ball-polytope and let* $X, Y \subseteq C_{\mathbf{U}}$ *be closed non-empty sets such that* $X \cup Y = C_{\mathbf{U}}$*. If* $f : C_{\mathbf{U}} \to \mathbb{M}^d$ *is a contraction such that* $\mathrm{vol}_{\mathbb{M}^d}(\mathbf{U}_{X,f}) \le \mathrm{vol}_{\mathbb{M}^d}(\mathbf{U}_X)$, $\mathrm{vol}_{\mathbb{M}^d}(\mathbf{U}_{Y,f}) \le \mathrm{vol}_{\mathbb{M}^d}(\mathbf{U}_Y)$ *, and* $\mathrm{vol}_{\mathbb{M}^d}(\mathbf{U}_{X\cap Y,f}) = \mathrm{vol}_{\mathbb{M}^d}(U_{X\cap Y})$*, then*

$$\mathrm{vol}_{\mathbb{M}^d}(\mathbf{U}_f) \le \mathrm{vol}_{\mathbb{M}^d}(\mathbf{U}).$$

Finally, the following statement is a rather technical lemma, which is needed as well for the proof of Theorem 79.

Lemma 201 *Let* $\mathbf{U} \subset \mathbb{M}^d$ *be a ball-polytope and let* $X \subseteq C_{\mathbf{U}}$ *be a closed non-empty set and let* $Z \subseteq X$ *be the relative boundary of* X *in* $C_{\mathbf{U}}$*. If* $C_{\mathbf{U}_Z} \subseteq X$*, then* $C_{\mathbf{U}_X} = X$.

8.10.2 An extension theorem via piecewise isometries

Besides the central sets, the other core ingredient for the proof of Theorem 79 is generated by piecewise isometries. For proofs of the following extension theorem via piecewise isometries, we refer the interested reader to the relevant references mentioned in Subsection 2.4 of [124].

Definition 44 *A map* $f : \mathbb{M}^d \to \mathbb{M}^d$ *is called a* piecewise isometry, *if the map* f *is continuous and there exists a locally finite triangulation of* $\mathbb{M}^d$ *such that for any simplex* $\mathbf{T}$ *of the triangulation, the restriction of* f *to* $\mathbf{T}$ *is an isometry.*

Theorem 202 *Any contraction* $f : X \to \mathbb{M}^d$ *of a finite set* $X \subset \mathbb{M}^d$ *can be extended to a piecewise isometry on* $\mathbb{M}^d$*.*

8.10.3 Deriving Theorem 79 from the preliminary results

We follow the presentation of Gorbovickis' proof published in Subsection 2.4 of [124]. Assume that there exists a counterexample to the statement of Theorem 79. According to Theorem 202, the corresponding rearrangement of the centers of the disks can be extended to a piecewise isometry $f : \mathbb{M}^d \to \mathbb{M}^d$. We fix this piecewise isometry f together with the associated triangulation of $\mathbb{M}^d$. Without loss of generality we may assume that every interior point of the union of the disks in the initial configuration is also an interior point of some disk. (Namely, if an interior point is not in the interior of some disk, then it lies in the intersection of at least two boundary circles. There are finitely many such points and we can always place new small disks that cover them.) We decrease the radii of the disks slightly, if necessary, so that the union of the disks in the initial configuration of the counterexample is a simply connected ball-polytope. Let us denote this ball-polytope by $\mathbf{U} \subseteq \mathbb{M}^d$. Since the perturbation of the radii can be arbitrarily small, we may assume that the new configuration of the disks still provides a counterexample to the statement of Theorem 79. Then Lemma 197 implies that

$$\mathrm{vol}_{\mathbb{M}^d}(U_f) > \mathrm{vol}_{\mathbb{M}^d}(U). \tag{8.67}$$

Let $\mathcal{U}$ be the set of all simply connected ball-polytopes $\mathbf{U} \subseteq \mathbb{M}^d$ such that inequality (8.67) holds. As we just noticed, the set $\mathcal{U}$ is non-empty. It follows from Lemma 198 that for every $\mathbf{U} \in \mathcal{U}$, the central set $C_{\mathbf{U}}$ has the structure of a tree with straight edges. Every edge intersects only finitely many simplices from the triangulation associated to f, so it can be split into finitely many edges such that f restricted to each one of them is an isometry.

For $\mathbf{U} \in \mathcal{U}$, let $\Gamma(\mathbf{U})$ denote the tree structure on $C_{\mathbf{U}}$ such that f restricted to every edge of $\Gamma(\mathbf{U})$ is an isometry and $\Gamma(\mathbf{U})$ has the smallest possible number of edges. Let $|\Gamma(\mathbf{U})| \geq 0$ denote the number of edges in the graph $\Gamma(\mathbf{U})$.

We choose an element $\mathbf{U} \in \mathcal{U}$ with the minimal number $|\Gamma(\mathbf{U})|$. This number is strictly positive, since otherwise $\mathbf{U}$ is a ball and inequality (8.67) does not hold. Now since $|\Gamma(\mathbf{U})| > 0$ and $\Gamma(\mathbf{U})$ is a tree, it has an edge with a vertex of degree 1. Let us denote this edge by $Y \subseteq C_{\mathbf{U}}$ and define $X = \overline{C_{\mathbf{U}} \setminus Y}$, that is, let X be the closure of $C_{\mathbf{U}} \setminus Y$. Since $X \cap Y$ is a singleton, the sets $\mathbf{U}_{X\cap Y}$ and $\mathbf{U}_{X\cap Y,f}$ are congruent balls; hence, their volumes are equal, and since f is an isometry on Y, it follows from Proposition 200 that

$$\mathrm{vol}_{\mathbb{M}^d}(\mathbf{U}_{X,f}) > \mathrm{vol}_{\mathbb{M}^d}(\mathbf{U}_X).$$

Moreover, it follows from part (ii) of Lemma 198 and Lemma 199 that the set $\mathbf{U} \setminus \mathbf{U}_X$ has the form $\mathbf{B}_1 \setminus \mathbf{B}_2$, where $\mathbf{B}_1, \mathbf{B}_2 \subseteq \mathbb{M}^2$ are two closed disks centered at the vertices of the edge Y. The latter implies that $\mathbf{U}_X$ is a simply connected ball-polytope, hence $\mathbf{U}_X \in \mathcal{U}$.

Finally, since the relative boundary of X in $C_{\mathbf{U}}$ is the set $X \cap Y$ that consists of one point, we have $C_{\mathbf{U}_{X\cap Y}} = X \cap Y \subseteq X$, and Lemma 201 implies that $C_{\mathbf{U}_X} = X$. Thus, $|\Gamma(\mathbf{U}_X)| < |\Gamma(\mathbf{U})|$, which contradicts the choice of the ball-polytope $\mathbf{U}$. This completes the proof of Theorem 79 also in the case when $\mathbb{M}^2 = \mathbb{E}^2$.

9

Proofs on Volumetric Bounds for Contact Numbers

Summary. In this chapter we present selected proofs of some of the theorems in Chapter 4. In particular, in Section 9.1 we prove estimates for $c(n,3)$ and $c_{\text{fcc}}(n)$. In Section 9.2 we treat the general case, i.e., we examine the contact numbers of translative packings of an arbitrary convex body in $\mathbb{E}^d$. In Sections 9.3 and 9.4 we establish a connection between contact numbers of ball packings and the densities of packings of soft balls, and prove a Blichfeldt-type estimate for the latter quantities, respectively. In Section 9.5 we solve the maximum packing density problem for soft Euclidean disks. In Section 9.6 we prove a Rogers-type bound for soft packings in $\mathbb{E}^3$. Then we turn our attention to totally separable packings. In Section 9.7 we determine the values of $c_{\text{sep}}(n,2)$ and prove bounds for $c_{\text{sep}}(n,3)$. In Section 9.8 we prove estimates for $c_{\mathbb{Z}}(n,d)$, while in Section 9.9 we examine the general problem of finding $c_{\text{sep}}(n,d)$. In Section 9.10 we determine the Hadwiger numbers of strictly convex and smooth convex bodies in $\mathbb{E}^d$ for $d \leq 4$, and apply these results to prove estimates for $c_{\text{sep}}(\mathbf{K},n,d)$ with $d \leq 4$. In Section 9.11 we construct topological disks with arbitrarily large Hadwiger numbers, and in Sections 9.12 and 9.13 we prove upper bounds for the Hadwiger numbers of starlike disks in the centrally symmetric and in the general case, respectively.

9.1 Proof of Theorem 87

9.1.1 An upper bound for sphere packings: Proof of (i)

The proof presented in this section follows the proof of (i) of Theorem 1.1 in [50] and as such it is based on the recent breakthrough results of Hales [136]. The details are as follows.

Let $\mathbf{B}$ denote the (closed) unit ball centered at the origin $\mathbf{o}$ of $\mathbb{E}^3$ and let $\mathcal{P} := \{\mathbf{c}_1 + \mathbf{B}, \mathbf{c}_2 + \mathbf{B}, \dots, \mathbf{c}_n + \mathbf{B}\}$ denote the packing of n unit balls with centers $\mathbf{c}_1, \mathbf{c}_2, \dots, \mathbf{c}_n$ in $\mathbb{E}^3$ having the largest number $C(n)$ of touching pairs among all packings of n unit balls in $\mathbb{E}^3$. ($\mathcal{P}$ might not be uniquely determined up to congruence in which case $\mathcal{P}$ stands for any of those extremal packings.)

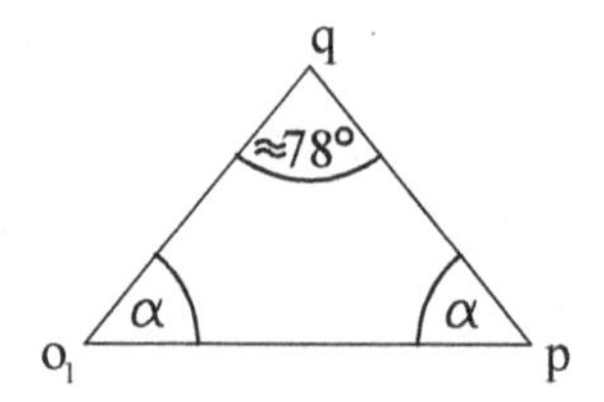

FIGURE 9.1
The isosceles triangle $\text{conv}\{\mathbf{o}_1, \mathbf{p}, \mathbf{q}\}$.

Now, let $\hat{r} := 1.58731$. The following statement shows the main property of $\hat{r}$ that is needed for our proof of Theorem 87.

Theorem 203 *Let $\mathbf{B}_1, \mathbf{B}_2, \dots, \mathbf{B}_{13}$ be 13 different members of a packing of unit balls in $\mathbb{E}^3$. Assume that each ball of the family $\mathbf{B}_2, \mathbf{B}_3, \dots, \mathbf{B}_{13}$ touches $\mathbf{B}_1$. Let $\hat{\mathbf{B}}_i$ be the closed ball concentric with $\mathbf{B}_i$ having radius $\hat{r}$, $1 \leq i \leq 13$. Then the boundary $\text{bd}(\hat{\mathbf{B}}_1)$ of $\hat{\mathbf{B}}_1$ is covered by the balls $\hat{\mathbf{B}}_2, \hat{\mathbf{B}}_3, \dots, \hat{\mathbf{B}}_{13}$, that is,*

$$\text{bd}(\hat{\mathbf{B}}_1) \subset \cup_{j=2}^{13} \hat{\mathbf{B}}_j \ .$$

Proof. Let $\mathbf{o}_i$ be the center of the unit ball $\mathbf{B}_i$, $1 \leq i \leq 13$ and assume that $\mathbf{B}_1$ is tangent to the unit balls $\mathbf{B}_2, \mathbf{B}_3, \dots, \mathbf{B}_{13}$ at the points $\mathbf{t}_j \in \text{bd}(\mathbf{B}_j) \cap \text{bd}(\mathbf{B}_1), 2 \leq j \leq 13$.

Let α denote the measure of the angles opposite to the equal sides of the isosceles triangle $\text{conv}\{\mathbf{o}_1, \mathbf{p}, \mathbf{q}\}$ with $\text{dist}(\mathbf{o}_1, \mathbf{p}) = 2$ and $\text{dist}(\mathbf{p}, \mathbf{q}) = \text{dist}(\mathbf{o}_1, \mathbf{q}) = \hat{r}$, where $\text{dist}(\cdot, \cdot)$ denotes the Euclidean distance between the corresponding two points (Figure 9.1). Clearly, $\cos\alpha = \frac{1}{\hat{r}}$ with $\alpha < \frac{\pi}{3}$.

Lemma 204 *Let $\mathbf{T}$ be the convex hull of the points $\mathbf{t}_2, \mathbf{t}_3, \dots, \mathbf{t}_{13}$. Then the radius of the circumscribed circle of each face of the convex polyhedron $\mathbf{T}$ is less than $\sin\alpha$.*

Proof. Let F be an arbitrary face of $\mathbf{T}$ with vertices $\mathbf{t}_j, j \in I_F \subset \{2, 3, \dots, 13\}$ and let $\mathbf{c}_F$ denote the center of the circumscribed circle of F. Clearly, the triangle $\text{conv}\{\mathbf{o}_1, \mathbf{c}_F, \mathbf{t}_j\}$ is a right triangle with a right angle at $\mathbf{c}_F$ and with an acute angle of measure β_F at $\mathbf{o}_1$ for all $j \in I_F$. We have to show that $\beta_F < \alpha$. We prove this by contradiction. Namely, assume that $\alpha \leq \beta_F$. Then either $\frac{\pi}{3} < \beta_F$ or $\alpha \leq \beta_F \leq \frac{\pi}{3}$. First, let us take a closer look of the case $\frac{\pi}{3} < \beta_F$. Reflect the point $\mathbf{o}_1$ about the plane of F and label the point obtained by $\mathbf{o}_1'$.

Clearly, the triangle $\text{conv}\{\mathbf{o}_1, \mathbf{o}_1', \mathbf{o}_j\}$ is a right triangle with a right angle at $\mathbf{o}_1'$ and with an acute angle of measure β_F at $\mathbf{o}_1$ for all $j \in I_F$. Then reflect the point $\mathbf{o}_1$ about $\mathbf{o}_1'$ and label the obtained point by $\mathbf{o}_1''$ furthermore, let

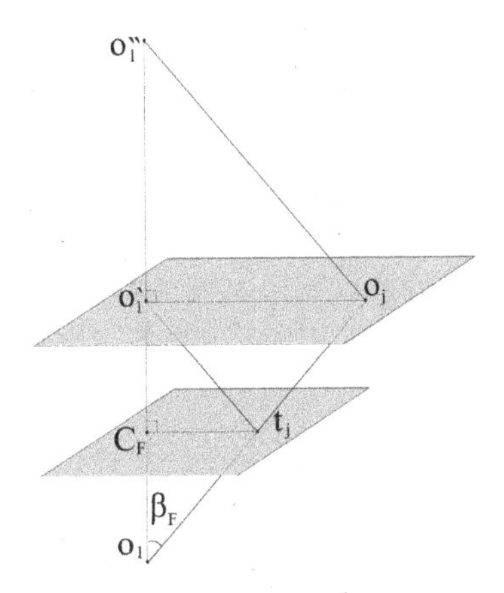

FIGURE 9.2
The plane reflections to obtain $\mathbf{o}_1'$ and $\mathbf{o}_1''$.

$\mathbf{B}_1''$ denote the unit ball centered at $\mathbf{o}_1''$ (Figure 9.2). As $\frac{\pi}{3} < \beta_F$ therefore $\mathrm{dist}(\mathbf{o}_1, \mathbf{o}_1'') < 2$ and so, one can simply translate $\mathbf{B}_1''$ along the line $\mathbf{o}_1\mathbf{o}_1''$ away from $\mathbf{o}_1$ to a new position, say $\mathbf{B}_1'''$, such that it is tangent to $\mathbf{B}_1$. However, this would mean that $\mathbf{B}_1$ is tangent to 13 non-overlapping unit balls namely, to $\mathbf{B}_1''', \mathbf{B}_2, \mathbf{B}_3, \dots, \mathbf{B}_{13}$, clearly contradicting the well-known fact ([220]) that this number cannot be larger than 12. Thus, we are left with the case when $\alpha \leq \beta_F \leq \frac{\pi}{3}$. By repeating the definitions of $\mathbf{o}_1'$, $\mathbf{o}_1''$, and $\mathbf{B}_1''$, the inequality $\beta_F \leq \frac{\pi}{3}$ implies in a straightforward way that the 14 unit balls $\mathbf{B}_1, \mathbf{B}_1'', \mathbf{B}_2, \mathbf{B}_3, \dots, \mathbf{B}_{13}$ form a packing in $\mathbb{E}^3$. Moreover, the inequality $\alpha \leq \beta_F$ yields that $\mathrm{dist}(\mathbf{o}_1, \mathbf{o}_1'') \leq 4\cos\alpha = \frac{4}{\hat{r}} = 2.51998... < 2.52$. Finally, notice that the latter inequality contradicts the following recent result of Hales [136].

Theorem 205 *Let* $\mathbf{B}_1, \mathbf{B}_2, \dots, \mathbf{B}_{14}$ *be* 14 *different members of a packing of unit balls in* $\mathbb{E}^3$. *Assume that each ball of the family* $\mathbf{B}_2, \mathbf{B}_3, \dots, \mathbf{B}_{13}$ *touches* $\mathbf{B}_1$. *Then the distance between the centers of* $\mathbf{B}_1$ *and* $\mathbf{B}_{14}$ *is at least* 2.52.

This completes the proof of Lemma 204. □

Now, we are ready to prove Theorem 203. First, we note that by projecting the faces F of $\mathbf{T}$ from the center point $\mathbf{o}_1$ onto the sphere $\mathrm{bd}(\hat{\mathbf{B}}_1)$ we get a tiling of $\mathrm{bd}(\hat{\mathbf{B}}_1)$ into spherically convex polygons $\hat{F}$. Thus, it is sufficient to show that if F is an arbitrary face of $\mathbf{T}$ with vertices $\mathbf{t}_j, j \in I_F \subset \{2, 3, \dots, 13\}$, then its central projection $\hat{F} \subset \mathrm{bd}(\hat{\mathbf{B}}_1)$ is covered by the closed balls $\hat{\mathbf{B}}_j, j \in I_F \subset \{2, 3, \dots, 13\}$. Second, in order to achieve this, it is sufficient to prove that the projection $\hat{\mathbf{c}}_F$ of the center $\mathbf{c}_F$ of the circumscribed circle of F from the center point $\mathbf{o}_1$ onto the sphere $\mathrm{bd}(\hat{\mathbf{B}}_1)$ is covered by each of the closed balls $\hat{\mathbf{B}}_j, j \in I_F \subset \{2, 3, \dots, 13\}$. Indeed, if in the triangle $\mathrm{conv}\{\mathbf{o}_1, \mathbf{o}_j, \hat{\mathbf{c}}_F\}$ the measure of the angle at $\mathbf{o}_1$ is denoted by β_F, then Lemma 204 implies in a straighforward way that $\beta_F < \alpha$. Hence, based on $\mathrm{dist}(\mathbf{o}_1, \mathbf{o}_j) = 2$ and $\mathrm{dist}(\mathbf{o}_1, \hat{\mathbf{c}}_F) = \hat{r}$, a simple comparison of the triangle $\mathrm{conv}\{\mathbf{o}_1, \mathbf{o}_j, \hat{\mathbf{c}}_F\}$ with the triangle $\mathrm{conv}\{\mathbf{o}_1, \mathbf{p}, \mathbf{q}\}$ yields that $\mathrm{dist}(\mathbf{o}_j, \hat{\mathbf{c}}_F) < \hat{r}$ holds for all $j \in I_F \subset \{2, 3, \dots, 13\}$, finishing the proof of Theorem 203. □

Next, let us take the union $\bigcup_{i=1}^{n} (\mathbf{c}_i + \hat{r}\mathbf{B})$ of the closed balls $\mathbf{c}_1 + \hat{r}\mathbf{B}, \mathbf{c}_2 + \hat{r}\mathbf{B}, \dots, \mathbf{c}_n + \hat{r}\mathbf{B}$ of radii $\hat{r}$ centered at the points $\mathbf{c}_1, \mathbf{c}_2, \dots, \mathbf{c}_n$ in $\mathbb{E}^3$.

Theorem 206

$$\frac{n\mathrm{vol}_3(\mathbf{B})}{\mathrm{vol}_3\left(\bigcup_{i=1}^{n} (\mathbf{c}_i + \hat{r}\mathbf{B})\right)} < 0.7547,$$

where $\mathrm{vol}_3(\cdot)$ *refers to the* 3*-dimensional volume of the corresponding set.*

Proof. First, partition $\bigcup_{i=1}^{n} (\mathbf{c}_i + \hat{r}\mathbf{B})$ into truncated Voronoi cells as follows. Let $\mathbf{P}_i$ denote the Voronoi cell of the packing $\mathcal{P}$ assigned to $\mathbf{c}_i + \mathbf{B}$, $1 \le i \le n$, that is, let $\mathbf{P}_i$ stand for the set of points of $\mathbb{E}^3$ that are not farther away from $\mathbf{c}_i$ than from any other $\mathbf{c}_j$ with $j \neq i, 1 \le j \le n$. Then, recall the well-known fact (see for example, [106]) that the Voronoi cells $\mathbf{P}_i$, $1 \le i \le n$ just introduced form a tiling of $\mathbb{E}^3$. Based on this, it is easy to see that the truncated Voronoi cells $\mathbf{P}_i \cap (\mathbf{c}_i + \hat{r}\mathbf{B})$, $1 \le i \le n$ generate a tiling of the non-convex container $\bigcup_{i=1}^{n} (\mathbf{c}_i + \hat{r}\mathbf{B})$ for the packing $\mathcal{P}$ (Figure 9.3). Second, as $\sqrt{2} < \hat{r}$; therefore, the following very recent result of Hales [136] (see Lemma 9.13 on p. 228) applied to the truncated Voronoi cells $\mathbf{P}_i \cap (\mathbf{c}_i + \hat{r}\mathbf{B})$, $1 \le i \le n$ implies the inequality of Theorem 206 in a straightforward way.

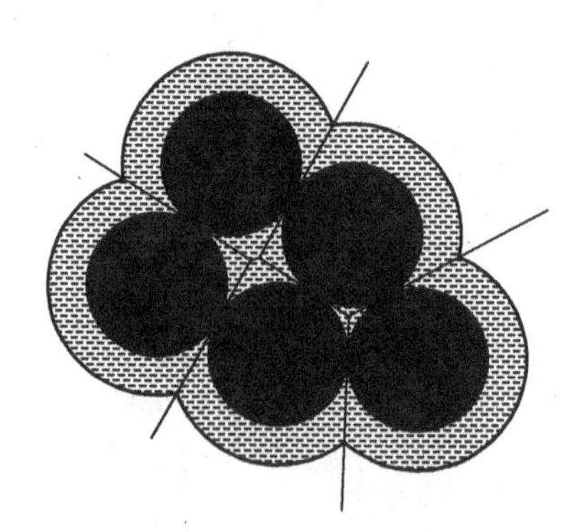

FIGURE 9.3
Voronoi cells of a packing with black $\mathbf{c}_i + \mathbf{B}$'s and dashed $\mathbf{c}_i + \hat{r}\mathbf{B}$'s.

Theorem 207 *Let* $\mathcal{F}$ *be an arbitrary (finite or infinite) family of non-overlapping unit balls in* $\mathbb{E}^3$ *with the unit ball* $\mathbf{B}$ *centered at the origin* $\mathbf{o}$ *of* $\mathbb{E}^3$

belonging to $\mathcal{F}$. Let $\mathbf{P}$ stand for the Voronoi cell of the packing $\mathcal{F}$ assigned to $\mathbf{B}$. Let $\mathbf{Q}$ denote a regular dodecahedron circumscribed $\mathbf{B}$ (having circumradius $\sqrt{3}\tan\frac{\pi}{5} = 1.2584...$). Finally, let $r := \sqrt{2} = 1.4142...$ and let $r\mathbf{B}$ denote the ball of radius r centered at the origin $\mathbf{o}$ of $\mathbb{E}^3$. Then

$$\frac{\mathrm{vol}_3(\mathbf{B})}{\mathrm{vol}_3(\mathbf{P})} \leq \frac{\mathrm{vol}_3(\mathbf{B})}{\mathrm{vol}_3(\mathbf{P} \cap r\mathbf{B})} \leq \frac{\mathrm{vol}_3(\mathbf{B})}{\mathrm{vol}_3(\mathbf{Q})} < 0.7547.$$

This finishes the proof of Theorem 206. □

The well-known isoperimetric inequality [194] applied to $\bigcup_{i=1}^{n}(\mathbf{c}_i + \hat{r}\mathbf{B})$ yields:

Lemma 208

$$36\pi \mathrm{vol}_3^2\left(\bigcup_{i=1}^{n}(\mathbf{c}_i + \hat{r}\mathbf{B})\right) \leq \mathrm{svol}_2^3\left(\mathrm{bd}\left(\bigcup_{i=1}^{n}(\mathbf{c}_i + \hat{r}\mathbf{B})\right)\right),$$

where $\mathrm{svol}_2(\cdot)$ refers to the 2-dimensional surface volume of the corresponding set.

Thus, Theorem 206 and Lemma 208 generate the following inequality.

Corollary 209

$$15.159805 n^{\frac{2}{3}} < 15.15980554... n^{\frac{2}{3}} = \frac{4\pi}{(0.7547)^{\frac{2}{3}}} n^{\frac{2}{3}}$$

$$< \mathrm{svol}_2\left(\mathrm{bd}\left(\bigcup_{i=1}^{n}(\mathbf{c}_i + \hat{r}\mathbf{B})\right)\right).$$

Now, assume that $\mathbf{c}_i + \mathbf{B} \in \mathcal{P}$ is tangent to $\mathbf{c}_j + \mathbf{B} \in \mathcal{P}$ for all $j \in T_i$, where $T_i \subset \{1, 2, \ldots, n\}$ stands for the family of indices $1 \leq j \leq n$ for which $\mathrm{dist}(\mathbf{c}_i, \mathbf{c}_j) = 2$. Then let $\hat{S}_i := \mathrm{bd}(\mathbf{c}_i + \hat{r}\mathbf{B})$ and let $\hat{\mathbf{c}}_{ij}$ be the intersection of the line segment $\mathbf{c}_i\mathbf{c}_j$ with $\hat{S}_i$ for all $j \in T_i$. Moreover, let $C_{\hat{S}_i}(\hat{\mathbf{c}}_{ij}, \frac{\pi}{6})$ (resp., $C_{\hat{S}_i}(\hat{\mathbf{c}}_{ij}, \alpha)$) denote the open spherical cap of $\hat{S}_i$ centered at $\hat{\mathbf{c}}_{ij} \in \hat{S}_i$ having angular radius $\frac{\pi}{6}$ (resp., α with $0 < \alpha < \frac{\pi}{2}$ and $\cos\alpha = \frac{1}{\hat{r}}$). Clearly, the family $\{C_{\hat{S}_i}(\hat{\mathbf{c}}_{ij}, \frac{\pi}{6}), j \in T_i\}$ consists of pairwise disjoint open spherical caps of $\hat{S}_i$; moreover,

$$\frac{\sum_{j \in T_i} \mathrm{svol}_2\left(C_{\hat{S}_i}(\hat{\mathbf{c}}_{ij}, \frac{\pi}{6})\right)}{\mathrm{svol}_2\left(\cup_{j \in T_i} C_{\hat{S}_i}(\hat{\mathbf{c}}_{ij}, \alpha)\right)} = \frac{\sum_{j \in T_i} \mathrm{Sarea}\left(C(\mathbf{u}_{ij}, \frac{\pi}{6})\right)}{\mathrm{Sarea}\left(\cup_{j \in T_i} C(\mathbf{u}_{ij}, \alpha)\right)}, \tag{9.1}$$

where $\mathbf{u}_{ij} := \frac{1}{2}(\mathbf{c}_j - \mathbf{c}_i) \in \mathbb{S}^2 := \mathrm{bd}(\mathbf{B})$ and $C(\mathbf{u}_{ij}, \frac{\pi}{6}) \subset \mathbb{S}^2$ (resp., $C(\mathbf{u}_{ij}, \alpha) \subset \mathbb{S}^2$) denotes the open spherical cap of $\mathbb{S}^2$ centered at $\mathbf{u}_{ij}$ having angular radius

$\frac{\pi}{6}$ (resp., α) and where Sarea$(\cdot)$ refers to the spherical area measure on $\mathbb{S}^2$. Now, Molnár's density bound (Satz I in [185]) implies that

$$\frac{\sum_{j\in T_i} \mathrm{Sarea}\left(C(\mathbf{u}_{ij},\frac{\pi}{6})\right)}{\mathrm{Sarea}\left(\cup_{j\in T_i} C(\mathbf{u}_{ij},\alpha)\right)} < 0.89332 \ . \tag{9.2}$$

In order to estimate

$$\mathrm{svol}_2\left(\mathrm{bd}\left(\bigcup_{i=1}^{n}(\mathbf{c}_i+\hat{r}\mathbf{B})\right)\right)$$

from above, let us assume that m members of $\mathcal{P}$ have 12 touching neighbors in $\mathcal{P}$ and k members of $\mathcal{P}$ have at most 9 touching neighbors in $\mathcal{P}$. Thus, $n-m-k$ members of $\mathcal{P}$ have either 10 or 11 touching neighbors in $\mathcal{P}$. (Here we have used the well-known fact that $\tau_3 = 12$, that is, no member of $\mathcal{P}$ can have more than 12 touching neighbors.) Without loss of generality we may assume that $4 \le k \le n-m$.

First, we note that $\mathrm{Sarea}\left(C(\mathbf{u}_{ij},\frac{\pi}{6})\right) = 2\pi(1-\cos\frac{\pi}{6}) = 2\pi(1-\frac{\sqrt{3}}{2})$ and $\mathrm{svol}_2\left(C_{\hat{S}_i}(\hat{\mathbf{c}}_{ij},\frac{\pi}{6})\right) = 2\pi(1-\frac{\sqrt{3}}{2})\hat{r}^2$. Second, recall Theorem 203 according to which if a member of $\mathcal{P}$, say $\mathbf{c}_i+\mathbf{B}$, has exactly 12 touching neighbors in $\mathcal{P}$, then $\hat{S}_i \subset \bigcup_{j\in T_i}(\mathbf{c}_j+\hat{r}\mathbf{B})$. These facts together with (9.1) and (9.2) imply the following estimate.

Corollary 210 $\mathrm{svol}_2\left(\mathrm{bd}\left(\bigcup_{i=1}^{n}(\mathbf{c}_i+\hat{r}\mathbf{B})\right)\right) < \frac{24.53902}{3}(n-m-k)+24.53902k$.

Proof.

$$\mathrm{svol}_2\left(\mathrm{bd}\left(\bigcup_{i=1}^{n}(\mathbf{c}_i+\hat{r}\mathbf{B})\right)\right)$$

$$< \left(4\pi\hat{r}^2 - \frac{10\cdot 2\pi(1-\frac{\sqrt{3}}{2})\hat{r}^2}{0.89332}\right)(n-m-k) + \left(4\pi\hat{r}^2 - \frac{3\cdot 2\pi(1-\frac{\sqrt{3}}{2})\hat{r}^2}{0.89332}\right)k$$

$$< 7.91956(n-m-k) + 24.53902k < \frac{24.53902}{3}(n-m-k) + 24.53902k \ .$$

□

Hence, Corollaries 209 and 210 yield in a straightforward way that

$$1.85335n^{\frac{2}{3}} - 3k < n-m-k \ . \tag{9.3}$$

Finally, as the number $C(n)$ of touching pairs in $\mathcal{P}$ is obviously at most

$$\frac{1}{2}\left(12n-(n-m-k)-3k\right),$$

therefore (9.3) implies that

$$C(n) \le \frac{1}{2}\left(12n-(n-m-k)-3k\right) < 6n - 0.926675n^{\frac{2}{3}} < 6n - 0.926n^{\frac{2}{3}},$$

finishing the proof of (i) in Theorem 87.

9.1.2 An upper bound for the fcc lattice: Proof of (ii)

Although the idea of the proof of (ii) is similar to that of (i), they differ in the combinatorial counting part (see (9.9)) as well as in the density estimate for packings of spherical caps of angular radii $\frac{\pi}{6}$ (see (9.8)). Moreover, the proof of (ii) is based on the new parameter value $\bar{r} := \sqrt{2}$ (replacing $\hat{r} = 1.81383$). The details are as follows.

First, recall that if Λ_{fcc} denotes the face-centered cubic lattice with shortest non-zero lattice vector of length 2 in $\mathbb{E}^3$ and we place unit balls centered at each lattice point of Λ_{fcc}, then we get the fcc lattice packing of unit balls, labelled by $\mathcal{P}_{fcc}$, in which each unit ball is touched by 12 others such that their centers form the vertices of a cuboctahedron. (Recall that a cuboctahedron is a convex polyhedron with 8 triangular faces and 6 square faces having 12 identical vertices, with 2 triangles and 2 squares meeting at each, and 24 identical edges, each separating a triangle from a square. As such, it is a quasiregular polyhedron, i.e., an Archimedean solid, being vertex-transitive and edge-transitive.) Second, it is well known (see [106] for more details) that the Voronoi cell of each unit ball in $\mathcal{P}_{fcc}$ is a rhombic dodecahedron (the dual of a cuboctahedron) of volume $\sqrt{32}$ (and of circumradius $\sqrt{2}$). Thus, the density of $\mathcal{P}_{fcc}$ is $\frac{\frac{4\pi}{3}}{\sqrt{32}} = \frac{\pi}{\sqrt{18}}$.

Now, let $\mathbf{B}$ denote the unit ball centered at the origin $\mathbf{o} \in \Lambda_{fcc}$ of $\mathbb{E}^3$ and let $\mathcal{P} := \{\mathbf{c}_1 + \mathbf{B}, \mathbf{c}_2 + \mathbf{B}, \dots, \mathbf{c}_n + \mathbf{B}\}$ denote the packing of n unit balls with centers $\{\mathbf{c}_1, \mathbf{c}_2, \dots, \mathbf{c}_n\} \subset \Lambda_{fcc}$ in $\mathbb{E}^3$ having the largest number $C_{fcc}(n)$ of touching pairs among all packings of n unit balls being a subpacking of $\mathcal{P}_{fcc}$. ($\mathcal{P}$ might not be uniquely determined up to congruence in which case $\mathcal{P}$ stands for any of those extremal packings.) As the Voronoi cell of each unit ball in $\mathcal{P}_{fcc}$ is contained in a ball of radius $\bar{r} = \sqrt{2}$, therefore, based on the corresponding decomposition of $\bigcup_{i=1}^{n}(\mathbf{c}_i + \bar{r}\mathbf{B})$ into truncated Voronoi cells, we get that

$$\frac{n\mathrm{vol}_3(\mathbf{B})}{\mathrm{vol}_3(\bigcup_{i=1}^{n}(\mathbf{c}_i + \bar{r}\mathbf{B}))} < \frac{\pi}{\sqrt{18}} = 0.7404\ldots\ . \tag{9.4}$$

As a next step we apply the isoperimetric inequality ([194]):

$$36\pi\mathrm{vol}_3^2\left(\bigcup_{i=1}^{n}(\mathbf{c}_i + \bar{r}\mathbf{B})\right) \leq \mathrm{svol}_2^3\left(\mathrm{bd}\left(\bigcup_{i=1}^{n}(\mathbf{c}_i + \bar{r}\mathbf{B})\right)\right)\ . \tag{9.5}$$

Thus, (9.4) and (9.5) yield in a straightforward way that

$$15.3532\ldots n^{\frac{2}{3}} = 4\sqrt[3]{18\pi}n^{\frac{2}{3}} < \mathrm{svol}_2\left(\mathrm{bd}\left(\bigcup_{i=1}^{n}(\mathbf{c}_i + \bar{r}\mathbf{B})\right)\right)\ . \tag{9.6}$$

Now, assume that $\mathbf{c}_i + \mathbf{B} \in \mathcal{P}$ is tangent to $\mathbf{c}_j + \mathbf{B} \in \mathcal{P}$ for all $j \in T_i$, where $T_i \subset \{1, 2, \dots, n\}$ stands for the family of indices $1 \leq j \leq n$ for which $\mathrm{dist}(\mathbf{c}_i, \mathbf{c}_j) = 2$. Then let $\bar{S}_i := \mathrm{bd}(\mathbf{c}_i + \bar{r}\mathbf{B})$ and let $\bar{\mathbf{c}}_{ij}$ be the intersection of the line segment $\mathbf{c}_i\mathbf{c}_j$ with $\bar{S}_i$ for all $j \in T_i$. Moreover, let $C_{\bar{S}_i}(\bar{\mathbf{c}}_{ij}, \frac{\pi}{6})$ (resp., $C_{\bar{S}_i}(\bar{\mathbf{c}}_{ij}, \frac{\pi}{4})$) denote the open spherical cap of $\bar{S}_i$ centered at $\bar{\mathbf{c}}_{ij} \in \bar{S}_i$ having angular radius $\frac{\pi}{6}$ (resp., $\frac{\pi}{4}$). Clearly, the family $\{C_{\bar{S}_i}(\bar{\mathbf{c}}_{ij}, \frac{\pi}{6}), j \in T_i\}$ consists of pairwise disjoint open spherical caps of $\bar{S}_i$; moreover,

$$\frac{\sum_{j\in T_i} \mathrm{svol}_2\left(C_{\bar{S}_i}(\bar{\mathbf{c}}_{ij}, \frac{\pi}{6})\right)}{\mathrm{svol}_2\left(\cup_{j\in T_i} C_{\bar{S}_i}(\bar{\mathbf{c}}_{ij}, \frac{\pi}{4})\right)} = \frac{\sum_{j\in T_i} \mathrm{Sarea}\left(C(\mathbf{u}_{ij}, \frac{\pi}{6})\right)}{\mathrm{Sarea}\left(\cup_{j\in T_i} C(\mathbf{u}_{ij}, \frac{\pi}{4})\right)}, \tag{9.7}$$

where $\mathbf{u}_{ij} = \frac{1}{2}(\mathbf{c}_j - \mathbf{c}_i) \in \mathbb{S}^2$ and $C(\mathbf{u}_{ij}, \frac{\pi}{6}) \subset \mathbb{S}^2$ (resp., $C(\mathbf{u}_{ij}, \frac{\pi}{4}) \subset \mathbb{S}^2$) denotes the open spherical cap of $\mathbb{S}^2$ centered at $\mathbf{u}_{ij}$ having angular radius $\frac{\pi}{6}$ (resp., $\frac{\pi}{4}$). Now, the geometry of the cuboctahedron representing the 12 touching neighbors of an arbitrary unit ball in $\mathcal{P}_{fcc}$ implies in a straightforward way that

$$\frac{\sum_{j\in T_i} \mathrm{Sarea}\left(C(\mathbf{u}_{ij}, \frac{\pi}{6})\right)}{\mathrm{Sarea}\left(\cup_{j\in T_i} C(\mathbf{u}_{ij}, \frac{\pi}{4})\right)} \leq 6(1 - \frac{\sqrt{3}}{2}) = 0.8038... \tag{9.8}$$

with equality when 12 spherical caps of angular radius $\frac{\pi}{6}$ are packed on $\mathbb{S}^2$.

Finally, as $\mathrm{svol}_2\left(C(\mathbf{u}_{ij}, \frac{\pi}{6})\right) = 2\pi(1 - \cos\frac{\pi}{6})$ and $\mathrm{svol}_2\left(C_{\bar{S}_i}(\bar{\mathbf{c}}_{ij}, \frac{\pi}{6})\right) = 2\pi(1 - \frac{\sqrt{3}}{2})\bar{r}^2$, therefore, (9.7) and (9.8) yield that

$$\mathrm{svol}_2\left(\mathrm{bd}\left(\bigcup_{i=1}^{n} \mathbf{c}_i + \bar{r}\mathbf{B}\right)\right) \leq 4\pi\bar{r}^2 n - \frac{1}{6(1-\frac{\sqrt{3}}{2})} 2\left(2\pi\left(1 - \frac{\sqrt{3}}{2}\right)\bar{r}^2\right) C_{fcc}(n)$$

$$= 8\pi n - \frac{4\pi}{3} C_{fcc}(n)\ . \tag{9.9}$$

Thus, (9.6) and (9.9) imply that

$$4\sqrt[3]{18\pi}\, n^{\frac{2}{3}} < 8\pi n - \frac{4\pi}{3} C_{fcc}(n)\ . \tag{9.10}$$

From (9.10) the inequality $C_{fcc}(n) < 6n - \frac{3\sqrt[3]{18\pi}}{\pi} n^{\frac{2}{3}} = 6n - 3.665\ldots n^{\frac{2}{3}}$ follows in a straightforward way for all $n \geq 2$. This completes the proof of (ii) in Theorem 87.

9.1.3 Octahedral unit sphere packings: Proof of (iii)

It is rather easy to show that for any positive integer $k \geq 2$ there are $n(k) := \frac{2k^3+k}{3} = \frac{k(2k^2+1)}{3}$ lattice points of the face-centered cubic lattice Λ_{fcc} such that their convex hull is a regular octahedron $\mathbf{K} \subset \mathbb{E}^3$ of edge length $2(k-1)$ having exactly k lattice points along each of its edges (see Figure 9.4 for $k = 4$).

Now, draw a unit ball around each lattice point of $\Lambda_{fcc} \cap \mathbf{K}$ and label the packing of the $n(k)$ unit balls obtained in this way by $\mathcal{P}_{fcc}(k)$. It is easy to check that if the center of a unit ball of $\mathcal{P}_{fcc}(k)$ is a relative interior point

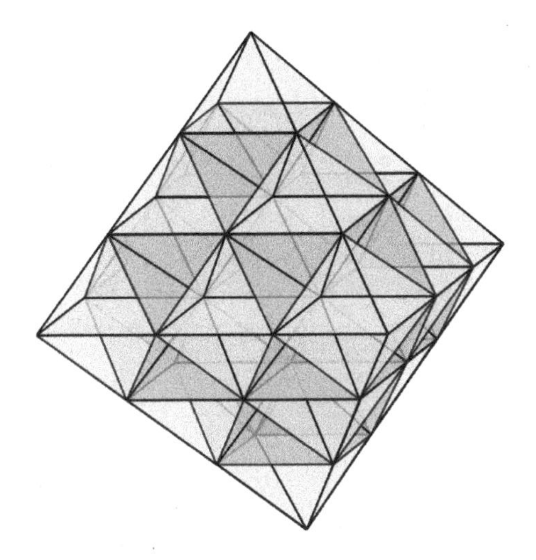

FIGURE 9.4
The Octahedral Construction for $k = 4$.

of an edge (resp., of a face) of $\mathbf{K}$, then the unit ball in question has 7 (resp., 9) touching neighbors in $\mathcal{P}_{fcc}(k)$. Last but not least, any unit ball of $\mathcal{P}_{fcc}(k)$ whose center is an interior point of $\mathbf{K}$ has 12 touching neighbors in $\mathcal{P}_{fcc}(k)$. Next we note that the number of lattice points of Λ_{fcc} lying in the relative interior of the edges (resp., faces) of $\mathbf{K}$ is $12(k-2) = 12k - 24$ (resp., $8\left(\frac{1}{2}(k-3)^2 + \frac{1}{2}(k-3)\right) = 4(k-3)^2 + 4(k-3)$). Furthermore, the number of lattice points of Λ_{fcc} in the interior of $\mathbf{K}$ is equal to $\frac{2}{3}(k-2)^3 + \frac{1}{3}(k-2)$. Thus, the contact number $C\left(\mathcal{P}_{fcc}(k)\right)$ of the packing $\mathcal{P}_{fcc}(k)$ is equal to

$$\frac{12}{2}\left(\frac{2}{3}(k-2)^3 + \frac{1}{3}(k-2)\right) + \frac{9}{2}\left(4(k-3)^2 + 4(k-3)\right) + \frac{7}{2}\left(12k-24\right) + \frac{24}{2}$$
$$= 4k^3 - 6k^2 + 2k.$$

As a result, we get that

$$C\left(\mathcal{P}_{fcc}(k)\right) = 6n(k) - 6k^2. \tag{9.11}$$

Finally, as $\frac{2k^3}{3} < n(k)$, therefore, $6k^2 < \sqrt[3]{486}n^{\frac{2}{3}}(k)$ and so, (9.11) implies (iii) of Theorem 87 in a straightforward way.

9.2 Proof of Theorem 88

Let $\{\mathbf{c}_1+\mathbf{K}, \mathbf{c}_2+\mathbf{K}, \dots, \mathbf{c}_n+\mathbf{K}\}$ be an arbitrary packing of $n>1$ translates of the convex body $\mathbf{K}$ in $\mathbb{E}^d$, $d \geq 3$ with $C_n = \{\mathbf{c}_1, \mathbf{c}_2, \dots, \mathbf{c}_n\}$. Clearly, $(\mathbf{c}_1+\mathbf{K}) \cup (\mathbf{c}_2+\mathbf{K}) \cup \dots \cup (\mathbf{c}_n+\mathbf{K}) = C_n+\mathbf{K}$.

Recall that the *Minkowski symmetrization* of the convex body $\mathbf{K}$ in $\mathbb{E}^d$ denoted by $\mathbf{K_o}$ is defined by $\mathbf{K_o} := \frac{1}{2}(\mathbf{K}+(-\mathbf{K})) = \frac{1}{2}(\mathbf{K}-\mathbf{K}) = \frac{1}{2}\{\mathbf{x}-\mathbf{y} : \mathbf{x}, \mathbf{y} \in \mathbf{K}\}$. Clearly, $\mathbf{K_o}$ is an $\mathbf{o}$-symmetric d-dimensional convex body. Minkowski [184] showed that if $\mathcal{P} = \{\mathbf{x}_1+\mathbf{K}, \mathbf{x}_2+\mathbf{K}, \dots, \mathbf{x}_n+\mathbf{K}\}$ is a packing of translates of $\mathbf{K}$, then $\mathcal{P}_\mathbf{o} = \{\mathbf{x}_1+\mathbf{K_o}, \mathbf{x}_2+\mathbf{K_o}, \dots, \mathbf{x}_n+\mathbf{K_o}\}$ is a packing as well. Moreover, the contact graphs of $\mathcal{P}$ and $\mathcal{P}_\mathbf{o}$ are the same. Thus, it is sufficient to give an upper bound for the number of touching pairs in the packing $\mathcal{K} = \{\mathbf{c}_1+\mathbf{K_o}, \mathbf{c}_2+\mathbf{K_o}, \dots, \mathbf{c}_n+\mathbf{K_o}\}$.

Notice that if $\mathbf{c}_i+\mathbf{K_o}$ is tangent to $H(\mathbf{K_o})$ members of the packing $\mathcal{K}$, then

$$\mathbf{c}_i+2\mathbf{K_o} \subset \bigcup_{j\neq i, 1\leq j\leq n} (\mathbf{c}_j+2\mathbf{K_o}).$$

Hence, if m denotes the number of members of $\mathcal{K}$ that are touched by exactly $H(\mathbf{K_o})$ members of $\mathcal{K}$, then the $(d-1)$-dimensional surface volume $\mathrm{svol}_{d-1}(\mathrm{bd}(C_n+2\mathbf{K_o}))$ of the boundary $\mathrm{bd}(C_n+2\mathbf{K_o})$ of the (non-convex) set $C_n+2\mathbf{K_o}$ must satisfy the inequality

$$\mathrm{svol}_{d-1}(\mathrm{bd}(C_n+2\mathbf{K_o})) \leq (n-m)2^{d-1}\mathrm{svol}_{d-1}(\mathrm{bd}\mathbf{K_o}). \tag{9.12}$$

On the other hand, the isoperimetric inequality applied to $C_n+2\mathbf{K_o}$ yields

$$\mathrm{iq}(\mathbf{B}^d) \leq \mathrm{iq}(C_n+2\mathbf{K_o}) = \frac{(\mathrm{svol}_{d-1}(\mathrm{bd}(C_n+2\mathbf{K_o})))^d}{(\mathrm{vol}_d(C_n+2\mathbf{K_o}))^{d-1}} \tag{9.13}$$

Finally, Lemma 235 with $\rho = 1$ implies the following inequality in a straightforward way:

$$\frac{\mathrm{vol}_d(C_n+\mathbf{K_o})}{\mathrm{vol}_d(C_n+2\mathbf{K_o})} \leq \delta(\mathbf{K_o}) \tag{9.14}$$

Hence, (9.12), (9.13), and (9.14) yield

$$\frac{1}{2^{d-1}[\delta(\mathbf{K_o})]^{\frac{d-1}{d}}} \left[\frac{\mathrm{iq}(\mathbf{B}^d)}{\mathrm{iq}(\mathbf{K_o})}\right]^{\frac{1}{d}} n^{\frac{d-1}{d}} \leq n-m. \tag{9.15}$$

Finally, notice that the convex hull of C_n, $n>1$ must have at least two vertices in $\mathbb{E}^d$, say, $\mathbf{c}_i$ and $\mathbf{c}_j$. Then, it is obvious that the number of members of $\mathcal{K}$ that are tangent to $\mathbf{c}_i+\mathbf{K_o}$ (resp., $\mathbf{c}_j+\mathbf{K_o}$) is at most $h(\mathbf{K_o})$. From this and (9.15), the first inequality of Theorem 88 follows in a straightforward way.

To prove the second inequality, we note that $\delta(\mathbf{K_o}) \leq 1$, and $(\mathrm{iq}(\mathbf{B}^d))^{1/d} = d\mathrm{vol}_d(\mathbf{B}^d)$. Next, according to Ball's reverse isoperimetric inequality [16], for any **o**-symmetric convex body $\mathbf{K_o}$, there is a non-degenerate affine map $T : \mathbb{E}^d \to \mathbb{E}^d$ with $\mathrm{iq}(T\mathbf{K_o}) \leq (2d)^d$. Finally, notice that $c(\mathbf{K_o}, n, d) = c(T\mathbf{K_o}, n, d)$, and the inequality follows in a straightforward way. This completes the proof of Theorem 88.

9.3 Proof of Theorem 92

First, we show that there exists $\lambda'_{d,n} > 0$ such that for every λ satisfying $0 < \lambda < \lambda'_{d,n}$, $\delta_d(n, \lambda)$ is generated by a packing of n unit balls in $\mathbb{E}^d$ possessing the largest contact number $c(n, d)$ for the given n. Our proof is by contradiction and starts by assuming that the claim is not true. Then there exists a sequence $\lambda_1 > \lambda_2 > \cdots > \lambda_m > \cdots > 0$ of positive reals with $\lim_{m\to+\infty} \lambda_m = 0$ such that the unit ball packing $\mathcal{P}(\lambda_m) := \{\mathbf{c}_i(\lambda_m) + \mathbf{B}^d \mid 1 \leq i \leq n \text{ with } \|\mathbf{c}_j(\lambda_m) - \mathbf{c}_k(\lambda_m)\| \geq 2 \text{ for all } 1 \leq j < k \leq n\}$ that generates $\delta_d(n, \lambda_m)$ has a contact number $c(\mathcal{P}(\lambda_m))$ satisfying

$$c(\mathcal{P}(\lambda_m)) \leq c(n, d) - 1 \tag{9.16}$$

for all $m = 1, 2, \ldots$. Clearly, by assumption

$$\mathrm{vol}_d(\mathbf{P}^n_{\lambda_m}) \geq \mathrm{vol}_d(\mathbf{P}(\lambda_m)) \tag{9.17}$$

must hold for every packing $\mathcal{P}^n = \{\mathbf{c}_i + \mathbf{B}^d \mid 1 \leq i \leq n \text{ with } \|\mathbf{c}_j - \mathbf{c}_k\| \geq 2 \text{ for all } 1 \leq j < k \leq n\}$ of n unit balls in $\mathbb{E}^d$ and for all $m = 1, 2, \ldots$, where

$$\mathbf{P}^n_{\lambda_m} = \bigcup_{i=1}^{n} (\mathbf{c}_i + (1 + \lambda_m)\mathbf{B}^d) \text{ and } \mathbf{P}(\lambda_m) := \bigcup_{i=1}^{n} (\mathbf{c}_i(\lambda_m) + (1 + \lambda_m)\mathbf{B}^d).$$

By choosing convergent subsequences if necessary, one may assume that $\lim_{m\to+\infty} \mathbf{c}_i(\lambda_m) = \mathbf{c}'_i \in \mathbb{E}^d$ for all $1 \leq i \leq n$. Clearly, $\mathcal{P}' := \{\mathbf{c}'_i + \mathbf{B}^d \mid 1 \leq i \leq n\}$ is a packing of n unit balls in $\mathbb{E}^d$. Now, let $\mathcal{P}'' := \{\mathbf{c}''_i + \mathbf{B}^d \mid 1 \leq i \leq n\}$ be a packing of n unit balls in $\mathbb{E}^d$ with maximum contact number $c(n, d)$. Finally, let $2 + 2\lambda'$ be the smallest distance between the centers of non-touching pairs of unit balls in the packings $\mathcal{P}'$ and $\mathcal{P}''$. Thus, if $0 < \lambda_m < \lambda'$ and m is sufficiently large, then the number of overlapping pairs in the ball arrangement $\{\mathbf{c}_i(\lambda_m) + (1 + \lambda_m)\mathbf{B}^d \mid 1 \leq i \leq n\}$ is at most $c(n, d)$. On the other hand, the number of overlapping pairs in the ball arrangement $\{\mathbf{c}''_i + (1+\lambda_m)\mathbf{B}^d \mid 1 \leq i \leq n\}$ is $c(n, d)$. Hence, (9.16) implies in a straightforward way that $\mathrm{vol}_d(\mathbf{P}(\lambda_m)) > \mathrm{vol}_d(\bigcup_{i=1}^n (\mathbf{c}''_i + (1 + \lambda_m)\mathbf{B}^d))$, a contradiction to (9.17). This completes our proof on the existence of $\lambda'_{d,n} > 0$.

Second, we turn to the proof of the existence of the packing $\widehat{\mathcal{P}}^n$ of n unit balls in $\mathbb{E}^d$ with the extremal property stated in Theorem 92. According to the first part of our proof, for every λ satisfying $0 < \lambda < \lambda'_{d,n}$ there exist a packing $\mathcal{P}(\lambda) := \{\mathbf{c}_i(\lambda)+\mathbf{B}^d \mid 1 \le i \le n \text{ with } \|\mathbf{c}_j(\lambda)-\mathbf{c}_k(\lambda)\| \ge 2 \text{ for all } 1 \le j < k \le n\}$ of n unit balls in $\mathbb{E}^d$ with contact number $c(\mathcal{P}(\lambda)) = c(n,d)$ such that

$$\mathrm{vol}_d(\mathbf{P}^n_\lambda) \ge \mathrm{vol}_d(\mathbf{P}(\lambda)) \tag{9.18}$$

holds for every packing $\mathcal{P}^n = \{\mathbf{c}_i+\mathbf{B}^d \mid 1 \le i \le n \text{ with } \|\mathbf{c}_j-\mathbf{c}_k\| \ge 2 \text{ for all } 1 \le j < k \le n\}$ of n unit balls in $\mathbb{E}^d$, where

$$\mathbf{P}^n_\lambda = \bigcup_{i=1}^{n}(\mathbf{c}_i + (1+\lambda)\mathbf{B}^d) \text{ and } \mathbf{P}(\lambda) := \bigcup_{i=1}^{n}(\mathbf{c}_i(\lambda) + (1+\lambda)\mathbf{B}^d).$$

Now, if we assume that $\widehat{\mathcal{P}}^n$ does not exist, then clearly we must have a sequence $\lambda_1 > \lambda_2 > \cdots > \lambda_m > \cdots > 0$ of positive reals with $\lim_{m\to+\infty}\lambda_m = 0$ and with unit ball packings $\mathcal{P}(\lambda_m) := \{\mathbf{c}_i(\lambda_m) + \mathbf{B}^d \mid 1 \le i \le n \text{ with } \|\mathbf{c}_j(\lambda_m) - \mathbf{c}_k(\lambda_m)\| \ge 2 \text{ for all } 1 \le j < k \le n\}$ in $\mathbb{E}^d$ each with maximum contact number $c(\mathcal{P}(\lambda_m)) = c(n,d)$ such that we have (9.18), i.e.,

$$\mathrm{vol}_d(\mathbf{P}^n_{\lambda_m}) \ge \mathrm{vol}_d(\mathbf{P}(\lambda_m)) \tag{9.19}$$

for every packing $\mathcal{P}^n = \{\mathbf{c}_i + \mathbf{B}^d \mid 1 \le i \le n \text{ with } \|\mathbf{c}_j - \mathbf{c}_k\| \ge 2 \text{ for all } 1 \le j < k \le n\}$ of n unit balls in $\mathbb{E}^d$ and for all $m = 1, 2, \dots$. In particular, we must have

$$\mathrm{vol}_d(\mathbf{P}(\lambda_M)) \ge \mathrm{vol}_d(\mathbf{P}(\lambda_m)) \tag{9.20}$$

for all positive integers $1 \le m \le M$. Last but not least, by the non-existence of $\widehat{\mathcal{P}}^n$ we may assume about the sequence of the unit ball packings $\mathcal{P}(\lambda_m), m = 1, 2, \dots$ (resp., of volumes $\mathrm{vol}_d(\mathbf{P}(\lambda_m)), m = 1, 2, \dots$) that for every positive integer N there exist $m'' > m' \ge N$ with

$$\mathrm{vol}_d(\mathbf{P}(\lambda_{m''})) > \mathrm{vol}_d(\mathbf{P}(\lambda_{m'})). \tag{9.21}$$

Finally, let $2 + 2\lambda'_m$ be the smallest distance between the centers of non-touching pairs of unit balls in the packing $\mathcal{P}(\lambda_m)$, $m = 1, 2, \dots$. We claim that there exists a positive integer N' such that

$$0 < \lambda_m < \lambda'_m \text{ for all } m \ge N'. \tag{9.22}$$

Indeed, otherwise there exists a subsequence λ'_{m_i}, $i = 1, 2, \dots$ with $\lambda_{m_i} \ge \lambda'_{m_i} > 0$ for all $i = 1, 2, \dots$ and so, with $\lim_{i\to+\infty}\lambda'_{m_i} = 0$ implying the existence of a packing of n unit balls in $\mathbb{E}^d$ (via taking a convergent subsequence of the unit ball packings $\mathcal{P}(\lambda_{m_i})$, $i = 1, 2, \dots$ in $\mathbb{E}^d$) with contact number at least $c(n,d) + 1$, a contradiction.

Thus, (9.22) and $c(\mathcal{P}(\lambda_m)) = c(d,n)$ imply in a straightforward way that $\mathrm{vol}_d(\mathbf{P}(\lambda_{m''})) = \mathrm{vol}_d(\mathbf{P}(\lambda_{m'}))$ must hold for all $m'' > m' \ge N'$, a contradiction to (9.21). This completes our proof of Theorem 92.

9.4 Proof of Theorem 93

For simplicity, we set $\bar{\lambda} := 1+\lambda$. In the following proof of [43], we apply Blichfeldt's idea to $\mathcal{P}^n$ within the container $\bigcup_{i=1}^n (c_i + \bar{\lambda}\mathbf{B}^d)$ using the presentation of Blichfeldt's method in [105].

For $i = 1, 2, \ldots, n$, let $\mathbf{c}_i = (c_{i1}, c_{i2}, \ldots, c_{in})$. Clearly, if $i \neq j$, we have $\|\mathbf{c}_i - \mathbf{c}_j\|^2 \geq 4$, or equivalently, $\sum_{k=1}^d (c_{ik} - c_{jk})^2 \geq 4$. Summing up for all possible pairs of different indices, we obtain

$$2n(n-1) = 4\binom{n}{2} \leq n\sum_{i=1}^n \left(\sum_{j=1}^d c_{ij}^2\right) - \sum_{j=1}^d \left(\sum_{i=1}^n c_{ij}\right)^2,$$

which yields

$$2(n-1) \leq \sum_{i=1}^n \|\mathbf{c}_i\|^2. \tag{9.23}$$

We need the following definitions and lemma.

Definition 45 *The function*

$$\rho_\lambda(\mathbf{x}) = \begin{cases} 1 - \frac{1}{2}\|\mathbf{x}\|^2, & \text{if } \|\mathbf{x}\| \leq \bar{\lambda} \\ 0, & \text{if } \|\mathbf{x}\| > \bar{\lambda} \end{cases}$$

is called the Blichfeldt gauge function.

Lemma 211 *For any* $\mathbf{y} \in \mathbb{E}^d$, *we have*

$$\sum_{i=1}^n \rho_\lambda(\mathbf{y} - \mathbf{c}_i) \leq 1.$$

Proof. Without loss of generality, let $\mathbf{y}$ be the origin. Then, from (9.23), it follows that

$$\sum_{i=1}^n \rho_\lambda(\mathbf{c}_i) = \sum_{\|\mathbf{c}_i\| \leq \bar{\lambda}} \left(1 - \frac{1}{2}\|\mathbf{c}_i\|^2\right) \leq \sum_{i=1}^n \left(1 - \frac{1}{2}\|\mathbf{c}_i\|^2\right)$$

$$= n - \frac{1}{2}\sum_{i=1}^n \|\mathbf{c}_i\|^2 \leq n - \frac{1}{2}2 \cdot (n-1) = 1. \qquad \square$$

Definition 46 *Let*

$$I(\rho_\lambda) = \int_{\mathbb{E}^d} \rho_\lambda(\mathbf{x}) d\mathbf{x}, \quad \delta = \frac{n\kappa_d}{\mathrm{vol}_d(\bigcup_{i=1}^n (\mathbf{c}_i + \bar{\lambda}\mathbf{B}^d))}, \quad \Delta = \delta\frac{I(\rho_\lambda)}{\kappa_d}.$$

Clearly, Lemma 211 implies that $\Delta \leq 1$, and therefore $\delta \leq \frac{\kappa_d}{I(\rho_\lambda)}$, which yields that $\delta_d(n,\lambda) \leq \frac{\kappa_d}{I(\rho_\lambda)}$.

Now,

$$I(\rho_\lambda) = \int_{\mathbb{E}^d} \rho_\lambda(\mathbf{x}) d\mathbf{x} = \int_{\bar{\lambda}\mathbf{B}^d} \left(1 - \frac{1}{2}\|\mathbf{x}\|^2\right) d\mathbf{x} = \int_0^{\bar{\lambda}} \left(1 - \frac{1}{2}r^2\right) r^{d-1} d\kappa_d dr$$

$$= \kappa_d \left(\bar{\lambda}^d - \frac{d}{2(d+2)}\bar{\lambda}^{d+2}\right).$$

Hence, we have

$$\delta_d(n,\lambda) \leq \frac{1}{\bar{\lambda}^d \left(1 - \frac{d}{2d+4}\bar{\lambda}^2\right)} = \frac{2d+4}{\left(2 - \bar{\lambda}^2\right) d + 4} \bar{\lambda}^{-d},$$

and the assertion follows.

9.5 Proof of Theorem 94

The following proof follows closely the relevant parts of [43]. Let $\mathcal{P}^n = \left\{\mathbf{c}_i + \mathbf{B}^2 : i = 1, 2, \ldots, n\right\}$ be a packing of n unit disks in $\mathbb{E}^2$, and let $1 < \bar{\lambda} = 1 + \lambda < \frac{2}{\sqrt{3}}$.

Definition 47 *The* λ-intersection graph *of* $\mathcal{P}^n$ *is the graph* $G(\mathcal{P}^n)$ *with* $\{\mathbf{c}_i : i = 1, 2, \ldots, n\}$ *as vertices, and with two vertices connected by a line segment if their distance is at most* $2\bar{\lambda}$.

Note that since $1 < \bar{\lambda} < \frac{2}{\sqrt{3}}$, the λ-intersection graph of $\mathcal{P}^n$ is planar, but if $\bar{\lambda} > \frac{2}{\sqrt{3}}$, it is not necessarily so.

Definition 48 *The unbounded face of the* λ*-intersection graph* $G(\mathcal{P}^n)$ *is bounded by finitely many closed sequences of edges of* $G(\mathcal{P}^n)$. *We call the collection of these sequences the* boundary of $G(\mathcal{P}^n)$, *and denote the sum of the lengths of the edges in them by* $\mathrm{perim}(G(\mathcal{P}^n))$.

We remark that an edge of $G(\mathcal{P}^n)$ may appear more than once in the boundary of $G(\mathcal{P}^n)$ (for instance, if the boundary of the unbounded face contains a vertex of degree one). Such an edge contributes its length more than once to $\mathrm{perim}(G(\mathcal{P}^n))$.

We prove the following, stronger statement, which readily implies Theorem 94.

Theorem 212 *Let* $\mathcal{P}^n = \{\mathbf{c}_i + \mathbf{B}^2 : i = 1, 2, \ldots, n\}$ *be a packing of* n *unit disks in* $\mathbb{E}^2$, *and let* $1 < \bar{\lambda} < \frac{2}{\sqrt{3}}$. *Let* $A = \text{area}\left(\bigcup_{i=1}^n (\mathbf{c}_i + \bar{\lambda}\mathbf{B}^2)\right)$ *and* $P = \text{perim}(G(\mathcal{P}^n))$. *Then*

$$A \geq \left(\text{area}\left(\mathbf{H} \cap \bar{\lambda}\mathbf{B}^2\right)\right) n + \left(\bar{\lambda}^2 \arccos \frac{1}{\bar{\lambda}} - \sqrt{\bar{\lambda}^2 - 1}\right) P + \bar{\lambda}^2 \pi. \tag{9.24}$$

Proof. An elementary computation yields

$$\text{area}\left(\mathbf{H} \cap \bar{\lambda}\mathbf{B}^2\right) = \bar{\lambda}^2 \left(\pi - 6 \arccos \frac{1}{\bar{\lambda}}\right) + 6\sqrt{\bar{\lambda}^2 - 1}. \tag{9.25}$$

Let $\mathbf{C}$ denote the union of the bounded faces of the graph $G(\mathcal{P}^n)$. Consider the Voronoi decomposition of $\mathbb{E}^2$ by $\mathcal{P}^n$. Observe that as $\bar{\lambda} < \frac{2}{\sqrt{3}}$, no point of the plane belongs to more than two disks of the family $\{\mathbf{c}_i + \bar{\lambda}\mathbf{B}^2 : i = 1, 2, \ldots, n\}$. Thus, if $E = [\mathbf{c}_i, \mathbf{c}_j]$ is an edge of $G(\mathcal{P}^n)$, the midpoint $\mathbf{m}$ of E is a common point of the Voronoi cells of $\mathbf{c}_i + \mathbf{B}^2$ and $\mathbf{c}_j + \mathbf{B}^2$; more specifically, $\mathbf{m}$ is the point of the common edge of these cells, closest to both $\mathbf{c}_i$ and $\mathbf{c}_j$. Hence, following Rogers's method [208], we may partition $\mathbf{C}$ into triangles of the form $\mathbf{T} = \text{conv}\{\mathbf{c}_i, \mathbf{c}_i', \mathbf{c}_i''\}$, where $\mathbf{c}_i'$ is the point on an edge E of the Voronoi cell of $\mathbf{c}_i + \mathbf{B}^2$, closest to $\mathbf{c}_i$, and $\mathbf{c}_i''$ is an endpoint of E. We call these triangles *interior cells*, define the *center* of any such cell $\mathbf{T} = \text{conv}\{\mathbf{c}_i, \mathbf{c}_i', \mathbf{c}_i''\}$ as $\mathbf{c}_i$, and its *angle* as the angle $\angle(\mathbf{c}_i', \mathbf{c}_i, \mathbf{c}_i'')$. Furthermore, we define the *edge contribution* of an interior cell to be zero.

Now, let $[\mathbf{c}_i, \mathbf{c}_j]$ be an edge in the boundary of $G(\mathcal{P}^n)$, with outer unit normal vector $\mathbf{u}$ and midpoint $\mathbf{m}$. Then the sets $\left([\mathbf{c}_t, \mathbf{m}] + [\mathbf{o}, \bar{\lambda}\mathbf{u}]\right) \cap \left(\mathbf{c}_t + \bar{\lambda}\mathbf{B}^2\right)$, where $t \in \{i, j\}$, are called *boundary cells*, with center $\mathbf{c}_t$ (Figure 9.5). We define their angles as $\frac{\pi}{2}$, and their edge contributions as $\frac{1}{2}\|\mathbf{c}_i - \mathbf{c}_j\|$. Note that, even though no two interior cells overlap, this is not necessarily true for boundary cells: such a cell may have some overlap with interior as well as boundary cells.

The proof of Theorem 212 is based on Lemma 213.

Lemma 213 *Let* $\mathbf{T}$ *be an interior or boundary cell with center* $\mathbf{c}$, *edge contribution* x *and angle* α. *Then*

$$\text{area}\left(\mathbf{T} \cap \left(\mathbf{c} + \bar{\lambda}\mathbf{B}^2\right)\right)$$

$$\geq \frac{\bar{\lambda}^2 \left(\frac{\pi}{6} - \arccos \frac{1}{\bar{\lambda}}\right) + \sqrt{\bar{\lambda}^2 - 1}}{\frac{\pi}{3}} \alpha + \left(\bar{\lambda}^2 \arccos \frac{1}{\bar{\lambda}} - \sqrt{\bar{\lambda}^2 - 1}\right) x. \tag{9.26}$$

First, we show how Lemma 213 yields Theorem 212. Let the (interior and boundary) cells of $\mathcal{P}^n$ be $\mathbf{T}_j$, $j = 1, 2, \ldots, k$, with center $\mathbf{c}_j$, angle α_j and edge contribution x_j. Let $\mathbf{T}_j' = \mathbf{T}_j \cap (\mathbf{c}_j + \bar{\lambda}\mathbf{B}^2)$. Since the sum of the (signed) turning angles at the vertices of a simple polygon is equal to 2π, we have

$$A = \sum_{j=1}^k \text{area}(\mathbf{T}_j') + s\bar{\lambda}^2\pi,$$

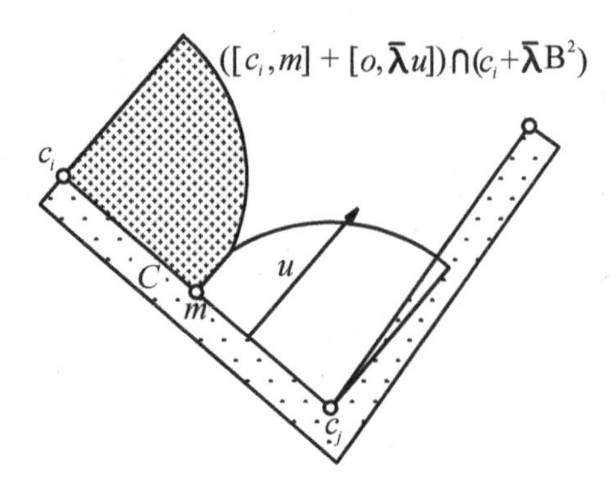

FIGURE 9.5
Boundary cells: the one with centre $\mathbf{c}_i$ is denoted by crosses, and int$\mathbf{C}$ is represented by dots.

where s is the number of components of the boundary of $G(\mathcal{P}^n)$. On the other hand,

$$\sum_{j=1}^{k} \alpha_j = 2\pi n, \quad \text{and} \quad \sum_{j=1}^{k} x_j = P.$$

Thus, summing up both sides in Lemma 213, and using the estimate $s \geq 1$ implies Theorem 212. □

Proof of Lemma 213. For simplicity, let $\mathbf{T}' = \mathbf{T} \cap \left(\mathbf{c} + \bar{\lambda}\mathbf{B}^2\right)$. First, we consider the case that $\mathbf{T}$ is a boundary triangle. Then $\alpha = \frac{\pi}{2}$, and an elementary computation yields that

$$\text{area}\left(\mathbf{T}'\right) = \frac{\bar{\lambda}^2}{2}\left(\frac{\pi}{2} - \arccos\frac{x}{\bar{\lambda}}\right) + \frac{x}{2}\sqrt{\bar{\lambda}^2 - x^2}. \tag{9.27}$$

Combining (9.26) and (9.27), it suffices to show that the function

$$f(x) = -\frac{\bar{\lambda}^2}{2}\arccos\frac{x}{\bar{\lambda}} + \frac{x}{2}\sqrt{\bar{\lambda}^2 - x^2} + \left(\frac{3}{2} - x\right)\left(\bar{\lambda}^2 \arccos\frac{1}{\bar{\lambda}} - \sqrt{\bar{\lambda}^2 - 1}\right)$$

is not negative for any $1 \leq x \leq \bar{\lambda} \leq \frac{2}{\sqrt{3}}$. Note that

$$f''(x) = \frac{-x}{\sqrt{\bar{\lambda}^2 - x^2}},$$

and hence, f is a strictly concave function of x, from which it follows that it is minimal either at $x = 1$ or at $x = \bar{\lambda}$.

Now, we have $f(1) = 0$, and $f(\bar{\lambda}) = \left(\frac{3}{2} - \bar{\lambda}\right)\left(\bar{\lambda}^2 \arccos\frac{1}{\bar{\lambda}} - \sqrt{\bar{\lambda}^2 - 1}\right)$.

Since $\bar{\lambda} \leq \frac{2}{\sqrt{3}} < \frac{3}{2}$, the first factor of $f(\bar{\lambda})$ is positive. On the other hand, comparing the second factor to (9.25), we can see that it is equal to $\frac{1}{6}\text{area}\left(\bar{\lambda}\mathbf{B}^2 \setminus \mathbf{H}\right) > 0$.

Second, let $\mathbf{T}$ be an interior cell triangle, which yields that $x = 0$. Observe that if $\mathbf{T} = \text{conv}\{\mathbf{c}, \mathbf{x}, \mathbf{y}\}$ is not a right triangle, then both $\mathbf{x}$ and $\mathbf{y}$ are vertices of the Voronoi cell of $\mathbf{c}+\mathbf{B}^2$, from which it follows that $\|\mathbf{x}-\mathbf{c}\|, \|\mathbf{y}-\mathbf{c}\| \geq \frac{2}{\sqrt{3}}$. In this case $\mathbf{T}'$ is a circle sector, and $\text{area}(\mathbf{T}') = \bar{\lambda}^2 \frac{\alpha}{2}$, which yields the assertion. Thus, we may assume that $\mathbf{T} = \text{conv}\{\mathbf{x}, \mathbf{y}\}$ has a right angle at $\mathbf{x}$, and that $\|\mathbf{x} - \mathbf{c}\| < \frac{2}{\sqrt{3}}$. Moving $\mathbf{y}$ towards $\mathbf{x}$ increases the ratio $\frac{\text{area}(\mathbf{T}')}{\alpha}$, and hence, we may assume that $\|\mathbf{y} - \mathbf{c}\| = \frac{2}{\sqrt{3}}$. Under these conditions, we have

$$\text{area}(\mathbf{T}') = \frac{\bar{\lambda}^2}{2}\left(\alpha - \arccos\frac{2\cos\alpha}{\sqrt{3}\bar{\lambda}}\right) + \frac{1}{\sqrt{3}}\cos\alpha\sqrt{\bar{\lambda}^2 - \frac{4}{3}\cos^2\alpha},$$

and, combining it with (9.26), it suffices to show that the function

$$g(\alpha) = -\frac{\bar{\lambda}^2}{2}\arccos\frac{2\cos\alpha}{\sqrt{3}\bar{\lambda}}$$

$$+\frac{1}{\sqrt{3}}\cos\alpha\sqrt{\bar{\lambda}^2 - \frac{4}{3}\cos^2\alpha} + \frac{\bar{\lambda}^2}{2}\arccos\frac{1}{\bar{\lambda}}\alpha - \frac{\alpha}{2}\sqrt{r^2 - 1}$$

is not negative if $1 \leq \bar{\lambda} \leq \frac{2}{\sqrt{3}}$ and $\arccos\frac{\sqrt{3}\bar{\lambda}}{2} \leq \alpha \leq \frac{\pi}{6}$. To do this, we may apply a computation similar to the one in case of a boundary triangle. □

9.6 Proof of Theorem 95

First of all, recall that $\bar{\lambda} = 1 + \lambda$, and let

$$\delta := \frac{\pi - 6\psi_0}{\pi - 6\psi_0 + (3\pi - 18\psi_0)\lambda - 18\psi_0\lambda^2 - (\pi + 6\psi_0)\lambda^3} < \sigma_3(\lambda).$$

Consider a unit ball packing $\mathcal{P}^n$ in $\mathbb{E}^3$, and let $\mathbf{V}$ be the Voronoi cell of some ball of $\mathcal{P}^n$, say $\mathbf{B}^3$. Let F be a face of $\mathbf{V}$, and denote the intersection of the conic hull of F with $\mathbf{V}$, $\mathbf{B}^3$ and $\text{bd}\mathbf{B}^3 = \mathbb{S}^2$ by $\mathbf{V}_F$, $\mathbf{B}_F$ and S_F, respectively. Furthermore, we set $\mathbf{V}'_F = \mathbf{V}_F \cap \left(\bar{\lambda}\mathbf{B}^3\right)$. To prove Theorem 95, it is sufficient to show that

$$\frac{\text{vol}_3(\mathbf{B}_F)}{\text{vol}_3(\mathbf{V}'_F)} \leq \delta. \tag{9.28}$$

Recall the well-known fact (cf. [208]) that the distance of any $(d - i)$-dimensional face of $\mathbf{V}$ from $\mathbf{o}$ is at least $\sqrt{\frac{2i}{i+1}}$. Thus, $\bar{\lambda} < \frac{2}{\sqrt{3}}$ yields that the intersection of $\operatorname{aff} F$ with $\bar{\lambda}\mathbf{B}^3$ is either contained in F, or disjoint from it. In the second case $\frac{\operatorname{vol}_3(\mathbf{B}_F)}{\operatorname{vol}_3(\mathbf{V}'_F)} = \frac{1}{\bar{\lambda}^3} < \delta$, and thus, we may assume that $\operatorname{aff} F \cap \left(\bar{\lambda}\mathbf{B}^3\right) \subset F$.

Let the distance of F and $\mathbf{o}$ be x, where $1 \leq x \leq \bar{\lambda} < \frac{2}{\sqrt{3}}$. An elementary computation yields that $\operatorname{vol}_3\left(\left(\bar{\lambda}\mathbf{B}_F\right) \setminus \mathbf{V}_F\right) = \pi\left(\frac{2}{3}\bar{\lambda}^3 - \bar{\lambda}^2 x + \frac{1}{3}x^3\right)$, from which it follows that

$$\frac{\operatorname{vol}_3(\mathbf{V}'_F)}{\operatorname{vol}_3(\mathbf{B}_F)} = \bar{\lambda}^3 - \frac{\pi\left(\frac{2}{3}\bar{\lambda}^3 - \bar{\lambda}^2 x + \frac{1}{3}x^3\right)}{\operatorname{vol}_3(\mathbf{B}_F)}. \tag{9.29}$$

First, we intend to minimize $\operatorname{vol}_3(\mathbf{B}_F)$, while keeping the value of x fixed. Recall the following lemma from [33].

Lemma 214 *Let F_i be an i-dimensional face of the Dirichlet-Voronoi cell of $\mathbf{p} + \mathbf{B}^d$, in a unit ball packing in $\mathbb{E}^d$. Let the distance of $\operatorname{aff} F_i$ from $\mathbf{p}$ be $R < \sqrt{2}$. If F_{i-1} is an $(i-1)$-dimensional face of F_i, then the distance of $\operatorname{aff} F_{i-1}$ from $\mathbf{p}$ is at least $\frac{2}{\sqrt{4-R^2}}$.*

This immediately yields that the distance of $\mathbf{o}$ from any sideline of F is at least $\frac{2}{\sqrt{4-x^2}}$, and from any vertex of F at least $\sqrt{\frac{4-x^2}{3-x^2}}$. By setting $H = \operatorname{aff} F$ and denoting the projection of $\mathbf{o}$ onto H by $\mathbf{c}$, we may rephrase this observation in the following way: F is a polygon in H, containing the circle C_1 with center $\mathbf{c}$ and radius $\sqrt{\frac{4}{4-x^2} - x^2} = \frac{2-x^2}{\sqrt{4-x^2}}$, such that each vertex of H is outside the circle C_1, with center c and radius $\frac{2-x^2}{\sqrt{3-x^2}}$. Observe that we have a similar condition for the projection of F onto the sphere $\mathbb{S}^2$. Thus, to minimize $\operatorname{vol}_3(\mathbf{B}_F)$, or equivalently, $\operatorname{Svol}_2(S_F) = 3\operatorname{vol}_3(\mathbf{B}_F)$, we may apply the following lemma from [**?**].

Lemma 215 (Hajós) *Let $0 < r < R < \frac{\pi}{2}$, and let C_r and C_R be two concentric circles on the sphere $\mathbb{S}^2$, of radii r and R, respectively. Let $\mathcal{P}$ denote the family of convex spherical polygons containing C_r, with no vertex contained in the interior of C_R. If $P \in \mathcal{P}$ has minimal spherical area over all the elements of $\mathcal{P}$, then each vertex of P lies on C_R, and each but at most one edge of P touches C_r.*

Such a polygon is called a *Hajós polygon* of the two circles. By Lemma 215, we may assume that F is a Hajós polygon, and compute $\operatorname{Svol}_2(S_F) = 3\operatorname{vol}_3(\mathbf{B}_F)$ under this condition.

Let $[\mathbf{p}, \mathbf{q}]$ be an edge of H that touches C_1, and let $\mathbf{m}$ be the midpoint of $[\mathbf{p}, \mathbf{q}]$. Let the angles of the triangle $T = \operatorname{conv}\{\mathbf{p}, \mathbf{m}, \mathbf{c}\}$, at $\mathbf{p}, \mathbf{m}$ $\mathbf{c}$, be β, $\gamma = \frac{\pi}{2}$ and α, respectively. Let T' be the central projection of T onto $\mathbb{S}^2$ from o, and denote the angles of T' by α', β', γ', according to the notation in T. We compute $\operatorname{Svol}_2(T') = \alpha' + \beta' + \gamma' - \pi$. First, we observe that, by the properties of the projection, we have $\alpha' = \alpha$, and $\gamma' = \gamma = \frac{\pi}{2}$. Since

$\|\mathbf{p}-\mathbf{c}\| = \frac{2-x^2}{\sqrt{3-x^2}}$ and $\|\mathbf{m}-\mathbf{c}\| = \frac{2-x^2}{\sqrt{4-x^2}}$, an elementary computation yields $\|\mathbf{p}-\mathbf{m}\| = \frac{2-x^2}{\sqrt{(3-x^2)(4-x^2)}}$, and

$$\alpha' = \arctan \frac{1}{\sqrt{3-x^2}}.$$

In the following, we use Lemma 216.

Lemma 216 *Let H denote the tangent plane of the unit sphere $\mathbb{S}^2$ at some point $\mathbf{p} \in \mathbb{S}^2$. Let $T = \operatorname{conv}\{\mathbf{p}_1, \mathbf{p}_2, \mathbf{p}_3\}$ with $\mathbf{p}_1 = \mathbf{p}$. For $i = 1, 2, 3$, let ϕ_i be the angle of T at $\mathbf{p}_i$, and $\mathbf{p}_i'$ be the central projection of $\mathbf{p}_i$ on $\mathbb{S}^2$ from o. Furthermore, let T' be the central projection of T, with $\mathbf{p}_i'$ and ϕ_i' and d_i' being the projections of p_i and ϕ_i', and the spherical length of the side of T' opposite of $\mathbf{p}_i'$, respectively. Then*

$$\tan \phi_2 = \tan \phi_2' \cos d_3', \quad \textit{and} \quad \tan \phi_3 = \tan \phi_3' \cos d_2'.$$

Proof. Let $\mathbf{q}$ be the orthogonal projection of $\mathbf{p}_1$ onto the line containing $\mathbf{p}_2$ and $\mathbf{p}_3$, and let $\mathbf{q}'$ be the central projection of $\mathbf{q}$ onto $\mathbb{S}^2$. Observe that the spherical angle $\mathbf{p}_1'\mathbf{q}'\mathbf{p}_2'\angle$ is a right angle. Thus, from the spherical law of cosines for angles, it follows that

$$1 = \tan(\mathbf{q}'\mathbf{p}_1'\mathbf{p}_2'\angle) \tan \phi_2' \cos d_3'.$$

Now, we have $\mathbf{q}'\mathbf{p}_1'\mathbf{p}_2'\angle = \mathbf{q}\mathbf{p}_1\mathbf{p}_2\angle = \frac{\pi}{2} - \phi_2$, from which the first equality readily follows. The second one can be proven in a similar way. □

From Lemma 216, we readily obtain that $\tan \beta = \tan \beta' \cos \arctan \frac{\|\mathbf{p}-\mathbf{c}\|}{x}$, which yields

$$\beta' = \arctan \frac{\sqrt{4-x^2}}{x}.$$

Thus,

$$\operatorname{Svol}_2(T') = \arctan \frac{1}{\sqrt{3-x^2}} + \arctan \frac{\sqrt{4-x^2}}{x} - \frac{\pi}{2}. \tag{9.30}$$

Now, if $1 \leq x \leq \frac{2}{\sqrt{3}}$, then $\frac{\pi}{6} < \phi_0 \leq \alpha' \leq 0.659058 < \frac{\pi}{4}$. Thus, F has either five or six edges, depending on the values of x. More specifically, if $1 \leq x < \sqrt{\frac{10-2\sqrt{5}}{5}} = 1.051462\ldots$, then F has six, and otherwise five edges. Using this, $\operatorname{vol}_3(\mathbf{B}_F) = \frac{1}{3}\operatorname{Svol}_2(S_F)$ can be computed similarly to $\operatorname{Svol}_2(T')$, which yields that if $1 \leq x \leq \sqrt{\frac{10-2\sqrt{5}}{5}}$, then

$$\operatorname{vol}_3(\mathbf{B}_F) = \frac{10}{3} \arctan \frac{\sqrt{4-x^2}}{x}$$

$$-\frac{2}{3}\operatorname{arccot} \frac{x\sqrt{3-x^2} \tan\left(5 \arctan\left(\frac{1}{\sqrt{3-x^2}}\right)\right)}{\sqrt{4-x^2}} - \frac{2}{3}\pi.$$

Let us denote the expression on the right by $f(x)$. We may observe that if $\sqrt{\frac{10-2\sqrt{5}}{5}} < x \leq \frac{2}{\sqrt{3}}$, then the area of the sixth triangle appears with a negative sign in $f(x)$, which yields, using a geometric observation, that in this case $\mathrm{vol}_3(\mathbf{B}_F) > f(x)$.

Let

$$F(x, \bar{\lambda}) = f(x) - C\pi \left(\frac{2}{3}\bar{\lambda}^3 - \bar{\lambda}^2 x + \frac{1}{3}x^3 \right),$$

where $C = \frac{f(1)}{\pi\left(\frac{2}{3}\bar{\lambda}^3 - \bar{\lambda}^2 + \frac{1}{3}\right)}$. Note that $F(1, \bar{\lambda}) = 0$ for every value of $\bar{\lambda}$. Thus, by (9.28), (9.29) and the inequality $\mathrm{vol}_3(\mathbf{B}_F) \geq f(x)$, it follows that to prove Theorem 95, it is sufficient to show that $F(x, \lambda) \geq 0$ for every $1 \leq \bar{\lambda} < \frac{2}{\sqrt{3}}$ and $1 \leq x \leq \bar{\lambda}$. On the other hand, it is an elementary exercise to check that $\frac{\partial^2 F}{\partial x^2} < 0$ on this region, which yields that $F(x, \bar{\lambda})$ is minimal at $F(1, \bar{\lambda})$ or $F(\bar{\lambda}, \bar{\lambda})$. We may observe that $F(\bar{\lambda}, \bar{\lambda}) = f(\bar{\lambda})$ is greater than four times the value of the expression in (9.30) at $x = \bar{\lambda}$, which is positive. Thus, $F(x, \bar{\lambda})$ is not negative on the examined region, from which Theorem 95 follows.

9.7 Proof of Theorem 98

The following proof is an analogue of the proof of Theorem 1.1 in [34] and as such it is based on the proper modifications of the main (resp., technical) lemmas of [34]. Overall, the method discussed below turns out to be more efficient for totally separable unit ball packings than for unit ball packings in general. The more exact details are as follows.

Let $\mathcal{P} := \{\mathbf{c}_1 + \mathbf{B}^3, \mathbf{c}_2 + \mathbf{B}^3, \ldots, \mathbf{c}_n + \mathbf{B}^3\}$ denote the totally separable packing of n unit balls with centers $\mathbf{c}_1, \mathbf{c}_2, \ldots, \mathbf{c}_n$ in $\mathbb{E}^3$, which has the largest number, namely $c(n, 3)$, of touching pairs among all totally separable packings of n unit balls in $\mathbb{E}^3$. ($\mathcal{P}$ might not be uniquely determined up to congruence in which case $\mathcal{P}$ stands for any of those extremal packings.)

Lemma 217

$$\frac{\frac{4\pi}{3}n}{\mathrm{vol}_3\left(\bigcup_{i=1}^{n}\left(\mathbf{c}_i + \sqrt{3}\mathbf{B}^3\right)\right)} < 0.6401,$$

where $\mathrm{vol}_3(\cdot)$ *refers to the 3-dimensional volume of the corresponding set.*

Proof. First, partition $\bigcup_{i=1}^{n}\left(\mathbf{c}_i + \sqrt{3}\mathbf{B}^3\right)$ into truncated Voronoi cells as follows. Let $\mathbf{P}_i$ denote the Voronoi cell of the packing $\mathcal{P}$ assigned to $\mathbf{c}_i + \mathbf{B}^3$, $1 \leq i \leq n$, that is, let $\mathbf{P}_i$ stand for the set of points of $\mathbb{E}^3$ that are not farther away from $\mathbf{c}_i$ than from any other $\mathbf{c}_j$ with $j \neq i, 1 \leq j \leq n$. Then, recall the well-known fact (see for example, [106]) that the Voronoi cells $\mathbf{P}_i$, $1 \leq i \leq n$ just introduced form a tiling of $\mathbb{E}^3$. Based on this, it is easy to see that the truncated Voronoi cells $\mathbf{P}_i \cap (\mathbf{c}_i + \sqrt{3}\mathbf{B}^3)$, $1 \le i \le n$ generate a tiling of the non-convex container $\bigcup_{i=1}^{n} \left(\mathbf{c}_i + \sqrt{3}\mathbf{B}^3\right)$ for the packing $\mathcal{P}$. Second, we prove the following metric properties of the Voronoi cells introduced above.

Sublemma 218 *The distance between the line of an arbitrary edge of the Voronoi cell* $\mathbf{P}_i$ *and the center* $\mathbf{c}_i$ *is always at least* $\frac{3\sqrt{3}}{4} = 1.299\ldots$ *for any* $1 \le i \le n$.

Proof. It is easy to see that the claim follows from the following 2-dimensional statement: If $\{\mathbf{a}+\mathbf{B}^2, \mathbf{b}+\mathbf{B}^2, \mathbf{c}+\mathbf{B}^2\}$ is a totally separable packing of three unit disks with centers $\mathbf{a}, \mathbf{b}, \mathbf{c}$ in $\mathbb{E}^2$, then the circumradius of the triangle $\mathrm{conv}\{\mathbf{a}, \mathbf{b}, \mathbf{c}\}$ is at least $\frac{3\sqrt{3}}{4}$. We prove the latter statement as follows. If some side of the triangle $\mathrm{conv}\{\mathbf{a}, \mathbf{b}, \mathbf{c}\}$ has length at least $2\sqrt{2}$, then the circumradius of the triangle $\mathrm{conv}\{\mathbf{a}, \mathbf{b}, \mathbf{c}\}$ is at least $\sqrt{2} > \frac{3\sqrt{3}}{4} = 1.299\ldots$. So, without loss of generality we may assume that $2 < \|\mathbf{a}-\mathbf{b}\| < 2\sqrt{2}$, $2 < \|\mathbf{a}-\mathbf{c}\| < 2\sqrt{2}$, and $2 < \|\mathbf{b}-\mathbf{c}\| < 2\sqrt{2}$ and so $\mathrm{conv}\{\mathbf{a}, \mathbf{b}, \mathbf{c}\}$ is an acute triangle. Moreover, as the three unit disks with centers $\mathbf{a}, \mathbf{b}, \mathbf{c}$ form a totally separable packing, therefore, there must exist two unit disks say, $\mathbf{a}+\mathbf{B}^2$ and $\mathbf{b}+\mathbf{B}^2$ such that their two inner tangent lines are disjoint from the interior of the third unit disk $\mathbf{c}+\mathbf{B}^2$ separating the unit disks $\mathbf{a}+\mathbf{B}^2$, $\mathbf{c}+\mathbf{B}^2$ (resp., $\mathbf{b}+\mathbf{B}^2$, $\mathbf{c}+\mathbf{B}^2$) from $\mathbf{b}+\mathbf{B}^2$ (resp., $\mathbf{a}+\mathbf{B}^2$). Finally, if necessary then by properly translating $\mathbf{c}+\mathbf{B}^2$ and thereby decreasing the circumradius of the triangle $\mathrm{conv}\{\mathbf{a}, \mathbf{b}, \mathbf{c}\}$ one can assume that the two inner tangent lines of the unit disks $\mathbf{a}+\mathbf{B}^2$ and $\mathbf{b}+\mathbf{B}^2$ are tangent to the unit disk $\mathbf{c}+\mathbf{B}^2$ with $2 < \|\mathbf{a}-\mathbf{b}\| < 2\sqrt{2}$ and $2 < \|\mathbf{a}-\mathbf{c}\| = \|\mathbf{b}-\mathbf{c}\| < 2\sqrt{2}$. Now, if $x = \frac{1}{2}\|\mathbf{a}-\mathbf{b}\|$, then an elementary computation yields that the circumradius of the triangle $\mathrm{conv}\{\mathbf{a}, \mathbf{b}, \mathbf{c}\}$ is equal to $f(x) = \frac{x^3}{2\sqrt{x^2-1}}$ with $1 < x < \sqrt{2}$. Hence, $f'(x) = \frac{x^2(2x^2-3)}{2(x^2-1)\sqrt{x^2-1}}$ implies in a straightforward way that $f\left(\sqrt{\frac{3}{2}}\right) = \frac{3\sqrt{3}}{4}$ is a global minimum of $f(x)$ over $1 < x < \sqrt{2}$. This finishes the proof of Sublemma 218. □

Remark 219 *As one can see from the above proof, the lower bound of Sublemma 218 is a sharp one and it should be compared to the lower bound* $\frac{2}{\sqrt{3}} = 1.154\ldots$ *valid for any unit ball packing not necessarily totally separable in* $\mathbb{E}^3$. *(For more details on the lower bound* $\frac{2}{\sqrt{3}}$ *see for example the discussion on page 29 in [35].)*

Sublemma 220 *The distance between an arbitrary vertex of the Voronoi cell* $\mathbf{P}_i$ *and the center* $\mathbf{c}_i$ *is always at least* $\sqrt{2} = 1.414\ldots$ *for any* $1 \le i \le n$.

Proof. Clearly, the claim follows from the following statement: If $\mathcal{P}_4 = \{\mathbf{c}_1 + \mathbf{B}^3, \mathbf{c}_2 + \mathbf{B}^3, \mathbf{c}_3 + \mathbf{B}^3, \mathbf{c}_4 + \mathbf{B}^3\}$ is a totally separable packing of four unit balls with centers $\mathbf{c}_1, \mathbf{c}_2, \mathbf{c}_3, \mathbf{c}_4$ in $\mathbb{E}^3$, then the circumradius of the tetrahedron $\mathrm{conv}\{\mathbf{c}_1, \mathbf{c}_2, \mathbf{c}_3, \mathbf{c}_4\}$ is at least $\sqrt{2}$. We prove the latter claim by looking at the following two possible cases. $\mathcal{P}_4$ is a totally separable packing with plane H

separating either $\mathbf{c}_1+\mathbf{B}^3, \mathbf{c}_2+\mathbf{B}^3$ from $\mathbf{c}_3+\mathbf{B}^3, \mathbf{c}_4+\mathbf{B}^3$ (*Case 1*) or $\mathbf{c}_1+\mathbf{B}^3$ from $\mathbf{c}_2+\mathbf{B}^3, \mathbf{c}_3+\mathbf{B}^3, \mathbf{c}_4+\mathbf{B}^3$ (*Case 2*). In both cases, it is sufficient to show that if $\bigcup_{i=1}^4(\mathbf{c}_i+\mathbf{B}^3)\subset \mathbf{x}+r\mathbf{B}^3$ for some $\mathbf{x}\in\mathbb{E}^3$ and $r\in\mathbb{R}$, then $r\geq 1+\sqrt{2}$.
Case 1: Let $\mathbf{H}^+$ and $\mathbf{H}^-$ denote the two closed half spaces bounded by H with $\mathbf{c}_1+\mathbf{B}^3\cup\mathbf{c}_2+\mathbf{B}^3\subset H^+$ and $\mathbf{c}_3+\mathbf{B}^3\cup\mathbf{c}_4+\mathbf{B}^3\subset\mathbf{H}^-$. Without loss of generality, we may assume that $\mathbf{x}\in\mathbf{H}^-$. Now, if $\mathbf{c}_1'$ (resp., $\mathbf{c}_2'$) denotes the image of $\mathbf{c}_1$ (resp., $\mathbf{c}_2$) under the reflection in H, then clearly $\mathcal{P}'=\{\mathbf{c}_1+\mathbf{B}^3, \mathbf{c}_2+\mathbf{B}^3, \mathbf{c}_1'+\mathbf{B}^3, \mathbf{c}_2'+\mathbf{B}^3\}$ is a packing of four unit balls in $\mathbf{x}+r\mathbf{B}^3$ symmetric about H. Then using the symmetry of $\mathcal{P}'$ with respect to H it is easy to see that $r\geq 1+\sqrt{2}$.
Case 2: Let $\mathbf{H}^+$ and $\mathbf{H}^-$ denote the two closed half spaces bounded by H with $\mathbf{c}_1+\mathbf{B}^3\subset\mathbb{H}^+$ and $\mathbf{c}_2+\mathbf{B}^3\cup\mathbf{c}_3+\mathbf{B}^3\cup\mathbf{c}_4+\mathbf{B}^3\subset\mathbf{H}^-$. If one assumes that $r-1<\sqrt{2}$, then using $\mathbf{c}_1\in(\mathbf{x}+(r-1)\mathbf{B}^3)\cap\mathbb{H}^+$ and $\{\mathbf{c}_2,\mathbf{c}_3,\mathbf{c}_4\}\subset(\mathbf{x}+(r-1)\mathbf{B}^3)\cap\mathbf{H}^-$ it is easy to see that the triangle $\mathrm{conv}\{\mathbf{c}_2,\mathbf{c}_3,\mathbf{c}_4\}$ is contained in a disk of radius less than $2\sqrt{\sqrt{2}-1}=1.287\ldots$. On the other hand, as the unit balls $\mathbf{c}_2+\mathbf{B}^3, \mathbf{c}_3+\mathbf{B}^3, \mathbf{c}_4+\mathbf{B}^3$ form a totally separable packing, therefore, the proof of Sublemma 218 implies that the radius of any disk containing the triangle $\mathrm{conv}\{\mathbf{c}_2,\mathbf{c}_3,\mathbf{c}_4\}$ must be at least $\frac{3\sqrt{3}}{4}=1.299\ldots$, a contradiction. □

Remark 221 *As one can see from the above proof, the lower bound of Sublemma 220 is a sharp one and it should be compared to the lower bound* $\sqrt{\frac{3}{2}}=1.224\ldots$ *valid for any unit ball packing not necessarily totally separable in* $\mathbb{E}^3$. *(For more details on the lower bound* $\sqrt{\frac{3}{2}}$ *see for example the discussion on page 29 in [35].)*

Now, we are ready to prove Sublemma 223. As the method used is well-known we give only an outline of the major steps of its proof, which originates from Rogers ([208]). In fact, what we need here is a truncated version of Rogers's method that has been introduced by Böröczky [56] (also for spherical and hyperbolic spaces). We recommend the interested reader to look up the relevant details in [56]. First we need to recall the notion of an orthoscheme.

Definition 49 *The i-dimensional simplex* $Y=\mathrm{conv}\{\mathbf{o},\mathbf{y}_1,\ldots,\mathbf{y}_i\}\subset\mathbb{E}^d$ *with vertices* $\mathbf{y}_0=\mathbf{o},\mathbf{y}_1,\ldots,\mathbf{y}_i$ *is called an i-dimensional orthoscheme if for each* $j, 0\leq j\leq i-1$ *the vector* $\mathbf{y}_j$ *is orthogonal to the linear hull* $\mathrm{lin}\{\mathbf{y}_k-\mathbf{y}_j \mid j+1\leq k\leq i\}$, *where* $1\leq i\leq d$.

Next we dissect each truncated Voronoi cell $\mathbf{P}_i\cap(\mathbf{c}_i+\sqrt{2}\mathbf{B}^3)$, $1\leq i\leq n$ into 3-dimensional, 2-dimensional, and 1-dimensional orthoschemes having pairwise disjoint relative interiors as follows. Namely, for each $\mathbf{x}\in\mathbf{P}_i\cap\mathrm{bd}(\mathbf{c}_i+\sqrt{2}\mathbf{B}^3)$ we assign an orthoscheme in the following well-defined way. (We note that due to Sublemma 220 the intersection $\mathbf{P}_i\cap\mathrm{bd}(\mathbf{c}_i+\sqrt{2}\mathbf{B}^3)$ is always non-empty.) We distinguish three cases. If $\mathbf{x}\in\mathrm{int}\mathbf{P}_i$, then the assigned orthoscheme is the line segment $\mathrm{conv}\{\mathbf{c}_i,\mathbf{x}\}$. If $\mathbf{x}$ is a relative interior point of some face F of $\mathbf{P}_i$,

then we assign to $\mathbf{x}$ the orthoscheme $\text{conv}\{\mathbf{c}_i, \mathbf{f}, \mathbf{x}\}$, where $\mathbf{f}$ is the orthogonal projection of $\mathbf{c}_i$ onto the plane of F. (We note that $\mathbf{f}$ lies in F). If $\mathbf{x}$ is a (relative interior) point of some edge E of $\mathbf{P}_i$ with E lying on the face F of $\mathbf{P}_i$, then we assign to $\mathbf{x}$ the orthoscheme $\text{conv}\{\mathbf{c}_i, \mathbf{f}, \mathbf{e}, \mathbf{x}\}$, where $\mathbf{e}$ (resp., $\mathbf{f}$) is the orthogonal projection of $\mathbf{c}_i$ onto the line of E (resp., onto the plane of F). We note that $\mathbf{e}$ (resp., $\mathbf{f}$) belongs to E (resp., F). This completes the process of dissecting $\mathbf{P}_i \cap (\mathbf{c}_i + \sqrt{2}\mathbf{B}^3)$ into orthoschemes.
As a next step we need to recall the so-called Lemma of Comparison of Rogers (for more details see, for example, page 33 in [35]).

Proposition 222 *Let* $\mathbf{W} := \text{conv}\{\mathbf{o}, \mathbf{w}_1, \ldots, \mathbf{w}_d\}$ *be a d-dimensional orthoscheme in* $\mathbb{E}^d$. *Moreover, let* $\mathbf{U} := \text{conv}\{\mathbf{o}, \mathbf{u}_1, \ldots, \mathbf{u}_d\}$ *be a d-dimensional simplex of* $\mathbb{E}^d$ *such that* $\|\mathbf{u}_i\| = \text{dist}(\mathbf{o}, \text{conv}\{\mathbf{u}_i, \mathbf{u}_{i+1}, \ldots, \mathbf{u}_d\})$ *for all* $1 \leq i \leq d$, *where* $\text{dist}(\cdot, \cdot)$ *refers to the usual Euclidean distance between two given sets. If* $1 \leq \|\mathbf{w}_i\| \leq \|\mathbf{u}_i\|$ *holds for all* $1 \leq i \leq d$, *then*

$$\frac{\text{vol}_d(\mathbf{U} \cap \mathbf{B}^d)}{\text{vol}_d(\mathbf{U})} \leq \frac{\text{vol}_d(\mathbf{W} \cap \mathbf{B}^d)}{\text{vol}_d(\mathbf{W})}.$$

Finally, let $\mathbf{W}^3 := \text{conv}\{\mathbf{o}, \mathbf{w}_1, \mathbf{w}_2, \mathbf{w}_3\}$ be the 3-dimensional orthoscheme with $\|\mathbf{w}_1\| = 1$, $\|\mathbf{w}_2\| = \frac{3\sqrt{3}}{4}$, and $\|\mathbf{w}_3\| = \sqrt{2}$. Clearly, Sublemmas 218, 220, and Proposition 222 imply that for any 3-dimensional orthoscheme $\mathbf{U}^3 := \text{conv}\{\mathbf{c}_i, \mathbf{f}, \mathbf{e}, \mathbf{x}\}$ of the above dissection of the truncated Voronoi cell $\mathbf{P}_i \cap (\mathbf{c}_i + \sqrt{2}\mathbf{B}^3)$ we have that

$$\frac{\text{vol}_3(\mathbf{U}^3 \cap \mathbf{B}^3)}{\text{vol}_d(\mathbf{U}^3)} \leq \frac{\text{vol}_3(\mathbf{W}^3 \cap \mathbf{B}^3)}{\text{vol}_d(\mathbf{W}^3)}.$$

As each 2-dimensional (resp., 1-dimensional) orthoscheme of the above dissection of $\mathbf{P}_i \cap (\mathbf{c}_i + \sqrt{2}\mathbf{B}^3)$ can be obtained as a limit of proper 3-dimensional orthoschemes, therefore, one can use the method of limiting density exactly the same way as it is described in [56] to obtain the following conclusion.

Sublemma 223

$$\frac{\text{vol}_3\left((\mathbf{P}_i \cap (\mathbf{c}_i + \sqrt{2}\mathbf{B}^3)) \cap (\mathbf{c}_i + \mathbf{B}^3)\right)}{\text{vol}_3(\mathbf{P}_i \cap (\mathbf{c}_i + \sqrt{2}\mathbf{B}^3))} = \frac{\frac{4\pi}{3}}{\text{vol}_3(\mathbf{P}_i \cap (\mathbf{c}_i + \sqrt{2}\mathbf{B}^3))}$$

$$\leq \frac{\text{vol}_3(\mathbf{W}^3 \cap \mathbf{B}^3)}{\text{vol}_3(\mathbf{W}^3)} < 0.6401.$$

Finally, as $\mathbf{P}_i \cap (\mathbf{c}_i + \sqrt{2}\mathbf{B}^3) \subset \mathbf{P}_i \cap (\mathbf{c}_i + \sqrt{3}\mathbf{B}^3)$, therefore, Sublemma 223 completes the proof of Lemma 217. □

As in Lemma 208, the well-known isoperimetric inequality ([194]) applied to $\bigcup_{i=1}^{n} (\mathbf{c}_i + \sqrt{3}\mathbf{B}^3)$ yields

Lemma 224

$$36\pi \ \mathrm{vol}_3\left(\bigcup_{i=1}^{n}\left(\mathbf{c}_i+\sqrt{3}\mathbf{B}^3\right)\right)^2 \leq \mathrm{svol}_2\left(\mathrm{bd}\left(\bigcup_{i=1}^{n}\left(\mathbf{c}_i+\sqrt{3}\mathbf{B}^3\right)\right)\right)^3,$$

where $\mathrm{svol}_2(\cdot)$ *refers to the surface area of the corresponding set.*

Thus, Lemma 217 and Lemma 224 generate the following inequality.

Corollary 225

$$\frac{4\pi}{(0.6401)^{\frac{2}{3}}}n^{\frac{2}{3}} < \mathrm{svol}_2\left(\mathrm{bd}\left(\bigcup_{i=1}^{n}\left(\mathbf{c}_i+\sqrt{3}\mathbf{B}^3\right)\right)\right).$$

Now, assume that $\mathbf{c}_i+\mathbf{B}^3 \in \mathcal{P}$ is tangent to $\mathbf{c}_j+\mathbf{B}^3 \in \mathcal{P}$ for all $j \in T_i$, where $T_i \subset \{1, 2, \dots, n\}$ stands for the family of indices $1 \leq j \leq n$ for which $\mathrm{dist}(\mathbf{c}_i, \mathbf{c}_j) = 2$. Then let $\hat{S}_i := \mathrm{bd}(\mathbf{c}_i + \sqrt{3}\mathbf{B})$ and let $\hat{\mathbf{c}}_{ij}$ be the intersection of the line segment $[\mathbf{c}_i, \mathbf{c}_j]$ with $\hat{S}_i$ for all $j \in T_i$. Moreover, let $C_{\hat{S}_i}(\hat{\mathbf{c}}_{ij}, \frac{\pi}{4})$ (resp., $C_{\hat{S}_i}(\hat{\mathbf{c}}_{ij}, \alpha)$) denote the open spherical cap of $\hat{S}_i$ centered at $\hat{\mathbf{c}}_{ij} \in \hat{S}_i$ having angular radius $\frac{\pi}{4}$ (resp., α with $0 < \alpha < \frac{\pi}{2}$ and $\cos\alpha = \frac{1}{\sqrt{3}}$). As $\mathcal{P}$ is totally separable, therefore, the family $\{C_{\hat{S}_i}(\hat{\mathbf{c}}_{ij}, \frac{\pi}{4}), j \in T_i\}$ consists of pairwise disjoint open spherical caps of $\hat{S}_i$; moreover,

$$\frac{\sum_{j\in T_i}\mathrm{svol}_2\left(C_{\hat{S}_i}(\hat{\mathbf{c}}_{ij}, \frac{\pi}{4})\right)}{\mathrm{svol}_2\left(\cup_{j\in T_i}C_{\hat{S}_i}(\hat{\mathbf{c}}_{ij}, \alpha)\right)} = \frac{\sum_{j\in T_i}\mathrm{Sarea}\left(C(\mathbf{u}_{ij}, \frac{\pi}{4})\right)}{\mathrm{Sarea}\left(\cup_{j\in T_i}C(\mathbf{u}_{ij}, \alpha)\right)}, \tag{9.31}$$

where $\mathbf{u}_{ij} := \frac{1}{2}(\mathbf{c}_j - \mathbf{c}_i) \in \mathbb{S}^2 := \mathrm{bd}(\mathbf{B}^3)$ and $C(\mathbf{u}_{ij}, \frac{\pi}{4}) \subset \mathbb{S}^2$ (resp., $C(\mathbf{u}_{ij}, \alpha) \subset \mathbb{S}^2$) denotes the open spherical cap of $\mathbb{S}^2$ centered at $\mathbf{u}_{ij}$ having angular radius $\frac{\pi}{4}$ (resp., α) and where $\mathrm{Sarea}(\cdot)$ refers to the spherical area measure on $\mathbb{S}^2$.

Lemma 226

$$\frac{\sum_{j\in T_i}\mathrm{Sarea}\left(C(\mathbf{u}_{ij}, \frac{\pi}{4})\right)}{\mathrm{Sarea}\left(\cup_{j\in T_i}C(\mathbf{u}_{ij}, \alpha)\right)} \leq 3\left(1 - \frac{1}{\sqrt{2}}\right) = 0.8786\dots .$$

Proof. By assumption $\mathcal{P}_i(\mathbb{S}^2) = \{C(\mathbf{u}_{ij}, \frac{\pi}{4}) \mid j \in T_i\}$ is a packing of spherical caps of angular radius $\frac{\pi}{4}$ in $\mathbb{S}^2$. Let $V_{ij}(\mathbb{S}^2)$ denote the Voronoi region of the packing $\mathcal{P}_i(\mathbb{S}^2)$ assigned to $C(\mathbf{u}_{ij}, \frac{\pi}{4})$, that is, let $V_{ij}(\mathbb{S}^2)$ stand for the set of points of $\mathbb{S}^2$ that are not farther away from $\mathbf{u}_{ij}$ than from any other $\mathbf{u}_{ik}$ with $k \neq j, k \in T_i$. Recall (see for example [106]) that the Voronoi regions $V_{ij}(\mathbb{S}^2)$, $j \in T_i$ are spherically convex polygons and form a tiling of $\mathbb{S}^2$. Moreover, it is easy to see that no vertex of $V_{ij}(\mathbb{S}^2)$ belongs to the interior of $C(\mathbf{u}_{ij}, \alpha)$ in $\mathbb{S}^2$. Thus, the Hajós Lemma (Hilfssatz 1 in [185], cf. also Lemma 215) implies that $\mathrm{Sarea}\left(V_{ij}(\mathbb{S}^2) \cap C(\mathbf{u}_{ij}, \alpha)\right) \geq \frac{2\pi}{3}$ because $\frac{2\pi}{3}$ is the spherical area of a regular spherical quadrilateral inscribed into $C(\mathbf{u}_{ij}, \alpha)$ with sides tangent to $C(\mathbf{u}_{ij}, \frac{\pi}{4})$. Hence,

$$\frac{\mathrm{Sarea}\left(C(\mathbf{u}_{ij}, \frac{\pi}{4})\right)}{\mathrm{Sarea}\left(V_{ij}(\mathbb{S}^2) \cap C(\mathbf{u}_{ij}, \alpha)\right)} \leq 3\left(1 - \frac{1}{\sqrt{2}}\right). \tag{9.32}$$

As the truncated Voronoi regions $V_{ij}(\mathbb{S}^2) \cap C(\mathbf{u}_{ij}, \alpha)$, $j \in T_i$ form a tiling of $\bigcup_{j \in T_i} C(\mathbf{u}_{ij}, \alpha)$, therefore, (9.32) finishes the proof of Lemma 226. □

Lemma 226 implies in a straightforward way that

$$\begin{aligned} \mathrm{svol}_2\left(\mathrm{bd}\left(\bigcup_{i=1}^{n}\left(\mathbf{c}_i + \sqrt{3}\mathbf{B}^3\right)\right)\right) &\leq 12\pi n - \frac{1}{3\left(1 - \frac{1}{\sqrt{2}}\right)} 12\pi\left(1 - \frac{1}{\sqrt{2}}\right) c(n,3) \\ &= 12\pi n - 4\pi c(n,3). \end{aligned} \tag{9.33}$$

Hence, Corollary 225 and (9.33) yield

$$\frac{4\pi}{(0.6401)^{\frac{2}{3}}} n^{\frac{2}{3}} < 12\pi n - 4\pi c(n,3),$$

from which it follows that $c(n,3) < 3n - \frac{1}{(0.6401)^{\frac{2}{3}}} n^{\frac{2}{3}} < 3n - 1.346 n^{\frac{2}{3}}$, finishing the proof of Theorem 98.

9.8 Proof of Theorem 99

The following short proof of Theorem 99 is taken from [51]. A union of finitely many axis parallel d-dimensional orthogonal boxes having pairwise disjoint interiors in $\mathbb{E}^d$ is called a *box-polytope.* One may call the following statement the isoperimetric inequality for box-polytopes, which together with its proof presented below is an analogue of the isoperimetric inequality for convex bodies derived from the Brunn–Minkowski inequality. (For more details on the latter, see for example [17].)

Lemma 227 *Among box-polytopes of given volume, the cubes have the least surface volume.*

Proof. Without loss of generality, we may assume that the volume $\mathrm{vol}_d(\mathbf{A})$ of the given box-polytope $\mathbf{A}$ in $\mathbb{E}^d$ is equal to 2^d, i.e., $\mathrm{vol}_d(\mathbf{A}) = 2^d$. Let $\mathbf{C}^d$ be an axis parallel d-dimensional cube of $\mathbb{E}^d$ with $\mathrm{vol}_d(\mathbf{C}^d) = 2^d$. Let the surface volume of $\mathbf{C}^d$ be denoted by $\mathrm{svol}_{d-1}(\mathrm{bd}\mathbf{C}^d)$. Clearly, $\mathrm{svol}_{d-1}(\mathrm{bd}\mathbf{C}^d) = d \cdot \mathrm{vol}_d(\mathbf{C}^d)$. On the other hand, if $\mathrm{svol}_{d-1}(\mathrm{bd}\mathbf{A})$ denotes the surface volume of the box-polytope $\mathbf{A}$, then it is rather straightforward to show that

$$\mathrm{svol}_{d-1}(\mathbf{A}) = \lim_{\varepsilon \to 0^+} \frac{\mathrm{vol}_d(\mathbf{A} + \varepsilon \mathbf{C}^d) - \mathrm{vol}_d(\mathbf{A})}{\varepsilon},$$

where "+" in the numerator stands for the Minkowski addition of the given sets. Using the Brunn–Minkowski inequality ([17]) we get that

$$\mathrm{vol}_d(\mathbf{A}+\varepsilon\mathbf{C}^d) \geq \left(\mathrm{vol}_d(\mathbf{A})^{\frac{1}{d}} + \mathrm{vol}_d(\varepsilon\mathbf{C}^d)^{\frac{1}{d}}\right)^d = \left(\mathrm{vol}_d(\mathbf{A})^{\frac{1}{d}} + \varepsilon \cdot \mathrm{vol}_d(\mathbf{C}^d)^{\frac{1}{d}}\right)^d.$$

Hence,

$$\begin{aligned} \mathrm{vol}_d(\mathbf{A} + \varepsilon\mathbf{C}^d) &\geq \mathrm{vol}_d(\mathbf{A}) + d \cdot \mathrm{vol}_d(\mathbf{A})^{\frac{d-1}{d}} \cdot \varepsilon \cdot \mathrm{vol}_d(\mathbf{C}^d)^{\frac{1}{d}} \\ &= \mathrm{vol}_d(\mathbf{A}) + \varepsilon \cdot d \cdot \mathrm{vol}_d(\mathbf{C}^d) \\ &= \mathrm{vol}_d(\mathbf{A}) + \varepsilon \cdot \mathrm{svol}_{d-1}(\mathrm{bd}\mathbf{C}^d) . \end{aligned}$$

So,

$$\frac{\mathrm{vol}_d(\mathbf{A} + \varepsilon\mathbf{C}^d) - \mathrm{vol}_d(\mathbf{A})}{\varepsilon} \geq \mathrm{svol}_{d-1}(\mathrm{bd}\mathbf{C}^d)$$

and therefore, $\mathrm{svol}_{d-1}(\mathrm{bd}\mathbf{A}) \geq \mathrm{svol}_{d-1}(\mathrm{bd}\mathbf{C}^d)$, finishing the proof of Lemma 227. □

Corollary 228 *For any box-polytope $\mathbf{P}$ of $\mathbb{E}^d$ the isoperimetric quotient of $\mathbf{P}$ is at least as large as the isoperimetric quotient of a cube, i.e.,*

$$\frac{\mathrm{svol}_{d-1}(\mathrm{bd}\mathbf{P})^d}{\mathrm{vol}_d(\mathbf{P})^{d-1}} \geq (2d)^d .$$

Now, let $\overline{\mathcal{P}} := \{\mathbf{c}_1 + \overline{\mathbf{B}}^d, \mathbf{c}_2 + \overline{\mathbf{B}}^d, \dots, \mathbf{c}_n + \overline{\mathbf{B}}^d\}$ denote the totally separable packing of n unit diameter balls with centers $\{\mathbf{c}_1, \mathbf{c}_2, \dots, \mathbf{c}_n\} \subset \mathbb{Z}^d$ having contact number $c_{\mathbb{Z}}(n, d)$ in $\mathbb{E}^d$. ($\overline{\mathcal{P}}$ might not be uniquely determined up to congruence in which case $\overline{\mathcal{P}}$ stands for any of those extremal packings.) Let $\mathbf{U}^d$ be the axis parallel d-dimensional unit cube centered at the origin $\mathbf{o}$ in $\mathbb{E}^d$. Then the unit cubes $\{\mathbf{c}_1 + \mathbf{U}^d, \mathbf{c}_2 + \mathbf{U}^d, \dots, \mathbf{c}_n + \mathbf{U}^d\}$ have pairwise disjoint interiors and $\mathbf{P} = \cup_{i=1}^n (\mathbf{c}_i + \mathbf{U}^d)$ is a box-polytope. Clearly, $\mathrm{svol}_{d-1}(\mathrm{bd}\mathbf{P}) = 2dn - 2c_{\mathbb{Z}}(n, d)$. Hence, Corollary 228 implies that

$$2dn - 2c_{\mathbb{Z}}(n, d) = \mathrm{svol}_{d-1}(\mathrm{bd}\mathbf{P}) \geq 2d\mathrm{vol}_d(\mathbf{P})^{\frac{d-1}{d}} = 2dn^{\frac{d-1}{d}} .$$

So, $dn - dn^{\frac{d-1}{d}} \geq c_{\mathbb{Z}}(n, d)$, finishing the proof of Theorem 99.

9.9 Proof of Theorem 100

Definition 50 *Let $\mathbf{B}^d = \{\mathbf{x} \in \mathbb{E}^d \mid \|\mathbf{x}\| \leq 1\}$ be the closed unit ball centered at the origin $\mathbf{o}$ in $\mathbb{E}^d$, where $\|\cdot\|$ refers to the standard Euclidean norm of $\mathbb{E}^d$.*

Let $R \geq 1$. We say that the packing

$$\mathcal{P}_{\text{sep}} = \{\mathbf{c}_i + \mathbf{B}^d \mid i \in I \text{ with } \|\mathbf{c}_j - \mathbf{c}_k\| \geq 2 \text{ for all } j \neq k \in I\}$$

*of (finitely or infinitely many) non-overlapping translates of $\mathbf{B}^d$ with centers $\{\mathbf{c}_i \mid i \in I\}$ is an R-*separable packing *in $\mathbb{E}^d$ if for each $i \in I$ the finite packing $\{\mathbf{c}_j + \mathbf{B}^d \mid \mathbf{c}_j + \mathbf{B}^d \subseteq \mathbf{c}_i + R\mathbf{B}^d\}$ is a totally separable packing (in $\mathbf{c}_i + R\mathbf{B}^d$). Finally, let $\delta_{\text{sep}}(R, d)$ denote the largest density of all R-separable unit ball packings in $\mathbb{E}^d$, i.e., let*

$$\delta_{\text{sep}}(R, d) = \sup_{\mathcal{P}_{\text{sep}}} \left(\limsup_{\lambda \to +\infty} \frac{\sum_{\mathbf{c}_i + \mathbf{B}^d \subset \mathbf{Q}_\lambda} \text{vol}_d(\mathbf{c}_i + \mathbf{B}^d)}{\text{vol}_d(\mathbf{Q}_\lambda)} \right) ,$$

where $\mathbf{Q}_\lambda$ denotes the d-dimensional cube of edge length 2λ centered at $\mathbf{o}$ in $\mathbb{E}^d$ having edges parallel to the coordinate axes of $\mathbb{E}^d$.

Remark 229 *For any $1 \leq R < 3$ we have that $\delta_{\text{sep}}(R, d) = \delta_d$, where δ_d stands for the supremum of the upper densities of all unit ball packings in $\mathbb{E}^d$.*

The following statement is a special case of Lemma 235 and it is the core part of the proof of Theorem 100 published in [51] and discussed below.

Lemma 230 *If $\{\mathbf{c}_i + \mathbf{B}^d \mid 1 \leq i \leq n\}$ is an R-separable packing of n unit balls in $\mathbb{E}^d$ with $R \geq 1$, $n \geq 1$, and $d \geq 2$, then*

$$\frac{n \text{vol}_d(\mathbf{B}^d)}{\text{vol}_d\left(\cup_{i=1}^n \mathbf{c}_i + 2R\mathbf{B}^d\right)} \leq \delta_{\text{sep}}(R, d) .$$

Next, let $\mathcal{P} = \{\mathbf{c}_1 + \mathbf{B}^d, \mathbf{c}_2 + \mathbf{B}^d, \ldots, \mathbf{c}_n + \mathbf{B}^d\}$ be a totally separable packing of n translates of $\mathbf{B}^d$ with centers at the points of $C_n = \{\mathbf{c}_1, \mathbf{c}_2, \ldots, \mathbf{c}_n\}$ in $\mathbb{E}^d$. Recall that any member of $\mathcal{P}$ is tangent to at most $2d$ members of $\mathcal{P}$ and if $\mathbf{c}_i + \mathbf{B}^d$ is tangent to $2d$ members, then the tangent points are the vertices of a regular cross-polytope inscribed in $\mathbf{c}_i + \mathbf{B}^d$ and therefore

$$\mathbf{c}_i + \sqrt{d}\mathbf{B}^d \subset \bigcup_{1 \leq j \leq n, j \neq i} \mathbf{c}_j + \sqrt{d}\mathbf{B}^d .$$

Thus, if m denotes the number of members of $\mathcal{P}$ that are tangent to $2d$ members in $\mathcal{P}$, then the $(d-1)$-dimensional surface volume $\text{svol}_{d-1}\left(\text{bd}(C_n + \sqrt{d}\mathbf{B}^d)\right)$ of the boundary $\text{bd}(C_n + \sqrt{d}\mathbf{B}^d)$ of the non-convex set $C_n + \sqrt{d}\mathbf{B}^d$ must satisfy the inequality

$$\text{svol}_{d-1}\left(\text{bd}(C_n + \sqrt{d}\mathbf{B}^d)\right) \leq (n - m) d^{\frac{d-1}{2}} \text{svol}_{d-1}\left(\text{bd}(\mathbf{B}^d)\right) . \tag{9.34}$$

Finally, the isoperimetric inequality ([194]) applied to $C_n + \sqrt{d}\mathbf{B}^d$ yields

$$\text{iq}(\mathbf{B}^d) = \frac{\text{svol}_{d-1}\left(\text{bd}(\mathbf{B}^d)\right)^d}{\text{vol}_d(\mathbf{B}^d)^{d-1}} = d^d \text{vol}_d(\mathbf{B}^d)$$

$$\leq \mathrm{iq}(C_n + \sqrt{d}\mathbf{B}^d) = \frac{\mathrm{svol}_{d-1}\left(\mathrm{bd}(C_n + \sqrt{d}\mathbf{B}^d)\right)^d}{\mathrm{vol}_d(C_n + \sqrt{d}\mathbf{B}^d)^{d-1}}, \tag{9.35}$$

where $\mathrm{iq}(\cdot)$ stands for the isoperimetric quotient of the given set. As $d \geq 4$, $\mathcal{P}$ is a $\frac{\sqrt{d}}{2}$-separable packing (in fact, it is an R-separable packing for all $R \geq 1$) and therefore (9.34), (9.35), and Lemma 230 imply in a straightforward way that

$$n - m \geq \frac{\mathrm{svol}_{d-1}\left(\mathrm{bd}(C_n + \sqrt{d}\mathbf{B}^d)\right)}{d^{\frac{d-1}{2}}\mathrm{svol}_{d-1}\left(\mathrm{bd}(\mathbf{B}^d)\right)} = \frac{\mathrm{svol}_{d-1}\left(\mathrm{bd}(C_n + \sqrt{d}\mathbf{B}^d)\right)}{d^{\frac{d+1}{2}}\mathrm{vol}_d(\mathbf{B}^d)}$$

$$\geq \frac{\mathrm{iq}(\mathbf{B}^d)^{\frac{1}{d}}\mathrm{vol}_d(C_n + \sqrt{d}\mathbf{B}^d)^{\frac{d-1}{d}}}{d^{\frac{d+1}{2}}\mathrm{vol}_d(\mathbf{B}^d)}$$

$$\geq \frac{\mathrm{iq}(\mathbf{B}^d)^{\frac{1}{d}}}{d^{\frac{d+1}{2}}\mathrm{vol}_d(\mathbf{B}^d)}\left(\frac{n\mathrm{vol}_d(\mathbf{B}^d)}{\delta_{\mathrm{sep}}(\frac{\sqrt{d}}{2}, d)}\right)^{\frac{d-1}{d}} = \frac{1}{d^{\frac{d-1}{2}}\delta_{\mathrm{sep}}(\frac{\sqrt{d}}{2}, d)^{\frac{d-1}{d}}}n^{\frac{d-1}{d}}.$$

Thus, the number of contacts in $\mathcal{P}$ is at most

$$\frac{1}{2}\left(2dn - (n-m)\right) \leq dn - \frac{1}{2d^{\frac{d-1}{2}}\delta_{\mathrm{sep}}(\frac{\sqrt{d}}{2}, d)^{\frac{d-1}{d}}}n^{\frac{d-1}{d}} < dn - \frac{1}{2d^{\frac{d-1}{2}}}n^{\frac{d-1}{d}},$$

finishing the proof of Theorem 100.

9.10 Proofs of Theorems 101, 102, and 103

9.10.1 Linearization, fundamental properties

First, in order to give a linearization of the problem, we consider a set of n pairs $(\mathbf{x}_1, f_1), \ldots, (\mathbf{x}_n, f_n)$ where $\mathbf{x}_i \in \mathbb{E}^d$ and f_i is a linear functional on $\mathbb{E}^d$ for all $1 \leq i \leq n$, and we define the following conditions that they may satisfy.

$$f_i(\mathbf{x}_i) = 1 \quad \text{and} \quad f_i(\mathbf{x}_j) \in [-1, 0] \text{ holds for all } 1 \leq i, j \leq n, i \neq j. \tag{Lin}$$

$$f_i(\mathbf{x}_j) = -1, \text{ if and only if } \mathbf{x}_j = -\mathbf{x}_i \text{ holds for all } 1 \leq i, j \leq n, i \neq j. \tag{StrictC}$$

$$f_i(\mathbf{x}_j) = -1, \text{ if and only if } f_j = -f_i \text{ holds for all } 1 \leq i, j \leq n, i \neq j. \tag{Smooth}$$

$$f_i(\mathbf{x}_i) = 1 \quad \text{and} \quad f_i(\mathbf{x}_j) \in (-1, 0] \text{ holds for all } 1 \leq i, j \leq n, i \neq j. \tag{OpenLin}$$

Lemma 231 *There is an* **o***-symmetric, strictly convex body* $\mathbf{K}$ *in* $\mathbb{E}^d$ *with* $H_{\text{sep}}(\mathbf{K}) \geq n$ *if and only if there is a set of* n *vector-linear functional pairs* $(\mathbf{x}_1, f_1), \ldots, (\mathbf{x}_n, f_n)$ *in* $\mathbb{E}^d$ *satisfying* (Lin) *and* (StrictC).

Similarly, there is an **o***-symmetric, smooth convex body* $\mathbf{K}$ *in* $\mathbb{E}^d$ *with* $H_{\text{sep}}(\mathbf{K}) \geq n$ *if and only if, there is a set of* n *vector-linear functional pairs* $(\mathbf{x}_1, f_1), \ldots, (\mathbf{x}_n, f_n)$ *in* $\mathbb{E}^d$ *satisfying* (Lin) *and* (Smooth).

Furthermore, the existence of an **o***-symmetric, smooth and strictly convex body with* $H_{\text{sep}}(\mathbf{K}) \geq n$ *is equivalent to the existence of* n *vector-linear functional pairs satisfying* (Lin), (StrictC) *and* (Smooth).

Proof. Let $\mathbf{K}$ be an **o**-symmetric convex body in $\mathbb{E}^d$. Assume that $2\mathbf{x}_1 + \mathbf{K}, 2\mathbf{x}_2 + \mathbf{K}, \ldots, 2\mathbf{x}_n + \mathbf{K}$ is a separable Hadwiger configuration of $\mathbf{K}$, where $\mathbf{x}_1, \ldots, \mathbf{x}_n \in \text{bd}\mathbf{K}$. For $1 \leq i \leq n$, let f_i be the linear functional corresponding to the separating hyperplane of $\mathbf{K}$ and $2\mathbf{x}_i + \mathbf{K}$ which is disjoint from the interior of all members of the family. That is, $f_i(\mathbf{x}_i) = 1$ and $-1 \leq f_i|_K \leq 1$.

Total separability yields that $f_i(\mathbf{x}_j) \in [-1, 1] \setminus (0, 1)$, for any $1 \leq i, j \leq n, i \neq j$. Suppose that $f_i(\mathbf{x}_j) = 1$. Then $2\mathbf{x}_i + \mathbf{K}$ and $2\mathbf{x}_j + \mathbf{K}$ both touch the hyperplane $H := \{x \in \mathbb{E}^d \ : \ f_i(x) = 1\}$ from one side, while $\mathbf{K}$ is on the other side of this hyperplane.

If $\mathbf{K}$ is strictly convex, then this is clearly not possible.

If $\mathbf{K}$ is smooth, then let S be a separating hyperplane of $2\mathbf{x}_i + \mathbf{K}$ and $2\mathbf{x}_j + \mathbf{K}$ which is disjoint from int$\mathbf{K}$. Since $\mathbf{K}$ is smooth, $\mathbf{K} \cap H \cap S = \emptyset$, and hence, $\mathbf{K}$ does not touch $2\mathbf{x}_i + \mathbf{K}$ or $2\mathbf{x}_j + \mathbf{K}$, a contradiction.

Thus, if $\mathbf{K}$ is strictly convex or smooth, then (Lin) holds.

If $\mathbf{K}$ is strictly convex (resp., smooth), then (StrictC) (resp., (Smooth)) follows immediately.

Next, assume that $(\mathbf{x}_1, f_1), \ldots, (\mathbf{x}_n, f_n)$ is a set of n vector-linear functional pairs satisfying (Lin) and (StrictC). We need to show that there is a strictly convex body $\mathbf{K}$ with $H_{\text{sep}}(\mathbf{K}) \geq n$. Consider the **o**-symmetric convex set $\mathbf{L} := \{\mathbf{x} \in \mathbb{E}^d \ : \ f_i(\mathbf{x}) \in [-1, 1] \text{ for all } 1 \leq i \leq n\}$, the intersection of n **o**-symmetric slabs.

Fix an $1 \leq i \leq n$. If there is no $j \neq i$ with $f_j(\mathbf{x}_i) = -1$, then $\mathbf{x}_i$ is in the relative interior of a facet of the polyhedral set $\mathbf{L}$, moreover, by (StrictC), no other point of the set $\{\pm\mathbf{x}_1, \ldots, \pm\mathbf{x}_n\}$ lies on that facet.

If there is a $j \neq i$ with $f_j(\mathbf{x}_i) = -1$, then $\mathbf{x}_i$ is in the intersection of two facets of $\mathbf{L}$, moreover, by (StrictC), no other point of the set $\{\pm\mathbf{x}_1, \ldots, \pm\mathbf{x}_n\}$ lies on the union of those two facets.

Thus, there is an **o**-symmetric, strictly convex body $\mathbf{K} \subset \mathbf{L}$ which contains each $\mathbf{x}_i$. Clearly, for $1 \leq i \leq n$, the hyperplane $\{\mathbf{x} \in \mathbb{E}^d \ : \ f_i(\mathbf{x}) = 1\}$ supports $\mathbf{K}$ at $\mathbf{x}_i$. It is an easy exercise to see that the family $2\mathbf{x}_1 + \mathbf{K}, 2\mathbf{x}_2 + \mathbf{K}, \ldots, 2\mathbf{x}_n + \mathbf{K}$ is a separable Hadwiger configuration of $\mathbf{K}$.

Next, assume that $(\mathbf{x}_1, f_1), \ldots, (\mathbf{x}_n, f_n)$ is a set of n vector-linear functional pairs satisfying (Lin) and (Smooth). To show that there is a smooth convex body $\mathbf{K}$ with $H_{\text{sep}}(\mathbf{K}) \geq n$, one may either copy the above proof and make the obvious modifications, or use duality: interchange the role of the $\mathbf{x}_i$s with that of the f_is, obtain a strictly convex body in the space of linear functionals, and then, by polarity obtain a smooth convex body in $\mathbb{E}^d$. We leave the details to the reader.

Finally, if (Lin), (StrictC) and (Smooth) hold, then in the above construction of a strictly convex body, we had that each point of the set $\{\pm\mathbf{x}_1, \ldots, \pm\mathbf{x}_n\}$ lies in the interior of a facet of $\mathbf{L}$, with no other point lying on the same facet. Thus, there is an $\mathbf{o}$-symmetric, smooth and strictly convex body $\mathbf{K} \subset \mathbf{L}$ which contains each $\mathbf{x}_i$. Clearly, we have $H_{\text{sep}}(\mathbf{K}) \geq n$. □

Note 1 *Let* $\mathbf{K}$ *be an* $\mathbf{o}$*-symmetric, strictly convex body in* $\mathbb{E}^d$*, and consider a separable Hadwiger configuration of* $\mathbf{K}$ *with* n *members. Then, by Lemma 231, we have a set of* n *vector-linear functional pairs satisfying* (Lin) *and* (StrictC).

If for each $1 \leq i \leq n$*, we have that* $-\mathbf{x}_i$ *is not in the set of vectors, then* (OpenLin) *is automatically satisfied. We remark that in this case, we may replace* $\mathbf{K}$ *with a strictly convex* and *smooth body.*

If for some $k \neq \ell$ *we have* $\mathbf{x}_\ell = -\mathbf{x}_k$*, then by* (Lin), $f_j(\mathbf{x}_k) = 0$ *for all* $j \in [n] \setminus \{k, \ell\}$*. Thus, if we remove* $(\mathbf{x}_k, f_k)$ *and* $(\mathbf{x}_\ell, f_\ell)$ *from the set of vector-linear functional pairs, then we obtain* $n-2$ *pairs that still satisfy* (Lin) *and* (StrictC)*, and the linear functionals lie in a* $(d-1)$*-dimensional linear hyperplane. Thus, we may consider the problem of bounding their maximum number,* $n-2$ *in* $\mathbb{E}^{d-1}$.

The same dimension reduction argument can be repeated when $\mathbf{K}$ *is smooth. Thus, in order to bound* $H_{\text{sep}}(\mathbf{K})$ *for smooth* or *strictly convex bodies, it is sufficient to consider smooth* and *strictly convex bodies, and bound* n *for which there are* n *vectors with linear functionals satisfying* (OpenLin).

We will rely on the following basic fact from convexity due to Steinitz [227] in its original form, and then refined later with the characterization of the case of equality, see [205].

Lemma 232 *Let* $\mathbf{x}_1, \ldots, \mathbf{x}_n$ *be points in* $\mathbb{E}^d$ *with* $\mathbf{o} \in \operatorname{int}\operatorname{conv}\{\mathbf{x}_1, \ldots, \mathbf{x}_n\}$*. Then there is a subset* $A \subseteq \{\mathbf{x}_1, \ldots, \mathbf{x}_n\}$ *of cardinality at most* $2d$ *with* $\mathbf{o} \in \operatorname{int}\operatorname{conv} A$.

Furthermore, if the minimal cardinality of such an A *is* $2d$*, then* A *consists of the endpoints of* d *line segments which span* $\mathbb{E}^d$*, and whose relative interiors intersect in* $\mathbf{o}$.

Proposition 233 *Let* $(\mathbf{x}_1, f_1), \ldots, (\mathbf{x}_n, f_n)$ *be vector-linear functional pairs in* $\mathbb{E}^d$ *satisfying* (Lin)*. Assume further that* $\mathbf{o} \in \operatorname{int}\operatorname{conv}\{\mathbf{x}_1, \ldots, \mathbf{x}_n\}$*. Then* $n \leq 2d$.

Moreover, if $n = 2d$*, then the points* $\mathbf{x}_1, \ldots, \mathbf{x}_n$ *are vertices of a cross-polytope with center* $\mathbf{o}$.

Proof. By (Lin), for any proper subset $A \subsetneq \{\mathbf{x}_1, \ldots, \mathbf{x}_n\}$, we have that the origin is not in the interior of $\operatorname{conv} A$. Thus, by Lemma 232, $n \leq 2d$.

Next, assume that $n = 2d$. Observe that it follows from (Lin) that if

$x_i = \lambda x_j$ for some $1 \leq i, j \leq n, i \neq j$ and $\lambda \in \mathbb{R}$, then $\lambda = -1$. Thus, combining the argument in the previous paragraph with the second part of Lemma 232 yields the second part of Proposition 233. □

Proposition 234 *Let* $(\mathbf{x}_1, f_1), \ldots, (\mathbf{x}_n, f_n)$ *be vector-linear functional pairs in* $\mathbb{E}^d$ *satisfying* (OpenLin). *Assume that* $\mathbf{o} \notin \operatorname{conv}\{\mathbf{x}_1, \ldots, \mathbf{x}_n\}$. *Then for any* $1 \leq k < \ell \leq n$, *the triangle* $\operatorname{conv}\{\mathbf{o}, \mathbf{x}_k, \mathbf{x}_\ell\}$ *is a face of the convex polytope* $\mathbf{P} := \operatorname{conv}\{\mathbf{x}_1, \ldots, \mathbf{x}_n, \mathbf{o}\}$.

Proof. By (OpenLin), we have that $f_i(\mathbf{x}_j) > -1$ for all $1 \leq i, j \leq n, i \neq j$. Suppose for a contradiction that $\operatorname{conv}\{\mathbf{x}_j : j \in \{1, 2, \ldots, n\} \setminus \{k, \ell\}\}$ contains a point of the form $\mathbf{x} = \lambda \mathbf{x}_k + \mu \mathbf{x}_\ell$ with $\lambda, \mu \geq 0, 0 < \lambda + \mu \leq 1$. By (OpenLin), we have $f_k(\mathbf{x}), f_\ell(\mathbf{x}) \leq 0$. Thus,

$$0 \geq f_k(\mathbf{x}) + f_\ell(\mathbf{x}) = \lambda(1 + f_\ell(\mathbf{x}_k)) + \mu(1 + f_k(\mathbf{x}_\ell)) > 0,$$

a contradiction. □

9.10.2 Proofs of Theorems 101 and 102

To prove part (i) of Theorem 102, we will use induction on d, the base case, $d = 1$ being trivial. By the dimension-reduction argument in Note 1, we may assume that there are n vector-linear functional pairs $(\mathbf{x}_1, f_1), \ldots, (\mathbf{x}_n, f_n)$ satisfying (OpenLin).

If $\mathbf{o} \notin \operatorname{conv}\{\mathbf{x}_1, \ldots, \mathbf{x}_n\}$, and $\mathbf{o} \notin \operatorname{conv}\{f_1, \ldots, f_n\}$, then, clearly, $n \leq h_{\mathsf{sep}}(d)$.

Thus, we may assume that $\mathbf{o} \in \operatorname{conv}\{\mathbf{x}_1, \ldots, \mathbf{x}_n\}$. We may also assume that $F = \operatorname{conv}\{\mathbf{x}_1, \ldots, \mathbf{x}_k\}$ is the face of the polytope $\operatorname{conv}\{\mathbf{x}_1, \ldots, \mathbf{x}_n\}$ that *supports* $\mathbf{o}$, that is the face which contains $\mathbf{o}$ in its relative interior. Let $H := \operatorname{span} F$. If H is the entire space $\mathbb{E}^d$, then $\mathbf{o} \in \operatorname{int}\operatorname{conv}\{\mathbf{x}_1, \ldots, \mathbf{x}_n\}$ and hence, $n \leq 2d$ follows from Proposition 233.

On the other hand, if H is a proper linear subspace of $\mathbb{E}^d$, then clearly, for any $i > k$, we have that f_i is identically zero on H.

Applying Proposition 233 on H for $\{\mathbf{x}_i : i \leq k\}$ with $\{f_i|_H : i \leq k\}$, we have

$$k \leq 2 \dim H. \tag{9.36}$$

Denote by $H^\perp$ the orthogonal complement of H, and by P the orthogonal projection of $\mathbb{E}^d$ onto $H^\perp$. It is not hard to see that P is one-to-one on the set $\{\mathbf{x}_i : i > k\}$. Moreover, the set of points $\{P\mathbf{x}_i : i > k\}$, with linear functionals $\{f_i|_{H^\perp} : i > k\}$ restricted to $H^\perp$, satisfy (OpenLin) in $H^\perp$.

Combining (9.36) with the induction hypothesis applied on $H^\perp$, we complete the proof of part (i) of Theorem 102.

For the three-dimensional bound in part (ii) of Theorem 102, suppose that $\mathbf{o} \notin \operatorname{conv}\{\mathbf{x}_1, \ldots, \mathbf{x}_4\} \in \mathbb{E}^3$. By Radon's lemma, the set $\{\mathbf{o}, \mathbf{x}_1, \ldots, \mathbf{x}_4\}$ admits a partition into two parts whose convex hulls intersect contradicting

Proposition 234. The same proof yields the two- and the four-dimensional statements, while the one-dimensional claim is trivial.

We use a projection argument to prove part (iii) of Theorem 102. Assume that $\mathbf{x}_1, \ldots, \mathbf{x}_n$ is a set of Euclidean unit vectors with $\langle \mathbf{x}_i, \mathbf{x}_j \rangle \in (-1, 0]$ for all $1 \le i, j \le n, i \ne j$. Furthermore, let $\mathbf{y}$ be a unit vector with $\langle \mathbf{y}, \mathbf{x}_i \rangle > 0$ for all $1 \le i \le n$. Consider the set of vectors $\mathbf{x}_i' := \mathbf{x}_i - \langle \mathbf{y}, \mathbf{x}_i \rangle \mathbf{y}$, $i = 1, \ldots, n$, all lying in the hyperplane $\mathbf{y}^\perp$. Now, for $1 \le i, j \le n, i \ne j$, we have

$$\langle \mathbf{x}_i', \mathbf{x}_j' \rangle = \langle \mathbf{x}_i, \mathbf{x}_j \rangle - \langle \mathbf{y}, \mathbf{x}_i \rangle \langle \mathbf{y}, \mathbf{x}_j \rangle < 0.$$

Thus, $\mathbf{x}_i', i = 1, \ldots, n$ form a set of n vectors in a $(d-1)$-dimensional space with pairwise obtuse angles. It is known [85, 157, 204], or may be proved using the same projection argument and induction on the dimension (projecting orthogonally to $(\mathbf{x}_n')^\perp$) that $n \le d$ follows. This completes the proof of Theorem 102.

Example 2 *By Lemma 231, it is sufficient to exhibit 6 vectors (with their convex hull not containing* $\mathbf{o}$ *in* $\mathbb{E}^5$*) and corresponding linear functionals satisfying* (OpenLin)*. Let the unit vectors* $\mathbf{v}_4, \mathbf{v}_5, \mathbf{v}_6$ *be the vertices of an equilateral triangle centered at* $\mathbf{o}$ *in the linear plane* $\operatorname{span}\{\mathbf{e}_4, \mathbf{e}_5\}$ *of* $\mathbb{E}^5$*. Let* $\mathbf{x}_i = \mathbf{e}_i$*, for* $i = 1, 2, 3$*, and let* $\mathbf{x}_i = (\mathbf{e}_1 + \mathbf{e}_2 + \mathbf{e}_3)/3 + \mathbf{v}_i$*, for* $i = 4, 5, 6$*. Observe that* $\mathbf{o} \notin \operatorname{conv}\{\mathbf{x}_1, \ldots, \mathbf{x}_6\}$*, as* $\langle \mathbf{e}_1 + \mathbf{e}_2 + \mathbf{e}_3, \mathbf{x}_i \rangle > 0$ *for* $i = 1, \ldots, 6$.

We define the following linear functionals.

$f_1(\mathbf{x}) = \langle \mathbf{e}_1 - \frac{\mathbf{e}_2+\mathbf{e}_3}{2}, \mathbf{x} \rangle$, $f_2(\mathbf{x}) = \langle \mathbf{e}_2 - \frac{\mathbf{e}_1+\mathbf{e}_3}{2}, \mathbf{x} \rangle$, $f_3(\mathbf{x}) = \langle \mathbf{e}_3 - \frac{\mathbf{e}_1+\mathbf{e}_2}{2}, \mathbf{x} \rangle$, *and* $f_i(\mathbf{x}) = \langle \mathbf{v}_i, \mathbf{x} \rangle$*, for* $i = 4, 5, 6$*. Clearly,* (OpenLin) *holds.*

Now, we are set for the proof of Theorem 101. First, we prove part (i). If the origin is in the interior of the convex hull of the translation vectors, then Proposition 233 yields $n \le 2d$ and the characterization of equality. In the case when $\mathbf{o} \notin \operatorname{intconv}\{\mathbf{x}_i\}$, Theorem 102 combined with Note 1 yields $n < 2d$.

The proof of part (ii) follows closely a classical proof of Danzer and Grünbaum [84] on the maximum size of an antipodal set in $\mathbb{E}^d$.

By Lemma 231 and Note 1, we may assume that $\mathbf{K}$ is an $\mathbf{o}$-symmetric smooth strictly convex body in $\mathbb{E}^d$. Assume that $2\mathbf{x}_1 + \mathbf{K}, 2\mathbf{x}_2 + \mathbf{K}, \ldots, 2\mathbf{x}_n + \mathbf{K}$ is a separable Hadwiger configuration of $\mathbf{K}$, where $\mathbf{x}_1, \ldots, \mathbf{x}_n \in \operatorname{bd}\mathbf{K}$. Let f_i denote the linear functional corresponding to the hyperplane that separates $\mathbf{K}$ from $2\mathbf{x}_i + \mathbf{K}$.

For each $1 \le i \le n$, let $\mathbf{K}_i$ be the set that we obtain by applying a homothety of ratio $1/2$ with center $\mathbf{x}_i$ on the set $\mathbf{K} \cap \{\mathbf{x} \in \mathbb{E}^d \ : \ f_i(\mathbf{x}) \ge 0\}$, that is,

$$\mathbf{K}_i := \frac{1}{2}\left(\mathbf{K} \cap \{\mathbf{x} \in \mathbb{E}^d \ : \ f_i(\mathbf{x}) \ge 0\}\right) + \frac{\mathbf{x}_i}{2}.$$

These sets are pairwise non-overlapping. In fact, it is easy to see that the following even stronger statement holds:

$$\left(\mu\mathbf{x}_i + \operatorname{int}\left(\frac{1}{2}\mathbf{K}\right)\right) \cap \left(\bigcup_{j \ne i} \mathbf{K}_j\right) = \emptyset$$

for any $\mu \geq 0$ and $1 \leq i \leq n$. On the other hand, $\mathrm{vol}_d(\mathbf{K}_i) = 2^{-(d+1)}\mathrm{vol}_d(\mathbf{K})$ by the central symmetry of $\mathbf{K}$, where $\mathrm{vol}_d(\cdot)$ stands for the d-dimensional volume of the given set. We remark that – unlike in the proof of the main result of [84] by Danzer and Grünbaum – the sets $\mathbf{K}_i$ are not translates of each other. Since each $\mathbf{K}_i$ is contained in $\mathbf{K}\setminus \mathrm{int}\left(\frac{1}{2}\mathbf{K}\right)$, we immediately obtain the bound $n \leq 2^{d+1} - 2$.

To decrease the bound further, replace $\mathbf{K}_1$ by

$$\widehat{\mathbf{K}}_1 := \mathbf{K} \cap \{\mathbf{x} \in \mathbb{E}^d \ : \ f_1(\mathbf{x}) \geq 1/2\}.$$

Now, $\widehat{\mathbf{K}}_1, \mathbf{K}_2, \ldots, \mathbf{K}_n$ are still pairwise non-overlapping, and are contained in $\mathbf{K} \setminus \mathrm{int}\left(\frac{1}{2}\mathbf{K}\right)$. The smoothness of $\mathbf{K}$ yields $\widehat{\mathbf{K}}_1 \supsetneq \mathbf{K}_1$, and hence, $\mathrm{vol}_d(\widehat{\mathbf{K}}_1) > 2^{-(d+1)}\mathrm{vol}_d(\mathbf{K})$. This completes the proof of part (ii) of Theorem 101.

9.10.3 Proof of Theorem 103

We recall Definition 31 in the following form.

Definition 51 *Let $\mathcal{P} := \{\mathbf{x}_i + \mathbf{K} \ : \ i \in I\}$ be a finite or infinite packing of translates of $\mathbf{K}$, and $\rho > 0$. We say that $\mathcal{P}$ is ρ-*separable *if for each $i \in I$ we have that the family $\{\mathbf{x}_j + \mathbf{K} \ : \ j \in I, \mathbf{x}_j + \mathbf{K} \subset \mathbf{x}_i + \rho\mathbf{K}\}$ is a totally separable packing of translates of $\mathbf{K}$. Let $\delta_{\mathrm{sep}}(\rho, \mathbf{K})$ denote the largest density of a ρ-separable packing of translates of $\mathbf{K}$, that is,*

$$\delta_{\mathrm{sep}}(\rho, \mathbf{K}) := \sup_{\mathcal{P}} \limsup_{\lambda \to \infty} \frac{\sum\limits_{i:\mathbf{x}_i+\mathbf{K}\subset[-\lambda,\lambda]^d} \mathrm{vol}_d(\mathbf{x}_i + \mathbf{K})}{(2\lambda)^d},$$

where the supremum is taken over all ρ-separable packings $\mathcal{P}$ of translates of $\mathbf{K}$.

We quote Lemma 1 of [45].

Lemma 235 *Let $\{\mathbf{x}_i + \mathbf{K} \ : \ 1 \leq i \leq n\}$ be a ρ-separable packing of translates of an $\mathbf{o}$-symmetric convex body $\mathbf{K}$ in $\mathbb{E}^d$ with $\rho \geq 1, n \geq 1$ and $d \geq 2$. Then*

$$\frac{n\mathrm{vol}_d(\mathbf{K})}{\mathrm{vol}_d\left(\bigcup\limits_{1\leq i\leq n} (\mathbf{x}_i + 2\rho\mathbf{K})\right)} \leq \delta_{\mathrm{sep}}(\rho, \mathbf{K}).$$

Proof. We use the method of the proof of the Lemma in [33] with proper modifications. The details are as follows. Assume that the claim is not true. Then there is an $\varepsilon > 0$ such that

$$\mathrm{vol}_d\left(\cup_{i=1}^n (\mathbf{c}_i + 2\rho\mathbf{C})\right) = \frac{n\mathrm{vol}_d(\mathbf{C})}{\delta_{\mathrm{sep}}(\rho, \mathbf{C})} - \varepsilon \tag{9.37}$$

Let $C_n = \{\mathbf{c}_i \mid i = 1, \ldots, n\}$ and let Λ be a packing lattice of $C_n + 2\rho\mathbf{C} =$

$\bigcup_{i=1}^{n}(\mathbf{c}_i+2\rho\mathbf{C})$ such that $C_n+2\rho\mathbf{C}$ is contained in a fundamental parallelotope of Λ say, in $\mathbf{P}$, which is symmetric about the origin. Recall that for each $\lambda > 0$, $\mathbf{W}_\lambda^d$ denotes the d-dimensional cube of edge length 2λ centered at the origin $\mathbf{o}$ in $\mathbb{E}^d$ having edges parallel to the coordinate axes of $\mathbb{E}^d$. Clearly, there is a constant $\mu > 0$ depending on $\mathbf{P}$ only, such that for each $\lambda > 0$ there is a subset L_λ of Λ with

$$\mathbf{W}_\lambda^d \subseteq L_\lambda + \mathbf{P} \text{ and } L_\lambda + 2\mathbf{P} \subseteq \mathbf{W}_{\lambda+\mu}^d\ . \tag{9.38}$$

The definition of $\delta_{\rm sep}(\rho, \mathbf{C})$ implies that for each $\lambda > 0$ there exists a ρ-separable packing of $m(\lambda)$ translates of $\mathbf{C}$ in $\mathbb{E}^d$ with centers at the points of $C(\lambda)$ such that

$$C(\lambda) + \mathbf{C} \subset \mathbf{W}_\lambda^d$$

and

$$\lim_{\lambda\to+\infty} \frac{m(\lambda)\mathrm{vol}_d(\mathbf{C})}{\mathrm{vol}_d(\mathbf{W}_\lambda^d)} = \delta_{\rm sep}(\rho, \mathbf{C})\ .$$

As $\lim_{\lambda\to+\infty} \frac{\mathrm{vol}_d(\mathbf{W}_{\lambda+\mu}^d)}{\mathrm{vol}_d(\mathbf{W}_\lambda^d)} = 1$ therefore there exist $\xi > 0$ and a ρ-separable packing of $m(\xi)$ translates of $\mathbf{C}$ in $\mathbb{E}^d$ with centers at the points of $C(\xi)$ and with $C(\xi) + \mathbf{C} \subset \mathbf{W}_\xi^d$ such that

$$\frac{\mathrm{vol}_d(\mathbf{P})\delta_{\rm sep}(\rho, \mathbf{C})}{\mathrm{vol}_d(\mathbf{P}) + \varepsilon} < \frac{m(\xi)\mathrm{vol}_d(\mathbf{C})}{\mathrm{vol}_d(\mathbf{W}_{\xi+\mu}^d)} \text{ and } \frac{n\mathrm{vol}_d(\mathbf{C})}{\mathrm{vol}_d(\mathbf{P}) + \varepsilon} < \frac{n\mathrm{vol}_d(\mathbf{C})\mathrm{card}(L_\xi)}{\mathrm{vol}_d(\mathbf{W}_{\xi+\mu}^d)}\ , \tag{9.39}$$

where card$(\cdot)$ refers to the cardinality of the given set. Now, for each $\mathbf{x} \in \mathbf{P}$ we define a ρ-separable packing of $\overline{m}(\mathbf{x})$ translates of $\mathbf{C}$ in $\mathbb{E}^d$ with centers at the points of

$$\overline{C}(\mathbf{x}) := \{\mathbf{x} + L_\xi + C_n\} \cup \{\mathbf{y} \in C(\xi) \mid \mathbf{y} \notin \mathbf{x} + L_\xi + C_n + \mathrm{int}(2\rho\mathbf{C})\}\ .$$

Clearly, (9.38) implies that $\overline{C}(\mathbf{x}) + \mathbf{C} \subset \mathbf{W}_{\xi+\mu}^d$. Now, in order to evaluate $\int_{\mathbf{x}\in\mathbf{P}} \overline{m}(\mathbf{x})d\mathbf{x}$, we introduce the function $\chi_{\mathbf{y}}$ for each $\mathbf{y} \in C(\xi)$ defined as follows: $\chi_{\mathbf{y}}(\mathbf{x}) = 1$ if $\mathbf{y} \notin \mathbf{x} + L_\xi + C_n + \mathrm{int}(2\rho\mathbf{C})$ and $\chi_{\mathbf{y}}(\mathbf{x}) = 0$ for any other $\mathbf{x} \in \mathbf{P}$. Based on the origin symmetric $\mathbf{P}$, it is easy to see that for any $\mathbf{y} \in C(\xi)$ one has $\int_{\mathbf{x}\in\mathbf{P}} \chi_{\mathbf{y}}(\mathbf{x})d\mathbf{x} = \mathrm{vol}_d(\mathbf{P}) - \mathrm{vol}_d(C_n + 2\rho\mathbf{C})$. Thus, it follows in a straightforward way that

$$\int_{\mathbf{x}\in\mathbf{P}} \overline{m}(\mathbf{x})d\mathbf{x} = \int_{\mathbf{x}\in\mathbf{P}} (n\mathrm{card}(L_\xi) + \sum_{\mathbf{y}\in C(\xi)} \chi_{\mathbf{y}}(\mathbf{x}))d\mathbf{x}$$

$$= n\mathrm{vol}_d(\mathbf{P})\mathrm{card}(L_\xi) + m(\xi)\big(\mathrm{vol}_d(\mathbf{P}) - \mathrm{vol}_d(C_n + 2\rho\mathbf{C})\big)\ .$$

Hence, there is a point $\mathbf{p} \in \mathbf{P}$ with

$$\overline{m}(\mathbf{p}) \geq m(\xi)\left(1 - \frac{\mathrm{vol}_d(C_n + 2\rho\mathbf{C})}{\mathrm{vol}_d(\mathbf{P})}\right) + n\mathrm{card}(L_\xi)$$

and so

$$\frac{\overline{m}(\mathbf{p})\mathrm{vol}_d(\mathbf{C})}{\mathrm{vol}_d(\mathbf{W}^d_{\xi+\mu})} \geq \frac{m(\xi)\mathrm{vol}_d(\mathbf{C})}{\mathrm{vol}_d(\mathbf{W}^d_{\xi+\mu})}\left(1 - \frac{\mathrm{vol}_d(C_n + 2\rho\mathbf{C})}{\mathrm{vol}_d(\mathbf{P})}\right) + \frac{n\mathrm{vol}_d(\mathbf{C})\mathrm{card}(L_\xi)}{\mathrm{vol}_d(\mathbf{W}^d_{\xi+\mu})}. \tag{9.40}$$

Now, (9.37) implies in a straightforward way that

$$\frac{\mathrm{vol}_d(\mathbf{P})\delta_{\mathrm{sep}}(\rho,\mathbf{C})}{\mathrm{vol}_d(\mathbf{P})+\varepsilon}\left(1 - \frac{\mathrm{vol}_d(C_n + 2\rho\mathbf{C})}{\mathrm{vol}_d(\mathbf{P})}\right) + \frac{n\mathrm{vol}_d(\mathbf{C})}{\mathrm{vol}_d(\mathbf{P})+\varepsilon} = \delta_{\mathrm{sep}}(\rho,\mathbf{C}). \tag{9.41}$$

Thus, (9.39), (9.40), and (9.41) yield that

$$\frac{\overline{m}(\mathbf{p})\mathrm{vol}_d(\mathbf{C})}{\mathrm{vol}_d(\mathbf{W}^d_{\xi+\mu})} > \delta_{\mathrm{sep}}(\rho,\mathbf{C})\ .$$

As $\overline{C}(\mathbf{p}) + \mathbf{C} \subset \mathbf{W}^d_{\xi+\mu}$ this contradicts the definition of $\delta_{\mathrm{sep}}(\rho,\mathbf{C})$, finishing the proof of Lemma 235. □

Lemma 236 *Let* $\mathbf{K}$ *be a smooth* $\mathbf{o}$*-symmetric convex body in* $\mathbb{E}^d$ *with* $d \in \{1,2,3,4\}$. *Then there is a* $\lambda > 0$ *such that for any separable Hadwiger configuration* $\{\mathbf{K}\} \cup \{\mathbf{x}_i + \mathbf{K}\ :\ i = 1,\ldots,2d\}$ *of* $\mathbf{K}$,

$$\lambda\mathbf{K} \subseteq \bigcup_{i=1}^{2d}(\mathbf{x}_i + \lambda\mathbf{K}). \tag{9.42}$$

holds. In particular, (9.42) *holds with* $\lambda = 2$ *when* $d = 2$.

Definition 52 *We denote the smallest* λ *satisfying* (9.42) *by* $\lambda_{\mathrm{sep}}(\mathbf{K})$, *and note that* $\lambda_{\mathrm{sep}}(\mathbf{K}) \geq 2$, *since otherwise* $\bigcup_{i=1}^{2d}(\mathbf{x}_i + \lambda\mathbf{K})$ *does not contain* $\mathbf{o}$.

Proof. Clearly, λ satisfies (9.42) if and only if for each boundary point $\mathbf{b} \in \mathrm{bd}(\mathbf{K})$ we have that at least one of the $2d$ points $\mathbf{b} - \frac{2}{\lambda}\mathbf{x}_i$ is in $\mathbf{K}$.

First, we fix a separable Hadwiger configuration of $\mathbf{K}$ with $2d$ members and show that for some $\lambda > 0$, (9.42) holds. By Theorem 101, we have that $\{\mathbf{x}_i\ :\ i = 1,\ldots,2d\}$ is an Auerbach basis of $\mathbf{K}$, and, in particular, the origin is in the interior of $\mathrm{conv}\{\mathbf{x}_i\ :\ i = 1,\ldots,2d\}$. It follows from the smoothness of $\mathbf{K}$ that for each boundary point $\mathbf{b} \in \mathrm{bd}(\mathbf{K})$ we have that at least one of the $2d$ rays $\{\mathbf{b} - t\mathbf{x}_i\ :\ t > 0\}$ intersects the interior of $\mathbf{K}$. The existence of λ now follows from the compactness of $\mathbf{K}$.

Next, since the set of Auerbach bases of $\mathbf{K}$ is compact (consider them as points in $\mathbf{K}^d$), it follows in a straightforward way that there is a $\lambda > 0$, for which (9.42) holds for all separable Hadwiger configurations of $\mathbf{K}$ with $2d$ members.

To prove the part concerning $d = 2$, we make use of the characterization of the equality case in part (i) of Theorem 101. An Auerbach basis of a planar $\mathbf{o}$-symmetric convex body $\mathbf{K}$ means that $\mathbf{K}$ is contained in an $\mathbf{o}$-symmetric parallelogram, the midpoints of whose edges are $\pm\mathbf{x}_1, \pm\mathbf{x}_2$, and $\pm\mathbf{x}_1, \pm\mathbf{x}_2 \in \mathbf{K}$. We leave it as an exercise to the reader that in this case, for each boundary point $\mathbf{b} \in \mathrm{bd}(\mathbf{K})$ we have that at least one of the 4 points $\mathbf{b} \pm \frac{\mathbf{x}_1}{2}, \mathbf{b} \pm \frac{\mathbf{x}_2}{2}$ is in $\mathbf{K}$. □

Finally, we are ready to state our main result, from which Theorem 103 immediately follows.

Theorem 237 *Let* $\mathbf{K}$ *be a smooth* $\mathbf{o}$*-symmetric convex body in* $\mathbb{E}^d$ *with* $d \in \{1, 2, 3, 4\}$. *Then*

$$c_{\mathrm{sep}}(\mathbf{K}, n, d) \leq$$

$$dn - \frac{n^{(d-1)/d}}{2\left[\lambda_{\mathrm{sep}}(\mathbf{K})\right]^{d-1}\left[\delta_{\mathrm{sep}}\left(\frac{\lambda_{\mathrm{sep}}(\mathbf{K})}{2}, \mathbf{K}\right)\right]^{(d-1)/d}}\left[\frac{\mathrm{iq}(\mathbf{B}^d)}{\mathrm{iq}(\mathbf{K})}\right]^{1/d}$$

$$\leq dn - \frac{n^{(d-1)/d}(\mathrm{vol}_d(\mathbf{B}^d))^{1/d}}{4\left[\lambda_{\mathrm{sep}}(\mathbf{K})\right]^{d-1}}$$

for all $n > 1$.

In particular, in the plane, we have

$$c_{\mathrm{sep}}(\mathbf{K}, n, 2) \leq 2n - \frac{\sqrt{\pi}}{8}\sqrt{n}$$

for all $n > 1$.

Proof. Let $\mathcal{P} = C + \mathbf{K}$ be a totally separable packing of translates of $\mathbf{K}$, where C denotes the set of centers $C = \{\mathbf{x}_1, \ldots, \mathbf{x}_n\}$. Assume that m of the n translates is touched by the maximum number, that is, by Theorem 101, $H_{\mathrm{sep}}(\mathbf{K}) = 2d$ others. By Lemma 236, we have

$$\mathrm{svol}_{d-1}\left(\mathrm{bd}\left(C + \lambda_{\mathrm{sep}}(\mathbf{K})\,\mathbf{K}\right)\right) \leq \tag{9.43}$$

$$(n-m)(\lambda_{\mathrm{sep}}(\mathbf{K}))^{d-1}\mathrm{svol}_{d-1}(\mathrm{bd}(\mathbf{K})).$$

By the isoperimetric inequality, we have

$$\mathrm{iq}(\mathbf{B}^d) \leq \mathrm{iq}(C + \lambda_{\mathrm{sep}}(\mathbf{K})\,\mathbf{K}) = \frac{\left(\mathrm{svol}_{d-1}\left(\mathrm{bd}\left(C + \lambda_{\mathrm{sep}}(\mathbf{K})\,\mathbf{K}\right)\right)\right)^d}{\left(\mathrm{vol}_d(C + \lambda_{\mathrm{sep}}(\mathbf{K})\,\mathbf{K})\right)^{d-1}}. \tag{9.44}$$

Combining (9.43) and (9.44) yields

$$n - m \geq \frac{(\mathrm{iq}(\mathbf{B}^d))^{1/d}\left[\mathrm{vol}_d(C + \lambda_{\mathrm{sep}}(\mathbf{K})\,\mathbf{K})\right]^{(d-1)/d}}{(\lambda_{\mathrm{sep}}(\mathbf{K}))^{d-1}\mathrm{svol}_{d-1}(\mathrm{bd}\mathbf{K})}.$$

The latter, by Lemma 235 is at least

$$\frac{(\mathrm{iq}(\mathbf{B}^d))^{1/d}\left[\frac{n\,\mathrm{vol}_d(\mathbf{K})}{\delta_{\mathrm{sep}}(\lambda_{\mathrm{sep}}(\mathbf{K})/2, \mathbf{K})}\right]^{(d-1)/d}}{(\lambda_{\mathrm{sep}}(\mathbf{K}))^{d-1}\mathrm{svol}_{d-1}(\mathrm{bd}\mathbf{K})}.$$

After rearrangement, we obtain the desired bound on n completing the proof of the first inequality in Theorem 237.

To prove the second inequality, we adopt the proof of [33, Corollary 1]. First, note that $\delta_{\mathrm{sep}}\left(\frac{\lambda_{\mathrm{sep}}(\mathbf{K})}{2}, \mathbf{K}\right) \leq 1$, and $(\mathrm{iq}(\mathbf{B}^d))^{1/d} = d\mathrm{vol}_d(\mathbf{B}^d)$. Next, according to Ball's reverse isoperimetric inequality (see Theorem 3 or [16]), for any convex body $\mathbf{K}$, there is a non-degenerate affine map $T : \mathbb{E}^d \to \mathbb{E}^d$ with $\mathrm{iq}(T\mathbf{K}) \leq (2d)^d$. Finally, notice that $c_{\mathrm{sep}}(\mathbf{K}, n, d) = c_{\mathrm{sep}}(T\mathbf{K}, n, d)$, and the inequality follows in a straightforward way.

The planar bound follows by substituting the value $\lambda_{\mathrm{sep}}(\mathbf{K}) = 2$ from Lemma 236. $\square$

9.10.4 Remarks

Lemma 236 does not hold for strictly convex but not smooth convex bodies. Indeed, in $\mathbb{E}^3$, consider the $\mathbf{o}$-symmetric polytope $\mathbf{P} := \mathrm{conv}\{\pm\mathbf{e}_1, \pm\mathbf{e}_2, \pm\mathbf{e}_3, \pm 0.9(\mathbf{e}_1 + \mathbf{e}_2 + \mathbf{e}_3)\}$ where the $\mathbf{e}_i$s are the standard basis vectors. The six translation vectors $\pm 2\mathbf{e}_1, \pm 2\mathbf{e}_2, \pm 2\mathbf{e}_3$ generate a separable Hadwiger configuration of $\mathbf{P}$. For the vertex $\mathbf{b} := 0.9(\mathbf{e}_1 + \mathbf{e}_2 + \mathbf{e}_3)$, we have that each of the 3 lines $\{\mathbf{b} + t\mathbf{e}_i \ : \ t \in \mathbb{R}\}$ intersect $\mathbf{P}$ in $\mathbf{b}$ only. Thus, there is a strictly convex $\mathbf{o}$-symmetric body $\mathbf{K}$ with the following properties. $\mathbf{P} \subset \mathbf{K}$, and $\pm\mathbf{e}_i$ is a boundary point of $\mathbf{K}$ for each $i = 1, 2, 3$, and at $\pm\mathbf{e}_i$, the plane orthogonal to $\mathbf{e}_i$ is a support plane of $\mathbf{K}$, and $\mathbf{b}$ is a boundary point of $\mathbf{K}$, and the 3 lines $\{\mathbf{b} + t\mathbf{e}_i \ : \ t \in \mathbb{R}\}$ intersect $\mathbf{K}$ in $\mathbf{b}$ only. For this strictly convex $\mathbf{K}$, we have $\lambda_{\mathrm{sep}}(\mathbf{K}) = \infty$.

Thus, it is natural to ask if in Theorem 103 smoothness can be replaced by strict convexity. We note that in our proof, Lemma 236 is the only place which does not carry over to this case.

The same construction of the polytope $\mathbf{P}$ shows that $\lambda_{\mathrm{sep}}(\mathbf{K})$ may be arbitrarily large for a three-dimensional smooth convex body $\mathbf{K}$. Indeed, if we take $\mathbf{K} := \mathbf{P} + \varepsilon\mathbf{B}^d$ with a small $\varepsilon > 0$, we obtain a smooth body for which, by the previous argument, $\lambda_{\mathrm{sep}}(\mathbf{K})$ is large.

Thus, it would be very interesting to see a lower bound on $f(\mathbf{K})$ of Theorem 103 which depends on d only.

9.11 Proof of Theorem 108

In the proof we use the following concepts and lemma.

Let S denote the sequence whose jth element s_j is the number of bits one needs to count from the right to reach the first non-zero value in the binary form of j; in other words, s_j is the smallest integer k such that 2^k is not a divisor of k. The first few elements of this sequence are

$$1, 2, 1, 3, 1, 2, 1, 4, 1, 2, 1, 3, 1, 2, 1, 5, 1, 2, 1, 3, 1, 2, 1, 4, 1, 2, 1, 3, 1, 2, 1, 6, \ldots$$

First, we prove a property of this sequence that we need in the construction of a topological disk with large Hadwiger number.

Lemma 238 *For any positive integer m, the sum of any m consecutive elements of S is minimal for the first m elements of S. In other words, for any non-negative integer r, we have*

$$\sum_{i=1}^{m} s_i \leq \sum_{i=r+1}^{r+m} s_i.$$

Proof. We prove by induction on m. If $m = 1$, then the assertion clearly holds, since $s_1 = 1 \leq s_r$ for all values of r. Assume that $m > 1$, and the statement holds for any positive integer smaller than m. If m is odd, then $s_m = 1$, and thus, by the induction hypothesis,

$$\sum_{i=1}^{m} s_i = 1 + \sum_{i=1}^{m-1} s_i \leq 1 + \sum_{i=r+1}^{r+m-1} s_i \leq s_{r+m} + \sum_{i=r+1}^{r+m-1} s_i = \sum_{i=r+1}^{r+m} s_i.$$

Consider the case that m is even. Observe that $s_i = 1$ if and only if i is odd, and if we remove all elements with an odd index, and decrease the value of every element with an even index, the obtained sequence coincides with S. Thus, if m is even, then

$$\sum_{i=1}^{m} s_i = \frac{m}{2} + \sum_{i=1}^{m/2} (s_i + 1) = m + \sum_{i=1}^{m/2} s_i \leq m + \sum_{i=r'+1}^{r'+m/2} s_i = \frac{m}{2} + \sum_{i=r'+1}^{r'+m/2} (s_i + 1)$$

$$= \sum_{i=r+1}^{r+m} s_i,$$

where $r' = \lfloor \frac{r}{2} \rfloor$. $\square$

To prove Theorem 108, for every positive integer m we construct a topological disk $\mathbf{K}_m$, and m translates of $\mathbf{K}_m$ which are pairwise non-overlapping, and each touches $\mathbf{K}_m$. To do it, first we introduce a 2-parameter family of topological disks.

Consider some arbitrary integers $k \geq 2$ and $m \geq 1$. To construct a topological disk $\mathbf{K}_m^k$ corresponding to these numbers, we use 2^m rectangles of size $k \times 1$, called *bars*, and $2^m - 1$ rectangles with unit base, called *connectors*. We denote the bars by $\mathbf{B}_1, \mathbf{B}_2, \ldots, \mathbf{B}_{2^m}$, and the connectors by $\mathbf{C}_1, \mathbf{C}_2, \ldots, \mathbf{C}_{2^m-1}$. Here, the height of the ith connector is s_i.

From these pieces, we build up $\mathbf{K}_m^k$ such that, starting with $\mathbf{B}_1$, in each step we attach the bottom side of $\mathbf{C}_i$ to the top side of the rightmost square of $\mathbf{B}_i$, and the left side of $\mathbf{B}_{i+1}$ to the right side of the uppermost square of $\mathbf{C}_i$. More formally, we set

$$\mathbf{B}_i = [(i-1)k, ik] \times [y_i, y_i + 1] \text{ and } \mathbf{C}_i = [ik-1, ik] \times [y_i + 1, y_{i+1} + 1],$$

where $y_i = \sum_{j=1}^{i-1} s_j$. Note that in this case the bottom, left corner of $\mathbf{B}_1$ is the origin.

Note that these disks can be constructed recursively as well by connecting two translates of $\mathbf{K}_{m-1}^k$ by $\mathbf{C}_{2^{m-1}}$.

First, we prove the next lemma.

Lemma 239 *Let* $\mathbf{F}$ *and* $\mathbf{F}'$ *be two translates of* $\mathbf{K}_m^k$*, where* $m, k \geq 2$*, such that the first bar* $\mathbf{B}_1'$ *of* $\mathbf{F}'$ *is obtained by translating some bar* $\mathbf{B}_r$ *of* $\mathbf{F}$ *with the vector* (x^*, y^*)*, where* $y^* \leq -1$*, and* $1 \leq x^* \leq k-1$*. Then* $\mathbf{F}$ *and* $\mathbf{F}'$ *do not overlap.*

Proof. Assume that $\mathbf{F} = \mathbf{K}_m^k$, and consider the vertical strip determined by the bar $\mathbf{B}_{r-1+i}$ for some $1 \leq i \leq 2^m - r + 1$. This strip may intersect only at most $\mathbf{B}_{i-1}'$, $\mathbf{B}_i'$ and $\mathbf{C}_{i-1}'$. The maximum y-coordinate of the points of $\mathbf{B}_{i-1}' \cup \mathbf{B}_i' \cup \mathbf{C}_{i-1}'$, measured from the bottom, left corner of $\mathbf{F}'$, is $y_i + 1$. Since the y-coordinate of this corner of $\mathbf{F}'$ is $y_r + y^* \leq y_r - 1$, the set $\mathbf{B}_{i-1}' \cup \mathbf{B}_i' \cup \mathbf{C}_{i-1}'$ lies below the line $y = y_r + y_i$.

On the other hand, the y-coordinate of the points of the bottom edge of $\mathbf{B}_{r-1+i}$ is y_{r-1+i}, from which

$$y_{r-1+i} - (y_r + y_i) = (y_{r-1+i} - y_r - y_i) = \sum_{j=r}^{r+i-2} s_j - \sum_{j=1}^{i-1} s_j \geq 0$$

follows. □

Now we return to the proof of Theorem 108. Let $k \geq m > 1$ be arbitrary. Consider the topological disk $\mathbf{X}_1 = \mathbf{K}_m^k$. We define the translates $\mathbf{X}_2, \mathbf{X}_3, \ldots, \mathbf{X}_m$ in the following way: The translate $\mathbf{X}_i$ is obtained from $\mathbf{X}_{i-1}$ in such a way that by translating the second instance of $\mathbf{K}_{m+1-i}^k$ from the left in $\mathbf{X}_{i-1}$ by one to the right and by one downward we obtain the first instance of $\mathbf{K}_{m+1-i}^k$ from the left in $\mathbf{X}_i$.

Let $1 \leq i < j \leq m$ arbitrary. Then the first instance of $\mathbf{K}_{m+1-i}$ from the left in $\mathbf{X}_j$ is obtained by translating some instance of this set in $\mathbf{X}_i$ to the right and downward by $j - i$. Since $1 \leq j - i \leq m \leq k$, Lemma 239 implies that $\mathbf{X}_i$ and $\mathbf{X}_j$ do not overlap.

Define $\mathbf{X}$ as the translate of $\mathbf{X}_i$ downward by $m+1$. To prove the theorem, it is sufficient to show that $\mathbf{X}_1, \mathbf{X}_2, \ldots, \mathbf{X}_m$ touch $\mathbf{X}$ (cf. Figure 9.6).

Let i be fixed. Let $\mathbf{K}$ be the last instance of $\mathbf{K}_{m+1-i}^k$ from the left in $\mathbf{X}$, and let $\mathbf{K}'$ be the first instance of $\mathbf{K}_{m+1-i}$ from the left in $\mathbf{X}_i$. Then $\mathbf{K}'$ is obtained from $\mathbf{K}$ by translating it first upward by $m+1$, then both to the right and downward by $i - 1$. Thus, $\mathbf{K}'$ is the translate of $\mathbf{K}$ by the vector

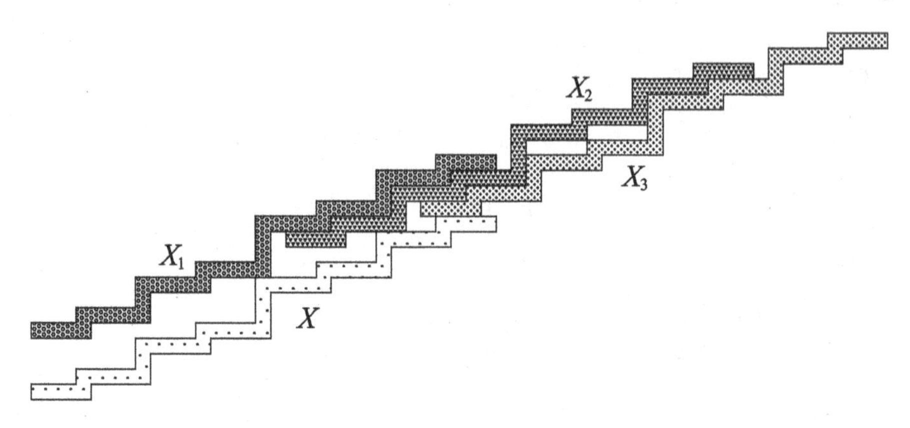

FIGURE 9.6
The translates of $\mathbf{K}_3^4$.

$(i-1, m+2-i)$. Since the height of the middle connector of $\mathbf{K}$ is $m+2-i$, this yields that $\mathbf{K}$ and $\mathbf{K}'$ touch. On the other hand, as $\mathbf{K}$ is the last, and $\mathbf{K}'$ is the first instance of $\mathbf{K}_{m+1-i}^k$ in $\mathbf{X}$ and $\mathbf{X}_i$, respectively, $\mathbf{X}$ and $\mathbf{X}_i$ do not overlap. □

9.12 Proof of Theorem 110

Let $\mathbf{S}$ be an $\mathbf{o}$-symmetric starlike disk. Let $\mathcal{F} = \{\mathbf{S}_i : i = 1, 2, \ldots, n\}$ be a family of translates of $\mathbf{S}$ such that $n = H(\mathbf{S})$ and, for $i = 1, 2, \ldots, n$, $\mathbf{S}_i = \mathbf{c}_i + \mathbf{S}$ touches $\mathbf{S}$ and does not overlap with any other element of $\mathcal{F}$. Let $\mathbf{K} = \text{conv}\mathbf{S}$, $X = \{\mathbf{c}_i : i = 1, 2, \ldots, n\}$, $\mathbf{C} = \text{conv}X$ and $\bar{\mathbf{C}} = \text{conv}\big(X \cup (-X)\big)$. Furthermore, let R_i denote the closed ray $R_i = \{\lambda \mathbf{c}_i : \lambda \in \mathbb{R} \text{ and } \lambda \geq 0\}$.

First, we prove a few lemmas.

Lemma 240 *The disk* $\mathbf{S}$ *is starlike relative to the origin* $\mathbf{o}$*. Furthermore,* $\mathbf{o} \in \text{int}\mathbf{S}$.

Proof. Let $\mathbf{S}$ be starlike relative to $\mathbf{p} \in \mathbf{S}$, and assume that $\mathbf{p} \neq \mathbf{o}$. By symmetry, S is starlike relative to $-\mathbf{p}$. Consider a point $\mathbf{q} \in \mathbf{S}$. Since $\mathbf{S}$ is starlike relative to $\mathbf{p}$ and $-\mathbf{p}$, the segments $[\mathbf{p}, \mathbf{q}]$ and $[-\mathbf{p}, \mathbf{q}]$ are contained in $\mathbf{S}$. Thus, any segment $[\mathbf{p}, \mathbf{r}]$, where $\mathbf{r} \in [-\mathbf{p}, \mathbf{q}]$, is contained in $\mathbf{S}$. In other words, we have $\text{conv}\{\mathbf{p}, -\mathbf{p}, \mathbf{q}\} \subset \mathbf{S}$, which yields that $[\mathbf{o}, \mathbf{q}] \subset \mathbf{S}$. The second assertion follows from the first and the symmetry of $\mathbf{S}$. □

Before the next lemma, recall that the *relative norm* of a convex body $\mathbf{K}$ is the norm induced by the central symmetrization of $\mathbf{K}$. In the following, we denote the distance of two points $\mathbf{x}, \mathbf{y}$ measured in the relative norm of $\mathbf{K}$ by $\text{dist}_{\mathbf{K}}(\mathbf{x}, \mathbf{y})$. An alternative definition of this concept is

$$\text{dist}_{\mathbf{K}}(\mathbf{x}, \mathbf{y}) = \frac{2||\mathbf{y} - \mathbf{x}||}{||\mathbf{p} - \mathbf{q}||},$$

where $[\mathbf{p}, \mathbf{q}]$ is a longest chord of $\mathbf{K}$ parallel to $[\mathbf{x}, \mathbf{y}]$, and $||\cdot||$ denotes Euclidean norm.

Lemma 241 *If* $\mathbf{x}+\mathbf{S}$ *and* $\mathbf{y}+\mathbf{S}$ *are non-overlapping translates of* $\mathbf{S}$, *then we have* $\text{dist}_{\mathbf{K}}(\mathbf{x}, \mathbf{y}) \geq 1$.

Proof. Without loss of generality, we may assume that $\mathbf{x} = \mathbf{o}$. Suppose that $\mathbf{y} \in \text{int}\mathbf{K}$. Note that there are points $\mathbf{p}, \mathbf{q} \in \mathbf{S}$ such that $\mathbf{y} \in \text{intconv}\{\mathbf{o}, \mathbf{p}, \mathbf{q}\}$. By the symmetry of $\mathbf{S}$, $[\mathbf{y} - \mathbf{p}, \mathbf{y}]$ and $[\mathbf{y} - \mathbf{q}, \mathbf{y}]$ are contained in $\mathbf{y} + \mathbf{S}$. Since $\mathbf{y} \in \text{intconv}\{\mathbf{o}, \mathbf{p}, \mathbf{q}\}$, the segments $[\mathbf{y}-\mathbf{p}, \mathbf{y}]$ and $[\mathbf{o}, \mathbf{q}]$ cross, which yields that $\mathbf{S}$ and $\mathbf{y} + \mathbf{S}$ overlap; a contradiction. Hence, $\mathbf{y} \notin \text{int}\mathbf{K}$. Since $\text{int}\mathbf{K}$ is the set of points in the plane whose distance from $\mathbf{o}$, in the norm with unit ball $\mathbf{K}$, is less than one, we have $\text{dist}_{\mathbf{K}}(\mathbf{o}, \mathbf{y}) \geq 1$. □

Remark 242 *The Hadwiger number* $H(\mathbf{S})$ *of* $\mathbf{S}$ *is at most twenty-four.*

Proof. Note that, for every value of i, $\mathbf{K}$ and $\mathbf{c}_i + \mathbf{K}$ either overlap or touch. Since $\mathbf{K}$ is $\mathbf{o}$-symmetric, it follows that $\mathbf{c}_i \in 2\mathbf{K}$, and $\mathbf{c}_i + \frac{1}{2}\mathbf{K}$ is contained in $\frac{5}{2}\mathbf{K}$. By Lemma 241, $\{\mathbf{c}_i + \frac{1}{2}\mathbf{K} : i = 1, 2, \dots, n\} \cup \{\frac{1}{2}\mathbf{K}\}$ is a family of pairwise non-overlapping translates of $\frac{1}{2}\mathbf{K}$. Thus, $n \leq 24$ follows from an area estimate. □

Lemma 243 *If* $j \neq i$, *then* $R_i \cap \text{int}\mathbf{S}_j = \emptyset$. *Furthermore,* $R_i \cap \mathbf{S}_j \subset (\mathbf{o}, \mathbf{c}_i)$.

Proof. Since $\mathbf{S}$ and $\mathbf{S}_i$ touch, there is a (possibly degenerate) parallelogram $\mathbf{P}$ such that $\text{bd}\mathbf{P} \subset (\mathbf{S} \cup \mathbf{S}_i)$ and $[\mathbf{o}, \mathbf{c}_i] \subset \mathbf{P}$ (cf. Figure 9.7). Note that if $\text{int}(\mathbf{x}+\mathbf{S})$ intersects neither $\mathbf{S}$ nor $\mathbf{S}_i$, then $\mathbf{x} \notin \mathbf{P}$ and $\text{int}(\mathbf{x}+\mathbf{S}) \cap (\mathbf{o}, \mathbf{c}_i) = \emptyset$.

If $\mathbf{S}_j \cap R_i = \emptyset$, we have nothing to prove. Let $\mathbf{S}_j \cap R_i \neq \emptyset$ and consider a point $\mathbf{c}_j + \mathbf{p} \in \mathbf{S}_j \cap R_i$. Since $\mathbf{o} \in \text{int}\mathbf{S}$, $\mathbf{c}_j + \mathbf{p} \neq \mathbf{o}$ and $\mathbf{c}_j + \mathbf{p} \neq \mathbf{c}_i$. By the previous paragraph, if $\mathbf{c}_j + \mathbf{p} \in (\mathbf{o}, \mathbf{c}_i)$, then $\mathbf{c}_j + \mathbf{p} \notin \text{int}\mathbf{S}_j$. Thus, we are left with the case that $\mathbf{c}_j + \mathbf{p} \in R_i \setminus [\mathbf{o}, \mathbf{c}_i]$. By symmetry, $\mathbf{c}_i - \mathbf{p} \in \mathbf{S}_i$. Note that $(\mathbf{c}_i, \mathbf{c}_i - \mathbf{p}) \cap (\mathbf{o}, \mathbf{c}_j) \neq \emptyset$, which yields that $\text{int}\mathbf{S}_i$ intersects $(\mathbf{o}, \mathbf{c}_j)$; a contradiction. □

Lemma 244 *We have* $\mathbf{o} \in \text{int}\mathbf{C}$, *and* $X \subset \text{bd}\mathbf{C}$.

Proof. Assume that $\mathbf{o} \notin \text{int}\mathbf{C}$. Note that there is a closed half plane $\mathbf{H}$, containing $\mathbf{o}$ in its boundary, such that $\mathbf{C} \subset \mathbf{H}$. Let $\mathbf{p}$ be a boundary point of $\mathbf{S}$ satisfying $\mathbf{S} \subset \mathbf{p} + \mathbf{H}$. Then, for $i = 1, 2, \dots, n$, we have $\mathbf{S}_i \subset \mathbf{p} + \mathbf{H}$. Observe that, for any value of i, $2\mathbf{p} + \mathbf{S}$ touches $\mathbf{S}$ and does not overlap $\mathbf{S}_i$. Thus, $\mathcal{F} \cup \{2\mathbf{p} + \mathbf{S}\}$ is a family of pairwise non-overlapping translates of $\mathbf{S}$

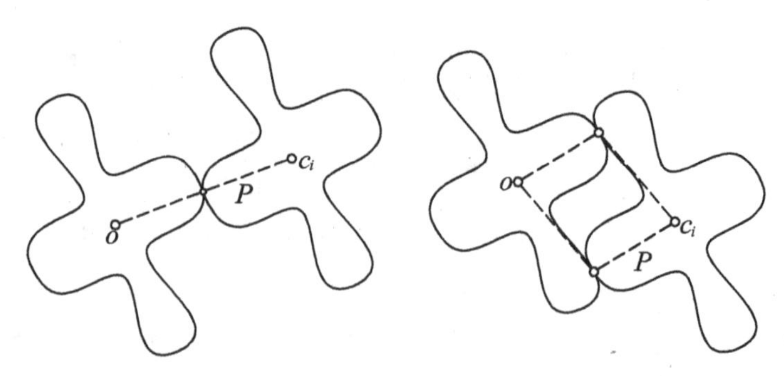

FIGURE 9.7
Touching pairs of translates of $\mathbf{S}$.

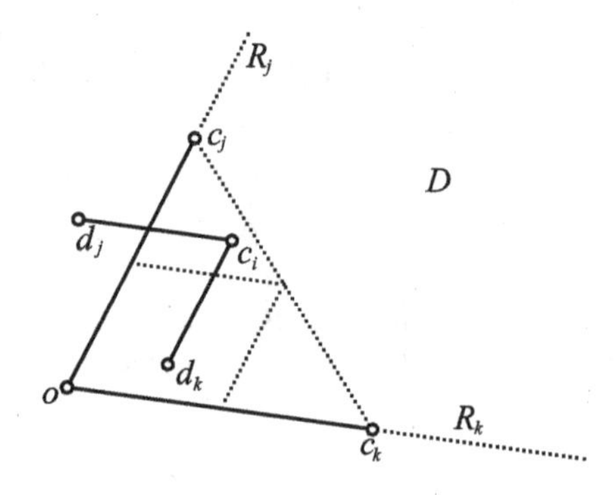

FIGURE 9.8
The angular domain bounded by $R_j \cup R_k$.

in which every element touches $\mathbf{S}$, which contradicts our assumption that $\text{card}\,\mathcal{F} = n = H(\mathbf{S})$.

Assume that $\mathbf{c}_i \notin \text{bd}\mathbf{C}$ for some i, and note that there are values j and k such that $\mathbf{c}_i \in \text{intconv}\{\mathbf{o}, \mathbf{c}_j, \mathbf{c}_k\}$. Since $\mathbf{S}_j$ and $\mathbf{S}_k$ touch $\mathbf{S}$, $\frac{1}{2}\mathbf{c}_j$ and $\frac{1}{2}\mathbf{c}_k$ are contained in $\mathbf{K}$. Observe that at least one of $\mathbf{d}_j = \mathbf{c}_i - \frac{1}{2}\mathbf{c}_j$ and $\mathbf{d}_k = \mathbf{c}_i - \frac{1}{2}\mathbf{c}_k$ is in the exterior of the closed, convex angular domain $\mathbf{D}$ bounded by $R_j \cup R_k$ (cf. Figure 9.8). Since $\mathbf{d}_j$ and $\mathbf{d}_k$ are points of $\mathbf{c}_i + \mathbf{K}$, we obtain $(\mathbf{c}_i + \mathbf{K}) \setminus \mathbf{D} \neq \emptyset$. On the other hand, Lemma 243 yields that $\mathbf{S}_i \subset \mathbf{D}$, hence, $\mathbf{c}_i + \mathbf{K} = \text{conv}\mathbf{S}_i \subset \mathbf{D}$; a contradiction. □

Remark 245 *The Hadwiger number $H(\mathbf{S})$ of $\mathbf{S}$ is at most sixteen.*

Proof. Gołab proved that the circumference of every centrally symmetric convex disk measured in its norm is at least six and at most eight (cf. Theorem 49 or [120]). Fáry and Makai proved that, in any norm, the circumferences of any

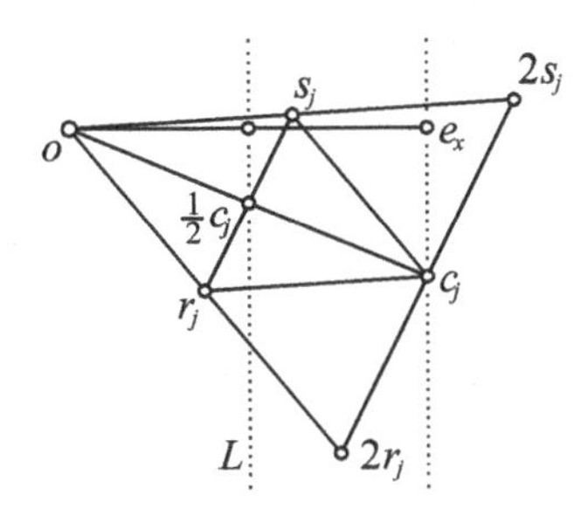

FIGURE 9.9
An illustration for the proof of Theorem 110.

convex disk $\mathbf{C}$ and its central symmetral $\frac{1}{2}(\mathbf{C}-\mathbf{C})$ are equal (cf. Theorem 50 or [101]). Thus, the circumference of $\mathbf{C}$ measured in the norm with unit ball $\frac{1}{2}(\mathbf{C}-\mathbf{C})$ is at most eight.

Since $\mathbf{C} \subset 2\mathbf{K}$, we have $\mathrm{dist}_{\mathbf{C}}(\mathbf{p},\mathbf{q}) \geq \mathrm{dist}_{2\mathbf{K}}(\mathbf{p},\mathbf{q}) = \frac{1}{2}\mathrm{dist}_{\mathbf{K}}(\mathbf{p},\mathbf{q})$ for any points $\mathbf{p},\mathbf{q}$. By Lemma 241, $\mathrm{dist}_{\mathbf{K}}(\mathbf{c}_i,\mathbf{c}_j) \geq 1$ for every $i \neq j$. Thus, $X = \{\mathbf{c}_i : i = 1,2,\ldots,n\}$ is a set of n points in the boundary of $\mathbf{C}$ at pairwise $\mathbf{C}$-distances at least $\frac{1}{2}$. Hence, $n \leq 16$. □

Now we are ready to prove our theorem. By [86], there is a parallelogram $\mathbf{P}$, circumscribed about $\bar{\mathbf{C}}$, such that the midpoints of the edges of $\mathbf{P}$ belong to $\bar{\mathbf{C}}$. Since the Hadwiger number of any affine image of $\mathbf{S}$ is equal to $H(\mathbf{S})$, we may assume that $\mathbf{P} = \{(\alpha,\beta) \in \mathbb{E}^2 : |\alpha| \leq 1 \text{ and } |\beta| \leq 1\}$. Note that the points $\mathbf{e}_x = (1,0)$ and $\mathbf{e}_y = (0,1)$ are in the boundary of $\bar{\mathbf{C}}$.

First, we show that there are two points $\mathbf{r}_x$ and $\mathbf{s}_x$ in $\mathbf{S}$, with x-coordinates ρ_x and σ_x, respectively, such that $\mathbf{e}_x \in \mathrm{conv}\{\mathbf{o}, 2\mathbf{r}_x, 2\mathbf{s}_x\}$ and $\rho_x + \sigma_x \geq 1$.

Assume that $\mathbf{e}_x = \mathbf{c}_i$ for some value of i. Since $\mathbf{S}$ and $\mathbf{S}_i$ touch, there is a (possibly degenerate) parallelogram $\mathbf{P}_i = \mathrm{conv}\{\mathbf{o},\mathbf{r}_x,\mathbf{s}_x,\mathbf{c}_i\}$ such that $\mathbf{c}_i = \mathbf{r}_x + \mathbf{s}_x$, $([\mathbf{o},\mathbf{r}_x] \cup [\mathbf{o},\mathbf{s}_x]) \subset \mathbf{S}$ and $([\mathbf{c}_i,\mathbf{r}_x] \cup [\mathbf{c}_i,\mathbf{s}_x]) \subset \mathbf{S}_i$ (cf. Figure 9.7). Observe that $\mathbf{c}_i \in \mathrm{conv}\{\mathbf{o}, 2\mathbf{r}_x, 2\mathbf{s}_x\}$ and $\rho_x + \sigma_x = 1$. If $\mathbf{e}_x = -\mathbf{c}_i$, we may choose $\mathbf{r}_x$ and $\mathbf{s}_x$ similarly.

Assume that $\mathbf{e}_x \in (\mathbf{c}_i,\mathbf{c}_j)$ for some values of i and j. Consider a parallelogram $\mathbf{P}_i = \mathrm{conv}\{\mathbf{o},\mathbf{r}_i,\mathbf{s}_i,\mathbf{c}_i\}$ such that $\mathbf{c}_i = \mathbf{r}_i + \mathbf{s}_i$, $([\mathbf{o},\mathbf{r}_i] \cup [\mathbf{o},\mathbf{s}_i]) \subset \mathbf{S}$ and $([\mathbf{c}_i,\mathbf{r}_i] \cup [\mathbf{c}_i,\mathbf{s}_i]) \subset \mathbf{S}_i$. Let L denote the line with equation $x = \frac{1}{2}$. We may assume that L separates $\mathbf{s}_i$ from $\mathbf{o}$. We define $\mathbf{r}_j$ and $\mathbf{s}_j$ similarly. If the x-axis separates the points $\mathbf{s}_i$ and $\mathbf{s}_j$, we may choose $\mathbf{s}_i$ and $\mathbf{s}_j$ as $\mathbf{r}_x$ and $\mathbf{s}_x$. If both $\mathbf{s}_i$ and $\mathbf{s}_j$ are contained in the open half plane, bounded by the x-axis and containing $\mathbf{c}_i$ or $\mathbf{c}_j$, say, $\mathbf{c}_i$, we may choose $\mathbf{r}_j$ and $\mathbf{s}_j$ as $\mathbf{r}_x$ and $\mathbf{s}_x$ (cf. Figure 9.9). If $\mathbf{e}_x$ is in $(-\mathbf{c}_i,\mathbf{c}_j)$ or $(-\mathbf{c}_i,-\mathbf{c}_j)$, we may apply a similar argument.

Analogously, we may choose points $\mathbf{r}_y$ and $\mathbf{s}_y$ in $\mathbf{S}$, with y-coordinates ρ_y and σ_y, respectively, such that $\mathbf{e}_y \in \mathrm{conv}\{\mathbf{o}, 2\mathbf{r}_y, 2\mathbf{s}_y\}$ and $\rho_y + \sigma_y \geq 1$. We may assume that $\rho_x \leq \sigma_x$ and that $\rho_y \leq \sigma_y$.

Let $\mathbf{Q}_1$, $\mathbf{Q}_2$, $\mathbf{Q}_3$ and $\mathbf{Q}_4$ denote the four closed quadrants of the coordinate

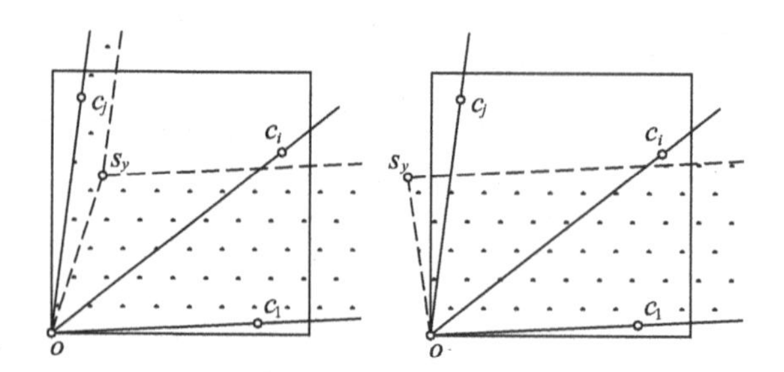

FIGURE 9.10
The y-coordinate of $\mathbf{c}_i$ is at least σ_j.

system in counterclockwise cyclic order. We may assume that $X \cap \mathbf{Q}_1 \neq \emptyset$, and that $\mathbf{Q}_1$ contains the points with non-negative x- and y-coordinates. We relabel the indices of the elements of $\mathcal{F}$ in a way that $R_1, R_2, \ldots, R_n$ are in counterclockwise cyclic order, and the angle between R_1 and the positive half of the x-axis, measured in the counterclockwise direction, is the smallest amongst all rays in $\{R_i : i = 1, 2, \ldots, n\}$.

If $\text{card}(\mathbf{Q}_i \cap X) \leq 3$ for each value of i, the assertion holds. Thus, we may assume that, say, $j = \text{card}(\mathbf{Q}_1 \cap X) > 3$. By Lemma 243, $[\mathbf{c}_i, \mathbf{c}_i - \mathbf{s}_y]$ does not cross the rays R_1 and R_j for $i = 2, 3, \ldots, j-1$. Thus, the y-coordinate of $\mathbf{c}_i$ is at least σ_y (cf. Figure 9.10, note that $\mathbf{c}_i$ is not contained in the dotted region). Similarly, the x-coordinate of $\mathbf{c}_i$ is at least σ_x for $i = 2, \ldots, j-1$. Thus, $\sigma_x \leq 1$ and $\sigma_y \leq 1$, which yield that $\rho_x \geq 0$ and $\rho_y \geq 0$. Since $\sigma_x \geq 1 - \rho_x$ and $\sigma_y \geq 1 - \rho_y$, each $\mathbf{c}_i$, with $2 \leq i \leq j-1$, is contained in the rectangle $\mathbf{T} = \{(\alpha, \beta) \in \mathbb{E}^2 : 1 - \rho_x \leq \alpha \leq 1 \text{ and } 1 - \rho_y \leq \beta \leq 1\}$.

Let $\mathbf{B} = \{(\alpha, \beta) \in \mathbb{E}^2 : |\alpha| \leq \rho_x \text{ and } |\beta| \leq \rho_y\}$. Note that if $\mathbf{S}$ and $\mathbf{p}+\mathbf{S}$ are non-overlapping and $\mathbf{x}, \mathbf{y} \in \mathbf{S}$, then the parallelogram $\text{conv}\{\mathbf{o}, \mathbf{x}, \mathbf{y}, \mathbf{x}+\mathbf{y}\}$ does not contain $\mathbf{p}$ in its interior. Thus, applying this observation with $\{\mathbf{x}, \mathbf{y}\} \subset \{\pm\mathbf{r}_x, \pm\frac{\rho_x}{\sigma_x}\mathbf{s}_x, \pm\mathbf{r}_y, \pm\frac{\rho_y}{\sigma_y}\mathbf{s}_x\}$, we obtain that $\mathbf{p} \notin \text{int}\mathbf{B}$ (cf. Figure 9.11, the dotted parallelograms show the region 'forbidden' for $\mathbf{p}$).

Furthermore, if $\mathbf{r}_x$ and $\mathbf{s}_x$ do not lie on the x-axis, and $\mathbf{r}_y$ and $\mathbf{s}_y$ do not lie on the y-axis, then the interiors of these parallelograms cover $\mathbf{B}$, apart from some points of $\mathbf{S}$, and thus, we have $\mathbf{p} \notin \mathbf{B}$. If $\mathbf{p}$ is on a vertical side of $\mathbf{B}$, then $\mathbf{r}_y$ or $\mathbf{s}_y$ lies on the y-axis (cf. Figure 9.12). Note that if $\mathbf{r}_y$ lies on the y-axis, then $\mathbf{e}_y \in \text{conv}\{\mathbf{o}, 2\mathbf{r}_y, 2\mathbf{s}_y\}$ yields $\rho_y \geq \frac{1}{2}$, or that also $\mathbf{s}_y$ lies on the y-axis. Thus, it follows in this case that $\frac{1}{2}\mathbf{e}_y \in \mathbf{S}$. Similarly, if $\mathbf{p}$ is on a horizontal side of $\mathbf{B}$, then $\frac{1}{2}\mathbf{e}_x \in \mathbf{S}$. We use this observation several times in the next three paragraphs.

Note that $\mathbf{T} = \left(1 - \frac{\rho_x}{2}, 1 - \frac{\rho_y}{2}\right) + \frac{1}{2}\mathbf{B}$. Since for any $2 \leq i < k \leq j-1$, $\mathbf{c}_i + \frac{1}{2}\mathbf{B}$ and $\mathbf{c}_k + \frac{1}{2}\mathbf{B}$ do not overlap, it follows that $\mathbf{c}_i$ and $\mathbf{c}_k$ lie on opposite sides of $\mathbf{T}$. By Lemma 4, we immediately obtain that $j \leq 5$.

Assume that $j = 5$. Then, we have $\text{card}(X \cap \mathbf{T}) = 3$, which implies that

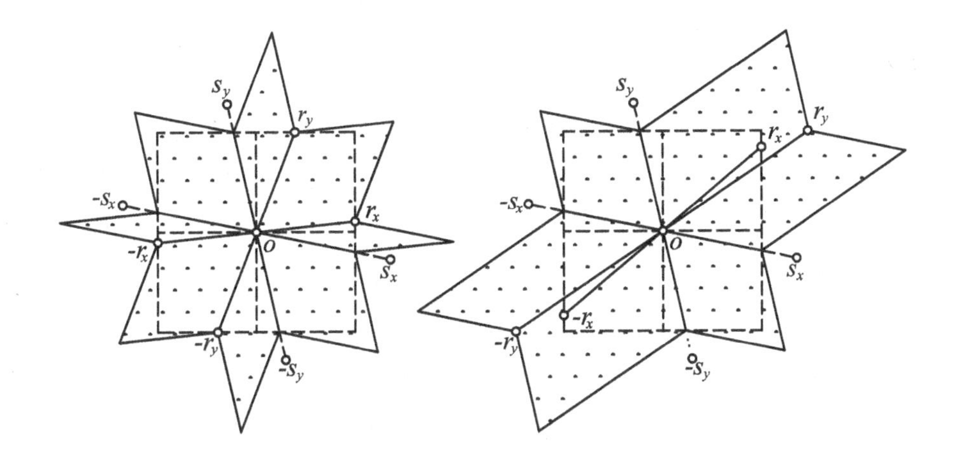

FIGURE 9.11
The regions 'forbidden' for $\mathbf{p}$.

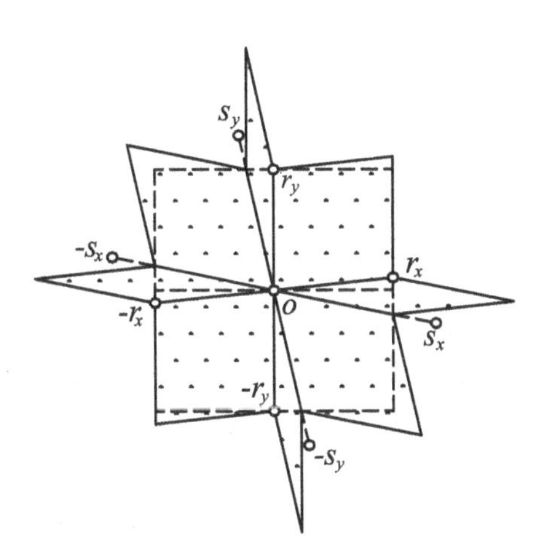

FIGURE 9.12
$\mathbf{p}$ is on a vertical side of $\mathbf{B}$.

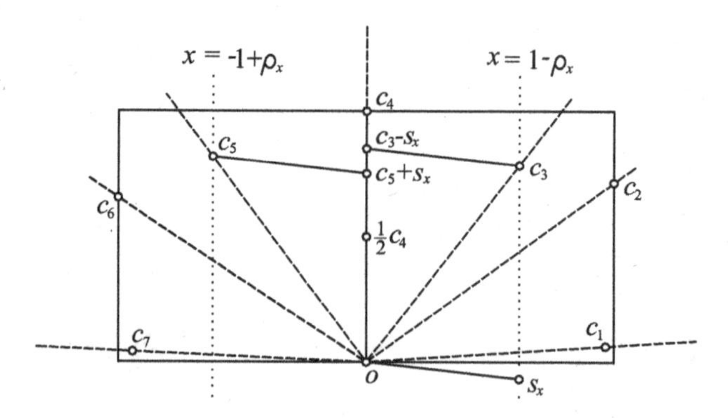

FIGURE 9.13
The case $j = 5$.

two points of $X \cap \mathbf{T}$ are consecutive vertices of $\mathbf{T}$. Without loss of generality, we may assume that $\mathbf{c}_4 = (1 - \rho_x, 1)$, $\mathbf{c}_3 = (1, 1)$ and $\mathbf{c}_2 = (\tau, 1 - \rho_y)$ for some $\tau \in [1 - \rho_y, 1]$. Since $\mathbf{c}_3 - \mathbf{c}_4$ lies on a vertical side of $\mathbf{B}$, we obtain that $\frac{1}{2}\mathbf{e}_y \in \mathbf{S}$. From the position of $\mathbf{c}_3 - \mathbf{c}_2$, we obtain similarly that $\frac{1}{2}\mathbf{e}_x \in \mathbf{S}$. Thus, if $\mathbf{c}_1$ is not on the x-axis or $\mathbf{c}_5$ is not on the y-axis, then $R_1 \cap \operatorname{int}\mathbf{S}_2 \neq \emptyset$ or $R_5 \cap \operatorname{int}\mathbf{S}_4 \neq \emptyset$, respectively; a contradiction. Hence, from $\frac{1}{2}\mathbf{e}_x, \frac{1}{2}\mathbf{e}_y \in \mathbf{S}$, it follows that $\mathbf{c}_1 = \mathbf{e}_x$ and $\mathbf{c}_5 = \mathbf{e}_y$. By Lemma 244, we have that $\mathbf{c}_2 = (1, 1-\rho_y)$, which yields that, for example, $\mathbf{S}_1$ and $\mathbf{S}_2$ overlap; a contradiction.

We are left with the case $j = 4$. We may assume that $\mathbf{c}_2$ and $\mathbf{c}_3$ lie, say, on the vertical sides of $\mathbf{T}$. Then we immediately have $\frac{1}{2}\mathbf{e}_y \in \mathbf{S}$. If $\mathbf{c}_4$ is not on the y-axis, then $R_4 \cap \operatorname{int}\mathbf{S}_3 \neq \emptyset$, and thus, it follows that $\mathbf{c}_4 = \mathbf{e}_y$. We show, by contradiction, that $\operatorname{card}((\mathbf{Q}_1 \cup \mathbf{Q}_2) \cap X) \leq 6$.

Assume that $\operatorname{card}((\mathbf{Q}_1 \cup \mathbf{Q}_2) \cap X) > 6$. Note that in this case $\operatorname{card}(\mathbf{Q}_2 \cap X) = 4$, and both $\mathbf{c}_5$ and $\mathbf{c}_6$ are either on the horizontal sides, or on the vertical sides of $\mathbf{T}' = (-2 + \rho_x, 0) + \mathbf{T}$. If they are on the horizontal sides, then $\frac{1}{2}\mathbf{e}_x \in \mathbf{S}$, $\mathbf{c}_5 = (-1, 1)$, $\mathbf{c}_7 = -\mathbf{e}_x$, and, by Lemma 244, $\mathbf{c}_6 = (-1, 1 - \rho_y)$. Thus, $\mathbf{S}_6$ overlaps both $\mathbf{S}_5$ and $\mathbf{S}_7$; a contradiction, and we may assume that $\mathbf{c}_5$ and $\mathbf{c}_6$ are on the vertical sides of $\mathbf{T}'$.

Since the y-coordinate of $\mathbf{c}_2$ is at least $\frac{1}{2}$, and since $\left(\mathbf{c}_3, \mathbf{c}_3 - \frac{1}{2}\mathbf{e}_y\right)$ does not intersect the ray R_2, we obtain that the y-coordinate of $\mathbf{c}_3$ is at least $\frac{3}{4}$. Similarly, the y-coordinate of $\mathbf{c}_5$ is at least $\frac{3}{4}$. Note that $\mathbf{c}_3 - \mathbf{s}_x$ and $\mathbf{c}_5 + \mathbf{s}_x$ are on the positive half of the y-axis. Then it follows from Lemma 243 that $\mathbf{c}_3 - \mathbf{s}_x$ and $\mathbf{c}_5 + \mathbf{s}_x$ lie on the open segment $(\mathbf{o}, \mathbf{c}_4)$ (cf. Figure 9.13). If $\mathbf{c}_3 - \mathbf{s}_x \notin \left(\frac{1}{2}\mathbf{c}_4, \mathbf{c}_4\right)$ or $\mathbf{c}_5 + \mathbf{s}_x \notin \left(\frac{1}{2}\mathbf{c}_4, \mathbf{c}_4\right)$, then we have $\mathbf{c}_5 + \mathbf{s}_x \notin (\mathbf{o}, \mathbf{c}_4)$ or $\mathbf{c}_3 - \mathbf{s}_x \notin (\mathbf{o}, \mathbf{c}_4)$, respectively. Thus, both $\mathbf{c}_5 + \mathbf{s}_x$ and $\mathbf{c}_3 - \mathbf{s}_x$ belong to $\left(\frac{1}{2}\mathbf{c}_4, \mathbf{c}_4\right)$, and a neighborhood of $\frac{1}{2}\mathbf{c}_4$ intersects $\mathbf{S}_4$ in a segment, which yields that $\mathbf{S}_4$ is not a disk; a contradiction.

Assume that $\operatorname{card}(\mathbf{Q}_4 \cap X) > 3$. Then $\operatorname{card}((\mathbf{Q}_1 \cup \mathbf{Q}_4) \cap X) > 6$ yields that

$\text{card}((\mathbf{Q}_3 \cup \mathbf{Q}_4) \cap X) \leq 6$, and the assertion follows. Thus, we may assume that $\text{card}(\mathbf{Q}_4 \cap X) \leq 3$.

Finally, assume that $\text{card}(\mathbf{Q}_3 \cap X) > 3$. Then we have $\text{card}((\mathbf{Q}_3 \cup \mathbf{Q}_4) \cap X) \leq 6$ or $\text{card}((\mathbf{Q}_2 \cup \mathbf{Q}_3) \cap X) \leq 6$. In the first case, we clearly have $\text{card}\, X \leq 12$. In the second case, by the argument used for $\mathbf{Q}_1 \cap X$, we obtain that $-\mathbf{e}_x \in X$ and $\text{card}(\mathbf{Q}_2 \cap X) \leq 3$, from which it follows that $\text{card}((\mathbf{Q}_1 \cup \mathbf{Q}_2 \cup \mathbf{Q}_3) \cap X) \leq 9$. Since $\text{card}(\mathbf{Q}_4 \cap X) \leq 3$, the assertion holds.

9.13 Proof of Theorem 111

The proof of Theorem 111 is a variant of the proof of Theorem 110. Again, let $\mathbf{S} \subset \mathbb{E}^2$ be a disk that is starlike relative to the origin, and let $\mathcal{F} = \{\mathbf{S}_i : i = 1, 2, \ldots, n\}$ be a family of pairwise non-overlapping translates of $\mathbf{S}$, with $n = H(\mathbf{S})$, such that each $\mathbf{S}_i = \mathbf{x}_i + \mathbf{S}$ touches $\mathbf{S}$. Let $\mathbf{K} = \text{conv}\mathbf{S}$, $\mathbf{K}_i = \text{conv}\mathbf{S}_i$ for $i = 1, 2, \ldots, n$, $X = \{\mathbf{x}_i : i = 1, 2, \ldots, n\}$, and $\mathbf{C} = \text{conv}X$. Furthermore, let $R_i = \{\lambda \mathbf{x}_i : \lambda \in \mathbb{R} \text{ and } \lambda \geq 0\}$.

First, we prove a few lemmas that we use in the proof.

Lemma 246 *We have* $\mathbf{o} \in \text{int}\mathbf{C}$, *and* $X \subset \text{bd}\mathbf{C}$.

Proof. Note that if $\mathbf{o} \in \text{bdconv}\,(X \cup \{\mathbf{o}\})$, then there is a supporting line $\bar{L}$ of $\mathbf{F} = \text{conv}\,(\mathbf{S} \cup (\bigcup_{i=1}^{n} \mathbf{S}_i))$ that passes through a point of $\mathbf{S}$. Thus, there is a translate of $\mathbf{S}$, on the other side of $\bar{L}$, that touches $\mathbf{S}$ and does not overlap $\mathbf{F}$. Since $n = H(\mathbf{S})$, we have a contradiction, which proves the first statement.

For contradiction, suppose that $\mathbf{x}_i \in \text{int}\mathbf{C}$ for some value of i. Note that if $i \neq j$, then $\mathbf{x}_i \notin [\mathbf{o}, \mathbf{x}_j]$. Thus, there are indices $j \neq k$ such that $\mathbf{x}_i \in \text{intconv}\{\mathbf{o}, \mathbf{x}_j, \mathbf{x}_k\}$. Since $H(\mathbf{S}') = H(\mathbf{S})$ for any affine image $\mathbf{S}'$ of $\mathbf{S}$, we may assume that $\mathbf{x}_j = \mathbf{e}_x$ and $\mathbf{x}_k = \mathbf{e}_y$.

Consider points $\mathbf{p} \in \mathbf{S}_j \cap \mathbf{S}$ and $\mathbf{q} \in \mathbf{S}_k \cap \mathbf{S}$, and note that $[\mathbf{o}, \mathbf{p}], [\mathbf{o}, \mathbf{q}], [\mathbf{o}, \mathbf{p} - \mathbf{x}_j], [\mathbf{o}, \mathbf{q} - \mathbf{x}_k] \subset \mathbf{S}$. Our aim is to show that for any such starlike disk $\mathbf{S}$, $\mathbf{S}_i$ overlaps $\mathbf{S}$, $\mathbf{S}_j$ or $\mathbf{S}_k$. In our examination, to help the reader follow the arguments, the segments in the figures belonging to $\mathbf{S}$, $\mathbf{S}_j$ or $\mathbf{S}_k$ are drawn with continuous lines, and all the other lines are dotted or dashed.

Observe that $\mathbf{x}_i$ is not contained in the open parallelograms $\mathbf{P}_j = \text{intconv}\{\mathbf{o}, \mathbf{x}_j, \mathbf{p}, \mathbf{x}_j - \mathbf{p}\}$, as otherwise the segment $[\mathbf{x}_i, \mathbf{x}_i + \mathbf{p}]$ crosses $[\mathbf{x}_j, \mathbf{p}]$, and thus, $\mathbf{S}_i$ and $\mathbf{S}_j$ overlap (note that this argument is valid also in the case that $\mathbf{p} \in [\mathbf{o}, \mathbf{x}_j]$). Similarly, $\mathbf{x}_i$ is not contained in $\mathbf{P}_k = \text{intconv}\{\mathbf{o}, \mathbf{x}_k, \mathbf{q}, \mathbf{x}_k - \mathbf{q}\}$, since otherwise $\mathbf{S}_i$ and $\mathbf{S}_k$ overlap. We set $\mathbf{T} = [\mathbf{o}, \mathbf{x}_j, \mathbf{x}_k]$ and $\mathbf{Q} = (\text{int}\mathbf{T}) \setminus (\mathcal{P}_j \cup \mathcal{P}_k)$. So far, we have that $\mathbf{x}_i \in \mathbf{Q}$.

Let $f : \mathbb{E}^2 \to \mathbb{R}$ be defined by $f((\alpha, \beta)) = \alpha + \beta$. We show that $0 \leq f(\mathbf{p}) \leq 1$ and $0 \leq f(\mathbf{q}) \leq 1$.

For contradiction, suppose first that $f(\mathbf{p}) < 0$ or $f(\mathbf{q}) < 0$. Without loss of generality, we may assume that $f(\mathbf{p}) < 0$ and $f(\mathbf{p}) \leq f(\mathbf{q})$

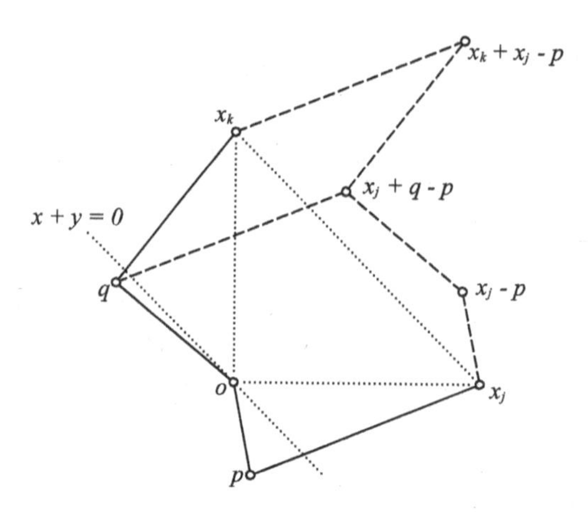

FIGURE 9.14
An illustration for Lemma 246.

(cf. Figure 9.14), which yields that $\mathbf{Q} \subseteq [\mathbf{o}, \mathbf{x}_j - \mathbf{p}) + ([\mathbf{o}, \mathbf{q}] \cup [\mathbf{q}, \mathbf{x}_k])$. If $\mathbf{x}_i \in ([\mathbf{o}, \mathbf{x}_j - \mathbf{p}) + [\mathbf{o}, \mathbf{q}))$, then $[\mathbf{x}_i, \mathbf{x}_i + \mathbf{x}_j - \mathbf{p}]$ crosses $[\mathbf{o}, \mathbf{q}]$, and thus, $\mathbf{S}_i$ overlaps $\mathbf{S}$; a contradiction. Similarly, if $\mathbf{x}_i \in ([\mathbf{o}, \mathbf{x}_j - \mathbf{p}) + (\mathbf{q}, \mathbf{x}_k])$, then $[\mathbf{x}_i, \mathbf{x}_i + \mathbf{x}_j - \mathbf{p}]$ crosses $[\mathbf{q}, \mathbf{x}_k]$, and $\mathbf{S}_i$ overlaps $\mathbf{S}_k$. Finally, if $\mathbf{x}_i \in [\mathbf{q}, \mathbf{q} + \mathbf{x}_j - \mathbf{p})$, then $\mathbf{q}$ lies in the relative interior of a segment in $\mathbb{S}_i$, from which it readily follows that $\mathbb{S}_i$ is not a disk; a contradiction.

Next, suppose that $f(\mathbf{p}) > 1$ or $f(\mathbf{q}) > 1$. Without loss of generality, we may assume that $f(\mathbf{p}) > 1$ and that $0 \leq f(\mathbf{q}) \leq f(\mathbf{p})$. Then $\mathbf{Q} \subseteq [\mathbf{o}, \mathbf{p}) + ([\mathbf{o}, \mathbf{x}_k - \mathbf{q}] \cup [\mathbf{x}_k - \mathbf{q}, \mathbf{x}_k])$. From here, the assertion follows by an argument similar to the one in the previous paragraph.

In the following, we denote the line with equation $x + y = 1$ by L.

Case 1: both the y-coordinate of $\mathbf{p}$ and the x-coordinate of $\mathbf{q}$ are negative. Without loss of generality, we may assume that $f(\mathbf{q}) \geq f(\mathbf{p})$. Then, since f is linear, we have that $f(\mathbf{x}_j + \mathbf{q} - \mathbf{p}) \geq 1$, or in other words, that L separates $\mathbf{x}_j + \mathbf{q} - \mathbf{p}$ from the origin. Note that $\mathbf{Q}$ is covered by the union of the sets $\mathbf{U}_1 = [\mathbf{x}_j, \mathbf{x}_j - \mathbf{p}) + [\mathbf{o}, \mathbf{q})$, $\mathbf{U}_2 = [\mathbf{o}, \mathbf{q}) + [\mathbf{o}, \mathbf{x}_j - \mathbf{p})$, $\mathbf{U}_3 = [\mathbf{x}_k, \mathbf{q}) + [\mathbf{o}, \mathbf{x}_j - \mathbf{p})$, $[\mathbf{q}, \mathbf{x}_j + \mathbf{q} - \mathbf{p})$ and $[\mathbf{x}_j - \mathbf{p}, \mathbf{x}_j + \mathbf{q} - \mathbf{p})$ (cf. the left-hand side of Figure 9.15). If $\mathbf{x}_i \in \mathbf{U}_1$, then $[\mathbf{x}_i, \mathbf{x}_i + \mathbf{p}]$ and $[\mathbf{x}_j, \mathbf{x}_j + \mathbf{q}]$ cross, and thus, $\mathbf{S}_i$ and $\mathbf{S}_j$ overlap; a contradiction. If $\mathbf{x}_i \in \mathbf{U}_2$ or $\mathbf{x}_i \in \mathbf{U}_3$, then $[\mathbf{x}_i, \mathbf{x}_i + \mathbf{x}_j - \mathbf{p}]$ crosses $[\mathbf{o}, \mathbf{q}]$ or $[\mathbf{q}, \mathbf{x}_k]$, respectively, and thus, $\mathbf{S}_i$ overlaps $\mathbf{S}$ or $\mathbf{S}_k$. If $\mathbf{x}_i \in [\mathbf{q}, \mathbf{x}_j + \mathbf{q} - \mathbf{p})$, then $\mathbf{S}$ and $\mathbf{S}_k$ touch each other in a relative interior point of $[\mathbf{x}_i, \mathbf{x}_i - \mathbf{p}]$, which yields that $\mathbf{S}_i$ is not a disk; a contradiction. Finally, if $\mathbf{x}_i \in [\mathbf{x}_j - \mathbf{p}, \mathbf{x}_j + \mathbf{q} - \mathbf{p})$, then $\mathbf{S}_i$ meets the segments $[\mathbf{o}, \mathbf{q})$ and $[\mathbf{x}_j, \mathbf{x}_j + \mathbf{q})$ from different sides. Since $[\mathbf{x}_j, \mathbf{x}_j + \mathbf{q})$ is the translate of $[\mathbf{o}, \mathbf{q})$ in $\mathbf{S}_j$, from this it follows that $\mathbf{S}$ is not a disk; a contradiction.

Case 2: either the y-coordinate of $\mathbf{p}$ or the x-coordinate of $\mathbf{q}$ is negative.

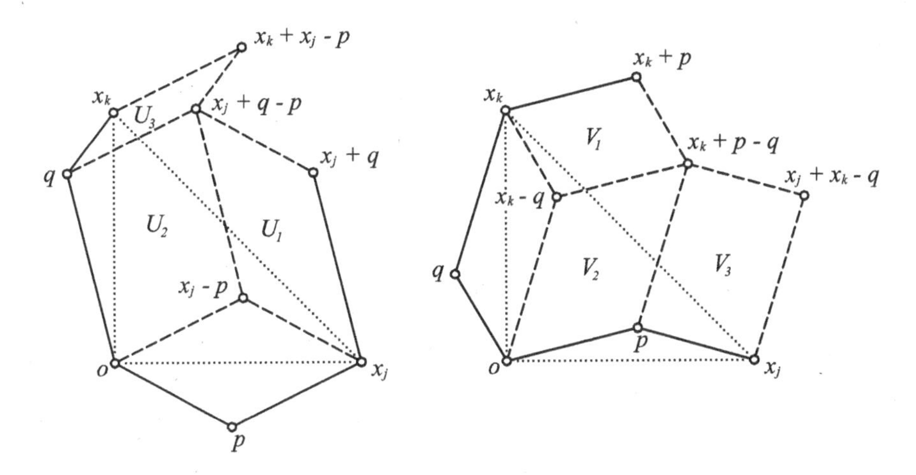

FIGURE 9.15
Illustrations for Cases 1 and 2 of Lemma 246.

Without loss of generality, we may assume that the y-coordinate of $\mathbf{p}$ is non-negative and that the x-coordinate of $\mathbf{q}$ is negative. First, we examine the case that $f(\mathbf{p}) \geq f(\mathbf{q})$, which yields that L separates $\mathbf{o}$ and $\mathbf{x}_k + \mathbf{p} - \mathbf{q}$ (cf. the right-hand side of Figure 9.15). Then $\mathbf{Q}$ is covered by the union of the sets $\mathbf{V}_1 = [\mathbf{x}_k, \mathbf{x}_k - \mathbf{q}) + [\mathbf{o}, \mathbf{p})$, $\mathbf{V}_2 = [\mathbf{o}, \mathbf{x}_k - \mathbf{q}) + [\mathbf{o}, \mathbf{p})$, $\mathbf{V}_3 = [\mathbf{x}_j, \mathbf{p}) + [\mathbf{o}, \mathbf{x}_k - \mathbf{q})$, $[\mathbf{p}, \mathbf{x}_k + \mathbf{p} - \mathbf{q})$ and $[\mathbf{x}_k - \mathbf{q}, \mathbf{x}_j + \mathbf{p} - \mathbf{q})$. If $\mathbf{x}_i \in \mathbf{V}_1$, $\mathbf{x}_i \in \mathbf{V}_2$ or $\mathbf{x}_i \in \mathbf{V}_3$, then $\mathbf{S}_i$ overlaps $\mathbf{S}_k$, $\mathbf{S}$ or $\mathbf{S}_j$, respectively. If $\mathbf{x}_i \in [\mathbf{p}, \mathbf{x}_k + \mathbf{p} - \mathbf{q})$ or $\mathbf{x}_i \in [\mathbf{x}_k - \mathbf{q}, \mathbf{x}_j + \mathbf{p} - \mathbf{q})$, then $\mathbf{S}$ is not a disk.

If $f(\mathbf{p}) \leq f(\mathbf{q})$, then the assertion follows by a similar argument.
Case 3: both the y-coordinate of $\mathbf{p}$ and the x-coordinate of $\mathbf{q}$ are non-negative. The proof in this case is similar to the proof in the previous two cases, hence we omit it. □

With reference to Lemma 246, we may relabel the indices of the elements of $\mathcal{F}$ in a way that $\mathbf{x}_1, \mathbf{x}_2, \ldots, \mathbf{x}_n = \mathbf{x}_0$ are in counterclockwise order on bd$\mathbf{C}$.

Lemma 247 *Consider points* $\mathbf{w}_i \in \mathbf{S} \cap \mathbf{S}_i$ *for* $i = 1, 2, \ldots, n$. *Then* $\mathbf{w}_1, \mathbf{w}_2, \ldots, \mathbf{w}_n$ *are in this counterclockwise order around* $\mathbf{o}$.

Proof. Note that as $\mathbf{o} \in$ int$\mathbf{C}$, and the points $\mathbf{x}_1, \mathbf{x}_2, \ldots, \mathbf{x}_n$ are in this counterclockwise order on bd$\mathbf{C}$, they are in the same order around $\mathbf{o}$. We define the points $\bar{\mathbf{x}}_i$ as follows: If $\mathbf{w}_i \in$ int$\mathbf{C}$, then $\bar{\mathbf{x}}_i = \mathbf{x}_i$, and otherwise it is the intersection point of $[\mathbf{o}, \mathbf{w}_i]$ and bd$\mathbf{C}$. Let $\bar{R}_i = \{\lambda \bar{\mathbf{x}}_i : \lambda \in \mathbb{R} \text{ and } \lambda \geq 0\}$, and let $\mathbf{Q}_i = $ intconv$(R_i \cup \bar{R}_i)$.

First, we show that if $\mathbf{x}_j \in \mathbf{Q}_i$ for some $j \neq i$, then $\mathbf{x}_j \notin$ conv$\{\mathbf{o}, \mathbf{w}_i, \mathbf{x}_i\}$, $\mathbf{w}_i \in$ conv$\{\mathbf{o}, \mathbf{w}_j, \mathbf{x}_j\}$ and $\mathbf{w}_j \notin$ int$\mathbf{C}$. Consider some $i \neq j$ with $\mathbf{x}_j \in \mathbf{Q}_i$. Then $\mathbf{w}_i \notin$ int$\mathbf{C}$, as otherwise $R_i = \bar{R}_i$. If $\mathbf{x}_j \in$ intconv$\{\mathbf{o}, \mathbf{w}_i, \mathbf{x}_i\}$, then $[\mathbf{x}_j, \mathbf{x}_j + \mathbf{w}_i]$

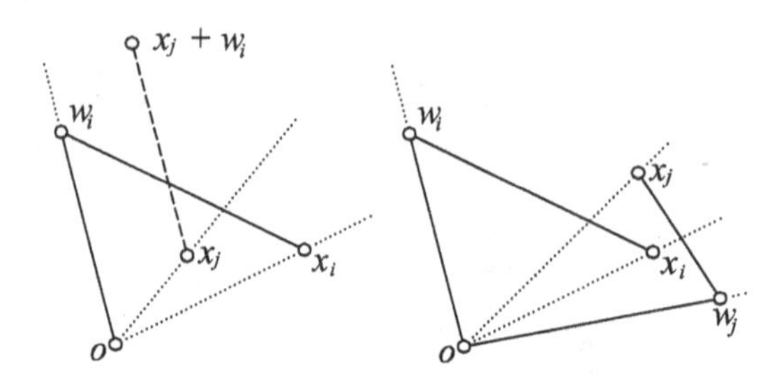

FIGURE 9.16
An illustration for Lemma 247.

crosses $[\mathbf{x}_i, \mathbf{w}_i]$, and $\mathbf{S}_i$ and $\mathbf{S}_j$ overlap; a contradiction. If $\mathbf{x}_j \in (\mathbf{w}_i, \mathbf{x}_i)$, then $[\mathbf{x}_j, \mathbf{x}_j + (\mathbf{w}_i - \mathbf{x}_i)] \subset \mathbf{S}_j$, which, since this segment is the translate of $[\mathbf{x}_i, \mathbf{w}_i]$ by $\mathbf{x}_j - \mathbf{x}_i$ and since their relative interiors intersect, yields that $\mathbf{S}$ is not a disk; a contradiction. If $\mathbf{x}_j \notin \operatorname{conv}\{\mathbf{o}, \mathbf{w}_i, \mathbf{x}_i\}$, then $[\mathbf{w}_j, \mathbf{x}_j] \cap \bar{R}_i \neq \emptyset$, as otherwise $[\mathbf{o}, \mathbf{w}_j]$ crosses $[\mathbf{w}_i, \mathbf{x}_i]$ or $\mathbf{x}_i \in \operatorname{intconv}\{\mathbf{o}, \mathbf{w}_j, \mathbf{x}_j\}$ (cf. Figure 9.16). Thus, in this case $\mathbf{w}_i \in \operatorname{conv}\{\mathbf{o}, \mathbf{w}_j, \mathbf{x}_j\}$, which, as $\mathbf{x}_j \in \operatorname{bd}\mathbf{C}$, yields that $\mathbf{w}_j \notin \operatorname{int}\mathbf{C}$.

Next, we show that $\bar{\mathbf{x}}_1, \bar{\mathbf{x}}_2, \ldots, \bar{\mathbf{x}}_n$ are in this counterclockwise order around $\mathbf{o}$. To do this, it suffices to show that there are no values of $i \neq j$ such that $\bar{\mathbf{x}}_i, \bar{\mathbf{x}}_j$ and $\bar{\mathbf{x}}_{i+1}$ are in this counterclockwise order around $\mathbf{o}$. Suppose for contradiction that there are such values.

First we consider the case that $\mathbf{x}_i$, $\mathbf{w}_j$ and $\mathbf{x}_{i+1}$ are in this counterclockwise order around $\mathbf{o}$. Since $\mathbf{x}_i$, $\mathbf{x}_j$ and $\mathbf{x}_{i+1}$ are *not* in this counterclockwise order, we have $\mathbf{x}_i \in \mathbf{Q}_j$ or $\mathbf{x}_{i+1} \in \mathbf{Q}_j$, say $\mathbf{x}_i \in \mathbf{Q}_j$. Then, clearly, $\mathbf{x}_{i+1} \notin \mathbf{Q}_j$, and, by the argument in the second paragraph of this proof, we have $\mathbf{x}_i \notin \operatorname{conv}\{\mathbf{o}, \mathbf{x}_j, \mathbf{w}_j\}$, $\mathbf{w}_j \notin \operatorname{int}\mathbf{C}$ and $\mathbf{w}_j \in \operatorname{conv}\{\mathbf{o}, \mathbf{w}_i, \mathbf{x}_i\}$. Thus, $\mathbf{w}_i \notin \operatorname{int}\mathbf{C}$, which yields that $\mathbf{x}_i, \mathbf{w}_i$ and $\mathbf{x}_{i+1}$ are in this counterclockwise order. Since $\mathbf{x}_j, \mathbf{w}_i$ and $\mathbf{w}_{i+1}$ are in this counterclockwise order, it follows that so are $\bar{\mathbf{x}}_j, \bar{\mathbf{x}}_i$ and $\bar{\mathbf{x}}_{i+1}$.

Now we examine the case that $\mathbf{x}_i$, $\mathbf{w}_j$ and $\mathbf{x}_{i+1}$ are not in this counterclockwise order. Then, since neither are $\mathbf{x}_i$, $\mathbf{x}_j$ and $\mathbf{x}_{i+1}$, we have that $\mathbf{w}_j \in \mathbf{Q}_i$ or $\mathbf{w}_j \in \mathbf{Q}_{i+1}$, say $\mathbf{w}_j \in \mathbf{Q}_i$. From this, we obtain that $\mathbf{w}_j \in \operatorname{conv}\{\mathbf{o}, \mathbf{w}_i, \mathbf{x}_i\}$, and as $\mathbf{x}_j \notin \operatorname{conv}\{\mathbf{o}, \mathbf{w}_i, \mathbf{x}_i\}$, we have that $[\mathbf{w}_j, \mathbf{x}_j]$ intersects both $[\mathbf{o}, \mathbf{x}_i]$ and $[\mathbf{o}, \mathbf{x}_{i+1}]$. Since $\mathbf{x}_i, \mathbf{x}_j \notin \operatorname{conv}\{\mathbf{o}, \mathbf{w}_{i+1}, \mathbf{x}_{i+1}\}$, this implies that $\mathbf{w}_j \in \operatorname{conv}\{\mathbf{o}, \mathbf{w}_{i+1}, \mathbf{x}_{i+1}\}$ and $\mathbf{w}_{i+1} \in \operatorname{conv}\{\mathbf{o}, \mathbf{w}_i, \mathbf{x}_i\}$. From this, it readily follows that $\mathbf{w}_i, \mathbf{w}_j, \mathbf{w}_{i+1} \notin \operatorname{int}\mathbf{C}$, and thus, $\bar{\mathbf{x}}_i$, $\bar{\mathbf{x}}_{i+1}$ and $\bar{\mathbf{x}}_j$ are in this counterclockwise order around $\mathbf{o}$.

We have shown that $\bar{\mathbf{x}}_1, \bar{\mathbf{x}}_2, \ldots, \bar{\mathbf{x}}_n$ are in this counterclockwise order around $\mathbf{o}$. Since these points are in $\operatorname{bd}\mathbf{C}$ and they can be connected to $\mathbf{o}$ by mutually non-crossing polygonal curves in $\operatorname{int}\mathbf{C}$, their counterclockwise order around $\mathbf{o}$ is the same as that of the points $\mathbf{w}_1, \mathbf{w}_2, \ldots, \mathbf{w}_n$. $\square$

We need the next lemma of A. Bezdek to prove Lemma 249 (cf. Lemma 3 in [28]).

Lemma 248 *For any* $i = 1, 2, \ldots, n$, $\text{int}\mathbf{K}_i$ *contains at most one element of* $X \setminus \{\mathbf{x}_i\}$.

We call $\mathbf{S}_i$ and $\mathbf{S}_j$ *separated*, if $\mathbf{x}_i \notin \text{int}\mathbf{K}_j$, and $\mathbf{x}_j \notin \text{int}\mathbf{K}_i$.

Lemma 249 *There is a subfamily* $\mathcal{F}'$ *of* $\mathcal{F}$, *of cardinality at least* $\lfloor \frac{n-2}{2} \rfloor$, *such that any two elements of* $\mathcal{F}'$ *are separated.*

Proof. For $i = 1, 2, \ldots, n$, we choose points $\mathbf{w}_i \in \mathbf{S} \cap \mathbf{S}_i$, and set $\Gamma_i = [\mathbf{o}, \mathbf{w}_i] \cup [\mathbf{w}_i, \mathbf{x}_i]$. By Lemma 247, the points $\mathbf{w}_1, \mathbf{w}_2, \ldots, \mathbf{w}_n$ are in counterclockwise order around $\mathbf{o}$.

By Lemma 248, $\text{int}\mathbf{K}_i$ contains at most one point of X different from $\mathbf{x}_i$. Hence, if $X \cap \text{int}\mathbf{K}_i \subset \{\mathbf{x}_{i-1}, \mathbf{x}_i, \mathbf{x}_{i+1}\}$ for every value of i, the assertion immediately follows with $\mathcal{F}' = \{S_{2m} : m = 1, 2, \ldots, \lfloor n/2 \rfloor\}$. Thus, it suffices to show that $X \cap \text{int}\mathbf{K}_i \not\subset \{\mathbf{x}_{i-1}, \mathbf{x}_i, \mathbf{x}_{i+1}\}$ for at most two values of i, as in this case, after removing these elements of $\mathcal{F}$, we may choose the elements of $\mathcal{F}'$ like in the previous case.

Consider the case that $\mathbf{x}_j \in \text{int}\mathbf{K}_i$ for some $j \notin \{i-1, i, i+1\}$. Without loss of generality, let $i = 2$. Since $\mathbf{o} \in \text{int}\mathbf{C}$, we have that the line $L_2 = R_2 \cup (-R_2)$ separates $\mathbf{x}_1$ and $\mathbf{x}_3$. Without loss of generality, we may assume that $\mathbf{x}_j$ and $\mathbf{x}_3$ lie in the same closed half plane $\mathbf{H}$ bounded by L_2, which yields that $L_3 = R_3 \cup (-R_3)$ separates $\mathbf{x}_2$ and $\mathbf{x}_j$.

Since $\mathbf{x}_j \in \text{int}\mathbf{K}_2$, there are points $\mathbf{p}, \mathbf{q} \in \mathbf{S}$ such that $\mathbf{x}_j \in \text{intconv}\{\mathbf{x}_2, \mathbf{x}_2 + \mathbf{p}, \mathbf{x}_2 + \mathbf{q}\}$. Note that by Lemma 248, we have that $\mathbf{x}_3 \notin \text{intconv}\{\mathbf{x}_2, \mathbf{x}_2 + \mathbf{p}, \mathbf{x}_2 + \mathbf{q}\}$. For contradiction, suppose that $\mathbf{o} \notin \text{intconv}\{\mathbf{x}_2, \mathbf{x}_2 + \mathbf{p}, \mathbf{x}_2 + \mathbf{q}\}$. Considering the cases that the line, passing through $\mathbf{x}_2 + \mathbf{p}$ and $\mathbf{x}_2 + \mathbf{q}$, separates $\mathbf{x}_j$ from $\mathbf{o}$, $\mathbf{x}_3$ or neither, it readily follows that at least one of $[\mathbf{x}_2, \mathbf{x}_2 + \mathbf{p}]$ or $[\mathbf{x}_2, \mathbf{x}_2 + \mathbf{q}]$ crosses both $[\mathbf{o}, \mathbf{x}_3]$ and the ray emanating from $\mathbf{x}_3$ and passing through $\mathbf{x}_j$. Since Γ_3 does not cross $[\mathbf{x}_2, \mathbf{x}_2 + \mathbf{p}]$ and $[\mathbf{x}_2, \mathbf{x}_2 + \mathbf{q}]$, we obtain that $\mathbf{x}_j \in \text{conv}\{\mathbf{o}, \mathbf{x}_3, \mathbf{w}_3\}$ or $\mathbf{x}_2 \in \text{conv}\{\mathbf{o}, \mathbf{x}_3, \mathbf{w}_3\}$, which, like in the proof of Lemma 247, immediately yields that $\mathbf{S}_j$ or $\mathbf{S}_2$ overlaps $\mathbf{S}_3$; a contradiction. Hence, we obtain that $\mathbf{o} \in \text{intconv}\{\mathbf{x}_2, \mathbf{x}_2 + \mathbf{p}, \mathbf{x}_2 + \mathbf{q}\}$. Without loss of generality, we may choose our notation so that $\mathbf{q} \in \mathbf{H}$, which implies that R_3 crosses the segment $[\mathbf{x}_2, \mathbf{x}_2 + \mathbf{q}]$.

Let $\mathbf{Q} = \text{int}\,(\text{conv}\{\mathbf{x}_2, \mathbf{o}, \mathbf{x}_2 + \mathbf{p}\} \cup \text{conv}\{\mathbf{x}_2, \mathbf{o}, \mathbf{x}_2 + \mathbf{q}\})$, and consider the case that $\mathbf{w}_1 \in \mathbf{Q}$. Since $\mathbf{S}_1$ and $\mathbf{S}$ do not overlap, we have that $\mathbf{x}_1 \notin \text{intconv}\{\mathbf{q}, \mathbf{p} + \mathbf{q}, -\mathbf{q}, \mathbf{p} - \mathbf{q}\}$. By Lemma 248, $\mathbf{x}_1 \notin \text{intconv}\{\mathbf{x}_2, \mathbf{x}_2 + \mathbf{p}, \mathbf{x}_2 + \mathbf{q}\}$, from which it readily follows that $\mathbf{x}_1 = \alpha\mathbf{p} + \beta\mathbf{q}$ with $\alpha \geq 1$. Thus, the segment $[\mathbf{x}_2, \mathbf{x}_2 + \mathbf{q}]$ crosses $[\mathbf{o}, \mathbf{w}_1 - \mathbf{x}_1]$ (cf. Figure 9.17). As $[\mathbf{o}, \mathbf{w}_1 - \mathbf{x}_1] \subset \mathbf{S}$, it follows that $\mathbf{S}$ and $\mathbf{S}_2$ overlap; a contradiction. We may show similarly that $\mathbf{w}_3 \notin \mathbf{Q}$.

We obtained that, from $\mathbf{x}_j \in \text{int}\mathbf{K}_2$ and $\mathbf{x}_j \notin \{\mathbf{x}_1, \mathbf{x}_2, \mathbf{x}_3\}$, it follows that the angle $\angle(\mathbf{w}_1, \mathbf{o}, \mathbf{w}_3)$ measured from $[\mathbf{o}, \mathbf{w}_1]$ to $[\mathbf{o}, \mathbf{w}_3]$ in the counterclockwise direction is strictly greater than π. Note that $\angle(\mathbf{w}_k, \mathbf{o}, \mathbf{w}_{k+2}) \leq \pi$ if $k \notin \{n, 1, 2\}$, and that $\angle(\mathbf{w}_n, \mathbf{o}, \mathbf{w}_2) \leq \pi$ or $\angle(\mathbf{w}_2, \mathbf{o}, \mathbf{w}_4) \leq \pi$. Thus,

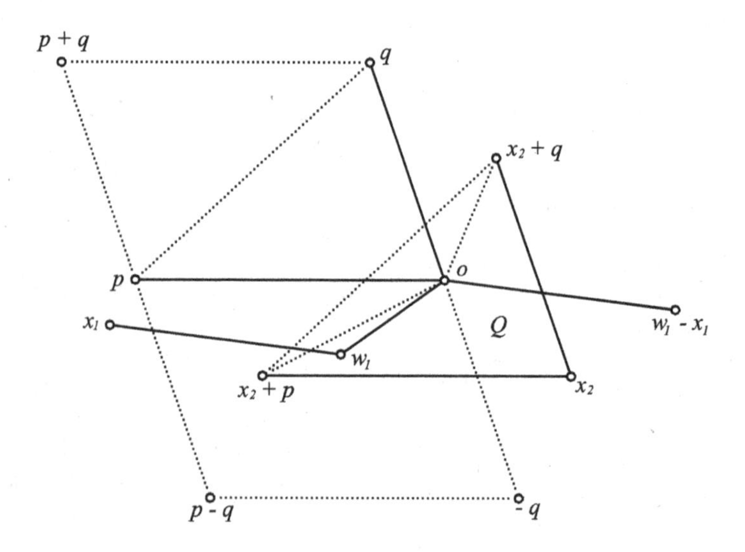

FIGURE 9.17
An illustration for Lemma 249.

$\angle(\mathbf{w}_{i-1}, \mathbf{o}, \mathbf{w}_{i+1}) > \pi$ holds for at most two values of i, and hence the assertion immediately follows. □

Now we are ready to prove Theorem 111. In the proof we use the notion of relative norm, described for example in Section 9.12, and apply an argument similar to the one in the proof of Remark 245.

Note that since $\mathbf{S}_i$ touches $\mathbf{S}$ for every i, $\mathbf{K}$ and $\mathbf{x}_i + \mathbf{K}$ intersect. Thus, $\mathrm{dist}_{\mathbf{K}}(\mathbf{o}, \mathbf{x}_i) = \mathrm{dist}_{\bar{\mathbf{K}}}(\mathbf{o}, \mathbf{x}_i) \leq 2$, where $\bar{\mathbf{K}} = \frac{1}{2}(\mathbf{K} - \mathbf{K})$. In other words, we have $X \subset 2\bar{\mathbf{K}} = \mathbf{K} - \mathbf{K}$, which yields that $\mathrm{dist}_{\mathbf{C}}(\mathbf{x}_i, \mathbf{x}_j) \geq \mathrm{dist}_{2\bar{\mathbf{K}}}(\mathbf{x}_i, \mathbf{x}_j) = \frac{1}{2}\mathrm{dist}_{\bar{\mathbf{K}}}(\mathbf{x}_i, \mathbf{x}_j)$ for any $i \neq j$.

By Lemma 249, we may choose a subfamily $\mathcal{F}'$ of at least $\lfloor \frac{n-2}{2} \rfloor$ pairwise separated elements of $\mathcal{F}$. Let X' denote the set of the translation vectors of the members of $\mathcal{F}'$. Note that if $\mathbf{y} + \mathbf{S}$ and $\mathbf{z} + \mathbf{S}$ are separated, then $\mathbf{y} + \frac{1}{2}\mathbf{K}$ and $\mathbf{z} + \frac{1}{2}\mathbf{K}$ are non-overlapping. In other words, we have $\mathrm{dist}_{\mathbf{K}}(\mathbf{y}, \mathbf{z}) \geq 1$ for any distinct $\mathbf{y}, \mathbf{z} \in X'$, which yields that $\mathrm{dist}_{\bar{\mathbf{C}}}(\mathbf{y}, \mathbf{z}) = \mathrm{dist}_{\mathbf{C}}(\mathbf{y}, \mathbf{z}) \geq \frac{1}{2}$, with $\bar{\mathbf{C}} = \frac{1}{2}(\mathbf{C} - \mathbf{C})$.

Gołab [120] proved that the circumference of every centrally symmetric convex disk measured in its norm is at least six and at most eight (cf. also Theorem 49). Fáry and Makai [101] proved that, in any norm, the circumferences of any convex disk $\mathbf{C}$ and its central symmetrization $\frac{1}{2}(\mathbf{C} - \mathbf{C})$ are equal (cf. also Theorem 50). Thus, the circumference of $\mathbf{C}$ measured in the norm with unit ball $\bar{\mathbf{C}}$ is at most eight. Since X' is a set of points in $\mathrm{bd}\mathbf{C}$ at pairwise $\bar{\mathbf{C}}$-distances at least $\frac{1}{2}$, we have $\lfloor \frac{n-2}{2} \rfloor \leq \mathrm{card}\, X' \leq 16$, from which the assertion immediately follows.

10

More Proofs on Volumetric Properties of Separable Packings

Summary. In this chapter we present selected proofs of some of the theorems in Chapter 5. Continuing our investigation of the contact numbers of totally separable translative packings, in Section 10.1 we determine $c_{\rm sep}(\mathbf{D}, n, 2)$ for any Birkhoff domain $\mathbf{D}$. In Section 10.2 we prove that Birkhoff domains are dense in the family of $\mathbf{o}$-symmetric, smooth strictly convex domains, and use this result in Section 10.3 to determine $c_{\rm sep}(\mathbf{K}, n, 2)$ for any $\mathbf{o}$-symmetric, smooth strictly convex region. In Section 10.4 we prove an Oler-type inequality for totally separable translative packings. Based on this result, we determine the value of $\delta_{\rm sep}(\mathbf{K})$ for any plane convex body $\mathbf{K}$ in Section 10.5, and prove sharp bounds for this quantity in the family of plane convex bodies in Section 10.6. We apply these results to find the minimum area of the convex hull of a totally separable packing of n translates of a plane convex body in Section 10.7, showing that the minimum is attained for example for linear packings. In Section 10.8 we prove that for ρ-separable packings, which can be regarded as local versions of totally separable packings, the same does not hold for the mean i-dimensional projection for any $i < d$ if n is suffficiently large, since in this case the minimal mean projection is attained if the convex hull is close to a Euclidean ball.

10.1 Proof of Theorem 113

Before getting to the main proof of this section, we take a detour to introduce some ideas that will be needed later. Consider the two-dimensional integer lattice $\mathbb{Z}^2$, which can also be thought of as an infinite plane tiling array of unit squares called lattice cells. For convenience, we imagine these squares to be centered at the integer points, rather than having their vertices at these points.

Definition 53 *Two lattice cells of $\mathbb{Z}^2$ are connected if they share an edge. A polyomino or n-omino is a collection of n lattice cells of $\mathbb{Z}^2$ such that from any cell we can reach any other cell through consecutive connected cells.*

Definition 54 *A packing of congruent unit diameter circular disks centered* *at the points of* $\mathbb{Z}^2$ *is called a digital circle packing [41, Section 6]. We denote the maximum contact number of such a packing of n circular disks by* $c_{\mathbb{Z}}(n, 2)$.

Recall that $c_{\text{sep}}(n, 2)$ stands for $c_{\text{sep}}(\mathbf{B}^2, n, 2)$. Clearly, every digital circle packing is totally separable and therefore $c_{\mathbb{Z}}(n, 2) \leq c_{\text{sep}}(n, 2)$. Consider a digital packing of n circular disks inscribed in the cells of an n-omino. Since each circular disk touches its circumscribing square at the midpoint of each edge and at no other point, it follows that the number of edges shared between the cells of the polyomino equals the contact number of the corresponding digital circle packing.

Through the rest of this section k, ℓ and ϵ are integers satisfying $\epsilon \in \{0, 1\}$ and $0 \leq k < \ell + \epsilon$. We note that any positive integer n can be uniquely expressed as $n = \ell(\ell + \epsilon) + k$ (as in [8]). We call this the *decomposition of* n.

Harary and Harborth [138] studied minimum-perimeter n-ominoes and Alonso and Cerf [8] characterized these in $\mathbb{Z}^2$. The latter also constructed a special class of minimum-perimeter polyominoes called basic polyominoes. Let $n = \ell(\ell + \epsilon) + k$. A *basic* n*-omino in* $\mathbb{Z}^2$ is formed by first completing a quasisquare $\mathbf{Q}_{\alpha \times \beta}$ (a rectangle whose dimensions differ by at most 1 unit) of dimensions $\alpha \times \beta$ with $\{\alpha, \beta\} = \{\ell, \ell + \epsilon\}$ and then attaching a strip $\mathbf{S}_{1 \times k}$ of dimensions $1 \times k$ (resp. $\mathbf{S}_{k \times 1}$ of dimensions $k \times 1$) to a vertical side of the quasisquare (resp. a horizontal side of the quasisquare). Here, we assume that the first dimension is along the horizontal direction. Moreover, we denote any of the resulting polyominoes by $\mathbf{Q}_{\alpha \times \beta} + \mathbf{S}_{1 \times k}$ (resp. $\mathbf{Q}_{\alpha \times \beta} + \mathbf{S}_{k \times 1}$). The results from [8, 138] indirectly show that $c_{\mathbb{Z}}(n, 2) = \lfloor 2n - 2\sqrt{n} \rfloor$. On the other hand, Bezdek, Szalkai and Szalkai [51] showed that $c_{\text{sep}}(n, 2) = \lfloor 2n - 2\sqrt{n} \rfloor$. Thus, $c_{\text{sep}}(n, 2) = c_{\mathbb{Z}}(n, 2)$ and therefore maximal contact digital packings of n circular disks are among maximal contact totally separable packings of n circular disks.

In order to make use of these ideas, we present analogues of polyominoes and digital circle packings in arbitrary normed planes.

Definition 55 *Let* $\mathbf{K_o}$ *be a smooth* $\mathbf{o}$*-symmetric convex domain in* $\mathbb{E}^2$ *and* $\mathbf{P}$ *any parallelogram (not necessarily of minimum area) circumscribing* $\mathbf{K_o}$ *such that* $\mathbf{K_o}$ *touches each side of* $\mathbf{P}$ *at its midpoint (and not at the corners of* $\mathbf{P}$ *as* $\mathbf{K_o}$ *is smooth). Let* $\mathbf{x}$ *and* $\mathbf{y}$ *be the midpoints of any two adjacent sides of* $\mathbf{P}$*. Then* $-\mathbf{x}$ *and* $-\mathbf{y}$ *are also points of* $\text{bd}\mathbf{K_o} \cap \text{bd}\mathbf{P}$*. It is easy to see that* $\{\mathbf{x}, \mathbf{y}\}$ *is an Auerbach basis of the normed plane* $(\mathbb{R}^2, \|\cdot\|_{\mathbf{K_o}})$*. We call the lattice* $\mathcal{L}_{\mathbf{P}}$ *in* $(\mathbb{R}^2, \|\cdot\|_{\mathbf{K_o}})$ *with fundamental cell* $\mathbf{P}$*, an Auerbach lattice of* $\mathbf{K_o}$ *as we can think of* $\mathcal{L}_{\mathbf{P}}$ *as being generated by the Auerbach basis* $\{\mathbf{x}, \mathbf{y}\}$ *of* $(\mathbb{R}^2, \|\cdot\|_{\mathbf{K_o}})$.

On the other hand, any Auerbach basis $\{\mathbf{x}, \mathbf{y}\}$ of a smooth $\mathbf{o}$-symmetric convex domain $\mathbf{K_o}$ generates an Auerbach lattice $\mathcal{L}_{\mathbf{P}}$ of $\mathbf{K_o}$, with fundamental cell determined by the unique lines supporting $\mathbf{K_o}$ at $\mathbf{x}$, $\mathbf{y}$, $-\mathbf{x}$ and $-\mathbf{y}$, respectively. In the sequel, we will use $\mathcal{L}_{\mathbf{P}}$ to denote the tiling of $\mathbb{R}^2$ by translates of $\mathbf{P}$ as well as the set of centers of the tiling cells. Indeed, the integer lattice $\mathbb{Z}^2$ is an Auerbach lattice of the circular disk $\mathbf{B}^2$.

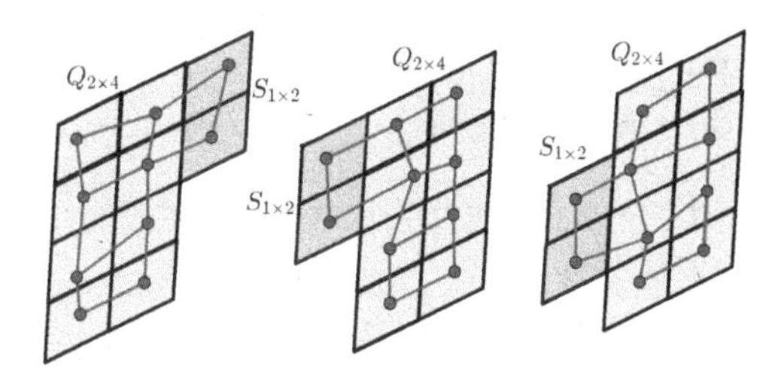

FIGURE 10.1
Some realizations of the basic polyomino $\mathbf{Q}_{2\times4}+\mathbf{S}_{1\times2}$ in some Auerbach lattice and the corresponding graphs.

Given an Auerbach lattice $\mathcal{L}_\mathbf{P}$ of a smooth $\mathbf{o}$-symmetric convex domain $\mathbf{K_o} \subseteq \mathbb{E}^2$ corresponding to the Auerbach basis $\{\mathbf{x}, \mathbf{y}\}$, we define polyominoes in $\mathcal{L}_\mathbf{P}$ as in Definition 53. We also define *basic n-ominoes in* $\mathcal{L}_\mathbf{P}$ on the same lines as in $\mathbb{Z}^2$ with the first dimension along $\mathbf{x}$, while the second dimension along $\mathbf{y}$. The *left* and *right* rows of an $\mathcal{L}_\mathbf{P}$-polyomino $\mathbf{F}$ are defined along $\mathbf{x}$-direction, while the *top* and *bottom* rows are defined along $\mathbf{y}$-direction in the natural way. The *base-lines* of $\mathbf{F}$ are the four sides of a minimal area parallelogram containing $\mathbf{F}$ (and having sides parallel to the sides of the cells of $\mathbf{F}$) and are designated (in a natural way) as the top, bottom, right and left base-line of $\mathbf{F}$. The *graph of* $\mathbf{F}$, denoted by $G(\mathbf{F})$, has a vertex for each cell of $\mathbf{F}$, with two vertices adjacent if and only if the corresponding cells share a side. Figure 10.1 shows some basic polyominoes and their graphs in some Auerbach lattice. We refer to the translates of $\mathbf{K_o}$ centered at the lattice points of $\mathcal{L}_\mathbf{P}$ (inscribed in the cells of $\mathcal{L}_\mathbf{P}$) as $\mathcal{L}_\mathbf{P}$*-translates* of $\mathbf{K_o}$. Any packing of such translates will be called an $\mathcal{L}_\mathbf{P}$*-packing* of $\mathbf{K_o}$. Since $\mathcal{L}_\mathbf{P}$ is a linear image of $\mathbb{Z}^2$, the results of [8, 138] also hold for $\mathcal{L}_\mathbf{P}$-polyominoes.

Lemma 250 *Let* $\mathbf{K_o}$ *be a smooth* $\mathbf{o}$*-symmetric convex domain,* $n = \ell(\ell+\epsilon)+k$ *be the decomposition of a positive integer* n *and* $\mathbf{F}$ *be an* n*-omino in an Auerbach lattice* $\mathcal{L}_\mathbf{P}$ *of* $\mathbf{K_o}$.

(i) If $\mathcal{P}$ *is a packing of* n *translates of* $\mathbf{K_o}$ *inscribed in the cells of* $\mathbf{F}$*, then* $G(\mathbf{F})$ *is the contact graph of* $\mathcal{P}$ *and therefore, the number of edges in* $G(\mathbf{F})$ *is equal to the contact number* $c(\mathcal{P})$ *of* $\mathcal{P}$.

(ii) If in addition $\mathbf{F}$ *is a minimum-perimeter (or basic) n-omino, then* $c(\mathcal{P}) = \lfloor 2n - 2\sqrt{n} \rfloor$.

Proof. Since $\mathbf{K_o}$ is smooth, no $\mathcal{L}_\mathbf{P}$-translate of $\mathbf{K_o}$ meets the cell of $\mathcal{L}_\mathbf{P}$ containing it at a corner of the cell. Also any $\mathcal{L}_\mathbf{P}$-translates of $\mathbf{K_o}$ touches the cell containing it at the midpoints of the four sides of the cell. Therefore, two $\mathcal{L}_\mathbf{P}$-translates of $\mathbf{K_o}$ touch if and only if the cells of $\mathcal{L}_\mathbf{P}$ containing them share a side. This proves (i).

Statement (ii) now follows from (i) and [8, 138]. □

We now show that $c_{\text{sep}}(\mathbf{D}, n, 2) = \lfloor 2n - 2\sqrt{n} \rfloor$, for any smooth B-domain

$\mathbf{D}$. The existence of a B-measure plays a key role in the following proof as it provides us a Euclidean-like angle measure to work with. The proof also heavily relies on the $\mathcal{L}_{\mathbf{P}}$-packing ideas discussed above.

Remark 251 *As $\mathbf{D}$ is a smooth B-domain, therefore by part (i) of Theorem 101, we have that $H_{\text{sep}}(\mathbf{D}) = 4$.*

First, we establish the lower bound whose proof neither uses smoothness nor the B-measure. Consider an Auerbach lattice $\mathcal{L}_{\mathbf{P}}$ of $\mathbf{D}$ corresponding to an Auerbach basis $\{\mathbf{x}, \mathbf{y}\}$ of $(\mathbb{R}^2, \|\cdot\|_{\mathbf{D}})$. Then $\mathcal{L}_{\mathbf{P}} = T(\mathbb{Z}^2)$, for some linear transformation $T : \mathbb{R}^2 \to \mathbb{R}^2$. Now for any $n \geq 2$, consider a maximal contact digital packing $\mathcal{C}$ of n circular disks and let $\mathbf{F}$ be the corresponding polyomino in $\mathbb{Z}^2$. Then $T(\mathbf{F})$ is an $\mathcal{L}_{\mathbf{P}}$-polyomino with n cells. Let $\mathcal{P}$ be the packing of $\mathcal{L}_{\mathbf{P}}$-translates of $\mathbf{D}$ inscribed in the cells of $T(\mathbf{F})$, then by Lemma 250, the contact number of $\mathcal{P}$ is at least as large as the contact number of $\mathcal{C}$. Thus $c_{\text{sep}}(\mathbf{D}, n, 2) \geq \lfloor 2n - 2\sqrt{n} \rfloor$.

Our goal is to show the reverse inequality

$$c_{\text{sep}}(\mathbf{D}, n, 2) \leq \left\lfloor 2n - 2\sqrt{n} \right\rfloor . \tag{10.1}$$

for any smooth B-domain $\mathbf{D}$ in $\mathbb{E}^2$ and $n \geq 2$. Since (10.1) clearly holds for $n \leq 3$, for proving it, we proceed by induction on n, the number of translates in the packing. For the sake of brevity, we write $c_{\text{sep}}(n) = c_{\text{sep}}(\mathbf{D}, n, 2)$. Suppose (10.1) is true for totally separable packings of up to $n-1$ translates of $\mathbf{D}$. Let G denote the contact graph of a maximal contact totally separable packing $\mathcal{P}$ of $n \geq 4$ translates of $\mathbf{D}$. Since $n \geq 2$ and $c_{\text{sep}}(n-1)+1 = \lfloor 2(n-1) - 2\sqrt{n-1} \rfloor + 1 \leq \lfloor 2n - 2\sqrt{n} \rfloor$, we can assume without loss of generality that every vertex of G has degree at least 2.

Proposition 252 *If G is not 2-connected, then the number of edges of G is at most $\lfloor 2n - 2\sqrt{n} \rfloor$.*

Proof. To prove Proposition 252 suppose that G has two subgraphs G_1 and G_2 with only one vertex in common and with n_1 and n_2 vertices, respectively. Then $n_1, n_2 \geq 2$ and $n_1 + n_2 = n + 1$, and by induction the number of edges of G is at most $(2n_1 - 2\sqrt{n_1}) + (2n_2 - 2\sqrt{n_2})$, which is easy to estimate from above by $2n - 2\sqrt{n}$. This completes the proof of Proposition 252. □

Thus, by Proposition 252 we may assume that G is a 2-connected planar graph with $c_{\text{sep}}(n)$ edges having minimum vertex degree at least 2 and so, every face of G – including the external one – is bounded by a cycle. Thus G is bounded by a simple closed polygonal curve P. Let v denote the number of vertices of P. By Remark 251, the degree of each vertex in G is 2, 3 or 4. For $j \in \{2, 3, 4\}$, let v_j be the number of vertices of P of degree j. By definition of B-domains, there exists a B-measure m in $(\mathbf{R}^2, \|\cdot\|_{\mathbf{D}})$ so that using the total separability of our packing, the internal angle of P at a vertex of degree j is at least $\frac{(j-1)\pi}{2}$. Since the internal angle sum formula holds for angular measures, the sum of these angles will be $(v-2)\pi$. Clearly $v = v_2 + v_3 + v_4$, and thus we get the inequality

$$v_2 + 2v_3 + 3v_4 \leq 2v - 4. \tag{10.2}$$

Now let g_j be the number of internal faces of G that have j sides. By total separability and smoothness, $j \geq 4$. It follows from Euler's polyhedral formula that

$$n - c_{\text{sep}}(n) + g_4 + g_5 + \ldots = 1. \tag{10.3}$$

In the process of adding up the number of sides of the internal faces of G, every edge of P is counted once and all the other edges are counted twice. Therefore,

$$4(g_4 + g_5 + \ldots) \leq 4g_4 + 5g_5 + \ldots = v + 2(c_{\text{sep}}(n) - v). \tag{10.4}$$

This, together with (10.3), implies that $4(1-n+c_{\text{sep}}(n)) \leq v+2(c_{\text{sep}}(n)-v)$, and thus we obtain

$$2c_{\text{sep}}(n) - 3n + 4 \leq n - v. \tag{10.5}$$

From G, delete the vertices of P along with the edges that are incident to them. From the definition of $c_{\text{sep}}(n-v)$, we get $c_{\text{sep}}(n) - v - (v_3 + 2v_4) \leq c_{\text{sep}}(n-v)$, which together with (10.2) implies that $c_{\text{sep}}(n) \leq c_{\text{sep}}(n-v)+2v-4$. By induction hypothesis, $c_{\text{sep}}(n-v) \leq 2(n-v) - 2\sqrt{n-v}$, and so

$$c_{\text{sep}}(n) \leq (2n-4) - 2\sqrt{n-v}. \tag{10.6}$$

Using (10.5) it follows that $c_{\text{sep}}(n) \leq (2n-4) - 2\sqrt{2c_{\text{sep}}(n) - 3n + 4}$, and so

$$c_{\text{sep}}(n)^2 - 4nc_{\text{sep}}(n) + (4n^2 - 4n) \geq 0.$$

Finally, since the solutions of the quadratic equation $x^2-4nx+(4n^2-4n) = 0$ are $x = 2n \pm 2\sqrt{n}$ and $c_{\text{sep}}(n) < 2n$, therefore, it follows that $c_{\text{sep}}(n) \leq 2n - 2\sqrt{n}$.

10.2 Proof of Theorem 116

Let $\mathbf{K_o}$ be a smooth $\mathbf{o}$-symmetric strictly convex domain and $\varepsilon > 0$ be sufficiently small. We describe the construction of a smooth strictly convex A-domain $\mathbf{A}$ with the property that $h(\mathbf{A}', \mathbf{K_o}) \leq \varepsilon$, for some image $\mathbf{A}' = T(\mathbf{A})$ of $\mathbf{A}$ under an invertible linear transformation $T : \mathbb{E}^2 \to \mathbb{E}^2$. Let $\mathbf{K}'_{\mathbf{o}} := T^{-1}(\mathbf{K_o})$. We note that $\mathbf{K}'_{\mathbf{o}}$ is a smooth $\mathbf{o}$-symmetric strictly convex domain in $\mathbb{E}^2$. It is sufficient to show that $\text{bd}\mathbf{A}$ lies in the annulus $\text{bd}\mathbf{K}'_{\mathbf{o}} + \varepsilon\mathbf{B}^2$ (see Figure 10.2)

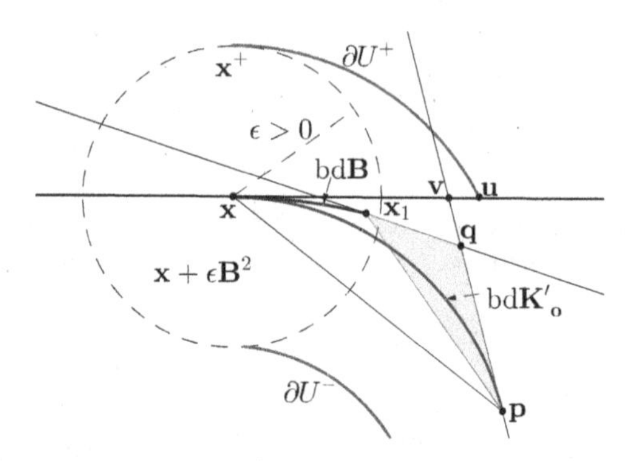

FIGURE 10.2
Replacing a part of the boundary of the smooth **o**-symmetric strictly convex domain $\mathbf{K}'_{\mathbf{o}}$ in the proof of Theorem 116 by a circular arc. The circular arc connecting $\mathbf{x}$ and $\mathbf{x}_1$ is on bd$\mathbf{B}$, while the one connecting $\mathbf{x}$ and $\mathbf{p}$ belongs to the boundary of $\mathbf{K}'_{\mathbf{o}}$. The arcs ∂U^+ and ∂U^- represent the outer and inner boundary of the annulus bd$\mathbf{K}'_{\mathbf{o}} + \varepsilon\mathbf{B}^2$, respectively. The construction is independent of whether $\mathbf{x}_1 \in \mathbf{K}'_{\mathbf{o}}$ or not.

and the length of bd$\mathbf{A} \cap$ bd$\mathbf{K}'_{\mathbf{o}}$ can be made arbitrarily close to the length of bd$\mathbf{K}'_{\mathbf{o}}$ during the construction.

We choose T so that the minimal area parallelogram $\mathbf{P}$ containing $\mathbf{K}'_{\mathbf{o}}$ is a square. Let $\mathbf{x}, \mathbf{y}, -\mathbf{x}, -\mathbf{y} \in$ bd$\mathbf{K}'_{\mathbf{o}}$ be the midpoints of the sides of $\mathbf{P}$. Then $\{\mathbf{x}, \mathbf{y}\}$ is an Auerbach basis of $(\mathbb{R}^2, \|\cdot\|_{\mathbf{K}'_{\mathbf{o}}})$ [202] and, by strict convexity, bd$\mathbf{K}'_{\mathbf{o}}$ intersects bd$\mathbf{P}$ only at points $\mathbf{x}$, $-\mathbf{x}$, $\mathbf{y}$ and $-\mathbf{y}$. Let $\mathbf{B}$ be the circular disk centered at $\mathbf{o}$ that touches bd$\mathbf{P}$ at the points $\mathbf{x}$, $\mathbf{y}$, $-\mathbf{x}$, $-\mathbf{y}$. Without loss of generality, we may assume that the side of $\mathbf{P}$ passing through $\mathbf{x}$ is horizontal and $\mathbf{y}$ lies on the clockwise arc on bd$\mathbf{K}'_{\mathbf{o}}$ from $\mathbf{x}$ to $-\mathbf{x}$.

Let $\mathbf{U}_\varepsilon =$ bd$\mathbf{K}'_{\mathbf{o}} + \varepsilon\mathbf{B}^2$ be the ε-annular neighborhood of bd$\mathbf{K}'_{\mathbf{o}}$ with outer boundary curve ∂U^+ and inner boundary curve ∂U^-. Let $\mathbf{x}^+$ be the unique point of intersection of ∂U^+ and $\mathbf{x} + \varepsilon\mathbf{B}^2$. Moving clockwise along ∂U^+ starting from $\mathbf{x}^+$, let $\mathbf{u}$ be the first point where ∂U^+ intersects bd$\mathbf{P}$. Starting from $\mathbf{x}$ and moving along bd$\mathbf{K}'_{\mathbf{o}}$ clockwise, choose a point $\mathbf{p} \in$ bd$\mathbf{K}'_{\mathbf{o}}$ so that the tangent line supporting $\mathbf{K}'_{\mathbf{o}}$ at $\mathbf{p}$ intersects $(\mathbf{x}, \mathbf{u})$. This unique point of intersection is represented by $\mathbf{v}$ in Figure 10.2. Note that by the strict convexity of $\mathbf{K}'_{\mathbf{o}}$, such a point $\mathbf{p}$ necessarily exists (for example, choose the point on bd$\mathbf{K}'_{\mathbf{o}}$ directly below $\mathbf{u}$) and any such $\mathbf{p}$ can be replaced by any point on the open arc $(\mathbf{x}, \mathbf{p})_{\mathbf{K}'_{\mathbf{o}}}$. Now choose a point $\mathbf{x}_1 \in$ bd$\mathbf{B}$ close to $\mathbf{x}$ in clockwise direction so that the line tangent to $\mathbf{B}$ at $\mathbf{x}_1$ intersects $(\mathbf{p}, \mathbf{v})$. In Figure 10.2, $\mathbf{q}$ denotes this point of intersection. Again note that such a point $\mathbf{x}_1$ necessarily exists as the line supporting $\mathbf{B}$ at $\mathbf{x}$ is horizontal. Moreover, $\mathbf{x}_1$ can be replaced by any point on the open arc $(\mathbf{x}, \mathbf{x}_1)_{\mathbf{B}}$. Therefore, we may assume that $\mathbf{x}_1 \neq \mathbf{q}$ and so $\mathbf{p}$, $\mathbf{q}$ and $\mathbf{x}_1$ form a triangle Δ. Thus there exists a (actually, infinitely many) smooth strictly convex curve $S \subseteq \Delta$ with endpoints

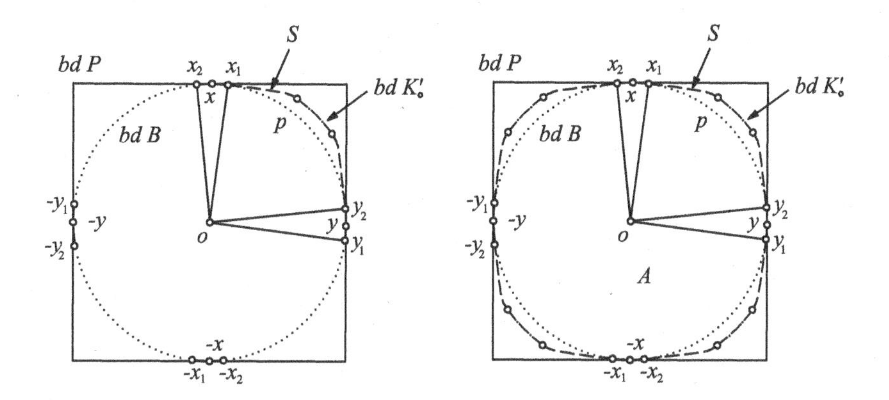

FIGURE 10.3
Construction of a smooth strictly convex A-domain approximating the smooth $\mathbf{o}$-symmetric strictly convex domain $\mathbf{K_o'}$ in the proof of Theorem 116. The circular arcs are drawn with solid lines, while the dashed-dotted arcs represent parts of the boundary of $\mathbf{K_o'}$. The dashed arcs represent smooth strictly convex connections.

$\mathbf{x}_1$ and $\mathbf{p}$ such that the convex domain

$$\mathbf{A}_1 = \text{conv}((\text{bd}\mathbf{K_o'} \backslash ([\mathbf{x}, \mathbf{p}]_{\mathbf{K_o'}} \cup [-\mathbf{x}, -\mathbf{p}]_{\mathbf{K_o'}})) \cup S \cup [\mathbf{x}, \mathbf{x}_1]_{\mathbf{B}} \cup (-S) \cup [-\mathbf{x}, -\mathbf{x}_1]_{\mathbf{B}})$$

obtained by replacing the antipodal boundary arcs $[\mathbf{x}, \mathbf{p}]_{\mathbf{K_o'}}$ and $[-\mathbf{x}, -\mathbf{p}]_{\mathbf{K_o'}}$ of $\mathbf{K_o'}$ with the antipodal circular arcs $[\mathbf{x}, \mathbf{x}_1]_{\mathbf{B}}$ and $[-\mathbf{x}, -\mathbf{x}_1]_{\mathbf{B}}$ and the smooth and strictly convex connecting curves S and $-S$ is a smooth $\mathbf{o}$-symmetric strictly convex domain with $\text{bd}\mathbf{A}_1 \subseteq \text{bd}\mathbf{K_o'} + \varepsilon\mathbf{B}^2$. Repeat the procedure for $\mathbf{A}_1$, but this time move counterclockwise along $\text{bd}\mathbf{A}_1$ starting from $\mathbf{x}$. The result is another smooth $\mathbf{o}$-symmetric strictly convex domain $\mathbf{A}_2$ with $\text{bd}\mathbf{A}_2 \subseteq \text{bd}\mathbf{K_o'} + \varepsilon\mathbf{B}^2$.

Let $[\mathbf{x}_1, \mathbf{x}_2]_{\mathbf{B}} \subseteq \text{bd}\mathbf{A}_2$ be the counterclockwise circular arc containing $\mathbf{x}$ (but not necessarily centered at $\mathbf{x}$) obtained in this way. We say that $[\mathbf{x}_1, \mathbf{x}_2]_{\mathbf{B}}$ is a *replacement arc* for $\mathbf{K_o'}$ at $\mathbf{x}$ (and therefore, $[-\mathbf{x}_1, -\mathbf{x}_2]_{\mathbf{B}}$ is a replacement arc for $\mathbf{K_o'}$ at $-\mathbf{x}$). Let $\mathbf{y}_1, \mathbf{y}_2 \in \text{bd}\mathbf{B}$ be such that $\mathbf{x}_1 \dashv_{\mathbf{B}} \mathbf{y}_1$ and $\mathbf{x}_2 \dashv_{\mathbf{B}} \mathbf{y}_2$ (and so $\mathbf{y}_1 \dashv_{\mathbf{B}} \mathbf{x}_1$, $\mathbf{y}_2 \dashv_{\mathbf{B}} \mathbf{x}_2$). By choosing $[\mathbf{x}_1, \mathbf{x}_2]_{\mathbf{B}}$ small enough, we can ensure that $[\mathbf{y}_1, \mathbf{y}_2]_{\mathbf{B}}$ is a replacement arc for $\mathbf{K_o'}$ at $\mathbf{y}$. Let $\mathbf{A}$ be the resulting convex domain as illustrated in Figure 10.3. Then $\mathbf{A}$ is a smooth strictly convex A-domain with circular pieces $[\mathbf{x}_1, \mathbf{x}_2]_{\mathbf{B}}$, $[-\mathbf{x}_1, -\mathbf{x}_2]_{\mathbf{B}}$, $[\mathbf{y}_1, \mathbf{y}_2]_{\mathbf{B}}$ and $[-\mathbf{y}_1, -\mathbf{y}_2]_{\mathbf{B}}$. Clearly, $h(\mathbf{A}, \mathbf{K_o'}) \leq \varepsilon$. Furthermore, we can make the length of $\text{bd}\mathbf{A} \cap \mathbf{K_o'}$ as close to the length of $\text{bd}\mathbf{K_o'}$ as we like. Therefore, $h(\mathbf{A}', \mathbf{K_o}) \leq \varepsilon$ and we can make the length of $\text{bd}\mathbf{A}' \cap \mathbf{K_o}$ as close to the length of $\text{bd}\mathbf{K_o}$ as we like.

10.3 Proof of Corollary 117

Let $\mathbf{K_o}$ be a smooth **o**-symmetric strictly convex domain. It is not hard to see that $c_{\mathrm{sep}}(\mathbf{K_o}, n, 2) \geq \lfloor 2n - 2\sqrt{n} \rfloor$ (for details see [42]). Let $\mathcal{P}$ be a maximal contact totally separable packing of n translates of $\mathbf{K_o}$ and $\mathcal{H}$ be a finite set of lines in $\mathbb{E}^2$ disjoint from the interiors of the translates in $\mathcal{P}$ such that any two translates are separated by at least one line in $\mathcal{H}$. We will construct a smooth strictly convex A-domain $\mathbf{A}$ such that $c_{\mathrm{sep}}(\mathbf{K_o}, n, 2) \leq c_{\mathrm{sep}}(\mathbf{A}', n, 2)$, for $\mathbf{A}' = T(\mathbf{A})$, where $T : \mathbb{E}^2 \to \mathbb{E}^2$ is a properly chosen invertible linear transformation.

Let $\mathbf{K_o} + \mathbf{c} \in \mathcal{P}$. By the strict convexity of $\mathbf{K_o}$, there exist finitely many points $\mathbf{c}_1 \ldots, \mathbf{c}_m \in \mathrm{bd}(\mathbf{K_o} + \mathbf{c})$ where $\mathbf{K_o} + \mathbf{c}$ touches other translates in $\mathcal{P}$ and the lines in $\mathcal{H}$. Then we call the points $\mathbf{c}_i - \mathbf{c} \in \mathrm{bd}\mathbf{K_o}$, $i = 1, \ldots, m$, the *contact positions* on $\mathbf{K_o}$ corresponding to $\mathbf{K_o} + \mathbf{x} \in \mathcal{P}$. Let $\mathrm{Con}(\mathcal{P})$ denote the set of all contact positions on $\mathbf{K_o}$ corresponding to all the translates in $\mathcal{P}$.

Let $\mathbf{x}$, $\mathbf{y}$, $\mathbf{P}$ and $\mathbf{B}$ be as in the proof of Theorem 116. Using Theorem 116, construct a smooth strictly convex A-domain $\mathbf{A}$ with circular pieces $c = [\mathbf{x}_1, \mathbf{x}_2]_{\mathbf{A}} = [\mathbf{x}_1, \mathbf{x}_2]_{\mathbf{B}}$, $-c = [-\mathbf{x}_1, -\mathbf{x}_2]_{\mathbf{A}} = [-\mathbf{x}_1, -\mathbf{x}_2]_{\mathbf{B}}$, $c' = [\mathbf{y}_1, \mathbf{y}_2]_{\mathbf{A}} = [\mathbf{y}_1, \mathbf{y}_2]_{\mathbf{B}}$ and $-c' = [-\mathbf{y}_1, -\mathbf{y}_2]_{\mathbf{A}} = [-\mathbf{y}_1, -\mathbf{y}_2]_{\mathbf{B}}$.

Using Theorem 116, we can make $c = [\mathbf{x}_1, \mathbf{x}_2]_{\mathbf{A}}$ sufficiently small so that

$$\mathrm{Con}(\mathcal{P}) \subseteq (\mathrm{bd}\mathbf{A}' \cap \mathrm{bd}\mathbf{K_o}) \cup \{\pm T(\mathbf{x}), \pm T(\mathbf{y})\}. \tag{10.7}$$

(Recall that both $\mathbf{B}$ and $\mathbf{K}'_{\mathbf{o}} = T^{-1}(\mathbf{K_o})$ are supported by the sides of $\mathbf{P}$ at the points $\{\pm\mathbf{x}, \pm\mathbf{y}\}$.) Thus from (10.7), the arrangement obtained by replacing each translate in $\mathcal{P}$ by the corresponding translate of $\mathbf{A}'$ is a totally separable packing with at least $c_{\mathrm{sep}}(\mathbf{K_o}, n, 2)$ contacts.

10.4 Proof of Theorem 119

For any permissible closed polygonal curve Π in Theorem 119, we set

$$F(\Pi) = \frac{\mathrm{area}(\Pi^*)}{\mathrm{area}\,(\square(\mathbf{K}))} + \frac{M_{\mathbf{K}}(\Pi)}{4} + 1.$$

We prove the assertion by induction on n. Clearly, if $n = 1$, then $F(\Pi) = 0 + 0 + 1 = 1$, and Theorem 119 holds.

Assume that for any $n' < n$, Theorem 119 holds for any totally separable translative packing of $\mathbf{K}$ with n' elements and for any permissible polygonal curve associated to it. We prove that it holds for n element packings as well.

Let L be a line intersecting Π and separating the elements of $\mathcal{F}$. We present the proof for the case only when L intersects Π at exactly two points, as the proof in the other cases is similar. Let these intersection points be $\mathbf{p}$ and $\mathbf{q}$. Then $\mathbf{p}$ and $\mathbf{q}$ are points in the relative interiors of some edges $\mathbf{p} \in [\mathbf{p}_1, \mathbf{p}_2]$ and $\mathbf{q} \in [\mathbf{q}_1, \mathbf{q}_2]$ of Π whose vertices are not contained in $L + \text{int}\mathbf{K}$. For simplicity, we imagine L as a horizontal line, $\mathbf{p}$ to the left of $\mathbf{q}$, and $\mathbf{p}_1$ and $\mathbf{q}_1$ to be above L. Let L_1 and L_2 be the upper, respectively lower, line bounding $L + \mathbf{K}$. For $i = 1, 2$, let $\mathbf{p}_i'$ and $\mathbf{q}_i'$ be the intersection points of L_i with $[\mathbf{p}, \mathbf{p}_i]$ and $[\mathbf{q}, \mathbf{q}_i]$, respectively.

Observe that if $\phi : \mathbb{E}^2 \to \mathbb{E}^2$ is an (invertible) affine transformation, then the translates $\phi(\mathbf{x}_i + \mathbf{K})$, where $i = 1, 2 \ldots, n$ form a totally separable packing, the polygon $\phi(\Pi)$ satisfies the conditions in Theorem 119, and the quantity on the left-hand side of (5.5) remains invariant under ϕ. Thus, without loss of generality, we may assume that the parallelogram $\mathbf{P}_L$ circumscribed about $\mathbf{K}$ and having the property that its area is minimal among the circumscribed parallelograms, under the condition that it has a pair of sides parallel to L, is a square of edge length 2, implying that $\text{area}(\mathbf{P}_L) = 4$.

Observe that the lines L_1 and L_2 decompose Π into four components: one above L_1, one below L_2, and the last two ones being the segments $[\mathbf{p}_1', \mathbf{p}_2']$ and $[\mathbf{q}_1', \mathbf{q}_2']$. We define Π_1' as the union of the component above L_1 and the segment $[\mathbf{p}_1', \mathbf{q}_1']$, and we define Π_2' similarly. Clearly, these polygonal curves are permissible. Finally, for $i = 1, 2$, we let $\Pi_i'^* = \Pi_i' \cup \text{int}\Pi_i'$ (cf. Figure 10.4).

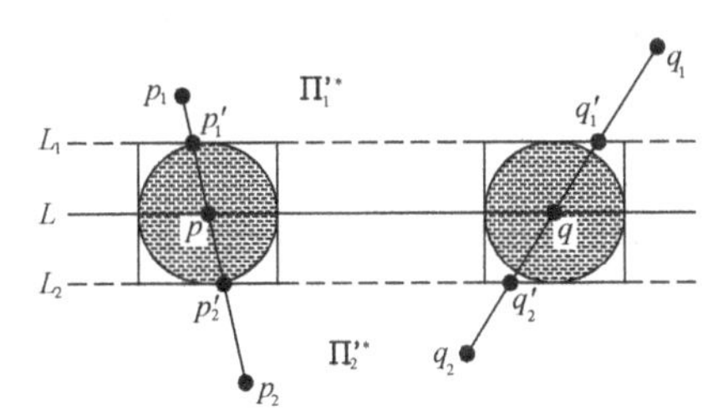

FIGURE 10.4
Notations in the proof of Theorem 119.

Then we have

$$\text{area}(\Pi^*) = \text{area}(\Pi_1'^*) + \text{area}(\Pi_2'^*) + \text{area}\left(\text{conv}\{\mathbf{p}_1', \mathbf{p}_2', \mathbf{q}_2', \mathbf{q}_1'\}\right) = \tag{10.8}$$

$$\text{area}(\Pi_1'^*) + \text{area}(\Pi_2'^*) + 2\|\mathbf{p} - \mathbf{q}\| = \text{area}(\Pi_1'^*) + \text{area}(\Pi_2'^*) + 2\|\mathbf{p} - \mathbf{q}\|_{\mathbf{K}}.$$

Furthermore, since the normed distance of L_1 and L_2 is two, we have

$$M_{\mathbf{K}}(\Pi) = M_{\mathbf{K}}(\Pi_1') + M_{\mathbf{K}}(\Pi_2') - \|\mathbf{p}_1' - \mathbf{q}_1'\|_{\mathbf{K}} - \|\mathbf{p}_2' - \mathbf{q}_2'\|_{\mathbf{K}} + \|\mathbf{p}_1' - \mathbf{p}_2'\|_{\mathbf{K}}$$

$$+\|\mathbf{q}_1' - \mathbf{q}_2'\|_{\mathbf{K}} \geq M_{\mathbf{K}}(\Pi_1') + M_{\mathbf{K}}(\Pi_2') - 2\|\mathbf{p} - \mathbf{q}\|_{\mathbf{K}} + 4. \tag{10.9}$$

Now we define a polygonal curve in $\Pi_1'^*$. Consider the points of X in the region $\mathbf{R}_1$ bounded by $[\mathbf{p}_1, \mathbf{p}_1']$, $[\mathbf{p}_1', \mathbf{q}_1']$, $[\mathbf{q}_1', \mathbf{q}_1]$ and $[\mathbf{q}_1, \mathbf{p}_1]$. Note that this region is a (not necessarily convex) quadrangle. If $\mathbf{R}_1$ is not convex, we assume, without loss of generality, that $\mathbf{p}_1 \in \mathrm{conv}\{\mathbf{p}_1', \mathbf{q}_1, \mathbf{q}_1'\}$. Consider the ray starting at $\mathbf{p}_1$ through $\mathbf{p}_1'$ and begin to rotate it counterclockwise until it hits the first point $\bar{\mathbf{p}}_1$ of X in $\mathbf{R}_1$. Then rotate this half line about $\bar{\mathbf{p}}_1$ counterclockwise until it hits the next point of X. Continuing this process we end up with a simple curve C_1 in $\mathbf{R}_1$, starting at $\mathbf{p}_1$ and ending at $\mathbf{q}_1$, which divides $\mathbf{R}_1$ into two connected components one of which contains all points of X in $\mathbf{R}_1$. We remark that if $\mathbf{R}_1$ is convex, then C_1 is a convex curve.

Let Π_1 denote the closed polygonal curve

$$(\Pi_1' \setminus ([\mathbf{p}_1, \mathbf{p}_1'] \cup [\mathbf{p}_1', \mathbf{q}_1'] \cup [\mathbf{q}_1', \mathbf{q}_1])) \cup C_1.$$

It is easy to see that Π_1 is a permissible polygonal curve whose vertices are points of X above L, and whose interior contains every other point of X above L. Let $\Pi_1^* = \Pi_1 \cup \mathrm{int}\Pi_1$. Clearly, $\mathrm{area}(\Pi_1^*) \leq \mathrm{area}(\Pi_1'^*)$. We show that $M_{\mathbf{K}}(\Pi_1) \leq M_{\mathbf{K}}(\Pi_1')$.

Case 1: $\mathbf{R}_1$ is convex. Note that in this case $C_1 \cup [\mathbf{p}_1, \mathbf{q}_1]$ is a convex region contained in the convex region $\mathbf{R}_1$, and thus, $M_{\mathbf{K}}(C_1) + \|\mathbf{p}_1 - \mathbf{q}_1\|_{\mathbf{K}} \leq M_{\mathbf{K}}(\mathrm{bd}\mathbf{R}_1)$, which readily implies our claim.

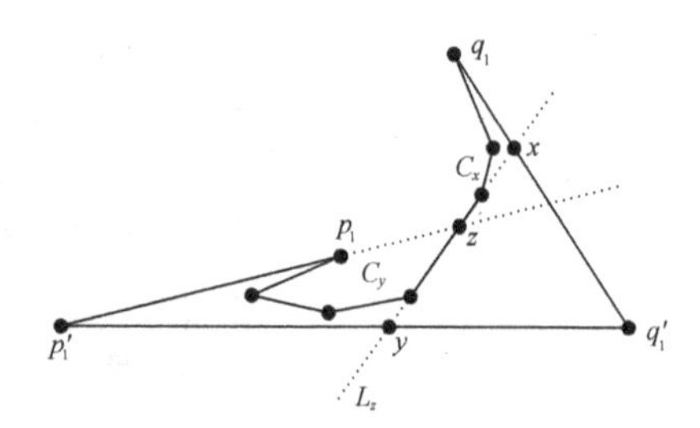

FIGURE 10.5
The points $\mathbf{p}_1, \mathbf{p}_1', \mathbf{q}_1, \mathbf{q}_1'$ are not in convex position as in Case 2.

Case 2: $\mathbf{R}_1$ is not convex. Then, according to our assumption, the line through $\mathbf{p}_1'$ and $\mathbf{p}_1$ intersects $[\mathbf{q}_1, \mathbf{q}_1']$ (cf. Figure 10.5). This line intersects C_1 at exactly one point $\mathbf{z}$, and there is a line $L_{\mathbf{z}}$ through $\mathbf{z}$ which supports C_1 at $\mathbf{z}$. Let $L_{\mathbf{z}}$ intersect $[\mathbf{q}_1, \mathbf{q}_1']$ at $\mathbf{x}$, and $[\mathbf{p}_1', \mathbf{q}_1']$ at $\mathbf{y}$. The point $\mathbf{z}$ decomposes C_1 into two convex polygonal curves $C_{\mathbf{x}}$ and $C_{\mathbf{y}}$ such that $\mathbf{p}_1 \in C_{\mathbf{y}}$ and $\mathbf{q}_1 \in C_{\mathbf{x}}$. Then we have

$$M_{\mathbf{K}}(C_1) = M_{\mathbf{K}}(C_{\mathbf{y}}) + M_{\mathbf{K}}(C_{\mathbf{x}})$$
$$\leq \|\mathbf{p}_1 - \mathbf{p}_1'\|_{\mathbf{K}} + \|\mathbf{p}_1' - \mathbf{y}\|_{\mathbf{K}} + \|\mathbf{y} - \mathbf{z}\|_{\mathbf{K}} + \|\mathbf{z} - \mathbf{x}\|_{\mathbf{K}} + \|\mathbf{x} - \mathbf{q}_1\|_{\mathbf{K}}$$
$$\leq \|\mathbf{p}_1 - \mathbf{p}_1'\|_{\mathbf{K}} + \|\mathbf{p}_1' - \mathbf{q}_1'\|_{\mathbf{K}} + \|\mathbf{q}_1' - \mathbf{q}_1\|_{\mathbf{K}}.$$

This implies that $M_{\mathbf{K}}(\Pi_1) \leq M_{\mathbf{K}}(\Pi_1')$.

To construct a permissible polygon Π_2 in $\Pi_2'^*$ with the same properties, we may apply an analogous process.

Thus, we have obtained two permissible polygons Π_1 and Π_2 associated to totally separable translative packings of $\mathbf{K}$, with strictly less elements than n, say k and $n-k$. Now, by (10.8), (10.9), area $(\square(\mathbf{K})) \leq 4$, area$(\Pi_i^*) \leq$ area$(\Pi_i'^*)$, $M_{\mathbf{K}}(\Pi_i) \leq M_{\mathbf{K}}(\Pi_i')$ (for $i = 1, 2$), and the induction hypothesis, we have

$$F(\Pi) \geq \frac{\text{area}(\Pi_1^*) + \text{area}(\Pi_2^*) + 2\|\mathbf{p} - \mathbf{q}\|_{\mathbf{K}}}{\text{area}\,(\square(\mathbf{K}))}$$

$$+\frac{M_{\mathbf{K}}(\Pi_1) + M_{\mathbf{K}}(\Pi_2) - 2\|\mathbf{p} - \mathbf{q}\|_{\mathbf{K}} + 4}{4} + 1$$

$$\geq F(\Pi_1) + F(\Pi_2) \geq k + n - k = n.$$

This completes the proof of Theorem 119.

10.5 Proof of Theorem 123

Clearly, $\delta_{sep}(\mathbf{K}) \geq \frac{\text{area}(\mathbf{K})}{\text{area}(\square(\mathbf{K}))}$. To show the opposite inequality, without loss of generality we may assume that $o \in \text{int}\mathbf{K}$.

Set $C = \min\{\mu > 0 : \mathbf{K} - \mathbf{K} \subseteq \mu\mathbf{K}\}$, and $C' = \frac{M_{\mathbf{K}}(\text{bd}\mathbf{K})}{4}$. Consider any totally separable packing $\mathcal{F}$ of translates of $\mathbf{K}$ in $\mathbb{E}^2$. For any $t > 0$, let $\mathcal{F}_t$ denote the subfamily of $\mathcal{F}$ consisting of the elements that intersect $t\mathbf{K}$, and let X_t denote the set of the translation vectors of the elements of $\mathcal{F}_t$ and n_t the cardinality of $\mathcal{F}_t$. Note that if $\mathbf{y} \in (\mathbf{x}+\mathbf{K}) \cap t\mathbf{K}$, then $\mathbf{x}+\mathbf{K} \subseteq \mathbf{y}+(\mathbf{K}-\mathbf{K})$ and therefore $\mathbf{x} + \mathbf{K} \subseteq (t + C)\mathbf{K}$, implying that $\bigcup \mathcal{F}_t \subseteq (t + C)\mathbf{K}$. On the other hand, by Theorem 119 and Remark 122, it follows that

$$n_t \leq \frac{\text{area}(\text{conv}(X_t))}{\text{area}(\square(\mathbf{K}))} + \frac{M_{\mathbf{K}}(\text{bdconv}(X_t))}{4} + 1$$

$$\leq \frac{\text{area}((t + C)\mathbf{K})}{\text{area}(\square(\mathbf{K}))} + \frac{M_{\mathbf{K}}(\text{bd}((t + C)\mathbf{K}))}{4} + 1 = (t+C)^2 \frac{\text{area}(\mathbf{K})}{\text{area}(\square(\mathbf{K}))} + (t+C)C' + 1.$$

This yields that

$$\frac{\text{area}((\bigcup \mathcal{F}) \cap t\mathbf{K})}{\text{area}(t\mathbf{K})} \leq \frac{\text{area}(\bigcup \mathcal{F}_t)}{\text{area}(t\mathbf{K})} = \frac{n_t\text{area}(\mathbf{K})}{\text{area}(t\mathbf{K})} = \frac{n_t}{t^2}$$

$$\leq \frac{(t + C)^2\text{area}(\mathbf{K})}{t^2\text{area}(\square(\mathbf{K}))} + \frac{tC' + CC' + 1}{t^2},$$

from which the claim follows by letting $t \to +\infty$.

10.6 Proof of Theorem 125

The right-hand side inequality in (5.7) is an immediate consequence of Theorem 123. We prove only the left-hand side inequality. Let $\mathbf{P}$ be a minimum area parallelogram circumscribed about $\mathbf{K}$. Without loss of generality, we may assume that $\mathbf{P}$ is the square $[0,1]^2$ in a suitable Cartesian coordinate system. Let the sides of $\mathbf{P}$ be S_1, S_2, S_3, S_4 in counterclockwise order such that the endpoints of S_1 are $(0,0)$ and $(1,0)$. Since $\mathbf{P}$ has minimum area, each side of $\mathbf{P}$ intersects $\mathbf{K}$.

We show that $S_2 \cap \mathbf{K}$ and $S_4 \cap \mathbf{K}$ contain points with equal y-coordinates. Suppose for contradiction that it is not so. Then $(1,0)+(S_4 \cap \mathbf{K})$ and $S_2 \cap \mathbf{K}$ are disjoint, implying that there is some point $\mathbf{p}_2 \in S_2$ separating these two sets. Set $\mathbf{p}_4 = (-1,0) + \mathbf{p}_2$. Then we may rotate the line of S_2 around $\mathbf{p}_2$, and the line of S_4 around $\mathbf{p}_4$ slightly, with the same angle, to obtain a parallelogram containing $\mathbf{K}$, with area equal to area$(\mathbf{P})$ and having two sides disjoint from $\mathbf{K}$, which contradicts our assumption that $\mathbf{P}$ has minimum area. Thus, there are some points $\mathbf{p}_4 = (0,t) \in S_4 \cap \mathbf{K}$ and $\mathbf{p}_2 = (1,t) \in S_2 \cap \mathbf{K}$ for some $t \in [0,1]$. We obtain similarly the existence of points $\mathbf{p}_1 = (s,0) \in S_1 \cap \mathbf{K}$ and $\mathbf{p}_3 = (s,1) \in S_3 \cap \mathbf{K}$. Hence, area$(\mathbf{K}) \geq$ area$(\mathrm{conv}\{\mathbf{p}_1, \mathbf{p}_2, \mathbf{p}_3, \mathbf{p}_4\}) = \frac{1}{2}$area$(\mathbf{P})$, which yields the left-hand side inequality in (5.7).

Now we examine the equality case. Note that, using the notations of the previous paragraph, $\frac{1}{2} = \delta_{sep}(\mathbf{K})$ implies that $\mathbf{K} = \mathrm{conv}\{\mathbf{p}_1, \mathbf{p}_2, \mathbf{p}_3, \mathbf{p}_4\}$. Consider the case that $s, t \in (0,1)$. Let $\mathbf{P}'$ be the parallelogram obtained by rotating the line of S_2 around $\mathbf{p}_2$ and the line of S_4 around $\mathbf{p}_4$, with the same small angle. Then $\mathbf{P}'$ is a parallelogram circumscribed about $\mathbf{K}$, having area equal to area$(\mathbf{P})$. Let the sides of $\mathbf{P}'$ be S_1', S_2', S_3', S_4' such that for $i = 1,2,3,4$, $\mathbf{p}_i \in S_i'$. Observe that $S_1' \cap \mathbf{K} = \{\mathbf{p}_1\}$, $S_3' \cap \mathbf{K} = \{\mathbf{p}_3\}$, and $[\mathbf{p}_1, \mathbf{p}_3]$ is not parallel to S_2' and S_4'. Thus, applying the argument in the previous paragraph, it follows that $\mathbf{P}'$ is not a minimum area circumscribed parallelogram, a contradiction. Thus, s or t is equal to 0 or 1, which implies that $\mathbf{K}$ is a triangle. This completes the proof of Theorem 125.

10.7 Proof of Theorem 126

We start with proving the following inequalities.

Lemma 253 *Let $\mathbf{K}$ be a convex domain in $\mathbb{E}^2$ and let $\mathbf{Q}$ be a convex polygon. Furthermore, let $A(\mathbf{Q}, \mathbf{K})$ denote the mixed area of $\mathbf{Q}$ and $\mathbf{K}$.*

(i) Then we have

$$\frac{12A(\mathbf{Q},\mathbf{K})}{\text{area}\,(\square(\mathbf{K}))} \geq M_{\mathbf{K}}(\text{bd}\mathbf{Q}).$$

Here, equality holds, for instance, if $\mathbf{Q} = \mathbf{K}$ *is a triangle.*

(ii) If $\mathbf{K}$ *or* $\mathbf{Q}$ *is centrally symmetric, then*

$$\frac{8A(\mathbf{Q},\mathbf{K})}{\text{area}\,(\square(\mathbf{K}))} \geq M_{\mathbf{K}}(\text{bd}\mathbf{Q}).$$

Furthermore, if $\mathbf{K}$ *is centrally symmetric, then equality holds for every convex polygon* $\mathbf{Q}$ *if and only if* bd$\mathbf{K}$ *is a Radon curve.*

Proof. Without loss of generality, we may assume that area$(\square(\mathbf{K})) = 1$.

Let k denote the number of sides of $\mathbf{Q}$, and for $i = 1, 2, \dots, k$, let l_i and $\mathbf{x}_i$ denote the (Euclidean) length and the outer unit normal vector of the ith side of $\mathbf{Q}$. Note that for every value of i, $w_{\mathbf{K}}(\mathbf{x}_i) = h_{\mathbf{K}}(\mathbf{x}_i) + h_{\mathbf{K}}(-\mathbf{x}_i)$, where $w_{\mathbf{K}}(\mathbf{x}_i)$ is the width of $\mathbf{K}$ in the direction of $\mathbf{x}_i$ and $h_{\mathbf{K}}(\mathbf{x}) = \sup\{\langle \mathbf{x}, \mathbf{k}\rangle : \mathbf{k} \in \mathbf{K}\}$ is the support function of $\mathbf{K}$ evaluated at $\mathbf{x} \in \mathbb{E}^2$ with "." standing for the standard inner product of $\mathbb{E}^2$. Furthermore, observe that

$$A(\mathbf{Q},\mathbf{K}) = \frac{1}{2}\sum_{i=1}^{k} l_i h_{\mathbf{K}}(\mathbf{x}_i). \tag{10.10}$$

First, we prove (i) for the case that $\mathbf{K}$ is centrally symmetric. Since a translation of $\mathbf{K}$ or $\mathbf{Q}$ does not change their mixed area, we may assume that $\mathbf{K}$ is $\mathbf{o}$-symmetric, which implies that $h_{\mathbf{K}}(\mathbf{x}_i) = \frac{1}{2}w_{\mathbf{K}}(\mathbf{x}_i)$ for every i. Let r_i be the Euclidean length of the radius of $\mathbf{K}$ in the direction of the ith side of $\mathbf{Q}$. Then the normed length of this side is $\frac{l_i}{r_i}$. On the other hand, since $2r_i w_{\mathbf{K}}(\mathbf{x}_i)$ is the area of a parallelogram circumscribed about $\mathbf{K}$ having minimum area under the condition that it has a side parallel to the ith side of $\mathbf{Q}$, therefore, for every value of i we have $r_i w_{\mathbf{K}}(\mathbf{x}_i) \geq \frac{1}{2}$. Combining these observations and (10.10), it follows that

$$A(\mathbf{Q},\mathbf{K}) = \frac{1}{4}\sum_{i=1}^{k} l_i w_{\mathbf{K}}(\mathbf{x}_i) = \frac{1}{4}\sum_{i=1}^{k} \frac{l_i}{r_i} r_i w_{\mathbf{K}}(\mathbf{x}_i) \geq \frac{1}{8}\sum_{i=1}^{k} \frac{l_i}{r_i} = \frac{1}{8}M_{\mathbf{K}}(\mathbf{Q}).$$

Here, equality holds for every convex polygon $\mathbf{Q}$ if and only if for any $\mathbf{v} \in \mathbb{S}^1 = \{\mathbf{x} \in \mathbb{E}^2 : |\mathbf{x}| = 1\}$, there is a minimum area parallelogram circumscribed about $\mathbf{K}$, which has a side parallel to $\mathbf{v}$. In other words, for any $\mathbf{v} \in \mathbb{S}^1$, we have that $l_K(v)w_{\mathbf{K}}(\mathbf{v}^{\perp})$ is independent of $\mathbf{v}$, where $l_{\mathbf{K}}(v)$ is the length of a longest chord of $\mathbf{K}$ in the direction of $\mathbf{v}$, and $w_K(\mathbf{v}^{\perp})$ is the width of $\mathbf{K}$ in the direction perpendicular to $\mathbf{v}$. The observation that this property is equivalent to the fact that bd$\mathbf{K}$ is a Radon curve can be found, for example, in the proof of Theorem 2 of [144] (see also the proof of Theorem 46).

Now consider the case that $\mathbf{Q}$ is $\mathbf{o}$-symmetric, but $\mathbf{K}$ is not necessarily. Note that in this case k is even, and for every i we have $l_{i+k/2} = l_i$, and $\mathbf{x}_{i+k/2} = -\mathbf{x}_i$. Thus, by (10.10)

$$A(\mathbf{Q}, \mathbf{K}) = \frac{1}{2}\sum_{i=1}^{k/2} l_i(h_{\mathbf{K}}(\mathbf{x}_i) + h_{\mathbf{K}}(-\mathbf{x}_i)) = \frac{1}{2}\sum_{i=1}^{k/2} l_i w_{\mathbf{K}}(\mathbf{x}_i) = \frac{1}{4}\sum_{i=1}^{k} l_i w_{\mathbf{K}}(\mathbf{x}_i).$$

From this equality, the statement follows by a similar argument using the relative norm of $\mathbf{K}$ whenever $\mathbf{K}$ is not centrally symmetric.

Finally, we prove (i) about the general case. Let $\bar{\mathbf{K}} = \frac{1}{2}(\mathbf{K} - \mathbf{K})$. Without loss of generality, we may assume that the origin $\mathbf{o}$ is the center of a maximum area triangle inscribed in $\mathbf{K}$. Then, clearly, $-\frac{1}{2}\mathbf{K} \subseteq \mathbf{K}$, from which a simple algebraic transformation yields that $\frac{2}{3}\bar{\mathbf{K}} \subseteq \mathbf{K}$. This implies that for any unit vector $\mathbf{x}$ we have

$$h_{\mathbf{K}}(\mathbf{x}) \geq \frac{2}{3} h_{\bar{\mathbf{K}}}(\mathbf{x}). \tag{10.11}$$

Then, by (10.10), we have

$$A(\mathbf{Q}, \mathbf{K}) \geq \frac{1}{3}\sum_{i=1}^{k} l_i h_{\bar{\mathbf{K}}}(\mathbf{x}_i) = \frac{2}{3} A(\mathbf{Q}, \bar{\mathbf{K}}).$$

Thus, our inequality readily follows from (253.2). The fact that here equality holds if $\mathbf{Q} = \mathbf{K}$ is a triangle can be shown by an elementary computation. □

First, we prove (ii). Note that bd$\mathbf{C}$ satisfies the conditions in Theorem 119, and thus (using Remark 122 if $\mathbf{K}$ is not centrally symmetric), we have

$$\frac{\operatorname{area}(\mathbf{C})}{\operatorname{area}(\square(\mathbf{K}))} + \frac{M_{\mathbf{K}}(\operatorname{bd}\mathbf{C})}{4} + 1 \geq n.$$

Thus, (253.2) of Lemma 253 yields that

$$\frac{\operatorname{area}(\mathbf{C})}{\operatorname{area}(\square(\mathbf{K}))} + \frac{2A(\mathbf{C}, \mathbf{K})}{\operatorname{area}(\square(\mathbf{K}))} + 1 \geq n.$$

From this, it follows that

$$\operatorname{area}\left(\operatorname{conv}\left(\bigcup_{i=1}^{n}(\mathbf{c}_i + \mathbf{K})\right)\right) = \operatorname{area}(\mathbf{C}+\mathbf{K}) = \operatorname{area}(\mathbf{C}) + 2A(\mathbf{C}, \mathbf{K}) + \operatorname{area}(\mathbf{K})$$

$$\geq (n-1)\operatorname{area}(\square(\mathbf{K})) + \operatorname{area}(\mathbf{K}).$$

Now we prove (126.1). In this case, Theorem 119 applied to bd$\mathbf{C}$ in the same way as above followed by (253.1) of Lemma 253 implies that

$$(n-1)\operatorname{area}(\square(\mathbf{K})) \leq \operatorname{area}(\mathbf{C}) + 3A(\mathbf{C}, \mathbf{K})$$

$$= \frac{3}{2}\text{area}(\mathbf{C}+\mathbf{K}) - \frac{1}{2}\text{area}(\mathbf{C}) - \frac{3}{2}\text{area}(\mathbf{K}).$$

This inequality yields

$$\text{area}(\mathbf{C}+\mathbf{K}) \geq \frac{2(n-1)}{3}\text{area}(\square(\mathbf{K})) + \text{area}(\mathbf{K}) + \frac{1}{3}\text{area}(\mathbf{C}),$$

finishing the proof of (126.1).

10.8 Proof of Theorem 130

We start recalling Lemma 235 in the following form.

Lemma 254 *Let $\{\mathbf{c}_i + \mathbf{C} \mid 1 \leq i \leq n\}$ be an arbitrary ρ-separable packing of n translates of the $\mathbf{o}$-symmetric convex body $\mathbf{C}$ in $\mathbb{E}^d$ with $\rho \geq 1$, $n \geq 1$, and $d \geq 2$. Then*

$$\frac{n\text{vol}_d(\mathbf{C})}{\text{vol}_d\left(\cup_{i=1}^{n}\mathbf{c}_i + 2\rho\mathbf{C}\right)} \leq \delta_{\text{sep}}(\rho, \mathbf{C}) \ .$$

Definition 56 *Let $d \geq 2$, $\rho \geq 1$, and let $\mathbf{K}$ (resp., $\mathbf{C}$) be a convex body (resp., an $\mathbf{o}$-symmetric convex body) in $\mathbb{E}^d$. Then let $\nu_{\mathbf{C}}(\rho, \mathbf{K})$ denote the largest n with the property that there exists a ρ-separable packing $\{\mathbf{c}_i + \mathbf{C} \mid 1 \leq i \leq n\}$ such that $\{\mathbf{c}_i \mid 1 \leq i \leq n\} \subset \mathbf{K}$.*

Lemma 255 *Let $d \geq 2$, $\rho \geq 1$, and let $\mathbf{K}$ (resp., $\mathbf{C}$) be a convex body (resp., an $\mathbf{o}$-symmetric convex body) in $\mathbb{E}^d$. Then*

$$\left(1 + \frac{2\rho R(\mathbf{C})}{r(\mathbf{K})}\right)^{-d} \frac{\text{vol}_d(\mathbf{C})\nu_{\mathbf{C}}(\rho, \mathbf{K})}{\delta_{\text{sep}}(\rho, \mathbf{C})} \leq \text{vol}_d(\mathbf{K}) \leq \frac{\text{vol}_d(\mathbf{C})\nu_{\mathbf{C}}(\rho, \mathbf{K})}{\delta_{\text{sep}}(\rho, \mathbf{C})}.$$

Proof. Observe that Lemma 254 and the containments $\mathbf{K} + 2\rho\mathbf{C} \subseteq \left(1 + \frac{2\rho R(\mathbf{C})}{r(\mathbf{K})}\right)\mathbf{K}$ yield the lower bound immediately.

We prove the upper bound. Let $0 < \varepsilon < \delta_{\text{sep}}(\rho, \mathbf{C})$. By the definition of $\delta_{\text{sep}}(\rho, \mathbf{C})$, if λ is sufficiently large, then there is a ρ-separable packing $\{\mathbf{c}_i + \mathbf{C} \mid 1 \leq i \leq n\}$ such that $C_n := \{\mathbf{c}_i \mid 1 \leq i \leq n\} \subset \mathbf{W}^d_\lambda$ and

$$\frac{n\text{vol}_d(\mathbf{C})}{\text{vol}_d(\mathbf{W}^d_\lambda)} \geq \delta_{\text{sep}}(\rho, \mathbf{C}) - \varepsilon. \tag{10.12}$$

Sublemma 256 *If $\mathbf{X}$ and $\mathbf{Y}$ are convex bodies in $\mathbb{E}^d$ and $\mathbf{C}$ is an $\mathbf{o}$-symmetric convex body in $\mathbb{E}^d$, then*

$$\nu_{\mathbf{C}}(\rho, \mathbf{Y}) \geq \frac{\text{vol}_d(\mathbf{Y})\nu_{\mathbf{C}}(\rho, \mathbf{X})}{\text{vol}_d(\mathbf{X} - \mathbf{Y})}. \tag{10.13}$$

Proof. Indeed, consider any finite point set $X := \{\mathbf{x}_1, \ldots, \mathbf{x}_N\} \subset \mathbf{X}$. Observe that the following are equivalent for a positive integer k:

- k is the maximum number a point of $\mathbf{X} - \mathbf{Y}$ is covered by the sets $\mathbf{x}_i - \mathbf{Y}$, $\mathbf{x}_i \in X$,
- k is the maximum number such that $\mathrm{card}((\mathbf{z} + \mathbf{Y}) \cap X) = k$ for some point $\mathbf{z} \in \mathbf{X} - \mathbf{Y}$.

Thus, $N\mathrm{vol}_d(\mathbf{Y}) \leq \mathrm{card}((\mathbf{z} + \mathbf{Y}) \cap X)\mathrm{vol}_d(\mathbf{X} - \mathbf{Y})$ for some $\mathbf{z} \in \mathbf{X} - \mathbf{Y}$. Hence, if $\{\mathbf{x}_i + \mathbf{C} \mid 1 \leq i \leq N\}$ is an arbitrary ρ-separable packing with $X = \{\mathbf{x}_1, \ldots, \mathbf{x}_N\} \subset \mathbf{X}$, then

$$\nu_{\mathbf{C}}(\rho, \mathbf{Y}) \geq \mathrm{card}((\mathbf{z} + \mathbf{Y}) \cap X) \geq \frac{\mathrm{vol}_d(\mathbf{Y})N}{\mathrm{vol}_d(\mathbf{X} - \mathbf{Y})},$$

which implies (10.13). $\square$

Applying (10.13) to $\mathbf{X} = \mathbf{W}^d_\lambda$ and $\mathbf{Y} = -\mathbf{K}$ and using (10.12), we obtain

$$\nu_{\mathbf{C}}(\rho, \mathbf{K}) \geq \frac{n\mathrm{vol}_d(\mathbf{K})}{\mathrm{vol}_d(\mathbf{W}^d_\lambda + \mathbf{K})} \geq \frac{\mathrm{vol}_d(\mathbf{K})}{\mathrm{vol}_d(\mathbf{W}^d_{\lambda + R(\mathbf{K})})} \cdot \frac{\mathrm{vol}_d(\mathbf{W}^d_\lambda)(\delta_{\mathrm{sep}}(\rho, \mathbf{C}) - \varepsilon)}{\mathrm{vol}_d(\mathbf{C})},$$

which finishes the proof of Lemma 255. $\square$

Definition 57 *Let $d \geq 2$, $n \geq 1$, $\rho \geq 1$, and let $\mathbf{C}$ be an $\mathbf{o}$-symmetric convex body in $\mathbb{E}^d$. Then let $R_{\mathbf{C}}(\rho, n)$ be the smallest radius $R > 0$ with the property that $\nu_{\mathbf{C}}(\rho, R\mathbf{B}^d) \geq n$.*

Clearly, for any $\varepsilon > 0$ we have $\nu_{\mathbf{C}}(\rho, (R_{\mathbf{C}}(\rho, n) - \varepsilon)\mathbf{B}^d) < n$, and thus, by Lemma 255 (for $\mathbf{K} = R_{\mathbf{C}}(\rho, n)\mathbf{B}^d$), we obtain

Corollary 257 *Let $d \geq 2$, $n \geq 1$, $\rho \geq 1$, and let $\mathbf{C}$ be an $\mathbf{o}$-symmetric convex body in $\mathbb{E}^d$. Then*

$$R_{\mathbf{C}}(\rho, n)^d \leq \frac{\mathrm{vol}_d(\mathbf{C})n}{\delta_{\mathrm{sep}}(\rho, \mathbf{C})\kappa_d} \leq (R_{\mathbf{C}}(\rho, n) + 2\rho R(\mathbf{C}))^d. \tag{10.14}$$

Lemma 258 *Let $n \geq \frac{4^d \delta_{\mathrm{sep}}(\rho, \mathbf{C})\rho^d R(\mathbf{C})^d}{r(\mathbf{C})^d}$ and $i = 1, 2, \ldots, d-1$. Then for $R = R_{\mathbf{C}}(\rho, n)$,*

$$M_i((R + \rho R(\mathbf{C}))\mathbf{B}^d) \leq$$

$$M_i(\mathbf{B}^d)\left(\frac{\mathrm{vol}_d(\mathbf{C})n}{\delta_{\mathrm{sep}}(\rho, \mathbf{C})\kappa_d}\right)^{\frac{i}{d}}\left(1 + \frac{2\delta_{\mathrm{sep}}(\rho, \mathbf{C})^{\frac{1}{d}}\rho R(\mathbf{C})}{r(\mathbf{C})} \cdot \frac{1}{n^{\frac{1}{d}}}\right)^i.$$

Proof. Set $t = R + 2\rho R(\mathbf{C})$. Then the first inequality in (10.14) yields that

$$R + \rho R(\mathbf{C}) \leq \frac{t - \rho R(\mathbf{C})}{t - 2\rho R(\mathbf{C})}\left(\frac{\mathrm{vol}_d(\mathbf{C})n}{\delta_{\mathrm{sep}}(\rho, \mathbf{C})\kappa_d}\right)^{\frac{1}{d}}.$$

Thus, by the second inequality in (10.14) and the condition that $n \geq \frac{4^d \delta_{\text{sep}}(\rho, \mathbf{C}) \rho^d R(\mathbf{C})^d}{r(\mathbf{C})^d} \geq \frac{4^d \delta_{\text{sep}}(\rho, \mathbf{C}) \rho^d R(\mathbf{C})^d \kappa_d}{\text{vol}_d(\mathbf{C})}$, we obtain that

$$\frac{t - \rho R(\mathbf{C})}{t - 2\rho R(\mathbf{C})} = 1 + \left(\frac{t}{\rho R(\mathbf{C})} - 2\right)^{-1} \leq 1 + \frac{2\delta_{\text{sep}}(\rho, \mathbf{C})^{\frac{1}{d}} \rho R(\mathbf{C}) \kappa_d^{\frac{1}{d}}}{\text{vol}_d(\mathbf{C})^{\frac{1}{d}}} \cdot \frac{1}{n^{\frac{1}{d}}}$$

$$\leq 1 + \frac{2\delta_{\text{sep}}(\rho, \mathbf{C})^{\frac{1}{d}} \rho R(\mathbf{C})}{r(\mathbf{C})} \cdot \frac{1}{n^{\frac{1}{d}}}. \qquad \square$$

In the proof of Theorem 130 that follows, we are going to use the following special case of the Alexandrov-Fenchel inequality ([217]): if $\mathbf{K}$ is a convex body in $\mathbb{E}^d$ satisfying $M_i(\mathbf{K}) \leq M_i(r\mathbf{B}^d)$ for given $1 \leq i < d$ and $r > 0$, then

$$M_j(\mathbf{K}) \leq M_j(r\mathbf{B}^d) \tag{10.15}$$

holds for all j with $i < j \leq d$. In particular, this statement for $j = d$ can be restated as follows: if $\mathbf{K}$ is a convex body in $\mathbb{E}^d$ satisfying $M_d(\mathbf{K}) = M_d(r\mathbf{B}^d)$ for given $d \geq 2$ and $r > 0$, then $M_i(\mathbf{K}) \geq M_i(r\mathbf{B}^d)$ holds for all i with $1 \leq i < d$.

Let $d \geq 2$, $1 \leq i \leq d-1$, $\rho \geq 1$, and let $\mathbf{Q}$ be the convex hull of the ρ-separable packing of n translates of the $\mathbf{o}$-symmetric convex body $\mathbf{C}$ in $\mathbb{E}^d$ such that $M_i(\mathbf{Q})$ is minimal and

$$n \geq \frac{4^d d^{4d}}{\delta_{\text{sep}}(\rho, \mathbf{C})^{d-1}} \cdot \left(\rho \frac{R(\mathbf{C})}{r(\mathbf{C})}\right)^d. \tag{10.16}$$

By the minimality of $M_i(\mathbf{Q})$ we have that

$$M_i(\mathbf{Q}) \leq M_i(R\mathbf{B}^d + \mathbf{C}) \leq M_i((R + \rho R(\mathbf{C}))\mathbf{B}^d) \tag{10.17}$$

with $R = R_{\mathbf{C}}(\rho, n)$. Note that (10.17) and Lemma 258 imply that

$$M_i(\mathbf{Q}) \leq \left(1 + \frac{2\delta_{\text{sep}}(\rho, \mathbf{C})^{\frac{1}{d}} \rho R(\mathbf{C})}{r(\mathbf{C})} \cdot \frac{1}{n^{\frac{1}{d}}}\right)^i M_i(\mathbf{B}^d) \left(\frac{\text{vol}_d(\mathbf{C})}{\delta_{\text{sep}}(\rho, \mathbf{C})\kappa_d}\right)^{\frac{i}{d}} \cdot n^{\frac{i}{d}}.$$

We examine the function $x \mapsto (1+x)^i$, where, by (10.16), we have $x \leq x_0 = \frac{1}{2d^4}$. The convexity of this function implies that $(1+x)^i \leq 1 + i(1+x_0)^{i-1}x$. Thus, from the inequality $\left(1 + \frac{1}{2d^4}\right)^{d-1} \leq \frac{33}{32} < 1.05$, where $d \geq 2$, the upper bound for $M_i(\mathbf{Q})$ in Theorem 130 follows.

On the other hand, in order to prove the lower bound for $M_i(\mathbf{Q})$ in Theorem 130, we start with the observation that (10.15) (based on (10.17)), (10.16), and Lemma 258 yield that

$$\text{svol}_{d-1}(\mathbf{Q}) \leq \text{svol}_{d-1}((R + \rho R(\mathbf{C}))\mathbf{B}^d)$$

$$\leq \kappa_d \left(\frac{n\text{vol}_d(\mathbf{C})}{\delta_{\text{sep}}(\rho, \mathbf{C})\kappa_d}\right)^{\frac{d-1}{d}} \left(1 + \frac{2\delta_{\text{sep}}(\rho, \mathbf{C})^{\frac{1}{d}} \rho R(\mathbf{C})}{r(\mathbf{C})} \cdot \frac{1}{n^{\frac{1}{d}}}\right)^{d-1}. \tag{10.18}$$

Thus, (10.18) together with the inequalities $\mathrm{svol}_{d-1}(\mathbf{Q})r(\mathbf{Q}) \geq \mathrm{vol}_d(\mathbf{Q})$ (cf. [195]) and $\mathrm{vol}_d(\mathbf{Q}) \geq n\mathrm{vol}_d(\mathbf{C})$ yield

$$r(\mathbf{Q}) \geq \left(1+\frac{2\delta_{\mathrm{sep}}(\rho,\mathbf{C})^{\frac{1}{d}}\rho R(\mathbf{C})}{r(\mathbf{C})}\cdot\frac{1}{n^{\frac{1}{d}}}\right)^{-(d-1)} \frac{\mathrm{vol}_d(\mathbf{C})^{\frac{1}{d}}\delta_{\mathrm{sep}}(\rho,\mathbf{C})^{\frac{d-1}{d}}}{d\kappa_d^{\frac{1}{d}}}\cdot n^{\frac{1}{d}}. \tag{10.19}$$

Applying the assumption (10.16) and $\mathrm{vol}_d(\mathbf{C}) \geq \kappa_d r(\mathbf{C})^d$ to (10.19), we get that

$$\begin{aligned} r(\mathbf{Q}) &\geq \left(1+\frac{1}{2d^4}\right)^{-(d-1)} \frac{\delta_{\mathrm{sep}}(\rho,\mathbf{C})^{\frac{d-1}{d}} r(\mathbf{C})}{d} n^{\frac{1}{d}} \\ &\geq \frac{4d^3}{(1+\frac{1}{2d^4})^{d-1}} R(\mathbf{C}) \geq 31R(\mathbf{C}). \end{aligned} \tag{10.20}$$

Let $\mathbf{P}$ denote the convex hull of the centers of the translates of $\mathbf{C}$ in $\mathbf{Q}$. Then, (10.20) implies

$$r(\mathbf{P}) \geq r(\mathbf{Q}) - R(\mathbf{C}) \geq \frac{30}{31} r(\mathbf{Q}) \geq \frac{8\delta_{\mathrm{sep}}(\rho,\mathbf{C})^{\frac{d-1}{d}} r(\mathbf{C})}{9d}\cdot n^{\frac{1}{d}}. \tag{10.21}$$

Hence, by (10.21) and Lemma 255,

$$\mathrm{vol}_d(\mathbf{Q}) \geq \mathrm{vol}_d(\mathbf{P}) \geq \left(1+\frac{9d\rho R(\mathbf{C})}{4\delta_{\mathrm{sep}}(\rho,\mathbf{C})^{\frac{d-1}{d}} r(\mathbf{C})}\cdot\frac{1}{n^{\frac{1}{d}}}\right)^{-d} \cdot \frac{n\mathrm{vol}_d(\mathbf{C})}{\delta_{\mathrm{sep}}(\rho,\mathbf{C})}, \tag{10.22}$$

which implies in a straightforward way that

$$\mathrm{vol}_d(\mathbf{Q}) \geq \left(1+\frac{9d\rho R(\mathbf{C})}{4\delta_{\mathrm{sep}}(\rho,\mathbf{C}) r(\mathbf{C})}\cdot\frac{1}{n^{\frac{1}{d}}}\right)^{-d} \cdot \frac{n\mathrm{vol}_d(\mathbf{C})}{\delta_{\mathrm{sep}}(\rho,\mathbf{C})}. \tag{10.23}$$

Note that (10.15) (see the restated version for $j = d$) implies that $M_i(\mathbf{Q}) \geq \left(\frac{\mathrm{vol}_d(\mathbf{Q})}{\kappa_d}\right)^{\frac{i}{d}} \kappa_i$. Then, replacing $\mathrm{vol}_d(\mathbf{Q})$ by the right-hand side of (10.23), and using the convexity of the function $x \mapsto (1+x)^{-i}$ for $x > -1$ yields the lower bound for $M_i(\mathbf{Q})$ in Theorem 130.

Finally, we prove the statement about the spherical shape of $\mathbf{Q}$, that is, the inequality (5.8). As in [58], let

$$\theta(d) = \frac{1}{2^{\frac{d+3}{2}}\sqrt{2\pi}\sqrt{d}(d-1)(d+3)} \min\left\{\frac{3}{\pi^2 d(d+2)2^d}, \frac{16}{(d\pi)^{\frac{d-1}{4}}}\right\}.$$

Using the inequality $\frac{\kappa_{d-1}}{\kappa_d} \geq \sqrt{\frac{d}{2\pi}}$ (cf. [23]) and (6) of [125], we obtain

$$\left(\frac{\mathrm{svol}_{d-1}(\mathbf{Q})}{\mathrm{svol}_{d-1}(\mathbf{B}^d)}\right)^d \left(\frac{\mathrm{vol}_d(\mathbf{B}^d)}{\mathrm{vol}_d(\mathbf{Q})}\right)^{d-1} - 1 \geq \theta(d)\cdot\left(1-\frac{r(\mathbf{Q})}{R(\mathbf{Q})}\right)^{\frac{d+3}{2}}$$

(see also (5) of [56]). Substituting (10.18) and (10.22) into this inequality, we obtain

$$\left(1+\frac{2\delta_{\text{sep}}(\rho,\mathbf{C})^{\frac{1}{d}}\rho R(\mathbf{C})}{r(\mathbf{C})}\cdot\frac{1}{n^{\frac{1}{d}}}\right)^{d(d-1)}\left(1+\frac{9d\rho R(\mathbf{C})}{4\delta_{\text{sep}}(\rho,\mathbf{C})^{\frac{d-1}{d}}r(\mathbf{C})}\cdot\frac{1}{n^{\frac{1}{d}}}\right)^{d(d-1)}$$

$$\geq\left(\frac{\text{svol}_{d-1}(\mathbf{Q})}{\text{svol}_{d-1}(\mathbf{B}^d)}\right)^{d}\left(\frac{\text{vol}_d(\mathbf{B}^d)}{\text{vol}_d(\mathbf{Q})}\right)^{d-1}.$$

By the assumptions $d\geq 2$ and (10.16), it follows that

$$4d^2(d-1)\frac{\rho R(\mathbf{C})}{\delta_{\text{sep}}(\rho,\mathbf{C})r(\mathbf{C})}\cdot\frac{1}{n^{\frac{1}{d}}}\geq\theta(d)\left(1-\frac{r(\mathbf{Q})}{R(\mathbf{Q})}\right)^{\frac{d+3}{2}}. \tag{10.24}$$

Note that by [199], $\frac{1}{\delta_{\text{sep}}(\rho,\mathbf{C})}\leq\frac{2^{\frac{3d}{2}}\cdot\sqrt{\binom{\frac{d(d+1)}{2}}{d}}}{(d+1)^{\frac{d}{2}}\pi^{\frac{d}{2}}\Gamma\left(\frac{d}{2}+1\right)}$. This and (10.24) implies (5.8), finishing the proof of Theorem 130.

11

Open Problems: An Overview

Summary. In this chapter we collect the open problems, questions and conjectures presented in the book. In particular, Sections 11.1-11.5 contain these problems from Chapters 1-5, respectively. For more information, the interested reader is referred to the appropriate place in the corresponding chapter, or the papers cited there.

11.1 Chapter 1

Theorem 4 by Lindelöf states that among polytopes in $\mathbb{E}^d$ with given outer facet normals, the ones with maximal isoperimetric ratio is circumscribed about a ball, implying also the same statement for polytopes with a given number of facets.

Question 12 *Can Theorem 4 be modified for convex polytopes in $\mathbb{S}^d$ or $\mathbb{H}^d$?*

Open Problem 1 *Prove or disprove that among simplices in $\mathbb{S}^d$ or $\mathbb{H}^d$ of a given volume, where $d > 2$, the regular ones have smallest surface volume.*

Open Problem 2 *Prove or disprove that among simplices in $\mathbb{S}^d$ or $\mathbb{H}^d$ circumscribed about a ball, where $d > 3$, the regular ones have smallest volume.*

Open Problem 3 *Among convex polytopes in $\mathbb{E}^d$, inscribed in $\mathbb{S}^{d-1}$ and having n vertices, which polytope has the largest volume?*

Recall that a simplicial convex polyhedron in $\mathbb{E}^3$ with n vertices is called medial, if the valence of each of its vertices is at least $\lfloor 6 - \frac{12}{n} \rfloor$ and at most $\lceil 6 - \frac{12}{n} \rceil$.

Conjecture 15 *Let $\mathbf{P}$ be a convex polyhedron with n vertices and inscribed in $\mathbf{B}^3$. If there is a medial polyhedron satisfying these properties, then $\mathrm{vol}_3(\mathbf{P})$ is maximal under these conditions for a medial polyhedron.*

Open Problem 4 *By Steiner symmetrization, it can be easily shown that the maximum volume of the intersection of a fixed ball in $\mathbb{E}^d$ and a variable simplex of given volume V is attained when the simplex is regular and concentric with the ball. Show that the above statement holds true in spherical and hyperbolic space as well.*

Open Problem 5 *Prove or disprove that if $\mathbf{S}$ is a simplex contained in*

$\mathbf{B}^d$, *and for some* $\mathbf{x} \in \mathbf{S}$, $\mathbf{x} + \rho\mathbf{B}^d$ *is a Euclidean ball whose interior is disjoint from the* k*-skeleton of* $\mathbf{S}$, *then* $\rho \le \sqrt{\frac{d-k}{d(k+1)}}$.

Question 25 *Does the statement in Problem 5 hold for spherical or hyperbolic simplices?*

Question 26 *Consider a ball* $\mathbf{S}$ *in* $\mathbb{S}^3$ *or in* $\mathbb{H}^3$. *What is the infimum and the supremum of volumes of the tetrahedra midscribed to* $\mathbf{S}$?

Open Problem 6 *Prove or disprove that every combinatorial class of convex polyhedra contains a Koebe polyhedron whose center of mass is the origin.*

For Theorem 29 and the notations in the next problem, see Section 1.4.

Open Problem 7 *Is it possible to prove variants of Theorem 29 if the weight functions* w_i *in (1.4) depend not only on* $\rho_T(\mathbf{C}_i)$, *but also on the radii of the other spherical caps as well?*

Open Problem 8 *As we already remarked in this section, Schramm [218] proved that if* $\mathbf{K}$ *is any smooth, strictly convex body in* $\mathbb{E}^3$, *then every combinatorial class of convex polyhedra contains a representative midscribed about* $\mathbf{K}$. *If* $\mathbf{K}$ *is symmetric to the origin, does this statement remain true with the additional assumption that the barycenter of the tangency points of this representative is the origin? Can the barycenter of the tangency points be replaced by other centers of the polyhedron?*

11.2 Chapter 2

Conjecture 35 *Let* $\mathbf{K}$ *be a convex body in* $\mathbb{E}^d$. *Then there is a point* $\mathbf{x} \in \mathbf{K}$ *such that*

$$\frac{\mathrm{vol}_d(\mathrm{conv}(\mathbf{K} \cup (2\mathbf{x} - \mathbf{K})))}{\mathrm{vol}_d(\mathbf{K})} \le \binom{d}{\lfloor \frac{d}{2} \rfloor}.$$

Recall that for any $i = 0, 1 \ldots, d-1$ and convex body $\mathbf{K} \subset \mathbb{E}^d$, we denote by $c_i(\mathbf{K})$ the maximal volume of the convex hull of $\mathbf{K}$ and one of its reflections about an i-dimensional subspace of $\mathbb{E}^d$ intersecting $\mathbf{K}$, divided by $\mathrm{vol}_d(\mathbf{K})$. We obtain the quantity $c^{co}(\mathbf{K})$ in the same way by replacing the reflected copy of $\mathbf{K}$ by any congruent copy of $\mathbf{K}$.

Conjecture 45 *Let* $d \ge 3$ *and* $1 < i < d-1$. *Prove that for any convex body* $\mathbf{K} \subset \mathbb{E}^d$, $c_i(\mathbf{K}) \ge 1 + \frac{2\kappa_{d-1}}{\kappa_d}$. *Is it true that equality holds only for Euclidean balls?*

If, for a convex body $\mathbf{K} \subset \mathbb{E}^d$, we have that $\mathrm{vol}_d(\mathrm{conv}((\mathbf{p}+\mathbf{K}) \cup (\mathbf{q}+\mathbf{K})))$ has the same value for any touching pair of translates, we say that $\mathbf{K}$ satisfies the translative constant volume property.

Conjecture 47 *Let* $d \ge 3$. *If some o-symmetric convex body* $\mathbf{K} \subset \mathbb{E}^d$ *satisfies the translative constant volume property, then* $\mathbf{K}$ *is an ellipsoid.*

Open Problem 9 *For any* $d \ge 3$ *and* $1 \le i \le d-1$, *find the least upper* *bound of $c_i(\mathbf{K})$ on the family of convex bodies in $\mathbb{E}^d$. Furthermore, find the least upper bound on $c^{co}(\mathbf{K})$ on the same family.*

Four variants of volume in a d-dimensional normed space with unit ball $\mathbf{M}$ appear in the literature, each of which is a scalar multiple of standard Lebesgue measure. In these variants, denoted by $\mathrm{vol}_{\mathbf{M}}^{Bus}(\cdot), \mathrm{vol}_{\mathbf{M}}^{HT}(\cdot), \mathrm{vol}_{\mathbf{M}}^{m}(\cdot)$, and $\mathrm{vol}_{\mathbf{M}}^{m*}(\cdot)$, the scalar multiple chosen in such a way that the volume of $\mathbf{M}$ is κ_d, the volume of $\mathbf{M}$ is $\frac{\mathrm{vol}_d(\mathbf{M})\mathrm{vol}_d(\mathbf{M}^\circ)}{\kappa_d}$, the volume of a largest cross-polytope inscribed in $\mathbf{M}$ is $\frac{2^d}{d!}$, and the volume of a smallest parallelotope circumscribed about $\mathbf{M}$ is 2^d, respectively.

Conjecture (Mahler Conjecture) 55 *Prove that among d-dimensional, $\mathbf{o}$-symmetric convex bodies, the minimum of $\mathrm{vol}_{\mathbf{M}}^{HT}(\mathbf{M})$ is attained, e.g., if $\mathbf{M}$ is a d-cube.*

For any $\tau \in \{HT, m, m^*\}$ and convex body $\mathbf{K}$, the quantity $c_{tr}^{\tau}(\mathbf{K})$ denotes the maximum of $\mathrm{vol}_{\mathbf{M}}^{\tau}(\mathrm{conv}(\mathbf{K} \cup (\mathbf{p} + \mathbf{K})))$ for any translate $\mathbf{p} + \mathbf{K}$ of $\mathbf{K}$ intersecting $\mathbf{K}$, where $\mathbf{M} = \frac{1}{2}(\mathbf{K} - \mathbf{K})$.

Open Problem 10 *Prove or disprove that if $c_{tr}^{HT}(\mathbf{K})$ is maximal for some plane convex body $\mathbf{K}$, then $\mathbf{K}$ is an affine image of $\mathbf{M}_0$.*

Open Problem 11 *For $d \geq 3$ and $\tau \in \{HT, m, m^*\}$, find the maximal values of $c_{tr}^{\tau}(\mathbf{K})$ over the family of d-dimensional convex bodies.*

Open Problem 12 *For $d \geq 3$ and $\tau \in \{Bus, HT, m, m^*\}$, find the minimal values of $c_{tr}^{\tau}(\mathbf{K})$ over the family of d-dimensional convex bodies.*

11.3 Chapter 3

Conjecture (Kneser-Poulsen Conjecture - Part 1) 63 *If $\mathbf{q} = (\mathbf{q}_1, \mathbf{q}_2, \ldots, \mathbf{q}_N)$ is a contraction of $\mathbf{p} = (\mathbf{p}_1, \mathbf{p}_2, \ldots, \mathbf{p}_N)$ in $\mathbb{E}^d$, $d \geq 3$, then*

$$\mathrm{vol}_d\left(\bigcup_{i=1}^{N} \mathbf{B}^d[\mathbf{p}_i, r_i]\right) \geq \mathrm{vol}_d\left(\bigcup_{i=1}^{N} \mathbf{B}^d[\mathbf{q}_i, r_i]\right).$$

Conjecture (Kneser-Poulsen Conjecture - Part 2) 64 *If $\mathbf{q} = (\mathbf{q}_1, \mathbf{q}_2, \ldots, \mathbf{q}_N)$ is a contraction of $\mathbf{p} = (\mathbf{p}_1, \mathbf{p}_2, \ldots, \mathbf{p}_N)$ in $\mathbb{E}^d$, $d \geq 3$, then*

$$\mathrm{vol}_d\left(\bigcap_{i=1}^{N} \mathbf{B}^d[\mathbf{p}_i, r_i]\right) \leq \mathrm{vol}_d\left(\bigcap_{i=1}^{N} \mathbf{B}^d[\mathbf{q}_i, r_i]\right).$$

Open Problem 13 *Prove the Kneser-Poulsen conjecture (i.e., Conjectures 63 and 64) for $d+4$ balls in $\mathbb{E}^d$, $d \geq 3$.*

Open Problem 14 *Let $1 \leq k \leq d$, $1 < N$, and $r > 0$. Prove or disprove that if $\mathbf{q} = (\mathbf{q}_1, \mathbf{q}_2, \ldots, \mathbf{q}_N)$ is a contraction of $\mathbf{p} = (\mathbf{p}_1, \mathbf{p}_2, \ldots, \mathbf{p}_N)$ in $\mathbb{E}^d$,* *then*

$$V_k\left(\bigcap_{i=1}^{N}\mathbf{B}^d[\mathbf{p}_i,r]\right)\leq V_k\left(\bigcap_{i=1}^{N}\mathbf{B}^d[\mathbf{q}_i,r]\right),$$

where $V_k(\cdot)$ denotes the k-th intrinsic volume.

Conjecture (Alexander Conjecture) 77 *Under an arbitrary contraction of the center points of finitely many congruent disks in the plane, the perimeter of the intersection of the disks cannot decrease.*

11.4 Chapter 4

Recall that $\deg_{\rm avr}(\mathcal{P}_r)$ denotes the average degree of the vertices of the contact graph of an arbitrary packing $\mathcal{P}_r$ of circles of radius $r>0$. It was proved in [31] that for packings $\mathcal{P}_r$ in $\mathbb{S}^2$, $\limsup_{r\to 0}\left(\sup_{\mathcal{P}_r}\deg_{\rm avr}(\mathcal{P}_r)\right)<5$.

Conjecture 86 *Let $\mathcal{P}_r$ be an arbitrary packing $\mathcal{P}_r$ of disks of radius $r>0$ in $\mathbb{H}^2$. Then*

$$\limsup_{r\to 0}\left(\sup_{\mathcal{P}_r}\deg_{\rm avr}(\mathcal{P}_r)\right)<5.$$

The next problem refers to the packing densities of packings of unit balls in $\mathbb{E}^d$ with respect to the union of balls of radius $1+\lambda$ centered at the centers of the elements of the packing.

Open Problem 15 *Determine $\delta_d(\lambda)$ for $d\geq 2$, $0<\lambda<\sqrt{\frac{2d}{d+1}}-1$.*

Recall that the maximum contact number of a totally separable packing of n unit balls in $\mathbb{E}^d$ is denoted by $c_{\rm sep}(n,3)$, and the maximum contact number of a packing of n unit diameter balls in $\mathbb{E}^d$ whose centers are points of $\mathbb{Z}^d$ is denoted by $c_{\mathbb{Z}}(n,d)$.

Open Problem 16 *Show that*

$$\lim_{n\to+\infty}\frac{3n-c_{\rm sep}(n,3)}{n^{\frac{2}{3}}}=3\ .$$

In particular, is it the case that $c_{\rm sep}(n,3)=c_{\mathbb{Z}}(n,3)$, for all positive integers n?

Recall that if $\mathbf{K}$ be a convex body in $\mathbb{E}^d$, then a family of translates of $\mathbf{K}$ that all touch $\mathbf{K}$ and, together with $\mathbf{K}$, form a totally separable packing in $\mathbb{E}^d$ is called a separable Hadwiger configuration of $\mathbf{K}$. Moreover, the separable Hadwiger number $H_{\rm sep}(\mathbf{K})$ of $\mathbf{K}$ is the maximum size of a separable Hadwiger configuration of $\mathbf{K}$.

Open Problem 17 *Determine the largest value of $H_{\rm sep}(\mathbf{K})$ for* **o***-symmetric smooth strictly convex bodies in $\mathbb{E}^d$, $d\geq 5$.*

Conjecture 106 *The Hadwiger number of any starlike disk is at most 8.*

Open Problem 18 *Is it true that for every positive integer k there is an* *integer $N(k)$ such that for any topological disk $\mathbf{S}$, if $(\text{conv}\mathbf{S}) \setminus \mathbf{S}$ has at most k connected components, then $H(\mathbf{S}) \leq N(k)$?*

11.5 Chapter 5

Corollary 117 states that for any **o**-symmetric smooth strictly convex domain in $\mathbb{E}^2$ and $n \geq 2$, $c_{\text{sep}}(\mathbf{K}, n, 2) = \lfloor 2n - 2\sqrt{n} \rfloor$.

Open Problem 19 *One may wonder whether $c_{\text{sep}}(\mathbf{K}, n, 2) = \lfloor 2n - 2\sqrt{n} \rfloor$ holds for any **o**-symmetric smooth convex domain $\mathbf{K}$ in $\mathbb{E}^2$ and $n \geq 2$.*

Open Problem 20 *Let $\mathbf{K}$ be a convex body in $\mathbb{E}^d$, $d \geq 3$. Prove or disprove that the highest density of totally separable translative packings of $\mathbf{K}$ in $\mathbb{E}^d$ is attained by the totally separable lattice packing of $\mathbf{K}$ generated by (any of) the smallest volume parallelotope (i.e., affine d-cube) circumscribed $\mathbf{K}$.*

Open Problem 21 *Let $\mathbf{C}$ be an **o**-symmetric convex body in $\mathbb{E}^d$, $d \geq 3$ and $n > 1$. Prove or disprove that the smallest volume of the convex hull of n translates of $\mathbf{C}$ forming a totally separable packing in $\mathbb{E}^d$ is obtained when the n translates of $\mathbf{C}$ form a sausage, that is, a linear packing.*

Recall that $\delta_{\text{sep}}(\mathbf{C}) \leq \delta_{\text{sep}}(\rho, \mathbf{C}) \leq \delta(\mathbf{C})$ hold for any **o**-symmetric convex body $\mathbf{C}$ in $\mathbb{E}^d$, $d \geq 2$ and for all $\rho \geq 1$.

Open Problem 22 *Let $\mathbf{C}$ be an **o**-symmetric convex body $\mathbf{C}$ in $\mathbb{E}^d$, $d \geq 2$. Then prove or disprove that there exists $\rho(\mathbf{C}) > 1$ such that for any $\rho \geq \rho(\mathbf{C})$ one has $\delta_{\text{sep}}(\mathbf{C}) = \delta_{\text{sep}}(\rho, \mathbf{C})$.*

Open Problem 23 *The nature of the question analogue to Theorem 130 on minimizing $M_d(\mathbf{Q}) = \text{vol}_d(\mathbf{Q})$ is very different. Namely, recall that Betke and Henk [24] proved L. Fejes Tóth's sausage conjecture for $d \geq 42$ according to which the smallest volume of the convex hull of n non-overlapping unit balls in $\mathbb{E}^d$ is obtained when the n unit balls form a sausage, that is, a linear packing (see also [25, 26]). As linear packings of unit balls are ρ-separable, therefore, the above theorem of Betke and Henk applies to ρ-separable packings of unit balls in $\mathbb{E}^d$ for all $\rho \geq 1$ and $d \geq 42$. On the other hand, the problem of minimizing the volume of the convex hull of n unit balls forming a ρ-separable packing in $\mathbb{E}^d$ remains an interesting open problem for $\rho \geq 1$ and $2 \leq d < 42$. Last but not least, the problem of minimizing $M_d(\mathbf{Q})$ for **o**-symmetric convex bodies $\mathbf{C}$ different from a ball in $\mathbb{E}^d$ seems to be wide open for $\rho \geq 1$ and $d \geq 2$.*

Bibliography

[1] V. È. Adler. Cutting of polygons. *Funktsional. Anal. i Prilozhen.*, 27(2):79–82, 1993.

[2] Hee-Kap Ahn, Peter Brass, and Chan-Su Shin. Maximum overlap and minimum convex hull of two convex polyhedra under translations. *Comput. Geom.*, 40(2):171–177, 2008.

[3] Arseniy V. Akopyan. Some remarks on the circumcenter of mass. *Discrete Comput. Geom.*, 51(4):837–841, 2014.

[4] M. Alexander, M. Fradelizi, and A. Zvavitch. Polytopes of maximal volume product. *arxiv*, 1708.07914, Aug 2017.

[5] R. Alexander. Lipschitzian mappings and total mean curvature of polyhedral surfaces i. *Trans. Amer. Math. Soc.*, 288:661–678, 1985.

[6] Ralph Alexander. The circumdisk and its relation to a theorem of Kirszbraun and Valentine. *Math. Mag.*, 57(3):165–169, 1984.

[7] Javier Alonso and Carlos Benítez. Orthogonality in normed linear spaces: a survey. I. Main properties. *Extracta Math.*, 3(1):1–15, 1988.

[8] Laurent Alonso and Raphaël Cerf. The three-dimensional polyominoes of minimal area. *Electron. J. Combin.*, 3(1):Research Paper 27, approx. 39, 1996.

[9] J. C. Álvarez Paiva and A. C. Thompson. Volumes on normed and Finsler spaces. In *A sampler of Riemann-Finsler geometry*, volume 50 of *Math. Sci. Res. Inst. Publ.*, pages 1–48. Cambridge Univ. Press, Cambridge, 2004.

[10] Juan Carlos Álvarez Paiva and Anthony Thompson. On the perimeter and area of the unit disc. *Amer. Math. Monthly*, 112(2):141–154, 2005.

[11] E. M. Andreev. Convex polyhedra in Lobačevskiĭ spaces. *Mat. Sb. (N.S.)*, 81 (123):445–478, 1970.

[12] E. M. Andreev. Convex polyhedra of finite volume in Lobačevskiĭ space. *Mat. Sb. (N.S.)*, 83 (125):256–260, 1970.

[13] Natalie Arkus, Vinothan N. Manoharan, and Michael P. Brenner. Deriving finite sphere packings. *SIAM J. Discrete Math.*, 25(4):1860–1901, 2011.

[14] Alex Baden, Keenan Crane, and Misha Kazhdan. Möbius registration. *Computer Graphics Forum*, 37(5):211–220, 2018.

[15] Vitor Balestro, Ákos G. Horváth, Horst Martini, and Ralph Teixeira. Angles in normed spaces. *Aequationes Math.*, 91(2):201–236, 2017.

[16] Keith Ball. Volume ratios and a reverse isoperimetric inequality. *J. London Math. Soc. (2)*, 44(2):351–359, 1991.

[17] Keith Ball. An elementary introduction to modern convex geometry. In *Flavors of geometry*, volume 31 of *Math. Sci. Res. Inst. Publ.*, pages 1–58. Cambridge Univ. Press, Cambridge, 1997.

[18] Imre Bárány and Zoltán Füredi. Computing the volume is difficult. *Discrete Comput. Geom.*, 2(4):319–326, 1987.

[19] Maria Belk and Robert Connelly. Realizability of graphs. *Discrete Comput. Geom.*, 37(2):125–137, 2007.

[20] Joel D. Berman and Kit Hanes. Volumes of polyhedra inscribed in the unit sphere E^3. *Math. Ann.*, 188:78–84, 1970.

[21] M. Bern and A. Sahai. Pushing disks together—the continuous-motion case. *Discrete Comput. Geom.*, 20(4):499–514, 1998.

[22] Marshall Bern and David Eppstein. Optimal Möbius transformations for information visualization and meshing. In *Algorithms and data structures (Providence, RI, 2001)*, volume 2125 of *Lecture Notes in Comput. Sci.*, pages 14–25. Springer, Berlin, 2001.

[23] U. Betke, P. Gritzmann, and J. M. Wills. Slices of L. Fejes Tóth's sausage conjecture. *Mathematika*, 29(2):194–201, 1982.

[24] U. Betke and M. Henk. Finite packings of spheres. *Discrete Comput. Geom.*, 19(2):197–227, 1998.

[25] U. Betke, M. Henk, and J. M. Wills. Sausages are good packings. *Discrete Comput. Geom.*, 13(3-4):297–311, 1995.

[26] Ulrich Betke, Martin Henk, and Jörg M. Wills. Finite and infinite packings. *J. Reine Angew. Math.*, 453:165–191, 1994.

[27] A. Bezdek. Locally separable circle packings. *Studia Sci. Math. Hungar.*, 18(2-4):371–375, 1983.

[28] A. Bezdek. On the Hadwiger number of a starlike disk. In *Intuitive geometry (Budapest, 1995)*, volume 6 of *Bolyai Soc. Math. Stud.*, pages 237–245. János Bolyai Math. Soc., Budapest, 1997.

[29] A. Bezdek, K. Kuperberg, and W. Kuperberg. Mutually contiguous translates of a plane disk. *Duke Math. J.*, 78(1):19–31, 1995.

[30] K. Bezdek. On Gromov's conjecture for uniform contractions. *arXiv*, 1810.11886, Oct 2018.

[31] K. Bezdek, R. Connelly, and G. Kertész. On the average number of neighbors in a spherical packing of congruent circles. In *Intuitive geometry (Siófok, 1985)*, volume 48 of *Colloq. Math. Soc. János Bolyai*, pages 37–52. North-Holland, Amsterdam, 1987.

[32] K. Bezdek, A. Deza, and Y. Ye. Selected open problems in discrete geometry and optimization. In *Discrete geometry and optimization*, volume 69 of *Fields Inst. Commun.* Springer, New York, 2013.

[33] Károly Bezdek. On the maximum number of touching pairs in a finite packing of translates of a convex body. *J. Combin. Theory Ser. A*, 98(1):192–200, 2002.

[34] Károly Bezdek. Contact numbers for congruent sphere packings in Euclidean 3-space. *Discrete Comput. Geom.*, 48(2):298–309, 2012.

[35] Károly Bezdek. *Lectures on sphere arrangements—the discrete geometric side*, volume 32 of *Fields Institute Monographs*. Springer, New York; Fields Institute for Research in Mathematical Sciences, Toronto, ON, 2013.

[36] Károly Bezdek. From r-dual sets to uniform contractions. *Aequationes Math.*, 92(1):123–134, 2018.

[37] Károly Bezdek and Peter Brass. On k^+-neighbour packings and one-sided Hadwiger configurations. *Beiträge Algebra Geom.*, 44(2):493–498, 2003.

[38] Károly Bezdek and Robert Connelly. Pushing disks apart—the Kneser-Poulsen conjecture in the plane. *J. Reine Angew. Math.*, 553:221–236, 2002.

[39] Károly Bezdek and Robert Connelly. The Kneser-Poulsen conjecture for spherical polytopes. *Discrete Comput. Geom.*, 32(1):101–106, 2004.

[40] Károly Bezdek, Robert Connelly, and Balázs Csikós. On the perimeter of the intersection of congruent disks. *Beiträge Algebra Geom.*, 47(1):53–62, 2006.

[41] Károly Bezdek and Muhammad Ali Khan. Contact numbers for sphere packings. *arXiv*, 1601.00145, Jan 2016.

[42] Károly Bezdek, Muhammad Ali Khan, and Michael Oliwa. Hadwiger and contact numbers of totally separable domains and their crystallization. *arXiv*, 1703.08568, Mar 2017.

[43] Károly Bezdek and Zsolt Lángi. Density bounds for outer parallel domains of unit ball packings. *Proc. Steklov Inst. Math.*, 288(1):209–225, 2015. Reprint of Tr. Mat. Inst. Steklova 88 (2015), 230–247.

[44] Károly Bezdek and Zsolt Lángi. Bounds for totally separable translative packings in the plane. *Discrete Comput. Geom.*, Oct 2017.

[45] Károly Bezdek and Zsolt Lángi. Minimizing the mean projections of finite ρ-separable packings. *Monatshefte für Mathematik*, Mar 2018.

[46] Károly Bezdek, Zsolt Lángi, Márton Naszódi, and Peter Papez. Ball-polyhedra. *Discrete Comput. Geom.*, 38(2):201–230, 2007.

[47] Károly Bezdek and Alexander E. Litvak. Packing convex bodies by cylinders. *Discrete Comput. Geom.*, 55(3):725–738, 2016.

[48] Károly Bezdek and Márton Naszódi. The Kneser–Poulsen Conjecture for Special Contractions. *Discrete Comput. Geom.*, 60(4):967–980, 2018.

[49] Károly Bezdek and Márton Naszódi. On contact graphs of totally separable packings in low dimensions. *Advances in Applied Mathematics*, 101:266 – 280, 2018.

[50] Károly Bezdek and Samuel Reid. Contact graphs of unit sphere packings revisited. *J. Geom.*, 104(1):57–83, 2013.

[51] Károly Bezdek, Balázs Szalkai, and István Szalkai. On contact numbers of totally separable unit sphere packings. *Discrete Math.*, 339(2):668–676, 2016.

[52] Garrett Birkhoff. Orthogonality in linear metric spaces. *Duke Math. J.*, 1(2):169–172, 1935.

[53] Wilhelm Blaschke. Konvexe Bereiche gegebener konstanter Breite und kleinsten Inhalts. *Math. Ann.*, 76(4):504–513, 1915.

[54] Wilhelm Blaschke. Über affine Geometrie VII: Neue Extremeigenschaften von Ellipse und Ellipsoid. *Ber. Verh. Sächs. Akad. Wiss. Leipzig Math.-naturw. Kl.*, 69:306–318, 1917.

[55] B. Bollobás. Area of the union of disks. *Elem. Math.*, 23:60–61, 1968.

[56] K. Böröczky. Packing of spheres in spaces of constant curvature. *Acta Math. Acad. Sci. Hungar.*, 32(3-4):243–261, 1978.

[57] K. Böröczky. On an extremum property of the regular simplex in S^d. In *Intuitive geometry (Siófok, 1985)*, volume 48 of *Colloq. Math. Soc. János Bolyai*, pages 117–121. North-Holland, Amsterdam, 1987.

[58] Károly Böröczky, Jr. Mean projections and finite packings of convex bodies. *Monatsh. Math.*, 118(1-2):41–54, 1994.

[59] Károly J. Böröczky and Endre J. Makai. Remarks on planar Blaschke-Santaló inequality. *arXiv*, 1411.3842, Nov 2014.

[60] Károly J. Böröczky and Imre Z. Ruzsa. Note on an inequality of Wegner. *Discrete Comput. Geom.*, 37(2):245–249, 2007.

[61] J. Bourgain and V. D. Milman. New volume ratio properties for convex symmetric bodies in $\mathbf{R}^n$. *Invent. Math.*, 88(2):319–340, 1987.

[62] Lewis Bowen. Circle packing in the hyperbolic plane. *Math. Phys. Electron. J.*, 6:Paper 6, 10, 2000.

[63] Rufus Bowen and Stephen Fisk. Generations of triangulations of the sphere. *Math. Comp.*, 21:250–252, 1967.

[64] Peter Boyvalenkov, Stefan Dodunekov, and Oleg Musin. A survey on the kissing numbers. *Serdica Math. J.*, 38(4):507–522, 2012.

[65] Herm Jan Brascamp and Elliott H. Lieb. Best constants in Young's inequality, its converse, and its generalization to more than three functions. *Advances in Math.*, 20(2):151–173, 1976.

[66] Peter Brass. Erdős distance problems in normed spaces. *Comput. Geom.*, 6(4):195–214, 1996.

[67] Peter Brass, William Moser, and János Pach. *Research problems in discrete geometry*. Springer, New York, 2005.

[68] Graham R. Brightwell and Edward R. Scheinerman. Representations of planar graphs. *SIAM J. Discrete Math.*, 6(2):214–229, 1993.

[69] Yu. D. Burago. *Mixed volume theory*, volume 10 of *Encyclopedia of Mathematics*. Springer Science+Business Media B.V., 1994.

[70] Yu. D. Burago and V. A. Zalgaller. *Geometric inequalities*, volume 285 of *Grundlehren der Mathematischen Wissenschaften [Fundamental Principles of Mathematical Sciences]*. Springer-Verlag, Berlin, 1988. Translated from the Russian by A. B. Sosinskiĭ, Springer Series in Soviet Mathematics.

[71] Herbert Busemann. The isoperimetric problem in the Minkowski plane. *Amer. J. Math.*, 69:863–871, 1947.

[72] Herbert Busemann. The isoperimetric problem for Minkowski area. *Amer. J. Math.*, 71:743–762, 1949.

[73] Vasilis Capoyleas. On the area of the intersection of disks in the plane. *Comput. Geom.*, 6(6):393–396, 1996.

[74] C. Carathéodory and E. Study. Zwei Beweise des Satzes, daßder Kreis unter allen Figuren gleichen Umfanges den größten Inhalt hat. *Math. Ann.*, 68(1):133–140, 1909.

[75] G. D. Chakerian. The isoperimetric problem in the Minkowski plane. *Amer. Math. Monthly*, 67:1002–1004, 1960.

[76] G. D. Chakerian. Sets of constant width. *Pacific J. Math.*, 19:13–21, 1966.

[77] Otfried Cheong and Mira Lee. The Hadwiger number of Jordan regions is unbounded. *Discrete Comput. Geom.*, 37(4):497–501, 2007.

[78] Andrew Cotton and David Freeman. The double bubble problem in spherical space and hyperbolic space. *Int. J. Math. Math. Sci.*, 32(11):641–699, 2002.

[79] B. Csikós. On the Hadwiger-Kneser-Poulsen conjecture. In *Intuitive geometry (Budapest, 1995)*, volume 6 of *Bolyai Soc. Math. Stud.*, pages 291–299. János Bolyai Math. Soc., Budapest, 1997.

[80] B. Csikós. On the volume of the union of balls. *Discrete Comput. Geom.*, 20(4):449–461, 1998.

[81] B. Csikós and M. Horváth. Two Kneser–Poulsen-type inequalities in planes of constant curvature. *Acta Math. Hungar.*, 155(1):158–174, 2018.

[82] Balázs Csikós. On the volume of flowers in space forms. *Geom. Dedicata*, 86(1-3):59–79, 2001.

[83] Balázs Csikós. A Schläfli-type formula for polytopes with curved faces and its application to the Kneser-Poulsen conjecture. *Monatsh. Math.*, 147(4):273–292, 2006.

[84] L. Danzer and B. Grünbaum. Über zwei Probleme bezüglich konvexer Körper von P. Erdős und von V. L. Klee. *Math. Z.*, 79:95–99, 1962.

[85] H. Davenport and Gy. Hajós. Problem 35. *Mat. Lapok*, 2:68, 1951.

[86] Mahlon M. Day. Polygons circumscribed about closed convex curves. *Trans. Amer. Math. Soc.*, 62:315–319, 1947.

[87] B. V. Dekster. The Jung theorem for spherical and hyperbolic spaces. *Acta Math. Hungar.*, 67(4):315–331, 1995.

[88] Michael B. Dillencourt and Warren D. Smith. Graph-theoretical conditions for inscribability and Delaunay realizability. *Discrete Math.*, 161(1-3):63–77, 1996.

[89] Gábor Domokos and Zsolt Lángi. The isoperimetric quotient of a convex body decreases monotonically under the Eikonal abrasion model. *Mathematika*, 65(1):119–129, 2019.

[90] Gábor Domokos, András Á. Sipos, Gyula M. Szabó, and Péter L. Várkonyi. Explaining the elongated shape of 'oumuamua by the eikonal abrasion model. *Research Notes of the AAS*, 1(1):50, 2017.

[91] C. H. Dowker. On minimum circumscribed polygons. *Bull. Amer. Math. Soc.*, 50:120–122, 1944.

[92] Nico Düvelmeyer. Angle measures and bisectors in Minkowski planes. *Canad. Math. Bull.*, 48(4):523–534, 2005.

[93] H. Edelsbrunner. The union of balls and its dual shape. *Discrete Comput. Geom.*, 13(3-4):415–440, 1995.

[94] F. Edler. Vervollständigung der Steinerschen elementargeometrischen Beweise für den satz, daß der Kreis grösseren Flächeninhalt besitzt, als jede andere ebene Figur gleich grossen Umfanges. *Nachrichten von der Königl. Gesellschaft der Wissenschaften und der Georg-Augusts-Universität zu Göttingen*, pages 73–80, 1882.

[95] P. Erdős. On sets of distances of n points. *Amer. Math. Monthly*, 53:248–250, 1946.

[96] P. Erdős. Problems and results in combinatorial geometry. In *Discrete geometry and convexity (New York, 1982)*, volume 440 of *Ann. New York Acad. Sci.*, pages 1–11. New York Acad. Sci., New York, 1985.

[97] Theodor Estermann. Über den Vektorenbereich eines konvexen Körpers. *Math. Z.*, 28(1):471–475, 1928.

[98] Andreas Fankhänel. On angular measures in Minkowski planes. *Beitr. Algebra Geom.*, 52(2):335–342, 2011.

[99] I. Fáry and L. Rédei. Der zentralsymmetrische Kern und die zentralsymmetrische Hülle von konvexen Körpern. *Math. Ann.*, 122:205–220, 1950.

[100] István Fáry. Sur la densité des réseaux de domaines convexes. *Bull. Soc. Math. France*, 78:152–161, 1950.

[101] István Fáry and Endre Makai, Jr. Isoperimetry in variable metric. *Studia Sci. Math. Hungar.*, 17(1-4):143–158, 1982.

[102] Herbert Federer. *Geometric measure theory*. Die Grundlehren der mathematischen Wissenschaften, Band 153. Springer-Verlag New York Inc., New York, 1969.

[103] G. Fejes Tóth. Ten-neighbour packing of equal balls. *Period. Math. Hungar.*, 12(2):125–127, 1981.

[104] G. Fejes Tóth and L. Fejes Tóth. On totally separable domains. *Acta Math. Acad. Sci. Hungar.*, 24:229–232, 1973.

[105] G. Fejes Tóth and W. Kuperberg. Blichfeldt's density bound revisited. *Math. Ann.*, 295(4):721–727, 1993.

[106] L. Fejes Tóth. *Regular Figures*. A Pergamon Press Book. The Macmillan Co., New York, 1964.

[107] L. Fejes Tóth. Packing of r-convex discs. *Studia Sci. Math. Hungar.*, 17(1-4):449–452, 1982.

[108] L. Fejes Tóth. On the densest packing of convex discs. *Mathematika*, 30(1):1–3, 1983.

[109] László Fejes Tóth. The isepiphan problem for n-hedra. *Amer. J. Math.*, 70:174–180, 1948.

[110] László Fejes Tóth. Some packing and covering theorems. *Acta Sci. Math. Szeged*, 12(Leopoldo Fejér et Frederico Riesz LXX annos natis dedicatus, Pars A):62–67, 1950.

[111] László Fejes Tóth. On circles covering a given circle. *Mat. Lapok*, 30(4):317–320, 1978/82.

[112] Stefan Felsner and Günter Rote. On Primal-Dual Circle Representations. In Jeremy T. Fineman and Michael Mitzenmacher, editors, *2nd Symposium on Simplicity in Algorithms (SOSA 2019)*, volume 69 of *OpenAccess Series in Informatics (OASIcs)*, pages 8:1–8:18. Schloss Dagstuhl–Leibniz-Zentrum fuer Informatik, Dagstuhl, Germany, 2018.

[113] Ferenc Fodor, Árpád Kurusa, and Viktor Vígh. Inequalities for hyperconvex sets. *Adv. Geom.*, 16(3):337–348, 2016.

[114] Dmitry Fuchs and Serge Tabachnikov. *Mathematical omnibus*. American Mathematical Society, Providence, RI, 2007. Thirty lectures on classic mathematics.

[115] Michael E. Gage. An isoperimetric inequality with applications to curve shortening. *Duke Math. J.*, 50(4):1225–1229, 1983.

[116] David Gale. Neighborly and cyclic polytopes. In *Proc. Sympos. Pure Math., Vol. VII*, pages 225–232. Amer. Math. Soc., Providence, R.I., 1963.

[117] Fuchang Gao, Daniel Hug, and Rolf Schneider. Intrinsic volumes and polar sets in spherical space. *Math. Notae*, 41:159–176 (2003), 2001/02. Homage to Luis Santaló. Vol. 1 (Spanish).

[118] Richard J. Gardner. The Brunn-Minkowski inequality. *Bull. Amer. Math. Soc. (N.S.)*, 39(3):355–405, 2002.

[119] Richard J. Gardner. *Geometric tomography*, volume 58 of *Encyclopedia of Mathematics and its Applications*. Cambridge University Press, New York, second edition, 2006.

[120] S. Gołab. Quelques problèmes mètriques de la géometrie de minkowski. *Trav. l'Acad. Mines Cracovie*, 6:1–79, 1932.

[121] M. Goldberg. The isoperimetric problem for polyhedra. *Tohoku Math. J.*, 40:226–236, 1934.

[122] Igors Gorbovickis. Strict Kneser-Poulsen conjecture for large radii. *Geom. Dedicata*, 162:95–107, 2013.

[123] Igors Gorbovickis. Kneser-Poulsen conjecture for a small number of intersections. *Contrib. Discrete Math.*, 9(1):1–10, 2014.

[124] Igors Gorbovickis. The central set and its application to the Kneser-Poulsen conjecture. *Discrete Comput. Geom.*, 59(4):784–801, 2018.

[125] H. Groemer and R. Schneider. Stability estimates for some geometric inequalities. *Bull. London Math. Soc.*, 23(1):67–74, 1991.

[126] Helmut Groemer. über die einlagerung von kreisen in einen konvexen bereich. *Math. Z.*, 73:285–294, 1960.

[127] Helmut Groemer. Abschätzungen für die Anzahl der konvexen Körper, die einen konvexen Körper berühren. *Monatsh. Math.*, 65:74–81, 1961.

[128] M. Gromov. Monotonicity of the volume of intersection of balls. In *Geometrical aspects of functional analysis (1985/86)*, volume 1267 of *Lecture Notes in Math.*, pages 1–4. Springer, Berlin, 1987.

[129] Peter M. Gruber. Error of asymptotic formulae for volume approximation of convex bodies in $\mathbb{E}^d$. *Monatsh. Math.*, 135(4):279–304, 2002. (Dedicated to Edmund Hlawka on the occasion of his 85th birthday).

[130] Branko Grünbaum. On a conjecture of H. Hadwiger. *Pacific J. Math.*, 11:215–219, 1961.

[131] Branko Grünbaum. *Convex polytopes*, volume 221 of *Graduate Texts in Mathematics*. Springer-Verlag, New York, second edition, 2003. Prepared and with a preface by Volker Kaibel, Victor Klee and Günter M. Ziegler.

[132] Alfred Haar. Der Massbegriff in der Theorie der kontinuierlichen Gruppen. *Ann. of Math. (2)*, 34(1):147–169, 1933.

[133] H. Hadwiger. Über Treffanzahlen bei translationsgleichen Eikörpern. *Arch. Math.*, 8:212–213, 1957.

[134] Charles J. A. Halberg, Jr., Eugene Levin, and E. G. Straus. On contiguous congruent sets in Euclidean space. *Proc. Amer. Math. Soc.*, 10:335–344, 1959.

[135] Thomas C. Hales. A proof of the Kepler conjecture. *Ann. of Math. (2)*, 162(3):1065–1185, 2005.

[136] Thomas C. Hales. *Dense sphere packings*, volume 400 of *London Mathematical Society Lecture Note Series*. Cambridge University Press, Cambridge, 2012. A blueprint for formal proofs.

[137] Richard S. Hamilton. Isoperimetric estimates for the curve shrinking flow in the plane. In *Modern methods in complex analysis (Princeton, NJ, 1992)*, volume 137 of *Ann. of Math. Stud.*, pages 201–222. Princeton Univ. Press, Princeton, NJ, 1995.

[138] Frank Harary and Heiko Harborth. Extremal animals. *J. Combinatorics Information Syst. Sci.*, 1(1):1–8, 1976.

[139] Heiko Harborth. Lösung zu Problem 664A. *Elem. Math.*, 29:14–15, 1974.

[140] Brian Hayes. The science of sticky spheres. *Am. Sci.*, 100:1442–1449, 2012.

[141] Martin Henk. Löwner-John ellipsoids. *Doc. Math.*, (Extra vol.: Optimization stories):95–106, 2012.

[142] M. Henze. *The Mahler Conjecture (diploma thesis)*. Otto-von-Guericke Universität Magdeburg, Magdeburg, 2008.

[143] Miranda Holmes-Cerfon. Enumerating rigid sphere packings. *SIAM Rev.*, 58(2):229–244, 2016.

[144] Ákos G. Horváth and Zsolt Lángi. On the volume of the convex hull of two convex bodies. *Monatsh. Math.*, 174(2):219–229, 2014.

[145] Ákos G. Horváth and Zsolt Lángi. Maximum volume polytopes inscribed in the unit sphere. *Monatsh. Math.*, 181(2):341–354, 2016.

[146] Robert S. Hoy, Jared Harwayne-Gidansky, and Corey S. O'Hern. Structure of finite sphere packings via exact enumeration: Implications for colloidal crystal nucleation. *Physical Review E*, 85:051403, 2012.

[147] Xiaojun Huang and Jinsong Liu. Characterizations of circle patterns and finite convex polyhedra in hyperbolic 3-space. *Math. Ann.*, 368(1-2):213–231, 2017.

[148] Gerhard Huisken. A distance comparison principle for evolving curves. *Asian J. Math.*, 2(1):127–133, 1998.

[149] G. A. Kabatjanskiĭ and V. I. Levensteĭn. Bounds for packings on the sphere and in space. *Problemy Peredači Informacii*, 14(1):3–25, 1978.

[150] Bernd Kawohl and Christof Weber. Meissner's mysterious bodies. *Math. Intelligencer*, 33(3):94–101, 2011.

[151] Béla Kerékjártó. *Les Fondements de la Géométrie. Tome II: Géométrie Projective.* Akadémiai Kiadó, Budapest, 1966.

[152] G. Kertész. On totally separable packings of equal balls. *Acta Math. Hungar.*, 51(3-4):363–364, 1988.

[153] Victor Klee. Facet-centroids and volume minimization. *Studia Sci. Math. Hungar.*, 21(1-2):143–147, 1986.

[154] Victor Klee and Stan Wagon. *Old and new unsolved problems in plane geometry and number theory*, volume 11 of *The Dolciani Mathematical Expositions.* Mathematical Association of America, Washington, DC, 1991.

[155] Martin Kneser. Einige Bemerkungen über das Minkowskische Flächenmass. *Arch. Math. (Basel)*, 6:382–390, 1955.

[156] Paul Koebe. Kontaktprobleme der konformen Abbildung. *Ber. Verh. Sächs. Akad. Leipzig*, 88:141–164, 1936.

[157] Wlodzimierz Kuperberg. Optimal arrangements in packing congruent balls in a spherical container. *Discrete Comput. Geom.*, 37(2):205–212, 2007.

[158] Y. S. Kupitz, H. Martini, and M. A. Perles. Ball polytopes and the Vázsonyi problem. *Acta Math. Hungar.*, 126(1-2):99–163, 2010.

[159] Zsolt Lángi. On the Hadwiger numbers of centrally symmetric starlike disks. *Beiträge Algebra Geom.*, 50(1):249–257, 2009.

[160] Zsolt Lángi. On the Hadwiger numbers of starlike disks. *European J. Combin.*, 32(8):1203–1212, 2011.

[161] Zsolt Lángi. On a normed version of a Rogers-Shephard type problem. *Israel J. Math.*, 212(1):203–217, 2016.

[162] Zsolt Lángi. Centering Koebe polyhedra via Möbius transformations. *arXiv*, 1804.07572, Apr 2018.

[163] Zsolt Lángi. A characterization of affinely regular polygons. *Aequationes Math.*, 92(6):1037–1049, Mar 2018.

[164] Zsolt Lángi, Márton Naszódi, and István Talata. Ball and spindle convexity with respect to a convex body. *Aequationes Math.*, 85(1-2):41–67, 2013.

[165] D. G. Larman and C. Zong. On the kissing numbers of some special convex bodies. *Discrete Comput. Geom.*, 21(2):233–242, 1999.

[166] Simon Larson. A bound for the perimeter of inner parallel bodies. *J. Funct. Anal.*, 271(3):610–619, 2016.

[167] Gary Lawlor. A new area-maximization proof for the circle. *The Mathematical Intelligencer*, 20(1):29–31, Mar 1998.

[168] Henry Lebesgue. Sur le problème des isopérimètres et sur les domaines de largeur constante. *Bull. Soc. Math. France, C. R.*, pages 72–76, 1914.

[169] Carl W. Lee. Regular triangulations of convex polytopes. In *Applied geometry and discrete mathematics*, volume 4 of *DIMACS Ser. Discrete Math. Theoret. Comput. Sci.*, pages 443–456. Amer. Math. Soc., Providence, RI, 1991.

[170] András Lengyel, Zsolt Gáspár, and Tibor Tarnai. The roundest polyhedra with symmetry constraints. *Symmetry*, 9(3):Paper No. 41, 15, 2017.

[171] L. Lindelöf. Propriétés générales des polyèdres qui, sous une étendue superficielle donnée, renferment le plus grand volume. *Math. Ann.*, 2(1):150–159, 1869.

[172] H. Maehara. The problem of thirteen spheres—a proof for undergraduates. *European J. Combin.*, 28(6):1770–1778, 2007.

[173] Kurt Mahler. Ein Minimalproblem für konvexe Polygone. *Mathematica (Zutphen)*, B7:118–127, 1938.

[174] Kurt Mahler. Ein übertragungsprinzip für konvexe Körper. *Časopis Pěst. Mat. Fys.*, 68:93–102, 1939.

[175] P. Mani. Automorphismen von polyedrischen Graphen. *Math. Ann.*, 192:279–303, 1971.

[176] Vinothan N. Manoharan. Colloidal matter: packing, geometry, and entropy. *Science*, 349(6251):942, 2015.

[177] Horst Martini and Zokhrab Mustafaev. Some applications of cross-section measures in Minkowski spaces. *Period. Math. Hungar.*, 53(1-2):185–197, 2006.

[178] Horst Martini and Konrad J. Swanepoel. Equiframed curves—a generalization of Radon curves. *Monatsh. Math.*, 141(4):301–314, 2004.

[179] Horst Martini and Konrad J. Swanepoel. Antinorms and Radon curves. *Aequationes Math.*, 72(1-2):110–138, 2006.

[180] William S. Massey. *Singular homology theory*, volume 70 of *Graduate Texts in Mathematics*. Springer-Verlag, New York-Berlin, 1980.

[181] M. Meyer, S. Reisner, and M. Schmuckenschläger. The volume of the intersection of a convex body with its translates. *Mathematika*, 40(2):278–289, 1993.

[182] Mathieu Meyer and Alain Pajor. On the Blaschke-Santaló inequality. *Arch. Math. (Basel)*, 55(1):82–93, 1990.

[183] Vitali D. Milman and Gideon Schechtman. *Asymptotic theory of finite-dimensional normed spaces*, volume 1200 of *Lecture Notes in Mathematics*. Springer-Verlag, Berlin, 1986. With an appendix by M. Gromov.

[184] Hermann Minkowski. Dichteste gitterförmige Lagerung kongruenter Körper. *Nachrichten von der Gesellschaft der Wissenschaften zu Göttingen, Mathematisch-Physikalische Klasse*, 1904:311–355, 1904.

[185] J. Molnár. Kreislagerungen auf Flächen konstanter Krümmung. *Math. Ann.*, 158:365–376, 1965.

[186] Frank Morgan and David L. Johnson. Some sharp isoperimetric theorems for Riemannian manifolds. *Indiana Univ. Math. J.*, 49(3):1017–1041, 2000.

[187] G. D. Mostow. *Strong rigidity of locally symmetric spaces*. Princeton University Press, Princeton, N.J.; University of Tokyo Press, Tokyo, 1973. Annals of Mathematics Studies, No. 78.

[188] Oleg R. Musin. The one-sided kissing number in four dimensions. *Period. Math. Hungar.*, 53(1-2):209–225, 2006.

[189] Oleg R. Musin. The kissing number in four dimensions. *Ann. of Math. (2)*, 168(1):1–32, 2008.

[190] Zokhrab Mustafaev. The ratio of the length of the unit circle to the area of the unit disc in Minkowski planes. *Proc. Amer. Math. Soc.*, 133(4):1231–1237, 2005.

[191] Nobuaki Mutoh. The polyhedra of maximal volume inscribed in the unit sphere and of minimal volume circumscribed about the unit sphere. In *Discrete and computational geometry*, volume 2866 of *Lecture Notes in Comput. Sci.*, pages 204–214. Springer, Berlin, 2003.

[192] A. M. Odlyzko and N. J. A. Sloane. New bounds on the number of unit spheres that can touch a unit sphere in n dimensions. *J. Combin. Theory Ser. A*, 26(2):210–214, 1979.

[193] N. Oler. An inequality in the geometry of numbers. *Acta Math.*, 105:19–48, 1961.

[194] Robert Osserman. The isoperimetric inequality. *Bull. Amer. Math. Soc.*, 84(6):1182–1238, 1978.

[195] Robert Osserman. Bonnesen-style isoperimetric inequalities. *Amer. Math. Monthly*, 86(1):1–29, 1979.

[196] F. Österreicher and J. Linhart. Packungen kongruenter Stäbchen mit konstanter Nachbarnzahl. *Elem. Math.*, 37(1):5–16, 1982.

[197] János Pach and Pankaj K. Agarwal. *Combinatorial geometry*. Wiley-Interscience Series in Discrete Mathematics and Optimization. John Wiley & Sons, Inc., New York, 1995. A Wiley-Interscience Publication.

[198] Arnau Padrol and Günter M. Ziegler. Six topics on inscribable polytopes. In *Advances in discrete differential geometry*, pages 407–419. Springer, [Berlin], 2016.

[199] A. Pełczyński and S. J. Szarek. On parallelepipeds of minimal volume containing a convex symmetric body in $\mathbf{R}^n$. *Math. Proc. Cambridge Philos. Soc.*, 109(1):125–148, 1991.

[200] Norbert Peyerimhoff. Simplices of maximal volume or minimal total edge length in hyperbolic space. *J. London Math. Soc. (2)*, 66(3):753–768, 2002.

[201] T. Pisanski, M. Kaufman, D. Bokal, E.C. Kirby, and A. Graovac. Isoperimetric quotient for fullerenes and other polyhedral cages. *J. Chem. Inf. Comput. Sci.*, 37:1028–1032, 1997.

[202] Anatolij M. Plichko. On the volume method in the study of Auerbach bases of finite-dimensional normed spaces. *Colloq. Math.*, 69(2):267–270, 1995.

[203] T. Poulsen. Problem 10. *Math. Scand.*, 2:346, 1954.

[204] R. A. Rankin. The closest packing of spherical caps in n dimensions. *Proc. Glasgow Math. Assoc.*, 2:139–144, 1955.

[205] John R. Reay. Generalizations of a theorem of Carathéodory. *Mem. Amer. Math. Soc. No.*, 54:50, 1965.

[206] Igor Rivin. A characterization of ideal polyhedra in hyperbolic 3-space. *Ann. of Math. (2)*, 143(1):51–70, 1996.

[207] C. A. Rogers. The closest packing of convex two-dimensional domains. *Acta Math.*, 86:309–321, 1951.

[208] C. A. Rogers. *Packing and covering*. Cambridge Tracts in Mathematics and Mathematical Physics, No. 54. Cambridge University Press, New York, 1964.

[209] C. A. Rogers and G. C. Shephard. The difference body of a convex body. *Arch. Math. (Basel)*, 8:220–233, 1957.

[210] C. A. Rogers and G. C. Shephard. Convex bodies associated with a given convex body. *J. London Math. Soc.*, 33:270–281, 1958.

[211] C. A. Rogers and G. C. Shephard. Some extremal problems for convex bodies. *Mathematika*, 5:93–102, 1958.

[212] J. Saint-Raymond. Sur le volume des corps convexes symétriques. In *Initiation Seminar on Analysis: G. Choquet-M. Rogalski-J. Saint-Raymond, 20th Year: 1980/1981*, volume 46 of *Publ. Math. Univ. Pierre et Marie Curie*, pages Exp. No. 11, 25. Univ. Paris VI, Paris, 1981.

[213] L. A. Santaló. An affine invariant for convex bodies of n-dimensional space. *Portugaliae Math.*, 8:155–161, 1949.

[214] Juan Jorge Schäffer. Inner diameter, perimeter, and girth of spheres. *Math. Ann. 173 (1967), 59–79; addendum, ibid.*, 173:79–82, 1967.

[215] Erhard Schmidt. über das isoperimetrische Problem im Raum von n Dimensionen. *Math. Z.*, 44(1):689–788, 1939.

[216] Erhard Schmidt. Beweis der isoperimetrischen Eigenschaft der Kugel im hyperbolischen und sphärischen Raum jeder Dimensionenzahl. *Math. Z.*, 49:1–109, 1943.

[217] Rolf Schneider. *Convex bodies: the Brunn-Minkowski theory*, volume 151 of *Encyclopedia of Mathematics and its Applications*. Cambridge University Press, Cambridge, expanded edition, 2014.

[218] Oded Schramm. How to cage an egg. *Invent. Math.*, 107(3):543–560, 1992.

[219] E. Schulte. Analogues of Steinitz's theorem about noninscribable polytopes. In *Intuitive geometry (Siófok, 1985)*, volume 48 of *Colloq. Math. Soc. János Bolyai*, pages 503–516. North-Holland, Amsterdam, 1987.

[220] K. Schütte and B. L. van der Waerden. Das Problem der dreizehn Kugeln. *Math. Ann.*, 125:325–334, 1953.

[221] Raimund Seidel. Exact upper bounds for the number of faces in d-dimensional Voronoïdiagrams. In *Applied geometry and discrete mathematics*, volume 4 of *DIMACS Ser. Discrete Math. Theoret. Comput. Sci.*, pages 517–529. Amer. Math. Soc., Providence, RI, 1991.

[222] Boris A. Springborn. A unique representation of polyhedral types. Centering via Möbius transformations. *Math. Z.*, 249(3):513–517, 2005.

[223] J. Steiner. Einfache Beweise der isoperimetrischen Hauptsätze. *J. Reine Angew. Math.*, 18:281–296, 1838.

[224] J. Steiner. Sur le maximum et le minimum des figures dans le plan, sur la sphère et dans l'espace en général. Premier mémoire. *J. Reine Angew. Math.*, 24:93–152, 1842.

[225] J. Steiner. Sur le maximum et le minimum des figures dans le plan, sur la sphère et dans l'espace en général. Second mémoire. *J. Reine Angew. Math.*, 24:189–250, 1842.

[226] J. Steiner. Systematische Entwicklung der Abhhngigkeit geometrischer Gestalten von einander, Reimer, Berlin, 1832. In *J. Steiner's Collected Works.* 1881.

[227] Ernst Steinitz. Bedingt konvergente Reihen und konvexe Systeme. *J. Reine Angew. Math.*, 143:128–176, 1913.

[228] Ernst Steinitz. *Polyeder und Raumeinteilungen*, volume vol III (Geometriae) of *Encyclopädie der matematischen Wissenschaften.* B.G. Teubner, Leipzig, 1922.

[229] Ernst Steinitz. Über isoperimetrische Probleme bei konvexen Polyedern. *J. Reine Angew. Math.*, 158:129–153, 1927.

[230] Ernst Steinitz and Hans Rademacher. *Vorlesungen über die Theorie der Polyeder.* Encyclopädie der matematischen Wissenschaften. Springer-Verlag, Berlin, 1934.

[231] Serge Tabachnikov and Emmanuel Tsukerman. Circumcenter of mass and generalized Euler line. *Discrete Comput. Geom.*, 51(4):815–836, 2014.

[232] I. Talata. Exponential lower bound for the translative kissing numbers of d-dimensional convex bodies. *Discrete Comput. Geom.*, 19(3, Special Issue):447–455, 1998. Dedicated to the memory of Paul Erdős.

[233] I. Talata. The translative kissing number of tetrahedra is 18. *Discrete Comput. Geom.*, 22(2):231–248, 1999.

[234] Ivor Thomas. *Selections illustrating the history of Greek mathematics. Vol. II. From Aristarchus to Pappus.* Harvard University Press, Cambridge, Mass; William Heinemann, Ltd., London, 1951.

[235] William P. Thurston. *Three-dimensional geometry and topology. Vol. 1*, volume 35 of *Princeton Mathematical Series*. Princeton University Press, Princeton, NJ, 1997. Edited by Silvio Levy.

[236] H.-W. van Wyk. *The Blaschke-Santaló inequality (diploma thesis)*. University of Pretoria, Pretoria, South Africa, 2007.

[237] Gerd Wegner. über endliche kreispackungen in der ebene. *Studia Sci. Math. Hungar.*, 21:1–28, 1986.

[238] Hans Zassenhaus. Modern developments in the geometry of numbers. *Bull. Amer. Math. Soc.*, 67:427–439, 1961.

[239] Günter M. Ziegler. *Lectures on polytopes*, volume 152 of *Graduate Texts in Mathematics*. Springer-Verlag, New York, 1995.

Index

(m, d)-scribable, 16
(m, d)-scribed polytopes, 3
A-domain, 68–70, 243, 245, 246
B-domain, 68, 69, 242
B-measure, 68, 69, 78, 242
R-separable packing, 213, 214
λ-intersection graph, 200
ρ-separable packing, 67, 75–77, 219, 220, 239, 253–255, 263
n-omino, 239–241
r-convex, 42, 159, 166–168
r-extreme normal vector, 89
r-tangential body, 89
'Oumuamua, 3, 8

affine isoperimetric inequalities, 5
Alexandrov's theorem, 7, 8
Alexandrov-Fenchel inequality, 4, 89, 176, 255
analytic contraction, 40, 41
angle graph, 106
angle measure, 68, 69, 242
angle of parallelism, 114
angular measure, 67, 68, 243
Archimedes' lemma, 19
associated $(d+1)$-dimensional convex body, 25, 131
Auerbach basis, 62, 69, 78, 221, 240–242, 244
Auerbach domain, 67, 68
Auerbach lattice, 78, 240–242

barycenter, 17, 20, 83, 113, 260
basic n-omino, 240, 241
basic polyomino, 240, 241
Benson's definition of volume, 29
Birkhoff domain, 67, 68, 239
Birkhoff measure, 67, 68
Birkhoff orthogonal, 67, 68, 142
Birkhoff orthogonality, 67, 78, 142
Blaschke's selection theorem, 143, 167
Blaschke-Lebesgue Theorem, 152
Blaschke-Santaló Inequality, 31, 42, 131, 142, 155, 159
Blichfeldt gauge function, 199
Blichfeldt's method, 57, 187, 199
box-polytope, 211, 212
Brunn-Minkowski inequality, 24, 34, 54, 135, 139, 145, 155, 157, 181, 211, 212
Busemann area, 157
Busemann volume, 29

Carathéodory's theorem, 132
Cauchy's surface area formula, 140
Cavalieri's principle, 145
Cayley-Menger determinant, 122
center of mass, 6, 18–20, 36, 119, 120, 125, 260
centering, 17, 18, 83
central projection, 17, 92, 189, 204, 205
central symmetrization, 30, 34, 36, 72, 138, 141, 152, 226, 238
Circle Packing Theorem, 3, 14
circulant matrix, 104
circumcenter of mass, 18, 19
circumscribed, 6, 8, 9, 11, 14, 16, 20, 21, 27–29, 36, 57, 61, 71–74, 83, 84, 89, 103, 141, 149, 150, 157, 188, 189, 191, 229, 247, 250, 251, 259, 261, 263
circumscribed polygon, 21

circumscribed polyhedra, 6, 17
circumscribed polytopes, 3, 6
coface, 99, 100
colloids, 51
contact graph, 49, 50, 52, 53, 61, 63, 196, 241, 242, 262
contact number, 49–55, 57, 58, 60, 63, 187, 195, 197, 198, 212, 239–242, 262
contact number problem, 49–51, 56, 57, 59, 67
continuous contraction, 37–39, 41, 45
contracted Gale diagram, 99–101
contraction, 37–41, 44–46, 173, 175, 182–184, 261, 262
Cramer's rule, 123
cyclic polytope, 12, 13, 21, 102

densest sphere packing problem, 50
density, 54–57, 67, 70, 73, 74, 76, 78, 192, 193, 209, 213, 219, 263
differentiable contraction, 174
digital contact number, 58, 59, 240
digital packing, 58, 59, 240, 242
discrete isoperimetric inequality, 5
double n-pyramid, 91
double pyramid, 9, 26, 92, 93
dual graph, 15, 105, 106, 111

edge graph, 15, 106, 108, 117–119
Eikonal abrasion model, 86, 87, 90
Eikonal equation, 7, 83
Eikonal wavefront model, 86, 87
equiframed curve, 36
Erdős-type distance problem, 52

face circle, 15, 19, 105, 106, 113, 114, 118, 119, 123
face-centered cubic lattice, 53, 193, 194
facial simplex, 10, 96, 98
facial tetrahedron, 90, 92
farthest point truncated Voronoi cell, 38, 160, 163
farthest point Voronoi cell, 38, 159, 160, 163
fcc lattice, 193
foliation, 116, 117, 120, 124
form body, 8

Gale diagram, 99, 101
Gale transform, 99
Gram matrix, 103, 104
Green's theorem, 149
Gromov's mass, 29, 131
Gromov's mass*, 29, 131, 150, 158

Haar measure, 29
Hadwiger number, 49, 54, 60, 61, 64–66, 187, 224, 227–229, 262
Hajós lemma, 204, 210
Hajós polygon, 204
Hausdorff distance, 70, 88, 89, 142, 143, 145
Holmes-Thompson volume, 29, 131, 145, 146, 150
homology group, 117
horosphere, 112, 120, 125, 128
hypersphere, 118

ideal point, 111–116, 118–121, 123, 124, 126–128
identity matrix, 103, 104
identity operator, 84
inner parallel body, 87, 88
inscribable, 16
inscribed, 9–14, 16, 19, 29, 35, 91, 92, 96, 97, 100, 101, 103, 104, 110, 148–150, 152, 153, 155–157, 211, 240–242, 252, 259, 261
inscribed polygon, 21
inscribed polyhedra, 17
inscribed polytopes, 3
integer lattice, 58, 60, 239, 240
intersection graph, 14, 15
isoperimetric inequality, 3–5, 24, 181, 191, 193, 196, 209, 211, 213, 222
isoperimetric polyomino, 59
isoperimetric problem, 3

isoperimetric quotient, 54, 212, 214
isoperimetric ratio, 3, 5–7, 21, 83, 259

John ellipsoid, 83, 84

kissing number, 49, 50, 55, 60, 66
kissing number problem, 49, 60
Kneser-Poulsen Conjecture, 37–39, 41, 44, 159, 182, 261
Koebe polyhedron, 3, 15, 17–20, 83, 113, 260

L'huilier's problem, 3, 9
Löwner-John ellipsoid, 103
Lebesgue measure, 4, 18, 29, 37, 42, 182, 261
lemma of comparison, 209
Lindelöf Condition, 3, 6, 16, 83, 259
linear parameter system, 24, 36, 131–133
locally separable, 78
longest chord, 34, 137, 138, 141, 152, 156, 227, 251

Möbius group, 113
Möbius transformation, 16–20, 105, 106, 111, 115, 118, 119, 121, 123, 125, 126
Mayer-Vietoris exact sequence, 117
medial polyhedron, 10, 93, 259
midscribed, 15, 17, 20, 83, 106, 260
Minkowski addition, 177, 212
Minkowski content, 4
Minkowski functional, 134
Minkowski lemma, 61, 196
Minkowski length, 71, 149
Minkowski perimeter, 5
Minkowski symmetrization, 61, 196
Minkowski's theorem on mixed volumes, 88
mixed area, 250, 251
mixed volume, 4, 76, 88, 89
moment curve, 12
moment lemma, 6
Mostow's rigidity theorem, 15

nearest point truncated Voronoi cell, 38, 160, 163
nearest point Voronoi cell, 38, 159, 160, 163
neighborly polytope, 102
normalized Gale diagram, 99

Oler's inequality, 67, 70–72, 239
one-sided Hadwiger number, 54, 62, 66
one-sided kissing number, 66
one-sided separable Hadwiger number, 62

permissible polygonal curve, 72, 246–249
Petty's projection inequality, 140
Picard-Lindelöf Theorem, 116, 126
Poincaré ball model, 111–114, 126
Poincaré half plane model, 115
Poincaré half space model, 115, 119, 123, 127, 128
polyhedral graph, 14, 15
polyomino, 58, 59, 239–242
primal-dual circle representation, 105–108, 110
projective ball model, 17
Property Z, 10–13, 91–93, 95–98, 100–102
pseudo-double pyramid, 26, 30, 34
pyramid, 97, 100

quadrangulation, 107, 109
quermassintegral, 4, 88

Radon arc, 78
Radon curve, 28, 29, 36, 142, 149–151, 153, 251
Radon domain, 78
Radon norm, 28, 29, 32, 131, 151, 153, 154
Radon partition, 101
Radon plane, 77, 78
Radon's lemma, 217
reduced angle graph, 107, 108
reflection body, 25, 131

relative norm, 23, 30, 32, 35, 71, 152, 157, 226, 227, 238, 252
Reuleaux triangle, 152
Reuleaux triangle in the norm of **M**, 152, 153, 156, 157
reverse isoperimetric inequality, 5, 83, 197, 223

sausage, 75, 77, 180, 263
self-assembly, 51
separable Hadwiger configuration, 61–63, 215, 216, 218, 221, 223, 262
separable Hadwiger number, 61, 66, 262
simplicial complex, 10
simplicial polyhedron, 19–21, 95, 97, 124, 125, 259
simplicial polytope, 11, 18, 19, 97, 99–101, 122
smooth contraction, 174
standard Gale diagram, 99
starlike ball, 64, 66
starlike disk, 64, 65, 187, 226, 233, 262
Steiner symmetrization, 4, 14, 21, 23, 24, 35, 131, 133, 135, 138, 139, 142–145, 155, 156, 259
stereographic projection, 14, 15, 106
strong contraction, 47
support function, 87, 251
surface area, 5, 6, 58, 210
surface volume, 3–5, 9, 21, 24, 40, 54, 76, 83, 88, 89, 140, 161, 191, 196, 211, 213, 259

topological ball, 64
topological disk, 64, 65, 187, 224, 225, 263
totally separable, 49, 50, 58–61, 63, 67, 70–76, 78, 79, 187, 206–208, 210, 212, 213, 219, 222, 239, 240, 242, 246, 247, 249, 262, 263
translation body, 25, 131
translative constant volume property, 28, 29, 131, 141, 260
translative kissing number, 49, 60
triakis tetrahedron, 16
triangulation, 18, 52, 90, 91, 96, 97, 184
trigonometric moment curve, 12
truncated Voronoi cell, 159, 161, 163, 190, 193, 206–209, 211
two-point symmetrization, 165–167

unconditional convex body, 47
uniform contraction, 42–44, 159, 177, 179

valence, 10, 21, 91, 93, 259
vertex circle, 15, 19, 105, 106, 113, 118
vertex-face incidence graph, 120, 122
volume product, 23, 150, 154, 155
volume ratio, 83, 84
Voronoi cell, 40, 159, 161, 163, 190, 191, 193, 201, 203, 204, 206, 207, 210
Voronoi decomposition, 201

width, 29, 30, 34, 72, 138, 141, 152, 177, 251